Health Informatics

"The World Health Organization Quality of Life (WHOQOL) assessment with its comprehensive 24 parameters is an outstanding but concomitantly daunting tool to measure the quality of life in individuals. Without a valid and reliable technology-based approach to this assessment tool, it is a set of parameters without the fullest impact on health with its measurements and interventions.

A technological and computational dyadic approach to the expansive WHOQOL tool would be the ideal visionary strategy. We should all applaud Sharon Wulfovich and Katarzyna Wac for their innovative philosophy espoused in the book Quantifying Quality of Life: Incorporating Daily Life into Medicine. This paradigm shift from a subjective, qualitative approach to a more objective, quantitative approach with human oversight renders this WHOQOL tool more accurate and efficient not to mention more continual and expedient. These aforementioned elements will be invaluable for the future of health in both individuals and populations.

We are in the midst of a global health crisis and its apocalyptic consequences. The erudite author Arundhati Roy wrote, "the pandemic is a portal, a gateway between one world and the next." This innovative approach to measuring the quality of health elucidated by Wulfovich and Wac is a very much welcomed part of this portfolio of tools we should bring along as we traverse through the portal into the new world."

—Anthony C. Chang, *Chief Intelligence and Innovation Officer,*
Children's Hospital of Orange County, CA, USA

"*Quantifying Quality of Life* is the first book of its kind by describing how to quantify both the short-term and long-term determinants of health and wellbeing over an individual's life course. With 5G broadband cellular networks on the horizon and the continual enthusiasm in the quantifying-self movement around the world, this book offers the methodologies and the tools for daily continuous assessment of the quality of life.

As the field gains more experience on how to quantify the quality of life by leveraging both technology and patient-reported outcomes, we will begin to see more relationships emerging between different domains and variables being measured, e.g. exercise and mental health; social relations and productivity; mobility and self-esteem, etc. This will help us gain new insights for building a better environment and society for all citizens.

This book is a remarkable first step towards understanding the science of an emerging and expanding field combining digital health, artificial intelligence, and Internet-of-Things. The editors and authors are to be congratulated for this pioneering work!"

—C. Jason Wang, *Director, Center for Policy,*
Outcomes, and Prevention, Associate Professor of Pediatrics, Medicine,
and Health Research and Policy, Co-chair, Mobile Health
and Other New Technologies, Center for Population Health Sciences,
Stanford University, USA

"The arrival of smartphones and technology-enabled tools and methods has sparked a veritable flood of what is called "Real-World Monitoring (RWM)." It allowed for creating unprecedented evidence data and patient-reported quality-of-life outcomes. Unlike traditional patient assessments via questionnaires or similar backward directed surveys, the new digital applications can enable objective, real-time data collection. This adds a completely new dimension to data quality and expressiveness. QoL Measurements and PRO's are relevant to all healthcare stakeholders including patients. Doctors are able to adapt and individualize therapies more easily and can communicate to patients well outside of their practice and routine examinations. Payers can finally move away from a "fee-for-service" to an outcomes-based remuneration model—which is an important requirement to implement connected care structures in an increasingly complex and costly care environment. Life Sciences firms as well as the MedTech and Health IT industries are in a position to comply with new approval and certification requirements from authorities, such as the FDA or the EMA, to include PRO's and not only traditional clinical data into their studies and applications. Finally, for regulators, solid and reliable outcomes data will be the basis to steer and manage healthcare systems much better than in the past—especially when it comes to placing innovation incentives and adapting processes or care structures. A multitude of health apps are available today and more of them are arriving each day which measure all sorts of personal health scores and figures.

However, monitoring isolated values without relating them to each other is of little use. Therefore, it is important to distinguish patient-reported QoL data—and this also includes data coming from digital tools and applications (called Technology Reported Outcomes, TechROs). These tools only make sense if they are incorporated into a coherent body image to avoid biases, misinterpretation, or false therapeutic conclusions. This book not only looks at different elements that make up a holistic QoL picture, but it also demonstrates scientific methods to assess them which will allow them to be put into a meaningful clinical context. This book is a useful guide for all healthcare stakeholders."

—Rainer Herzog, *Managing Director,*
Digital Health Partners, Germany

"Recently gathered insights indicate that conditions such as heart disease, stroke, and diabetes are essentially the pathophysiologic endpoints of a complex interplay of an individual's life experiences in five domains of influence: genetic predisposition, social circumstances, physical environment, behavior, and medical care.

It now becomes possible to measure QoL objectively and quantitatively using validated methods and to incorporate these into both everyday clinical care and research to measure the effect of various treatments and strategies in the real world. We have witnessed an influx of market adaption for digital health and quality of life technologies this past year from telemedicine and virtual clinical trials to remote patient monitoring and quality of life assessments. Digital health technologies are here to stay and will play a critical role in the way we treat and cure diseases while empowering patients to manage their self-care.

I was delighted to see the complex and rather challenging topic "Quantifying Quality of Life: Incorporating Daily Life into Medicine" being tackled by Prof Katarzyna Wac and Sharon Wulfovich. I found the QoL domains: physical health, psychological, social relationships, and environment to be meticulously addressed in this book with a state-of-the-art approach taken for the interpretation of the 24 related variables."

—Mathieu Ghadanfar, *President,*
M-Ghadanfar Consulting Life Sciences,
Cardiovascular Physician | Healthcare Executive
| Biotech and Pharmaceutical Executive

"*Quantifying Quality of Life: Incorporating Daily Life into Medicine* is an innovative and informative text that explores technology-enabled assessments of specific dimensions of the WHOQOL framework. An underlying construct running through the text is the significance of quality of life, and its core role within clinical, health, and societal decision-making. Within this construct, the text brings together critical thinking across key concepts and domains of the International Classification of Functioning, Disability, and Health (ICF) to consider the emergent role of technology in measurement. The authors guide the reader through a series of chapters exploring measurement of quality of life domains through technology and innovation both with active self-report PROs and passively collecting physiological, physical, biological, and contextual signals. The text is rich with deep thinking by leading critical thinkers across domains of the ICF. Original work is freshly presented in the book's chapters, which draw on case–control to umbrella review methodologies. The authors all succeed in compellingly presenting the background, state of the art, and suggesting the direction of future research activity. The book is abundant with fine details and expertise across WHO domains to inform the quality of life research. The book provides a unique insight into an easily accessible template for multidisciplinary students, clinicians, and academics."

—Prof. Helen Dawes, *Professor,*
Movement Science and Elizabeth Casson Trust Chair, Director,
Centre for Movement and Occupational Rehabilitation Sciences (MOReS),
Academic Director, Oxford Clinical Allied Technology
and Trial Services Unit (OxCATTS), Faculty of Health and Life Sciences,
Oxford Brookes University, Associate Research Fellow,
Department of Clinical Neurology,
University of Oxford,
Oxford Health Biomedical Research Centre, UK

"La transition vers l'hôpital numérique est en cours à une époque où l'on soigne de plus en plus à domicile, ou le patient prend le contrôle de sa santé. La médecine jusqu'ici essentiellement curative, notamment en Occident, se transforme elle aussi avec une mise en exergue du patient: une personnalisée, préventive, prédictive et participative (4P).

Nous sommes entrés dans cette nouvelle ère supportées par des technologies innovantes, comme le bigdata (analyse de données), le deep learning, la robotique, l'intelligence artificielle, le suivi de données personnelles, voir la génétique.

Nous pourrons ainsi anticiper, prévenir, modifier notre hygiène, non pas pour répondre à un modèle, mais à notre modèle personnalisé. Derrière cette personnalisation est en train de naître la médecine de trajectoires avec une prise en compte accrue de tout notre environnement (alimentation, qualité de l'air, pratique physique…).

Le patient devient donc désormais l'acteur principal et participe à sa santé de manière active au travers de ses soins mais surtout d'une meilleure connaissance de sa santé à l'aide principalement d'applications (comme des protocoles de suivi post-cancer), ou de capteurs de données régulières ou alarmistes (IOT) et d'un accès facilité à l'information médical. Cette vision est à la base de ce livre, et il fournit de nombreux exemples d'innovations technologiques potentielles menant à la réalisation de cette vision. Ce livre est le fruit d'un effort remarquable pour documenter l'état de l'art sur les 24 variables contribuant à la qualité de vie dans ces chapitres bien présentés, clairs et complémentaires. Il est écrit de manière à être accessible à tous, même aux patients. Je suis heureux de recommander cet excellent livre à tous!"

—Moïse Gerson, *A Patient, Digital Health Director and Co-Founder,*
Campus 2030, INNOVATING FOR THE PEOPLE, Switzerland

"Accurately assessing the quality of life (QoL) using established parameters, such as those set out by the WHO, is key to a revolution in the health and life insurance sectors. That is the use of real-time data to monitor risks and more accurately underwrite it using feedback loops. "Connected Insurance" is already in use in areas like the use of telematics for auto insurance and is similarly changing health and life insurance, helping insurers assist in preventing and managing chronic disease and acute incidents as well as encouraging healthier lifestyles in their customers. It has two important drivers. The first is that technologies for measurement, including mobile phone-based apps and wearables, are proliferating and improving and can allow real-time monitoring of many of the facets in the WHOQoL domains. The second is that accurate quantification of risk factors to scientific levels of accuracy (as demonstrated in these chapters) will provide increased statistical accuracy and confidence for underwriters of health and life risk.

Quantifying the Quality of Life is an important mosaic of exciting research that shows how facets of QoL can be quantified."

—Lawrence Reed, *Managing Director, IMCG Group, UK*

"*Quantifying Quality Of Life: Incorporating Daily Life into Medicine* is a must-read primer for anyone considering using patient-related outcomes with subjective and objective data. In health insurance, we are aware that healthy lifestyle choices like eating well, exercising, managing stress, and sufficient sleep can help mitigate the risks associated with lifestyle diseases. Therefore, we need to show clients how these different health factors are linked together. The power of ecosystems to develop a dynamic, personalized pricing approach that rewards good behaviors will potentially help to drive this forward. To achieve the next steps, it is essential to have valid, accurate, and reliable data for the QoL assessment for individuals. The book is full of important information for those of us who are on the journey to real patient-centricity by using relevant data outcomes."

—Roman Sauter, *Head, Sourcing & Procurement, Helsana, Switzerland*

This series is directed to healthcare professionals leading the transformation of healthcare by using information and knowledge. For over 20 years, Health Informatics has offered a broad range of titles: some address specific professions such as nursing, medicine, and health administration; others cover special areas of practice such as trauma and radiology; still other books in the series focus on interdisciplinary issues, such as the computer based patient record, electronic health records, and networked healthcare systems. Editors and authors, eminent experts in their fields, offer their accounts of innovations in health informatics. Increasingly, these accounts go beyond hardware and software to address the role of information in influencing the transformation of healthcare delivery systems around the world. The series also increasingly focuses on the users of the information and systems: the organizational, behavioral, and societal changes that accompany the diffusion of information technology in health services environments.

Developments in healthcare delivery are constant; in recent years, bioinformatics has emerged as a new field in health informatics to support emerging and ongoing developments in molecular biology. At the same time, further evolution of the field of health informatics is reflected in the introduction of concepts at the macro or health systems delivery level with major national initiatives related to electronic health records (EHR), data standards, and public health informatics.

These changes will continue to shape health services in the twenty-first century. By making full and creative use of the technology to tame data and to transform information, Health Informatics will foster the development and use of new knowledge in healthcare.

More information about this series at https://link.springer.com/bookseries/1114

Katarzyna Wac • Sharon Wulfovich

Editors

Quantifying Quality of Life

Incorporating Daily Life into Medicine

 Springer

Editors
Katarzyna Wac
Geneva School of Economics and
Management, Center for Informatics,
Quality of Life Technologies Lab
University of Geneva
Geneva, Switzerland

Sharon Wulfovich
School of Medicine
University of California San Diego
San Diego, CA, USA

Quality of Life Technologies Lab
Department of Computer Science
University of Copenhagen
Copenhagen, Denmark

Swiss National Science Foundation

This book is an open access publication.

ISSN 1431-1917   ISSN 2197-3741   (electronic)

ISBN 978-3-030-94211-3   ISBN 978-3-030-94212-0   (eBook)
https://doi.org/10.1007/978-3-030-94212-0

This Springer imprint is published by the registered company Springer Nature Switzerland AG
The registered company address is: Gewerbestrasse 11, 6330 Cham, Switzerland

Acknowledgements

The editors are deeply grateful to all the contributing authors and anonymous reviewers, whose time and dedication have made this book possible. The pre-press stage of this publication and its Gold Open Access were supported by the Swiss National Science Foundation (fund no. SNSF 10BP12_206013). Additionally, K. Wac's efforts were supported by internal grants from the University of Copenhagen (Denmark) and University of Geneva (Switzerland), as well as the swissuniversities P-13 project, H2020 WellCo Project (769765), AAL GUARDIAN project (AAL-2019-6-120-CP), and Data+ AI@CARE project.

Contents

Part I
Introduction

Chapter 1
Unfolding the Quantification of Quality of Life

Sharon Wulfovich, Jeppe Buur, and Katarzyna Wac

Introduction

There are many ways to define health. Health is defined by the World Health Organization (WHO) as "*a state of complete physical, mental and social well-being and not merely the absence of disease or infirmity.*" [1, 2] This definition has recently been challenged by a team of international experts who suggested that health be defined as "*the ability to adapt and self-manage in the face of social, physical, and emotional challenges.*" [3] Health contributes greatly to quality of life (QoL), and some authors suggest that health-related QoL and QoL can be used interchangeably [4]. However, QoL is more than health, as other factors including work capacity, social support, and the physical environment are also necessary for QoL [5–7]. QoL can be defined in multiple ways through a more global approach (from the psychological, economics, policy, or medical science perspective) [8], a categorical breakdown from an individual perspective (e.g., physical or psychological aspects), or a field-specific definition applied to individuals or specific populations (e.g., Liver QoL) [7, 9].

Across these different definitions, there is some agreement that QoL integrates an individual's multidimensional evaluation of their own life and total well-being [7]. Furthermore, an individual's QoL is not merely focused on the individual; it

S. Wulfovich
School of Medicine, University of California San Diego, La Jolla, CA, USA

J. Buur
Quality of Life Technologies Lab, Department of Computer Science, University of Copenhagen, Copenhagen, Denmark

K. Wac (✉)
Geneva School of Economics and Management, Center for Informatics, Quality of Life Technologies Lab, University of Geneva, Geneva, Switzerland
e-mail: katarzyna.wac@unige.ch

© The Author(s) 2022
K. Wac, S. Wulfovich (eds.), *Quantifying Quality of Life*, Health Informatics,
https://doi.org/10.1007/978-3-030-94212-0_1

encompasses the individual's physical and psychological state, the environment the individual is in, as well as the interaction between the two. The environment includes other individuals; nonmaterial things such as parks and roads; as well as water, air, and access to other resources.

Measuring an individual's QoL allows us to obtain a more holistic assessment of his or her state in the multiple contexts like disease progression (via symptoms), or treatment progress, and to put that in the context of clinical decision making. QoL or well-being has been indirectly assessed since the dawn of the field of medicine. Almost every doctor or physician informally asks the patient about his or her state using questions such as "how are you feeling right now?" or "how are your symptoms?"

With the need to systematically assess QoL in clinical decision making [10], there are two primary ways to capture this information: (1) asking people about different aspects of their lives following subjective self-reporting using validated patient-reported outcomes (PRO) [11] instruments [12]; examples and an overview of the current validated instruments for QoL assessment can be found in the studies of Gill [13] and Linton et al. [14]; and (2) leveraging technologies to objectively capture individuals' biological samples, physiological signals, behaviors, or interactions with the environment [4, 11].

One of the most widely used QoL assessment instruments is the WHO's Quality of Life instrument (WHOQOL), which is used as a framework for organizing this book. The WHOQOL defines QoL as *"individuals' perception of their position in life in the context of the culture and value systems in which they live and in relation to their goals, expectations, standards and concerns."* [15] The WHOQOL-BREF instrument assesses individuals' QoL across four domains: physical health, psychological health, social relationships, and environmental aspects [16]. These four large domains are further broken down into 24 subdomains, denoted by the WHO as *'facets'* [15] (Fig. 1.1). The subdomains embrace subjective and objective aspects of life, are mutually nonexclusive, and potentially intertwine [15]. For example, there is an influence of noise (i.e., environmental aspect) on sleep and rest (i.e., physical health).

The overarching assumption carried throughout this book is that within each of the QoL domains, there are specific daily behaviors that (a) can be accessed objectively through personal technologies or (b) enabled through the use of these technologies. A behavior is defined by the scientific community as "internally coordinated responses (actions or inactions) of whole living organisms (individuals or groups) to internal and/or external stimuli, excluding responses more easily understood as developmental changes," [17] or as "a comportment, or what someone does or how someone acts." [18] Behaviors can be assessed by means of, for example, their frequency, rate, duration, magnitude, and latency [19]. In the scope of this book, we focus specifically on external observable behaviors (or the lack thereof) that may be assessed using technologies. This assumption follows the definition of QoL Technologies (QoLT) as *"any technologies for assessment or improvement of the individual's QoL."* [20] The variety of designs of QoLT used to assess behaviors in daily life remains unknown, as does their influence on QoL. In this book we focus solely on the approaches using technology-enabled QoL

QoL Domains	QoL Subdomains
Physical Health	Activities of daily living
	Dependence on medicinal substances and medical aids
	Energy and fatigue
	Mobility
	Pain and discomfort
	Sleep and rest
	Work capacity
Psychological	Bodily image and appearance
	Negative feelings
	Positive feelings
	Self-esteem
	Spirituality/religion/personal beliefs
	Thinking, learning, memory and concentration
Social Relationships	Personal relationships
	Social support
	Sexual activity
Environment	Financial resources
	Freedom, physical safety and security
	Health and social care: accessibility and quality
	Home environment
	Opportunities for acquiring new information and skills
	Participation in and opportunities for recreation/leisure act.
	Physical environment (pollution / noise / traffic / climate)
	Transport

Fig. 1.1 WHOQOL Instrument Domains and Subdomains [16]

assessments. We therefore assume that QoLT enable behavior assessments and as a result the assessment of QoL in individuals [11].

This book presents QoLT leveraged for QoL assessment and draws from the WHOQOL, providing a way to categorize behaviors and QoL aspects. As a result, the WHOQOL instrument presented here serves as the organizational method for this book.

The remainder of this chapter is organized as follows: First, we present the WHOQOL instrument in detail (Sect. 2), and then we present the 'quantified-selfers', who leverage daily life technologies to assess their own behaviors and daily life (Sect. 3). Lastly, we conclude with a discussion further motivating the vision for this book (Sect. 4).

This book follows the WHOQOL instrument, and its chapters are organized along the WHOQOL subdomains. The following chapter discusses conclusive remarks and future directions for the field of QoL assessment. Finally, the last chapter discusses the current state of the subjective assessment of QoL by summarizing a set of validated instruments and scales for assessing daily life behaviors in the context of QoL, also organized along the variables in the WHOQOL instrument.

The WHOQOL Instrument

The WHO developed its first edition of an international QoL assessment approach in 1995 [15]. The development of the WHOQOL consisted of many stages: (i) QoL concept clarification; (ii) qualitative pilot; (iii) development pilot; and (iv) field test [15]. Due to the multidimensional essence of QoL, the WHOQOL divided QoL into six broad domains: (1) physical domain; (2) psychological domain; (3) level of independence; (4) social relationships; (5) environment; and (6) spirituality/religion/personal beliefs [15], embraced within the original 100-question instrument referred to as the WHOQOL-100 [21]. Later, the WHO developed a WHOQOL-BREF QoL assessment [16, 22], a shorter version of the original WHOQOL-100, which defines four broad domains: (1) physical health; (2) psychological health; (3) social relationships; and (4) environment [16, 22]. This shorter version, used as the assessment model throughout this book, was developed to minimize respondent burden and unnecessary detail when approaching QoL assessment in the general population [16, 22]. The WHOQOL-BREF instrument has been demonstrated to have "*good to excellent psychometric properties of reliability*" and to perform "*well in preliminary tests of validity.*" [22]

The WHOQOL-BREF, its four domains, and the 24 subdomains are outlined in Fig. 1.1. The paragraphs below provide a working definition for each of the four domains and 24 subdomains of the WHOQOL-BREF. We use the titles of the WHOQOL-BREF and have adapted the definitions of the WHOQOL User manual [23] accordingly, as the WHOQOL-BREF does not have its own manual.

Domain I: Physical Health
1. *Activities of Daily Living*: "a person's ability to perform usual daily living activities."
2. *Dependence on Medicinal Substances and Medical Aids*: "a person's dependence on medication or alternative medicines for supporting his/her physical and psychological well-being."
3. *Energy and Fatigue*: "the energy, enthusiasm and endurance that a person has in order to perform the necessary tasks of daily living, as well as other chosen activities such as recreation."
4. *Mobility*: "the person's view of his/her ability to get from one place to another, to move around the home, move around the workplace, or to and from transportation services."
5. *Pain and Discomfort*: "unpleasant physical sensation experienced by a person and the extent to which these sensations are distressing and interfere with life." The topics include pain control.
6. *Sleep and Rest*: problems getting enough sleep and rest.

7. *Work Capacity*: "a person's use of his or her energy for work." "Work" is defined as any major activity in which the person is engaged.

Domain II: Psychological Health

8. *Bodily Image and Appearance*: "the person's view of his/her body."
9. *Negative Feelings*: "how much a person experiences negative feelings, including despondency, guilt, sadness, tearfulness, despair, nervousness, anxiety and a lack of pleasure in life."
10. *Positive Feelings*: "how much a person experiences positive feelings of contentment, balance, peace, happiness, hopefulness, joy and enjoyment of the good things in life."
11. *Self-Esteem*: "how people feel about themselves."
12. *Spirituality/Religion/Personal Beliefs*: "examines the person's personal beliefs and how these affect quality of life."
13. *Thinking, Learning, Memory, and Concentration*: "a person's view of his/her thinking, learning, memory, concentration and ability to make decisions."

Domain III: Social Relationships

14. *Personal Relationships*: "the extent to which people feel the companionship, love and support they desire from the intimate relationship(s) in their life."
15. *Social Support*: "how much a person feels the commitment, approval, and availability of practical assistance from family and friends."
16. *Sexual Activity*: "a person's urge and desire for sex, and the extent to which the person is able to express and enjoy his/her sexual desire appropriately."

Domain IV: Environment

17. *Financial Resources*: "the person's view of how his/her financial resources (and other exchangeable resources) and the extent to which these resources meet the needs for a healthy and comfortable life style."
18. *Freedom, Physical Safety, and Security*: "the person's sense of safety and security from physical harm."
19. *Health and Social Care: Availability and Quality:* "the person's view of the health and social care in the near vicinity."
20. *Home Environment*: the "principal place where a person lives, and the way that this impacts on the person's life. The quality of the home would be assessed on the basis of being comfortable, as well as affording the person a safe place to reside."
21. *Opportunities for Acquiring New Information and Skills*: "a person's opportunity and desire to learn new skills, acquire new knowledge, and feel in touch with what is going on."
22. *Participation in and Opportunities for Recreation/Leisure Activities*: "a person's ability, opportunities and inclination to participate in leisure, pastimes and relaxation."
23. *Physical Environment (Pollution/Noise/Traffic/Climate)*: "the person's view of his/her environment. This includes the noise, pollution, climate and general aesthetic of the environment and whether this serves to improve or adversely affect quality of life."

24. *Transport*: "the person's view of how available or easy it is to find and use transport services to get around." [23]

The WHOQOL instrument provides a way to categorize behaviors and QoL aspects. As defined earlier, the WHOQOL instrument presented here serves as the organizational method for this book.

Learning from the 'Quantified-Self' Community

In this section, we present and discuss a subset of currently available technologies for the assessment of behaviors, health state, and—as a result—QoL. The "quantified self" is a term coined in 2007 by Gary Wolf and Kevin Kelly to accommodate actions such as lifelogging and self-tracking, in which motivated individuals use various analogues (e.g., paper and pencil) or digital, technology-enabled tools (e.g., Excel spreadsheets) and devices (e.g., wearables) for tracking certain aspects of their lives—be they in relation to physical health, mental health, social relationships, or even the environment surrounding them. This section presents a nonexhaustive view on the community, surveying individuals who actively participate in the quantified-self movement, as well as what they self-track and how they do it. Specifically, this chapter presents a qualitative study that examined the quantified-self community based on a curated set of self-tracked projects presented in video talks from quantified-self conferences and meetings (organized in the form of meetups) between 2015 and 2019. In total, 71 quantified-self projects were analyzed with the purpose of finding out who self-tracks, what they track, and how they track it. A variety of variables are categorized and analyzed, including the self-tracker's sex, domains of tracking (coded along the WHOQOL instrument dimensions and subdomains), and devices and tools used, among others. We then extrapolate upon the applicability of the tools, approaches, and lessons learned toward the larger public, for which we aim to quantify QoL.

This section is structured as follows: we first define the quantified-self movement and quantified-self community, then describe our research methods and results, and then analyze the outcomes, implications, and limitations. We end with a conclusive remarks section that summarizes the lessons learned within the section and links them to the chapter as a whole.

The Quantified-Self Movement

The quantified-self is a way of logging and measuring a variety of data about an individual and/or his/her surroundings, such as steps, calories eaten, or miles biked [24]. The quantified-self in its simplest form is a way of logging a variety of data for different reasons, be it for self-improvement, curiosity, or health benefits related to a specific tracking category. It is a practice that has developed a particularly rich design space with the introduction of personal digital technologies enabling self-tracking, such as smartphones, smartwatches, and intelligent wristbands that, among other devices, are now a part of many individuals' lives. Yet, the actual practice, emergence, and use of self-tracking as a method have been discussed for millennia [24]. While most individuals may or may not be aware of technology such as step counters in smartphones, the population of quantified-selfers purposely tracks an array of different variables of their lives, both quantitatively and qualitatively, with various goals, thus contributing to an enhanced understanding of their own behaviors, state, and potentially their own QoL.

The Quantified-Self Community

The quantified-self community arranges conferences and meet-ups where quantified-selfers have the opportunity to present their individual projects. Their motto is *"self-knowledge through numbers."* [25] This community provides a platform for individuals—in principle anyone—to present *what they did, how they did it,* and *what they learned*, from which other individuals can both learn and be inspired to shape their own projects [26]. Throughout the year, location-based group meet-ups are conducted within the quantified-self community. Furthermore, yearly (or bi-yearly) conferences are held in which individuals from all over the world participate [25]. The talks at the conferences are recorded, and the best of which (as selected by the community founders) are curated and published on the community website. These video recordings are the primary self-project materials leveraged for the analysis in this section.

Methods

This section provides insights into the methodological considerations of this study as well as a justification of methodological choices, as there will arguably be different ways to interpret and work with the self-tracking project material gathered by the quantified-self community in both current and future research.

The research in this section largely follows a qualitative research methodology that incorporates the basic principles of hermeneutics [27], which ensures that both the data and their interpretation are conducted cautiously. For this section, it is crucial to examine these results with a pre-understanding of self-tracking projects being both *inductive* and *deductive* in nature, and to acknowledge a potential confirmation bias in the self-tracking projects. As with most qualitative research, given this approach, it is difficult to generalize the results for the specified population, but the goal is not to develop a standing thesis about the quantified-self community;

rather, the goal is to identify current self-tracking trends and patterns within the sample [28].

The self-tracking projects analyzed in this section are in the form of video recordings of a talk related to the project, as presented by its author at a quantified-self conference. The employed video sampling method was *purposive sampling*. This represents several approaches within purposive sampling in qualitative research methods [28]. The video inclusion criteria included the following: the video material had to revolve around a quantified-self project and had to be selected by the community leaders to be uploaded to the quantified-self website, thus narrowing the analyzed examples to those presented within the curated content. Therefore, some of the examined cases are arguably extreme deviant cases, rather than typical self-tracking project cases.

This research was approached as a bottom-up gathering of data, from which 72 video presentations from quantified-self conferences (2015–2019) were examined and analyzed based on pre-existing themes and categorizations, as well as on themes growing from the material while trawling video presentations. Without adding search filters, the website was trawled from the top (the newest) to bottom, within the timeframe for the study (14 weeks in total). The website was updated twice (week 10 and week 14 of 2019) during the project, which means the order of the videos examined was disrupted at least these twice, and new video presentations were added during the study period. One video was deemed to be outside the scope of the project due to it having a vastly different goal to the others (i.e., educational), in which a use case of 'quantified-self' as an educational material was presented. This video was removed from the data set, thus making the actual data set consist of 71 videos. An overview of the 71 analyzed self-tracking projects (authors, titles, and years of publication) can be found in Appendix 1.

While it may be difficult to fully transform the words and personal experiences of self-trackers into quantitative evidence, we aimed to present here the qualitative approach that we employed; categorization and thematic analysis provided the opportunity to count self-tracking projects and partly quantify some of the material presented within each project [28]. The thematic analysis is based on examining project descriptions and identifying two kinds of codes to describe the projects: *descriptive* and *interpretive* codes [29]. The interpretive codes were defined beforehand (along the WHOQOL subdomains), and the descriptive codes were identified afterwards and noted within the dataset, as presented later in this section. The codes were agreed upon by two independent coders.

Results

First, the overall findings are presented per a WHOQOL variable (Sect. 3.4.1), and afterwards a deeper examination is presented of what is tracked and how it is done, starting at a macro level and then proceeding into a micro level analysis of different variables included in each of the projects (Sect. 3.4.2).

One-Dimensional Presentation of Data and Findings

Sex Self-tracker sample comprised n = 26 females (36%) and n = 45 males (64%).

QoL Domains Figure 1.2 presents the coded self-tracking project along the WHOQOL-BREF domains. Each project was assigned one main QoL domain, even if, as presented below, some projects in fact analyzed two or more different subdomains. As seen in Fig. 1.2, the largest domain that was tracked is the physical domain, embracing variables such as "exercise" and "sleep," as detailed later. The distribution of the tracked QoL domains is as follows: physical (n = 41), psychological (n = 26), social (n = 1), and environmental (n = 3).

QoL Subdomains Figure 1.3 presents the domain distribution with the subdomains. It is important to keep in mind that several projects are marked with more than one code, due to the projects sometime being cross-field examinations of parts of the self-tracker's life, or even holding variables from different domains against each-other, such as "location tracking" as a facilitation of "memory tracking of daily life activities," which in WHOQOL codes corresponds to "env-environ" (location) being tracked to keep track of "phy-adl" (activities) (appendix 2) [30]. A total of 84 codes were applied across 71 videos and are presented in Fig. 1.3 below.

Quantitative Vs. Qualitative Project Each project was assigned to one category (quantitative or qualitative) depending on its main goal. While the community is named after the term 'quantified self', qualitative studies are also present in the sample. Studies that have been defined as quantitative rely on datasets derived from smartphone apps (e.g., location) or wearables (e.g., steps). Studies that were defined

Fig. 1.2 Distribution of the WHOQOL Domains Tracked (N)

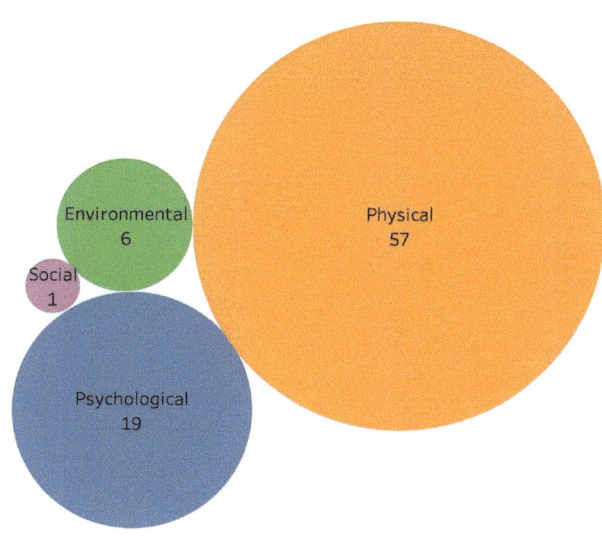

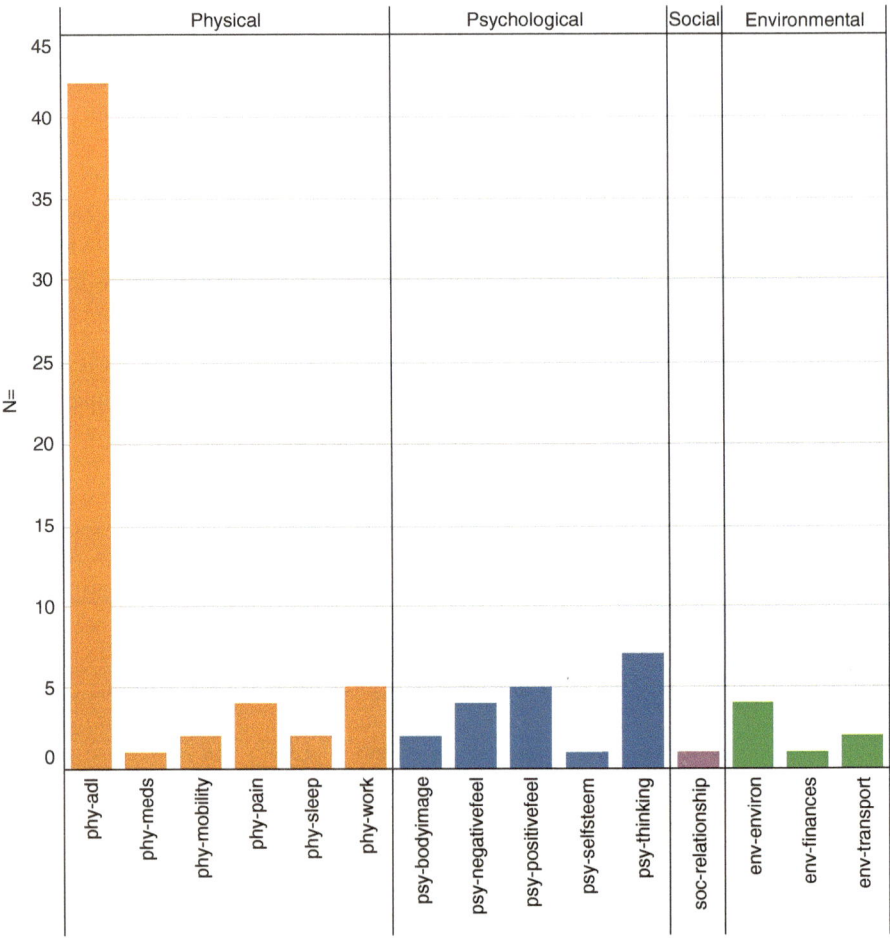

Fig. 1.3 Distribution of the WHOQOL Subdomains Tracked

as qualitative relied on journals/diaries or other kinds of self-reporting tools to describe feelings/emotions or other internal, difficult-to-observe states of individuals. Overall, within the sample projects, quantitative projects (n = 59) were more popular than qualitative projects (n = 12).

Manual Vs. Automatic Tracking Each self-tracking project has an element of tracking quantitative and/or qualitative data, and this tracking can be realized through manual (e.g., paper and pencil) or automatic means (e.g., smartphone phone loggers). Each project was assigned to one category (manual or automatic) depending on its main goal. The results reveal that the majority of projects include automatic logging (n = 42), whereas the projects with manual logging (via e.g., a spreadsheet) are less represented (n = 29).

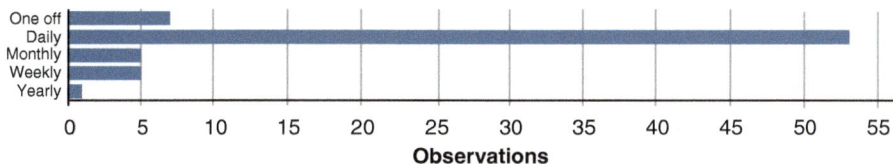

Fig. 1.4 Projects' Self-Tracking Frequency

Tracking Frequency Each self-tracking project has an element of tracking of some data at a specific frequency (from 'one-off', i.e., one observation, not repeated, to repeated 'daily' to less frequent), and it was assigned to a category depending on its main goal. Figure 1.4 presents the tracking frequency, whether it is daily (n = 53), weekly (n = 5), monthly (n = 5), yearly (n = 1) or one-off (n = 7). The daily group is the largest group, followed by one-off projects, weekly, monthly, and finally 'yearly' tracking.

Self-Tracking Project Duration It was difficult to analyze project duration because many projects contained no clear indication of their length and were thus coded as 'N/A' (n = 54). This is due to a variety of reasons, but most commonly it seemed that some projects did not focus on events in a given time duration, but rather on a number of certain events to be tracked in some (unspecified) observational period, selected as convenient, or even defined only post-experimentally by the individual. However, it can be noted that the most common durations range from 1 month (n = 3), 1 year (n = 7), and 3 years (n = 4) to 10 years (n = 3).

Observational Vs. Interventional Projects With regard to whether the project was an observation of an individual state or behavior, or explicit intervention (implying an implicit intention of change of an individual state or behavior), it was found that observations were the most common aim of individual self-tracking (n = 47). Interventions were documented within 22 projects. The last two projects led from observation to intervention on the state or behavior observed at first.

Self-Tracking Tools Used A total of 71 unique commercial and noncommercial devices and digital tools were identified through the course of this study. For the sake of simplicity, both actual devices such as wearables and smartphones (and their apps) were defined as a *"self-tracking tool."* Additionally, these seem to have increasing importance in the quantified-self community as well as in everyday life [24, 31]. Furthermore, several projects relied on data provided from companies such as 23andme and uBiome, which were also defined as a *tool* for self-tracking in this project. Several projects used multiple tools to gather their desired data. Figure 1.5 illustrates the distribution of the various tools across all 71 projects in a diagram, where the tools written in larger fonts correspond to the more common tools and those in smaller ones correspond to less common tools. The color coding is arbitrary.

Fig. 1.5 Tools Used in Self-Tracking Projects

The most commonly used tool was a spreadsheet (n = 7), which has many affordances with regards to data. It allows for data manipulation and statistical analysis and cooperates well, for example, with self-written analytics scripts (e.g., Python) and programs. Other popular tools were wearable devices such as Fitbit (n = 5). These provide basic biometric information, such as current heart rate and sleep schedule as well as an activity counter [32]. The Freestyle Libre (n = 5) was another popular tool in this sample, which is a continuous blood glucose monitor (CGM) essentially developed for diabetics to minimally invasively monitor their blood glucose levels. Due to its ease of use, availability for 'over-the-counter' sale, and affordable price, nondiabetics also use it [33].

Two-Dimensional Presentation of Data and Findings

Sex Vs. Self-Tracked WHOQOL Domain As seen in Fig. 1.6, the sex distribution analyzed against the WHOQOL domains illustrates that an imbalanced distribution exists for physical health tracking (with more male trackers) and an even level of the tracking of psychological aspects of life, even though male presenters represent the vast majority of the sample overall (n = 44). It is important to consider these results for the population analyzed within this section, rather than as results that can be generalized over a wider population of self-trackers.

Sex Vs. Self-Tracked Variable The top portion of Fig. 1.7 presents the distribution of all the self-tracking variables among the projects. The variables written in larger fonts correspond to the most common ones, and those in smaller fonts correspond to less common variables. The bottom portion of the figure presents the distribution of self-tracking variables (i.e., independent variables) sorted by the tracker's sex. There were n = 46 independent variables identified as categories describing the

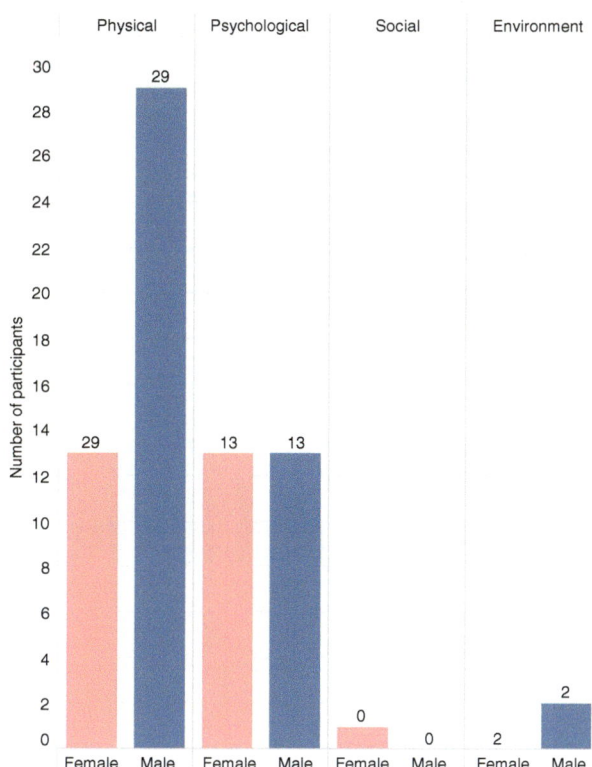

Fig. 1.6 QoL Main Domain Vs. Self-Trackers' Sex

self-tracking project and categorized under the four main WHOQOL domains. There were relatively few repeated variables, but "productivity" (n = 5) was the most observed variable for female trackers, whereas "sleep" (n = 4) was the most observed variable for male trackers. Furthermore, "menstrual cycle" (n = 3) was the second most tracked independent variable for female trackers, whereas "daily activities" (n = 3), "running" (n = 3), and "stress" (n = 3) shared this position for male trackers. The rest of the variables were unique for one or two projects (n = 40 projects).

Discussion

While the results are not generalizable, they do prove one point: the field and interest of the quantified-self projects and inputs to the community are highly diverse and represent a broad spectrum of self-tracking projects.

The distribution of self-trackers' sex is interesting to reflect upon. It suggests that female self-trackers would perhaps be more inclined to conduct a self-tracking study with an emphasis on psychological means of life, whereas male trackers seem

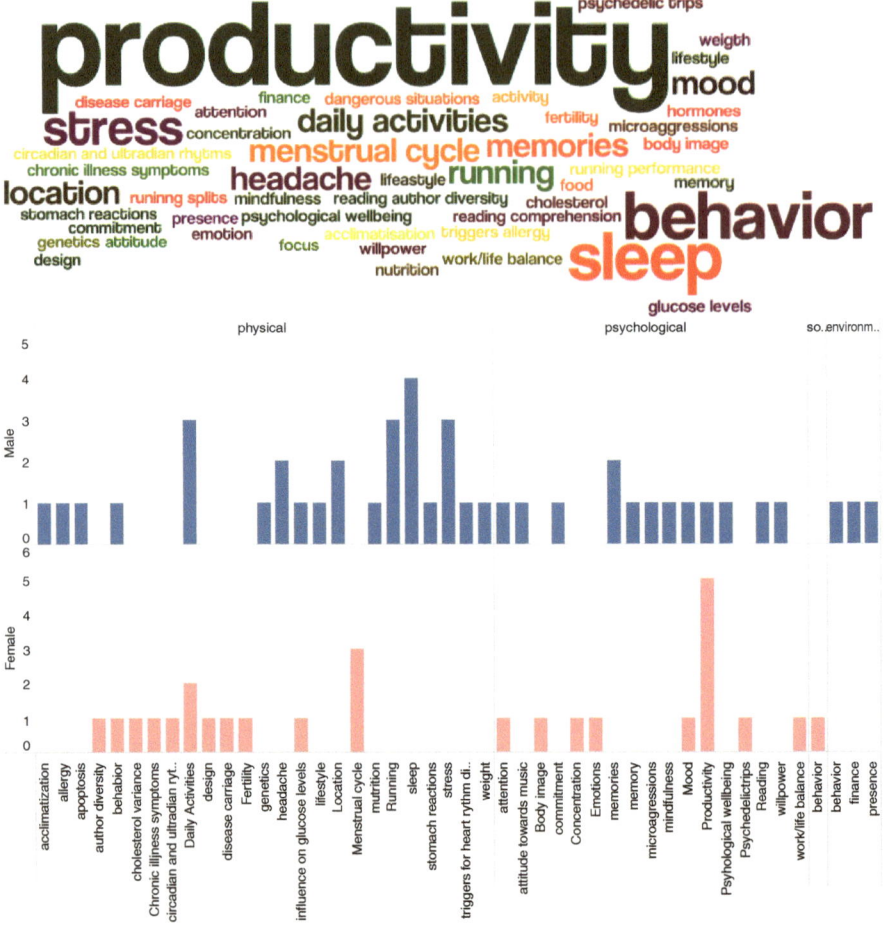

Fig. 1.7 Self-Tracked Independent Variables (top) Mapped by the Sex of Self-Trackers (bottom)

more inclined toward physical and environmental studies. While these results do not necessarily say anything absolute about the correlation of these variables, it is an interesting perspective on the distribution of sex vs. domain-related projects as well as personal self-interest.

As for the tools utilized within the self-tracking projects, they varied as greatly as the variables that were tracked. There was a relatively high number of observations for tools such as simple spreadsheets. The most tracked WHOQOL category was the physical health domain, while at the subdomain level it was "activities of daily life"—a broad category that accommodates numerous types of activities and diverse tracking approaches depending on the type and frequency of activity tracked. The self-tracker community is also interested in tracking variables that are not yet available for autonomous, pervasive tracking leveraging digital tools (e.g., moods and mental states). This can be seen when analyzing the lesser categories—the psychological health domain and more specifically the "thinking, learning, memory, and concentration" variable, which has high interest and includes self-tracking of, for example, mindfulness and willpower. Again, to date, these are almost impossible to track autonomously and pervasively in the daily life of individuals. Our results indicate that the projects are not necessarily dictated by which self-tracking tools exist, but rather by curiosity and personal interest in self-tracking.

The results acquired here also provide an interesting perspective on the quantified-self as a trend itself, since on the one hand we are living in an age where we are "datafying" ourselves at an increasing rate, while on the other hand data protection and privacy questions are arising with the digitalization of everyday means [24]. These questions will become even more urgent to tackle with the emergence of tools tracking psychological health [34, 35].

In addition, the acquired results bring into question the ultimate goal of self-tracking. The idea of the quantified-self stems from the idea of converting aspects of life into numbers and statistics, rather than (qualitative) writing in a daily journal to keep up with life. This can be described as the aspect of self-betterment, in which individuals seek to better understand themselves through numbers and analyses of everyday actions [24]. With roughly two out of three projects being observational in nature, this does not mean that the individuals involved do not seek behavior change, but it may be a distant goal rather than an immediate need. The immediate need focuses on understanding factors influencing one's own behavior and state. Few cases have provided evidence that the results of an observation could be transformed into an intervention, specifically when the results have been too crucial to ignore for the individual self-tracker. One such case was a male individual who felt upset with drivers looking at their personal devices while driving. He decided to investigate his own behavior while driving and set up a quantified-self project to help him reach his goal of spending even less time looking at his phone while driving. He hypothesized that the results would indicate that he was much better than other drivers; however, that was not the case, as he realized he spent up to 25% of his time while driving on his phone (up to around 25 minutes per a day along a 100-minute commute). This observation called for an immediate intervention, and this self-tracker ultimately bought a bike for smaller trips—which also implied he was getting some physical activity while commuting (and not using his phone) [36]. This is just one example of a self-tracking project—including meticulous observation and self-reflection—turning into an intervention.

Limitations

An array of limitations arose in this study, which stemmed from the methodological approach as well as the approach to data analysis. This section presents some of these limitations.

At first, the coding and categories examined in the study were predefined, partially based on the WHOQOL and only loosely based on existing literature. The other categorizations (the study duration and tools used) were agreed upon between two independent coders (having 90% agreement) as to what could prove to be interesting for the domain of QoLT, rather than what was found interesting in previous studies informing the QoL domain. This led to several categories yielding insufficient results or not covering relevant aspects of the research. Two examples are additional coding dimensions discussed along the "work/leisure" category (i.e., the professional or personal domain aim of the project) and the "chronic illness" category (i.e., if the self-tracker was a patient). It was proposed that, when identifying relevant categories, projects related to either work or leisure could yield interesting results; instead, it was almost impossible to define whether a project was solely based on or related to individuals' work or leisure activities, and most categorizations ended up being a mixture of the two, which were then omitted. The self-trackers approached their self-tracking projects—as well as their own lives—holistically, and the projects encompassed these two domains. It was also proposed that chronic illness of the self-tracker, if applicable to his or her condition, could prove interesting to examine, especially if it was explicitly stated to be a major part (and potentially part of the aim) of the project. It turned out that just five out of 71 projects were based on chronic conditions, thus making it difficult to examine this dimension thoroughly.

We only analyzed a small percentage of the whole set of self-tracked projects (using the video database of quantified-self talks) that could have been examined. Our material only provides a narrow view of the overall population and its recent projects and cannot be generalized. Overall, the nature of qualitative studies makes it difficult to replicate their results, as the qualitative understanding and perception of material may differ in the "eyes of the analyzer." [28]

This research does not derive or even suggest correlations between multiple variables, which could have proved to be an interesting aim on its own and should be considered in future work in the field, especially when larger datasets are acquired. Another limitation with regards to multivariate analysis is a lack of acquisition of basic information about the individuals studied. The only personal information collected is sex, which does not distinguish level of education, age, country of origin, cultural background, attitudes, or motivations for specific self-tracked variable(s), nor does it distinguish the level of digital literacy, which may be of importance when discussing the tools employed and use of data. These variables would have been paramount to include in an actual multivariate analysis, but it has not been possible to include them due to the structure of the datasets (i.e., the data were derived from videos).

Concluding Remarks on Self-Tracking and Quality of Life

While it is still too early to conclude anything that could be generalized to the population of the quantified-self community, which could then be applied to quantifying QoL, several valuable observations should be noted. Male individuals were still dominant in the sample of self-trackers, which was also presupposed [24]. The projects have mostly focused on the physical domain of QoL, whereas the social domain has been focused on the least. Tools enabling automatic tracking of variables have been more commonly used as a method for collecting data within the projects. At a more specific level, physical health—"activities of daily life" is the most tracked subdomain across all QoL domains, with the next most common being psychological health—"thinking, learning, memory, and concentration." Overall, individuals track many different categories of their lives and only a few variables are more dominant than others, namely "sleep," "stress," and "running" are the dominant variables for males, whereas "productivity" and "menstrual cycle" are for females. In total, 47 projects were observational, and thus had no inherent goal of changing the behavior of the individual, and 22 projects were designed to be behavioral interventions from the beginning. Two observational projects led to interventions. These behavioral interventions were self-designed and self-tracked and led to sustainable behavior change in most cases.

What we are able to derive from self-trackers is that their attitude, motivation, and overall curiosity-driven and personalized approach are likely to lead to effective change and improvement in the understanding of factors influencing the behavior or state, or to sustainable change in this behavior.

Future Outlook: Importance of Improving Quality of Life Quantification

Quality of life is a critical outcome in daily life and in medicine. Long-term QoL stems from behaviors and states that are repeated frequently; therefore, long-term QoL may be extrapolated through the quantification of these (short-term) behaviors/states. The quantified-self community's efforts illustrate that we can leverage various existing and emerging tools to observe and understand our own behaviors and states, and improve them through self-management as well as meticulously designed, highly personalized interventions. An integral part of future research on QoL technologies and their use in medicine is an interdisciplinary effort for achieving a user-centric and holistic approach, including physical, psychological, social, and environmental viewpoints. An interdisciplinary approach is required because assessment and management of behavior in medicine cannot be readily completed using solely one of the dimensions (e.g., physical) or by one systematic methodological approach (e.g., qualitative or quantitative). Holistic individual assessment and improvement research will bring new approaches to theory, design, methods, measurement, and data analysis specific to each dimension, thus deepening it while enabling breadth. Because of the technological and methodological advances required, such research is a long-term process rather than a short-term self-contained activity.

The QoLT field is in its nascent stage. This book presents an overview of the state-of-the-art methods and tools for quantifying daily life, health, and QoL state through QoLT across all the QoL domains and subdomains. An enhanced understanding of technology-enabled or -supported continuous assessment methods of behaviors and states will allow for an improved understanding and modeling of the short- as well as long-term health and QoL of individuals.

Appendices

Appendix 1

Quantified-Self Talks (Author Name: Title, Year)	Year
Steven Jonas: *Stressing out loud*	2013
Kendra Albert: *The great book project of 2013*	2014
Valerie Lanard: *Breaking the TV habit*	2015
Jim McCarter: *Effects of a year in ketosis*	2015
Ilyse Magy: *Know thy cycle, know thy self*	2016
Ellis Bartholomeus: *Draw a face a day*	2016
Robert Macdonell: *The data is in, I am a distracted driver*	2016
Ahnjili Zhuparris: *Menstrual cycles, 50 cents and right swipes*	2016
Randy Sargent: *Unlocking patterns with spectograms*	2016
Richard sprague: *Microbiome gut cleanse*	2016
Peter Torelli: *Narratives hidden in 20 years of personal financial data*	2016
Abe gong: *Changing sleep habits with unforgettable reminders*	2016
Mark Leavitt: *Daily HRV as a measure of health and willpower*	2016
Akhsar Kharebov: *A smart scale for healthy weight loss*	2016
Shelly Jang: *Can you see that I was falling in love?*	2016
Steven cartwright: *17 years of location tracking*	2016
Paul Lafontaine: *Using heart rate variability to analyze stress in conversation*	2016
Jon cousins: *Why I weighed my whiskers*	2016
Mark Wilson: *Three years of logging my inbox*	2016
Bethany Soule: *Extreme productivity*	2016
Jacek Smolicki: *Self-tracking as an artistic practice*	2016
Robby Macdonell: *The data is in, I am a distracted driver*	2016
Randy Sargent: *Unlocking patterns with spectograms*	2016
Thomas Christiansen: *Over-instrumented running: What I learned from doing too much*	2017
Ahnjili ZhuParris: *Finding my psychedelic sweet spot using R*	2017
Stephen cartwright: *Seeing my data in 3d*	2017

Quantified-Self Talks (Author Name: Title, Year)	Year
Whitney Erin Boesel: *My numbers sucked but I made this baby anyway*	2017
Kyril Potapov: *Tracking productivity for personal growth*	2017
Lillian Karabaic: *What if my life was the economy of A small country?*	2017
Sara Riggare: *Balancing neurotransmitters in neurological illness*	2017
Ellis Bartholomeus: *My health scars*	2017
Robin Weis: *Crying*	2017
Azure Grant: Hot stuff: *Body temperature and ovulatory cycles*	2017
Justin Lawler: *Taking on my osteoporosis*	2017
Azure Grant: *My biological rhythms in sickness and in health*	2018
Thomas Blomseth Christiansen: *Which grasses aggravate my allergies?*	2018
Mikey Sklar: *Three marathons on zero calories*	2018
Justin Lawler: *Tracking glucose as A person without diabetes*	2018
Madison Lukaczyk: *How work distractions affect my focus*	2018
Whitney Erin Boesel: *Cholesterol levels while nursing*	2018
Benjamin best: *My blood values from diet and other activities*	2018
Albara Alohali: *Running storytelling*	2018
Lydia Lutsyshyna: *Separating work and home*	2018
Benjamin Smarr: *Does my stomach anticipate my meals?*	2018
Shamay Agaron: *Tracking breathing to control my focus*	2018
Maggie Delano: *Quantifying my Phd: Pomodoros and productivity*	2018
Jessica Ching: *Learning an impossible form of exercise*	2018
Kyrill Potapov: *What Insidetracker taught me about my five-day fast*	2018
Daniel reeves: *Tracking my personal reliability*	2018
Fah Sathirapongsasuti: *Blood oxygen on Mt. Everest*	2018
Mad ball: *A self-study of my Child's genetic risk*	2018
Aaron Parecki: *Ten years of tracking my location*	2018
Aaron Yih: *Tracking across generations*	2018
Alec Rogers: *What I'm learning from my meditation app*	2018
Jordan Clark: *Quantifying the effects of microaggressions*	2018
Jakob Eg Larsen: *My headaches from tracking headaches*	2018
Todd Greco: *Building my external brain*	2018
Lillian Karabaic: *#100daysofqs: Daily art from data*	2018
Ralph Pethica: *Finding the optimal training zone*	2018
Anna Franziska Michel: *Using running and cycling data to inform my fashion*	2018
Eli Ricker: *Tracking what I do versus what I say I'll do*	2018
Shara Raqs: *Estrogen and invention*	2018
Stephen Maher: *A decade of tracking headaches*	2018
Valerie Lanard: *Learning from excuses*	2018
Eric Jain: *Four weeks of blood sugar tracking*	2019

Quantified-Self Talks (Author Name: Title, Year)	Year
Kyrill Potapov: *Finding my optimum Reading speed*	N/A
Rocio Chongtay: *Quantified brain and music for self-tuning*	N/A
Mark Drangsholt: *What causes my heart rhythm disorder*	N/A
Steven Jonas: *Memorizing my daybook*	N/A
Steven Jonas: *Spaced listening*	N/A
Ari Meisel: *Experiments in treating my Crohn's disease*	**N/A**

Appendix 2

WHOQOL Codes used for categorizing projects from the quantified-self community, following past work of Wac [37].

QoL Domain	Subdomains
'Phy': Physical health	Phy-adl, phy-meds, phy-energy, phy-mobility, phy-pain, phy-sleep, phy-work
'Psy':Psychological health	Psy-bodyimage, psy-negativefeel, psy-positivefeel, psy-selfesteem, psy-beliefs, psy-thinking
'Soc':Social relations	Soc-relationships, soc-support, soc-sex
'Env':Environment	Env-finances, env-freedom, env-healthcare, env-home, env-info, env-leisure, env-environ, env-transport

References

1. Constitution of the World Health Organization. In: Basic Documents. forty-fift.; 2006:1–18.
2. Callahan D. The WHO definition of "health.". Hast Cent Stud. 1973;1(3):77. https://doi.org/10.2307/3527467.
3. Huber M, Knottnerus JA, Green L, et al. How should we define health? BMJ. 2011;343(jul26 2):d4163. https://doi.org/10.1136/bmj.d4163.
4. Wilson IB, Cleary P. Linking clinical variables with health-related quality of life. A conceptual model of patient outcomes. JAMA J Am Med Assoc. 1995;273(1):59–65. https://doi.org/10.1001/jama.273.1.59.
5. Farquhar M. Elderly people's definitions of quality of life. Soc Sci Med. 1995;41(10):1439–46. https://doi.org/10.1016/0277-9536(95)00117-P.
6. Bowling A. What things are important in people's lives? A survey of the public's judgements to inform scales of health related quality of life. Soc Sci Med. 1995;41(10):1447–62. https://doi.org/10.1016/0277-9536(95)00113-L.
7. Wit M, Hajos T. Quality of life. In: Encyclopedia of behavioral medicine. New York: Springer; 2013. p. 1602–3. https://doi.org/10.1007/978-1-4419-1005-9_1196.

8. Alexandrova A. A philosophy for the science of Well-being, vol. 1. Oxford University Press; 2017. https://doi.org/10.1093/oso/9780199300518.001.0001.
9. Kalaitzakis E. Quality of life in liver cirrhosis. In: Handbook of disease burdens and quality of life measures. New York: Springer; 2010. p. 2239–54. https://doi.org/10.1007/978-0-387-78665-0_131.
10. Lavallee DC, Chenok KE, Love RM, et al. Incorporating patient-reported outcomes into health care to engage patients and enhance care. Health Aff. 2016;35(4):575–82. https://doi.org/10.1377/hlthaff.2015.1362.
11. Mayo NE, Figueiredo S, Ahmed S, Bartlett SJ. Montreal accord on patient-reported outcomes (PROs) use series–paper 2: terminology proposed to measure what matters in health. J Clin Epidemiol. 2017;89:119–24. https://doi.org/10.1016/j.jclinepi.2017.04.013.
12. Katz S, Ford A, Moskowitz R, Jackson B, Jaffe M. The index of ADL: a standardized measure of biological and psychosocial function. JAMA. 1963;185(12):914. https://doi.org/10.1001/jama.1963.03060120024016.
13. Gill TM. A critical appraisal of the quality of quality-of-life measurements. JAMA J Am Med Assoc. 1994;272(8):619–26. https://doi.org/10.1001/jama.272.8.619.
14. Linton M-J, Dieppe P, Medina-Lara A. Review of 99 self-report measures for assessing Well-being in adults: exploring dimensions of Well-being and developments over time. BMJ Open. 2016;6(7):e010641. https://doi.org/10.1136/bmjopen-2015-010641.
15. The World Health Organization quality of life assessment (WHOQOL): position paper from the World Health Organization. Soc Sci Med. 1995;41(10):1403–9. https://doi.org/10.1016/0277-9536(95)00112-K.
16. World Health Organization. WHOQOL-BREF: introduction, administration, Scoring and Generic Version of the Assessment, Field Trial Version; 1996. https://www.who.int/mental_health/media/en/76.pdf.
17. Levitis DA, Lidicker WZ, Freund G. Behavioural biologists do not agree on what constitutes behaviour. Anim Behav. 2009;78(1):103–10. https://doi.org/10.1016/j.anbehav.2009.03.018.
18. Rosenthal R, Rosnow R. Essentials of behavioral research. Methods and data analysis third. 2008;
19. Nock MK, Kurtz SMS. Direct behavioral observation in school settings: bringing science to practice. Cogn Behav Pract. 2005;12(3):359–70. https://doi.org/10.1016/S1077-7229(05)80058-6.
20. Wac K. Quality of life technologies. In: Encyclopedia of behavioral medicine. New York: Springer; 2019.
21. Skevington SM. Measuring quality of life in Britain: introducing the WHOQOL-100. J Psychosom Res. 1999;47(5):449–59. http://www.ncbi.nlm.nih.gov/pubmed/10624843
22. Skevington SM, Lotfy M, O'Connell KA. The World Health Organization's WHOQOL-BREF quality of life assessment: psychometric properties and results of the international field trial. A report from the WHOQOL group. Qual Life Res. 2004;13(2):299–310. https://doi.org/10.1023/B:QURE.0000018486.91360.00.
23. Organization WH. Programme on mental health. WHOQOL User Manual. 1998;
24. Lupton D. The quantified self. Polity. 2016;
25. Quantifiedself.com. Published 2019. Accessed November 8, 2019. https://quantifiedself.com
26. Quantified Self Meetups. Accessed November 8, 2019. https://quantifiedself.com/meetups/
27. Gilje N, Grimen H. Samfundsvidenskabernes Forudsætninger: Indføring i Samfundsvidenskabernes Videnskabsfilosofi. København: Hans Reitzels Forlag; 2002.
28. Bryman A. Social research methods. Oxford University Press; 2016.
29. King N, Horrocks C. Interviews in qualitative research. SAGE.
30. Show and Tell: ten years of tracking my location.
31. Papathanassopoulos S. Media perspectives for the 21st century. Routledge. 2011; https://doi.org/10.4324/9780203834077.

32. Fitbit. Accessed August 15, 2019. https://www.fitbit.com/dk/home
33. Freestyle Libre. Accessed August 15, 2019. https://www.freestylelibre.co.uk/libre/
34. Wulfovich S, Fiordelli M, Rivas H, Concepcion W, Wac K. "I Must Try Harder": Design Implications for Mobile Apps and Wearables Contributing to Self-Efficacy of Patients With Chronic Conditions. Front Psychol. 2019;10:2388. https://www.frontiersin.org/article/10.3389/fpsyg.2019.02388.
35. Swan M. Health 2050: the realization of personalized medicine through crowdsourcing, the quantified self, and the participatory biocitizen. J Pers Med. 2012;2(3):93–118. https://doi.org/10.3390/jpm2030093.
36. Show and Tell: The Data is In, I'm a Distracted Driver.
37. Wac K. From quantified self to quality of life. 2018;83-108 https://doi.org/10.1007/978-3-319-61446-5_7.

Part II
Physical Health

Chapter 2
Assessing Activity of Daily Living through Technology-Enabled Tools: Mobility and Nutrition Assessment: MiranaBot: A Nutrition Assessment Use Case

Mirana Randriambelonoro

Introduction

Developed countries are facing the challenge of ageing societies, lack of infrastructure for healthcare and high cost of care. Researchers have been attempting to answer these problems by using innovative technology to promote healthy ageing [1]. Rather than the absence of disease, "healthy ageing" is defined as a process that enables older people to continue to do the things that are important for them such as performing activity of daily living, maintaining social contact and conserving dignity [2–4].

Activities of Daily Living (ADLs) include individual's basic physical needs such as dressing, feeding, personal hygiene, continence and transferring [5]. To assess individual's level of independence, researchers and clinicians often measure the ability to perform ADLs' tasks through different approaches such as self-report and performance-based measures [6]. However, these methods which are commonly completed by caregivers are not completely free of bias and are sometimes subject to under or overestimation of the individual's true functioning [7].

Besides, the Quantified Self (QS) movement is gaining more and more attention. It is an emerging trend, which allows individuals to self-monitor or self-track their daily life activities, to analyze and self-reflect on their behaviors, and to bring potential change in their daily habits [8, 9]. Furthermore, with the recent development of monitoring technologies and artificial intelligence techniques, the possibilities to automatically distinguish between different ADLs and detect unexpected events such as falls, have become an interesting, important and potentially impactful topic [10]. How efficient these tools are, to assess and improve ADLs, needs further investigations.

M. Randriambelonoro (✉)
Division of eHealth and Telemedicine, Geneva University Hospitals and University of Geneva, Geneva, Switzerland
e-mail: Mirana.Randriambelonoro@unige.ch

© The Author(s) 2022   27
K. Wac, S. Wulfovich (eds.), *Quantifying Quality of Life*, Health Informatics,
https://doi.org/10.1007/978-3-030-94212-0_2

This chapter will present the standard validated scales for ADLs and the current researches on the use of technologies to assess one's ability to perform ADLs, mainly indoor-outdoor mobility and nutrition. To expand our overview on the topic, the focus will be on elderly as well as on younger and healthier individuals. We then follow with a nutrition use case assessment based on a conversational agent called "MiranaBot".

Conclusive remarks will emphasize the necessity to consider behavioral science along with cultural and environmental context to elaborate personalized monitoring and intervention strategies, which will be translated to innovative solutions to promote independence at home.

State of the Art

Activities of Daily Living (ADLs)

Created by Sidney Katz in 1950 [11, 12], the term "Activities of Daily Living" refers to the fundamental skills required to care for one-self and live independently. ADLs include dressing, feeding, personal hygiene, continence and transferring [5]. ADL dependence is often associated with lower quality of life, higher healthcare costs, higher risk of mortality and institutionalization [13–15]. Assessing individual's ADLs is therefore essential, especially for vulnerable population who may need assistance in performing these activities. In this chapter, the focus will be on indoor/ outdoor mobility and nutrition.

Mobility is defined as the ability to move oneself from one point (or state) to another inside and outside the home environment to maintain a certain level of independence [16, 17]. As activity restriction is associated with various physical deconditioning effects and decreased rates of social involvement, mobility is closely linked to health status and quality of life. Researchers has been attempting to propose a mobility model based on multiple factors such as cognitive, social, physical, environmental, financial as well as gender, culture and age, which all may influence individual's mobility in different way [16]. As people age, decreased mobility is often predictor of falls and physical disabilities [18]. Consequently, there is a growing interest in measuring mobility and determining the factors that influence mobility to strengthen the physical and functional capacities of older adults.

Nutrition capacity or the ability to eat is a part of the essential ADLs to maintain an independent life [5]. Food intake plays a crucial role in individual's life and momentary, as well as long-term health and quality of life. Research on nutrition science, human nutrition, preventive nutrition, clinical nutrition and public health nutrition has been specifically on the rise the past years [19], to treat and prevent medical conditions such as diabetes [20] or sarcopenia [21] and ensure a better quality of life [22]. Reliably measuring not only the quantity but also the quality and the frequency of food intake is therefore very important to picture one's nutritional habits and to be able to recommend adapted and personalized advice or treatment to the person.

Validated Scales for ADLs

Several methods, such as self-report, caregiver report, or direct observation, have been used to measure the degree of independence in ADLs. Self-reporting is easy to perform and are often used when direct observation is not possible, or if the person presents no cognitive deficiency. Nevertheless, the self-reports may be biased and less valid when ones have difficulties to evaluate their functional capacity [23, 24]. Performance-based assessments imply in-lab assessment of the individual's ability and skills for self-care using mock settings, role play and simulation of real-world environment. They may provide accurate information about the ability to execute the ADLs but typically require more qualified assessor to be performed, compared to self-reporting or informant reports. Table 2.1 presents four main validated scales widely used to measure ADLs in older adults or patients with cognitive or mobility impairment as well as their benefits and drawbacks.

Although these scales are validated and standardized, they are, first, often dependent of an informant or a caregiver, which may include biases and, second, mainly performed in the control settings of the hospital. Being able to automatically assess ADLs at the home environment would benefit individual's health in terms of disease prevention and treatment but would also enhance individual's quality of life and independence. In the following section, we will review the current researches on the use of technologies to assess one's ability to perform ADLs, mainly indoor-outdoor mobility and nutrition.

Technology-Enabled Tools to Assess Indoor-Outdoor Mobility

Over the last few years, Global Positioning System (GPS) technology, embedded in personal smartphone, has become the leading solution for outdoor positioning, and new technologies for indoor positioning and navigation exponentially expanded. Smart devices with embedded inertial sensors, radio beacons and image processing are just few examples of the systems deployed to assess indoor mobility. As most of the times, choosing one solution for any type of mobility assessment scenario is proven to be not possible, ongoing developments tend to combine different technologies to find the best balance between precision, cost, robustness, scalability and energy consumption [31]. With respect to the ADLs, such solutions enable assessment of the indoor-outdoor mobility. Some of them may enable recognition of activities such as transfer, use of stairs, dressing, meal preparation or use of rooms (bathroom, toilet and kitchen). Privacy concerns and environmental context recognition are often discussed in the research when choosing one or more systems for assessing mobility. In this section, we will review the current research on systems for mobility measurement by emphasizing two main categories: activity recognition and wireless indoor-outdoor positioning systems.

Table 2.1 Examples of validated scales for ADLs

Scale name	Recall period	Items	Scoring	Advantages	Disadvantages
Katz index of independence [25]	3 to 4 weeks	Bathing Dressing Toileting Transferring Continence Feeding	**Binary score:** Fully independent, dependent **3-point score:** Fully independent, some assistance, dependent **Total score:** 6: Fully independent 4: Moderately impaired 2: Severely impaired	Reliability Validity Accepted Standardized Useful for a first assessment	Insensitive to variation in low level of disabilities Not suitable for health survey
Older Americans resources and services (OARS) [26]	1 week	Feeding Dressing Grooming Walking Transferring bathing	**Score:** Coded from 0 (completely unable) to 2 (without help) **ADL rating:** From 1 (completely impaired) to 6 (excellent)	Reliability Validity Comprehensiveness	Biased towards overestimation of the capacity Necessary to have a reliable caregiver
Barthel ADL index [27, 28]	4 weeks	Feeding Bathing Grooming Dressing Toilet Transfers Mobility Stairs Bowels Bladder	**Score:** Performance rated by level of assistance needed, with each task. **Maximum score:** 100 points Independence in transferring and stairs weighted more heavily than other ADLs.	Performance-based measure Reliability Validity More nuanced picture of disability able to detect more subtle changes in functioning	Do not take into consideration individual factors or contextual differences
Functional Independence measure (FIM) [29, 30]	3 to 4 weeks	18 items to evaluate 6 functional areas. **Motor-FIM:** 13 items based on the Barthel index. **Cognitive-FIM:** 5 items	**Score scale:** 1 (total assistance) to 7 (complete independence) **Maximum score:** 126 **Minimum score:** 18	Performance-based measure Reliability Validity Comprehensiveness Include social cognition and communication	Not adapted to home patients who are impacted with the sensitive change in their surroundings

Activity Recognition

Activity recognition is focusing on distinguishing a person's activities (sedentary: sleeping, sitting /active: walking, using stairs, cooking, etc.) in a given space is now currently performed using two types of technologies: vision-based systems and sensor-based systems [32, 33].

The vision-based recognition focuses on the processing and evaluation of video data from cameras or low-cost integrated depth sensor. Cameras installed in fixed positions can be fitted to the environment. In this case, the goal is to locate a moving target in images that are taken by one or more cameras [31]. Commons examples include video-based exercise program which allow user, caregiver or family to track daily exercise. Ayase et al. [34] investigated a "multimedia fitness exercise progress notes" which forwarded the video recording of elderly exercise movements to a research center, which were used to identify the duration, the distances and the angle of the movement. Another study [35] tried to analyze sit-to-stand motion from monocular videos by comparing the images received with the motion of a human body model. Li et al. [36] proposed a system that could extract gait features by using images from two orthogonal viewing cameras. Although optical systems show better accuracy nowadays [37], they remain intrusive, costly and are often limited to specific environments in which variable lighting and other disturbances can be controlled. However, attentions are now given to mobile and low-cost solutions based on camera embedded in mobile phone [38, 39]. Kahlert and Ehrhardt [40] used a photo-based ambulatory assessment to measure out-of-home mobility and social participation of older adults. The use of video and image may however be recognized as potentially privacy-threatening.

Sensor-based recognition includes the use of pressure sensors, accelerometers, magnetic sensors and gyroscopes embedded within an environment, where the individual is active (or not). On one hand, smart floors and smart furniture are based on pressure sensitive sensors installed on or under the corresponding materials and focus either on load cell systems or pixelated surfaces. Load cell systems use separated sensors embedded on tiles and try to compute user location by sensing vertical force applied to the tiles [41, 42]. Pixelated surfaces on the contrary embed many sensors working as binary switches when someone is standing or moving on the surfaces [43–45]. Other furniture such as chairs and beds can also be embedded with pressure sensors to assess user mobility. Merilampi et al. [46] tested an exergame coupled with a smart chair where users were asked to stand up and sit to play the game. On the other hand, the emergence and progress of wearables technologies have opened a door to develop different tools to assess individual's mobility. Non-invasive off-the-shelf sensors such as wearables, smart clothes or smartphones have gained more and more attention for their potential to recognize indoor and outdoor activities while being continuously embedded on/around the user. Several reviews [47–49] discuss the potential of physical activity monitoring using accelerometry techniques to enable activity measurement of individuals in a free-living environment. Steps and activity intensity are often measured through an off-the-shelf sensors or a smartphone and used to estimate energy expenditure following a physical activity. Furthermore, Berglind et al. [50] showed in their study that smartphone apps can be as efficient as supervised exercise sessions (assessing the activities of the user) to improve body composition and cardiorespiratory fitness.

Wireless Indoor-Outdoor Positioning Systems

Through the past few years, several wireless technologies and techniques have been used and researched to assess indoor-outdoor positioning and navigation. Infrared, Ultrasound, Wi-Fi (Wireless Fidelity), RFID (Radio Frequency Identification), Bluetooth are mainly used indoor [31] while GPS (Global Positioning System) often represent outdoor solution [51].

All of them present advantages and drawbacks, which have to be considered while choosing the right solution. Infrared transmits infrared signals and compute the distance between the emitter and the receiver to locate the user. The main advantages are the absence of radio electromagnetic interference and an adjustable power of transmission. However, it can be costly to implement, less accurate and often requires a line of sight between transmitter and receiver to function properly [52]. Ultrasound systems use similar techniques by transmitting ultrasonic waves to assess the distance between fixed receivers and a mobile target. Synchronization of multiple receivers is then necessary. This system is relatively low-cost and could bypass indoors obstruction, but large-scale implementation remains complex and temperature variation could influence the speed of the sound [53]. Wi-Fi-based and Bluetooth-based systems are also known for being low-cost as it can localize every Wi-Fi compatible device without any extra installation, but it has lower accuracy and could induce large power consumption for smartphone [54, 55]. RFID system works without direct line of sight. Nevertheless, the intensity of the signal depends on the density of the obstacles in the building, and therefore precision is always reduced. However, RFID technology has more benefit such as better data rate, high security and compactness [56]. GPS-based applications are now widely used to track outdoor mobility. GPS-based navigation services are also often investigated in the context of facilitating visually impaired people mobility [57]. When combined with other devices such as accelerometers, the GPS system shows better accuracy in identifying the type of activity [58]. Weber and Porter [59] examined the feasibility of using GPS watch and accelerometers to monitor walking and community mobility in older adults and found promising results for monitoring community mobility patterns.

Technology-Enabled Tools to Assess Nutrition

Regarding techniques and technologies to assess nutrition which is important when assessing the meal preparation and feeding abilities of an individual, we also distinguish between vision-based systems, sensor-based systems and self-assessment via mobile applications.

Ghali et al. [60] designed a system using cameras to provide real-time feedback to stroke patients while performing daily activities such as making a cup of coffee. They used histogram-based recognition methods to compare the recorded movement to a target task model and identify key events. P-W Lo et al. [61] investigated a vision-based dietary assessment approach to estimate food volume by using mobile phone 3D camera, deep learning and deep sensing techniques. They concluded an efficiency in portion size estimation even under view occlusion (food items scanned only from the front view). Pettitt et al. [62] combined micro-camera images with food diaries and improved the accuracy of dietary assessment. Their system also provided valuable information on macronutrient intake and eating rate. Gemming and al [63]. used a wearable camera to capture and categorize the environmental and social context of self-identified eating episodes and found that these contexts could be assessed objectively by using wearable cameras. Most of the eating episodes were at home and indoors, seated at tables or sofas but also standing or at a desk.

Sensor-based systems include the uses of furniture embedded sensors such as smart fork, smart table or smart plate. Zhou et al. [64] developed an unobtrusive smart table surface and were able to distinguish between different eating actions, such as cutting, scooping and poking. They could also indicate the type of food taken and the way the meal is consumed. Qualitative study [65] evaluating smart fork to decrease eating has been conducted and demonstrated user awareness of eating rate. However, the incapacity of the fork to consider meal characteristics, were less appreciated. Huang et al. [66] prototyped a smart utensil which analyzes light spectra reflected by foods that differ for every food ingredient. They could recognize up to 20 types of aliments with 93% accuracy. Mertes et al. [67] proposed a novel plate system that can detect weight and location of individual bites during meals. They used a compartmented plate were filled with different type of food and initially weighed. Depending on the weight variation, they could detect the category of food eaten by the user.

The most common techniques to assess nutrition are through self-assessment via an application or a platform. Todays' systems are not only trying to measure food intake but based on nutrition habits, recommends healthy diet. Food recommender systems have received increasing attention to help people adopt healthier eating behavior. These systems focus on suggesting appropriate food items based on individuals' preferences and health conditions [68, 69]. Nevertheless, despite the extensive research and the large number of existing nutrition applications, food recommender systems are still facing many challenges in terms of nutrition habits tracking and delivery of the proper recommendations [70, 71]. People find it often difficult and time consuming to enter manually their food every day in the system. Many studies are also facing the uncertainty of the information given by the users as they may not know or tend to forget what they have eaten which makes it more challenging for the system to recommend the correct food item. Additionally, many studies have shown that the recommendation is not necessarily followed by a change in the behavior [72]. There is still a lack of understanding in how to incorporate efficiently behavior change techniques into a food recommender system.

Use Case: Nutrition Assessment

With the recent development in natural language understanding, conversational agents defined as a dialog system that supports human-like conversational interaction, have gained popularity over the past three years. In the nutrition domain, they allow to collect user data in an easy and user-friendly manner. Researchers have developed a Web-based prototype of a speech-controlled nutrition-logging system, which converts the entry spoken by the users into calories intake [73]. Others researched a chatbot that help people reduce their meat and increase fruits and vegetables consumption [74]. Users were able to set nutrition goals themselves and had a follow-up with the system every day. Although, only 11% could reach their objectives, more than half of the participants showed positive changes in their nutrition habits.

We developed a conversational agent called "MiranaBot" that helps the user to assess their nutritional patterns and to be aware of their eating habits in terms of variety and regularity.

Interviews, observations and focus groups, presented in the methodology below, were conducted to collect the requirements from users and caregivers for an efficient nutritional conversational agent.

Rather than focusing on food quantity and nutritional value, the system targets the variety of the individuals' diet. From a regular description of the individual's meals during a certain period, "MiranaBot" is able to assess the quality of their nutrition, identify the foods they need to consume less, and explain why and how to replace them. The system proposes personalized solutions tailored to the older adults' needs and context (Fig. 2.1).

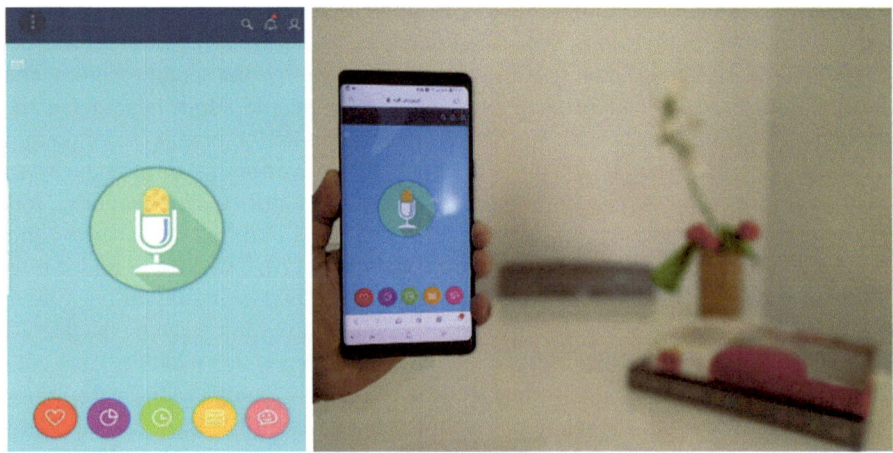

Fig. 2.1 Graphical User Interface—MiranaBot—Speech interface

Methodology

We conducted a qualitative study with 10 obese and diabetic patients, 2 nutritionists and 1 physician with a specific expertise on therapeutic education of chronically-ill patients from University of Geneva hospital in Switzerland. Nutrition and overall self-management with respect to nutrition, usually implying the need for nutritional behavior change, plays a vital role in the self-management of disease for these patients; their future health state depends on the daily nutrition patterns. Observation sessions, focus group and semi-structured interviews were performed to get insights from the different participants. The prototype was developed through an iterative process, following the agile development principles [75]. A first prototype was developed within 4 weeks and changes were quickly implemented according to the participants feedback and in collaboration with them.

To understand patient's current self-management with respect to nutrition, and identify barriers for adopting a healthy nutrition, we first observed 4 consultation sessions led by a physician at the therapeutic education center. In addition, we followed a group of 10 patients for one-week period, while they came to the center for a workshop to learn about their disease. To inquire about their needs and their issues, we conducted a focus group with the patients, while presenting them the first prototype of the application. We conducted a semi-structured interview with one physician and two nutritionists, who helped us to understand the real issues faced by the patients as well as the complexity of nutritional behavior change. The nutritionists participated in creating the nutritional database, we later used in the application.

We then selected the corresponding Behavior Change Techniques (BCTs) from the literature aiming at and answering and fulfilling the needs and requirements identified for the MiranaBot. These BCTs are then converted into a set of functionalities that we incorporated in MiranaBot.

Factors Influencing Nutrition Self-Management

Our observation followed by the discussion with the patients and the health professionals gave us insights on the barriers for patients to adopt a healthy nutrition as well as their needs and requirements for a successful behavior change. N1, N2 and P are used to refer to the nutritionists (N) and the physician (P). Patients are addressed as P1 to P10.

Barriers to Healthy Nutrition

Lack of regularity, lack of variety, false belief and hunger unawareness were identified as the main barriers to healthy nutrition.

Lack of Regularity

Nutritionists identified the lack of regularity in eating as the first issue commonly faced by their patients. N1 said *"Most of our patients are not eating three times a day. They are not taking breakfast, neither a lunch, which make them starve for a big quantity of food later during the day or at night."* P6 mentioned *"I don't really have time because of my job, so I'm just grabbing some food that I can easily find when I finish my round. It's easier like that for me."* P3 added *"I don't eat breakfast, I'm used to that. At lunch, I just grab some sandwiches, but I come back home, I think I deserve a big dinner. I'm usually very hungry at night."*

Lack of Variety

Health professionals also mentioned the lack of food variety as a recurring issue observed in chronically-ill patients. N1 stated *"We often share the optimal plate during our workshop because most of our patients have difficulty to bring variety to their food. They are often eating the same aliment every day"*. P4 said *"I have to admit that I'm always eating pasta and pizza, because I like them, and you know, it's not expensive."* P9 added *"I'm always buying the same thing when I do my groceries. I don't allow a lot of surprise about my food. I already have my personal habits."*

False Belief

Following a consultation with a patient, the physician shared with us that one of the most difficult issue they face as health professionals is patients' false beliefs. P said, *"As one of my patients heard that olive oil was good for the heart, he started to put a big quantity of olive oil in everything he eats."* During the session with her doctor, P2 shared *"You know doctor, I only eat fresh cheese every day. Whatever is the menu, I always have my fresh cheese. It is fresh, so for me that's healthy."*

Hunger Unawareness

Nutritionists and physician both mentioned the complexity of nutritional behavior change due to the different external and internal factors influencing people to eat. P shared with us *"Most of the time, when patients comes here, it is not only about food. It's about how they face the loss of their loved ones or the family problem they have. It's much more than a food addiction."* N2 stated *"It often happen that the patients do not eat because they feel hungry but rather to fill a void inside them."* P7 added *"You know, I just eat automatically because there is food there, I don't think too much about why."* P4 stated *"I'm not always feeling good after eating, that's how I somehow realize maybe that was too much."*

Needs and Requirements

Monitoring, education, empowerment and practicality were identified as the main needs and requirements and facilitators for nutrition behavior change approach.

Monitoring

Patients and health professionals both shared the need for a monitoring tool to raise self-awareness on patients' nutritional habits. P5 said *"It's true that I don't always remember the food I ate. So, I don't really realize how I eat. If a tool can help me know that, why not?"* P7 added *"The nutritionist asked me to describe what I ate on a diary, so that she can know how to help me. I guess it's important to monitor what we eat but it is not always easy."* N2 stated *"Just being aware of what and how they eat would already represents a big advancement because a lot of my patients do not even think of how much and why they eat. That's the real issue. It is essential for them to monitor their eating habits."*

Education

The plethora of nutritional advice and beliefs accessible on the web make it often difficult for patients to identify trustworthy information. Patients appreciated the one-week workshop organized by the hospital as it taught them trustworthy and practical tips to change their habits. P6 said *"You know before coming here, I thought drinking smoothie should be fine as it is just fruits. I did not know that fruits also contain sugar and drinking a lot of smoothie is really not healthy."* N1 mentioned *"During our session here, we teach our patients to always eat proteins, carbs and vegetable. We share simple tricks to measure portion. Our patients need that."*

Empowerment

From the perspective of the health professionals, there is no "one size fit all" solution for everyone. Their role remains to give the necessary education. It is then up to the patients to build their own strategy and apply it for changing their behavior. P4 effectively confirmed *"I won't like it if you tell me what I have to eat. I should be free to choose what I like as long as I follow your general recommendation."*

Practicality

Most of our participants expressed the need for a practical and simple solution. P8 shared *"I used a nutrition application before to record my food, but it was too time consuming, so I gave up."* N2 mentioned *"Specifying the quantity of your pasta and beef is not so obvious. Patients need a hands-on tool that would motivate them and make their life easy."*

Summary

Lack of regularity, lack of variety, false belief and hunger unawareness were identified as the main barriers to healthy nutrition; whereas monitoring, education, empowerment and practicality were identified as the main needs and requirements for nutrition behavior change. Considering these findings, we selected appropriate BCTs to include inside our systems, as described in the next sections.

MiranaBot: Behavior Change Theory and Techniques in Action

Multiple researches have been conducted to identify motivational strategies for behavior change, especially in the nutrition and physical activity domain [76, 77]. Studies grounded in behavior theory appeared to show a positive impact on patient

Table 2.2 Mapping of requirements/barriers and the proposed BCTs

Needs and requirements	Barriers to healthy nutrition	BCTs
Monitoring		Self-monitoring (eating habits history)
	Lack of regularity	Personalized visual feedback (visual cue showing the time of eating)
	Lack of variety	Personalized visual feedback (illustrating the type and the percentage of food eaten on plate)
Education	False belief	Personalized education (through the chatbot)
Empowerment		Goal setting (with information on goal progression)
	Hunger unawareness	Self-awareness (asking the reason of eating before each meal)
Practicality		Simple and user-friendly system (speech interface, visualization)

health behavior. Sawesi et al. [78] demonstrated a significant relationship between theory-based health behavior change intervention and patient engagement. Various systematic reviews and meta-analysis [79–81] showed that efficacy of dietary intervention is associated with well-defined behavior change techniques like: social support, increased contact frequency with the therapeutic team, goal setting, just-in-time feedback on behavior, review of goal progress and self-monitoring.

There are many models and theories on behavior change, some focusing on parts of the behavior change process [82], while others are more holistic, aiming to include all factors that can influence behavior change [83]. "MiranaBot" is based on the Self-regulation theory [84] and the COM-B (Capability, Opportunity, Motivation, Behavior) model [82]. Self-monitoring, personalized visual feedback, goal setting, self-awareness and personalized education were the behavior change techniques incorporated in the tool. Table 2.2 summarizes the mapping of the different requirements, barriers with the proposed BCTs (Fig. 2.2).

Self-Monitoring

"MiranaBot" allows the users to track their nutrition easily by asking them to briefly describe what they ate after each meal in terms of portions. The system will assess the portion size and initially train the users on the portion size to help them describe their meals as best as possible.

Personalized Visual Feedback

"MiranaBot" provides visual feedback every day regarding nutrition variety and regularity. Food variety is illustrated on a plate showing the ratio of proteins, carbohydrates and vegetables consumed by the user versus the optimal plate suggested by

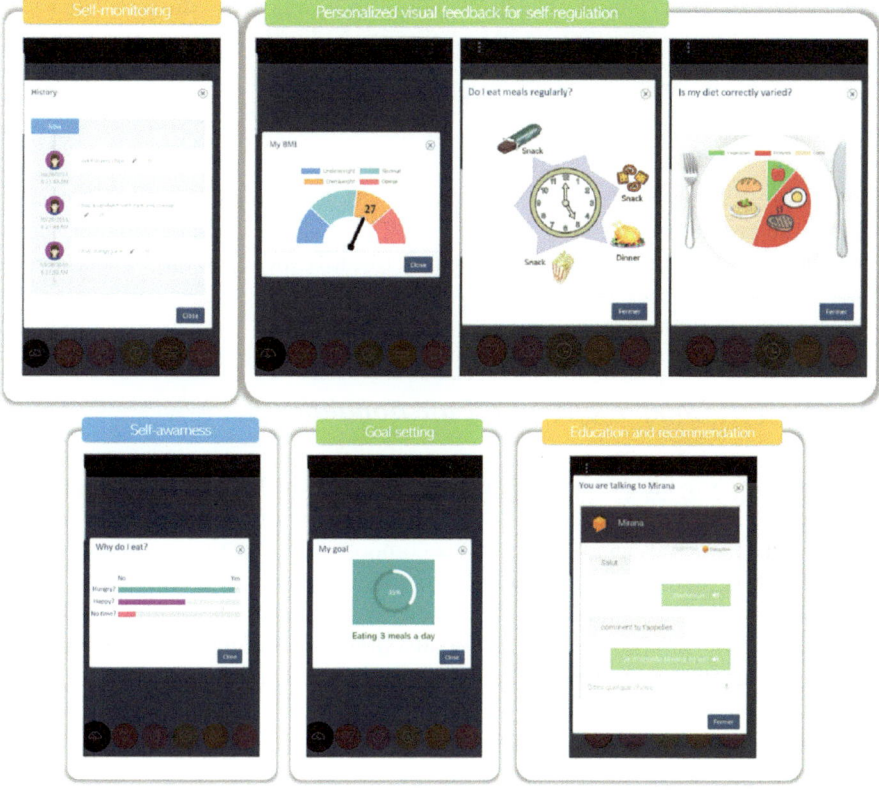

Fig. 2.2 Behavior change theories and techniques operationalized within MiranaBot

the Swiss Society of Nutrition [85]. A visual watch resuming the individual's eating times is intended to encourage self-reflection on their nutrition regularity.

Goal Setting

Following a training period analyzing the eating habits of the users, "MiranaBot" allows them to set personal goals, such as "eating three meals a day" or reducing "bread" and to define a timeline to reach their goals. Prompt reminders and useful information are provided to the users to help them reaching their goal.

Self-Awareness

Research has demonstrated that the reason to eat for most patients with chronic conditions are rather linked to psychological issues (emotions, stress, anxiety) or automatic behavior, than hunger itself [86, 87]. "MiranaBot" raises awareness of real physical hunger by asking users how they are feeling before and after each meal.

Personalized Education

"MiranaBot" is developed to be continuously available to answer users' questions during the process of change. Advice on food item alternatives, benefits of healthy nutrition behavior and promptly reminders towards their personal goals are frequently delivered.

MiranaBot: Components and Functionalities

"Mirana" refers to the overall system, providing "MiranaBot" service, as one of the components. Figure 2.3 illustrates the process and the different technical modules of the conversational agent. "Mirana" is composed of four main components: the speech to text module, the text analysis module, the natural conversation module and the graphical user interface. We denote as "nuggets" the results of "Mirana" analysis after a certain period of data collection from the individual; focusing on types of foods that may need to be replaced by others, to achieve healthier nutritional outcomes. For example, a nugget could be "bread", then the recommendation through the natural conversation module or the Graphic User Interface Module would be "let's reduce bread consumption by ¼ for a week. Here is what you can eat instead of bread".

Speech to Text Module

Users can easily describe, vocally, what they have eaten after each meal. This voice description is then converted in real time into text using DialogFlow, Google's human–computer interaction technologies based on natural language

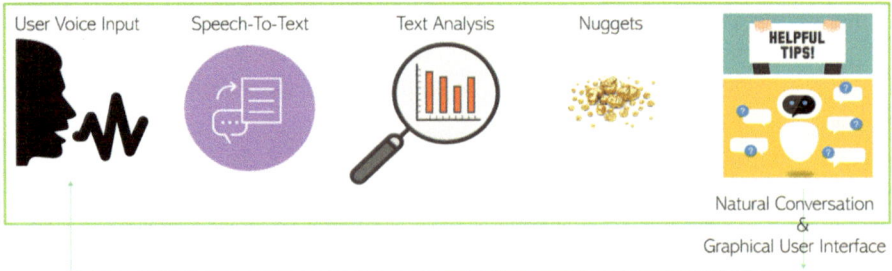

Fig. 2.3 Process and components of Mirana, including MiranaBot

conversations.[1] Data are stored in the secure server of the researcher's institution in Switzerland. MiranaBot is now available in three languages, namely French, English and Danish.

Text Analysis Module

MiranaBot analyses users' every day food data to provide them with daily and weekly visual feedbacks. MiranaBot also detects frequencies, varieties, and quantities (in terms of portions) of a specific food (or category of food). With the help of nutritionists and dieticians, we were able to build a database of food classified by categories. We have set rules for each category, which are then used by our algorithm to identify recurring patterns (e.g., eat too much bread = more than the normal consumption defined by the rules).

Natural Conversation Module

One important functionality of "Mirana" is its ability to discuss, have a conversation and answer the questions from the user. As mentioned previously, the conversational module is built upon DialogFlow developed by Google. Their module uses machine learning for small talk conversations to recognize user basic questions. In addition, we developed specific intents (or functions) related to our case to give "Mirana" the ability to answer questions about nutrition.

Graphical User Interface

To maximize the output provided to the user, MiranaBot uses the screen of the host device (phone, tablet, computer) to display graphical feedback in addition to the vocal answers. The regularity of the nutrition and the variety of the food eaten by the users are displayed on the screen to induce better self-reflection. While describing their food, the user can also check, in real time, if what they said was transcribed correctly. In case of error, they have the possibility to edit the entry through the graphical user interface. The user interface is also used to share articles, tips or to send useful notifications to the user.

---

[1] https://cloud.google.com/dialogflow

Conclusions

This chapter reviews existing validated scales to measure ADLs and different technology-enabled tools, methodologies and strategies currently used to assess the ability to perform ADLs, namely mobility and nutrition. Although different scales and techniques exist, we found that there is no one-size-fit-all solution to assess ADLs. Depending on the individual's conditions and environmental context, researchers and designers may choose one technology or even combine several of them to reach better accuracy. Each of the solutions has their own limitations. Despite the plethora of research in this area, most of the studies were conducted in controlled environment or in short time period. Further investigations are needed, in this promising area of Quality of Life Technologies (QoLT), specifically to assess ADLs.

In this chapter, we also described the development of a conversational agent to promote healthy nutrition for chronically-ill patients. We conducted a qualitative study with 13 patients and health professionals who helped us identify the main barriers as well as the needs and requirements for adopting a healthy nutrition behavior. Lack of regularity, lack of variety, false belief and hunger unawareness were identified as the main barriers to healthy nutrition; whereas monitoring, education, empowerment and practicality were identified as the main needs and requirements for nutrition behavior change. These findings allowed us to suggest the appropriate behavior change techniques to be leveraged in our systems, which are self-monitoring, personalized visual feedback, goal setting, self-awareness and personalized education. Finally, these were translated into a set of functionalities that build up to construct our final solution: MiranaBot. We envision that solutions similar to MiranaBot will be leveraged in the near future to enable easy to use and highly engaging assessment and potential improvement of the Activities of Daily Life and behaviors, leading to better life quality of individuals at all ages.

The QoLT we proposed here falls under the definition given by Wac [88], as a system for assessing or improving the individual's QoL relying on a hardware and software technology via a visual and an auditory interface. The primary aim of our QoLT here was mainly to assess nutritional habits. However, it also holds the potential to enhance individual's nutrition. As an extension, QoLT can also potentially be tools using prediction from current behavior to prevent undesired future behavior.

References

1. World Health Organization. Tracking universal health coverage: first global monitoring report. 2015. ISBN 978 92 4 156497 7.
2. World Health Organization. 10 facts on ageing and health. 2017. https://www.who.int/features/factfiles/ageing/en/ Accessed 01 June 2020.
3. Wiles JL, Leibing A, Guberman N, Reeve J, Allen RE. The meaning of "aging in place" to older people. Gerontologist. 2012;52(3):357–66. https://doi.org/10.1093/geront/gnr098.

4. Bacsu JR, Jeffery B, Johnson S, Martz D, Novik N, Abonyi S. Healthy aging in place: supporting rural seniors' health needs. Online J Rural Nurs Health Care. 2012;12(2):77–87. https://doi.org/10.14574/ojrnhc.v12i2.52.

5. Mlinac ME, Feng MC. Assessment of activities of daily living, self-care, and Independence. Arch Clin Neuropsychol. 2016;31(6):506–16. https://doi.org/10.1093/arclin/acw049.

6. Bravell ME, Zarit SH, Johansson B. Self-reported activities of daily living and performance-based functional ability: a study of congruence among the oldest old. Eur J Ageing. 2011;8(3):199–209. https://doi.org/10.1007/s10433-011-0192-6.

7. Miller LS, Brown CL, Mitchell MB, Williamson GM. Activities of daily living are associated with older adult cognitive status caregiver versus self-reports. J Appl Gerontol. 2013;32(1):3–30. https://doi.org/10.1177/0733464811405495.

8. Choe EK, Lee NB, Lee B, Pratt W, Kientz JA. Understanding quantified-selfers' practices in collecting and exploring personal data. In: Proceedings of the SIGCHI Conference on Human Factors in Computing Systems (CHI '14); 2014. p. 1143–52. https://doi.org/10.1145/2556288.2557372.

9. Wac K. From quantified self to quality of life. In: Rivas H, Wac K, editors. Digital health. Health informatics. Cham: Springer; 2018. https://doi.org/10.1007/978-3-319-61446-5_7.

10. De Falco I, De Pietro G, Sannino G. Evaluation of artificial intelligence techniques for the classification of different activities of daily living and falls. Neural Comput & Applic. 2020;32:747–58. https://doi.org/10.1007/s00521-018-03973-1.

11. Katz S. Assessing self-maintenance: activities of daily living, mobility, and instrumental activities of daily living. J Am Geriatr Soc. 1983;31(12):721–7. https://doi.org/10.1111/j.1532-5415.1983.tb03391.x.

12. Bieńkiewicz MM, Brandi ML, Goldenberg G, Hughes CM, Hermsdörfer J. The tool in the brain: apraxia in ADL. Behavioral and neurological correlates of apraxia in daily living. Front Psychol. 2014;5(353) https://doi.org/10.3389/fpsyg.2014.00353.

13. Millán-Calenti JC, Tubío J, Pita-Fernández S, González-Abraldes I, Lorenzo T, Fernández-Arruty T, et al. Prevalence of functional disability in activities of daily living (ADL), instrumental activities of daily living (IADL) and associated factors, as predictors of morbidity and mortality. Arch Gerontol Geriatr. 2010;50(3):306–10. https://doi.org/10.1016/j.archger.2009.04.017.

14. Gaugler JE, Duval S, Anderson KA, Kane RL. Predicting nursing home admission in the US: a meta-analysis. BMC Geriatr. 2007;7(1):13–26. https://doi.org/10.1186/1471-2318-7-13.

15. Ramos LR, Simoes EJ, Albert MS. Dependence in activities of daily living and cognitive impairment strongly predicted mortality in older urban residents in Brazil: a 2-year follow-up. J Am Geriatr Soc. 2001;49(9):1168–75. https://doi.org/10.1046/j.1532-5415.2001.49233.x.

16. Webber SC, Porter & M. M, Menec, V.H. Mobility in older adults: a comprehensive framework. The Gerontologist. 2010;50(4):443–50. https://doi.org/10.1093/geront/gnq013.

17. Patla AE, Shumway-Cook A. Dimensions of mobility: defining the complexity and difficulty associated with community mobility. J Aging Phys Activity. 1999;7(1):7–19. https://doi.org/10.1123/japa.7.1.7.

18. Tromp AM, Pluijm SMF, Smit JH, Deeg DJH, Bouter LM, Lips P. Fall-risk screening test: a prospective study on predictors for falls in community-dwelling elderly. J Clin Epidemiol. 2001;54(8):837–44. https://doi.org/10.1016/S0895-4356(01)00349-3.

19. Cederholm T, Barazzoni R, Austin P, Ballmer P, Biolo G, Bischoff SC, et al. ESPEN guidelines on definitions and terminology of clinical nutrition. Clin Nutr. 2017;36(1):49–64. https://doi.org/10.1016/j.clnu.2016.09.004.

20. Franz MJ, Bantle JP, Beebe CA, Brunzell JD, Chiasson J-L, Garg A, et al. Evidence-based nutrition principles and recommendations for the treatment and prevention of diabetes and related complications. Diabetes Care. 2002;25(1):148–98. https://10.2337/diacare.25.1.148

21. Beaudart C, Dawson A, Shaw SC, et al. Nutrition and physical activity in the prevention and treatment of sarcopenia: systematic review. Osteoporos Int. 2017;28:1817–33. https://doi.org/10.1007/s00198-017-3980-9.

22. Lindqvist C, Slinde F, Majeed A, Bottai M, Wahlin S. Nutrition impact symptoms are related to malnutrition and quality of life–a cross-sectional study of patients with chronic liver disease. Clin Nutr. 2020;39(6):1840–8. https://doi.org/10.1016/j.clnu.2019.07.024.

23. Jekel K, Damian M, Wattmo C, et al. Mild cognitive impairment and deficits in instrumental activities of daily living: a systematic review. Alz Res Therapy. 2015;7:17. https://doi.org/10.1186/s13195-015-0099-0.

24. Desai AK, Grossberg GT, Sheth DN. Activities of daily living in patients with dementia: clinical relevance, methods of assessment and effects of treatment. CNS Drugs. 2004; 18(13):853–75. https://doi.org/10.2165/00023210-200418130-00003.

25. Katz S, Ford AB, Moskowitz RW, Jackson BA, Jaffe MW. Studies of illness in the aged: the index of ADL: a standardized measure of biological and psychosocial function. JAMA. 1963;185(12):914–9. https://doi.org/10.1001/jama.1963.03060120024016.

26. George & Fillenbaum. Validity and reliability of OARS multidimensional functional assessment questionnaire in Iranian elderly. Iran Rehabil J. 1985;16 https://doi.org/10.32598/irj.16.2.169.

27. Ferrucci L, Koh C, Bandinelli S, Guralnik JM. Disability, functional status, and activities of daily living. In: Encyclopedia of gerontology (Second Edition). Elsevier; 2007. p. 427–36. https://doi.org/10.1016/B0-12-370870-2/00075-5.

28. Roedl J, Wilson LS, Fine J. A systematic review and comparison of functional assessments of community-dwelling elderly patients. J Am Assoc Nurse Pract. 2016;28(3):160–9. https://doi.org/10.1002/2327-6924.12273.

29. Adachi T. Advantages and disadvantages of the functional Independence measure for home care. In: Chino N, Melvin JL, editors. Functional evaluation of stroke patients. Tokyo: Springer; 1996. https://doi.org/10.1007/978-4-431-68461-9_10.

30. Grey N, Kennedy P. The functional Independence measure: a comparative study of clinician and self ratings. Paraplegia. 1993;31(7):457–61. https://doi.org/10.1038/sc.1993.74.

31. Mainetti L, Patrono L, Sergi I. A survey on indoor positioning systems. In: 22nd International Conference on Software, Telecommunications and Computer Networks (SoftCOM); 2014. p. 111–20. https://doi.org/10.1109/SOFTCOM.2014.7039067.

32. Heinz EA, Kunze KS, Gruber M, Bannach D. Using wearable sensors for real time recognition tasks in games of martial arts–an initial experiment. In: In Proceedings of the IEEE Symposium on Computational Intelligence and Games; 2006. p. 98–102. https://doi.org/10.1109/CIG.2006.311687.

33. Wang LK. Recognition of human activities using continuous autoencoders with wearable sensors. Sensors. 2016;16:189. https://doi.org/10.3390/s16020189.

34. Ayase R, Higashi T, Takayama S, Sagawa S, Ashida N. A method for supporting at-home fitness exercise guidance and at-home nursing Care for the Elders, video-based simple measurement system. In: In Proceedings of IEEE 10th International Conference on e-health Networking, Applications and Services (HealthCom); 2008. p. 182–6. https://doi.org/10.1109/HEALTH.2008.4600133.

35. Goffredo M, Schmid M, Conforto S, Carli M, Neri A, D'Alessio T. Markerless human motion analysis in gauss-Laguerre transform domain: an application to sit-to-stand in young and elderly people. IEEE Trans Inf Technol Biomed. 2009;13:207–16. https://doi.org/10.1109/TITB.2008.2007960.

36. Li Y, Miaou S, Hung CK, Sese JT. A gait analysis system using two cameras with orthogonal view. In: Proceedings of IEEE International Conference on Multimedia Technology (ICMT); 2011. p. 2841–4. https://doi.org/10.1109/ICMT.2011.6002046.

37. Mautz R, Tilch S. Survey of optical indoor positioning systems. In: International Conference Indoor Positioning Indoor Navigation (IPIN); 2011. p. 1–7. https://doi.org/10.1109/IPIN.2011.6071925.

38. Ausmeier B, Campbell T, Berman S. Indoor navigation using a Mobile phone. In: African Conf. Software Engineering and Applied Computing (ACSEAC); 2012. p. 109–15. https://doi.org/10.1109/ACSEAC.2012.26.

39. Elloumi W, et al. Indoor navigation assistance with a smartphone camera based on vanishing points. In: International Conference Indoor Positioning and Indoor Navigation (IPIN); 2013. p. 1–9. https://doi.org/10.1109/IPIN.2013.6817911.

40. Kahlert D, Ehrhardt N. Out-of-home mobility and social participation of older people: a photo-based ambulatory assessment study. Population Ageing. 2020; https://doi.org/10.1007/s12062-020-09278-3.

41. Addlesee MD, Jones A, Livesey F, Samaria F. The ORL active floor [sensor system]. IEEE Pers Commun. 1997;4(5):35–41. https://doi.org/10.1109/98.626980.

42. Schmidt A, Strohbach M, van Laerhoven K, Friday A, Gellersen HW. Context acquisition based on load sensing. In: Borriello G, Holmquist LE, editors. UbiComp 2002: ubiquitous computing. Lecture notes in computer science, vol. 2498; 2002. https://doi.org/10.1007/3-540-45809-3_26.

43. Middleton L, Buss AA, Bazin A, Nixon MS. A floor sensor system for gait recognition. In: Proceedings of the fourth IEEE workshop on automatic identification advanced technologies (AUTOID '05). IEEE Computer Society; 2005. p. 171–6. https://doi.org/10.1109/AUTOID.2005.2.

44. Paradiso J, Abler C, Hsiao K, Reynolds M. The magic carpet: physical sensing for immersive environments. In: CHI '97 extended abstracts on human factors in computing systems; 1997. https://doi.org/10.1145/1120212.1120391.

45. Richardson B, Leydon K, Fernström M, Paradiso J. Z-tiles: building blocks for modular, pressure-sensing floorspaces. Proceedings of CHI. 2004;1529-1532 https://doi.org/10.1145/985921.986107.

46. Merilampi, S., Mulholland, K., Ihanakangas, V., Ojala, J., Valo P. & Virkki, J. (2019). A smart chair physiotherapy Exergame for fall prevention–user experience study. IEEE 7th International Conference on Serious Games and Applications for Health (SeGAH), pp. 1–5, doi:https://doi.org/10.1109/SeGAH.2019.8882482.

47. Allet L, Knols RH, Shirato K, Bruin ED. Wearable Systems for Monitoring Mobility-Related Activities in chronic disease: a systematic review. Sensors. 2010;2010(10):9026–52. https://doi.org/10.3390/s101009026.

48. De Bruin ED, Hartmann A, Uebelhart D, Murer K, Zijlstra W. Wearable systems for monitoring mobility-related activities in older people: a systematic review. Clin Rehabil. 2008;22(10–11):878–95. https://doi.org/10.1177/0269215508090675.

49. Yang C-C, Hsu Y-L. A review of Accelerometry-based wearable motion detectors for physical activity monitoring. Sensors. 2010;2010(10):7772–88. https://doi.org/10.3390/s100807772.

50. Berglind D, Yacaman-Mendez D, Lavebratt C, Forsell Y. The effect of smartphone apps versus supervised exercise on physical activity, cardiorespiratory fitness, and body composition among individuals with mild-to-moderate mobility disability: randomized controlled trial. JMIR Mhealth Uhealth. 2020;8(2):e14615. https://doi.org/10.2196/14615.

51. Fillekes MP, Kim EK, Trumpf R, Zijlstra W, Giannouli E, Weibel R. Assessing older adults' daily mobility: a comparison of GPS-derived and self-reported mobility indicators. Sensors (Basel, Switzerland). 2019;19(20):4551. https://doi.org/10.3390/s19204551.

52. Elgala H, Mesleh R, Haas H. Indoor optical wireless communication: potential and state-of-the-art. IEEE Commun Mag. 2011;49(9):56–62. https://doi.org/10.1109/MCOM.2011.6011734.

53. Medina C, Segura JC, De la Torre Á. Ultrasound indoor positioning system based on a low-power wireless sensor network providing sub-centimeter accuracy. Sensors. 2013;2013(13):3501–26. https://doi.org/10.3390/s130303501.

54. Chen L, Wu E, Chen G. Intelligent fusion of Wi-fi and inertial sensor-based positioning Systems for Indoor Pedestrian Navigation. IEEE Sensors J. 2014;no. 99 https://doi.org/10.1109/JSEN.2014.2330573.

55. Faragher R, Harle R. Location fingerprinting with Bluetooth low energy beacons. IEEE Journal on Selected Areas in Communications. 2015;33(11):2418–28. https://doi.org/10.1109/JSAC.2015.2430281.

56. Di Giampaolo E. A passive-RFID based indoor navigation system for visually impaired people. In: 3rd International Symposium on Applied Sciences in Biomedical and Communication Technologies (ISABEL 2010); 2010. p. 1–5. https://doi.org/10.1109/ISABEL.2010.5702800.
57. Zeng L. A survey: outdoor mobility experiences by the visually impaired. In: Weisbecker A, Burmester M, Schmidt A, editors. Mensch und Computer 2015; 2015. p. S. 391–7.
58. Allahbakhshi H, Conrow L, Naimi B, Weibel R. Using accelerometer and GPS data for real-life physical activity type detection. Sensors. 2020;2020(20):588. https://doi.org/10.3390/s20030588.
59. Webber SC, Porter MM. Monitoring mobility in older adults using global positioning system (GPS) watches and accelerometers: a feasibility study. J Aging Phys Act. 2009;17(4):455–67. https://doi.org/10.1123/japa.17.4.455.
60. Ghali A, Cunningham AS, Pridmore TP. Object and event recognition for stroke rehabilitation. In: In Proceedings of Visual Communications and Image processing; 2003. p. 980–9. https://doi.org/10.1117/12.503470.
61. Lo FP, Sun Y, Qiu J, Lo BPL. Point2Volume: a vision-based dietary assessment approach using view synthesis. IEEE Transactions on Industrial Informatics. 2020;16(1):577–86. https://doi.org/10.1109/TII.2019.2942831.
62. Pettitt C, Liu J, Kwasnicki R, Yang G, Preston T, Frost G. A pilot study to determine whether using a lightweight, wearable micro-camera improves dietary assessment accuracy and offers information on macronutrients and eating rate. Br J Nutr. 2016;115(1):160–7. https://doi.org/10.1017/S0007114515004262.
63. Gemming L, Doherty A, Utter J, Shields E, Mhurchu CN. The use of a wearable camera to capture and categorise the environmental and social context of self-identified eating episodes. Appetite. 2015;92:118–25. https://doi.org/10.1016/j.appet.2015.05.019.
64. Zhou B, et al. Smart table surface: a novel approach to pervasive dining monitoring. In: 2015 IEEE International Conference on Pervasive Computing and communications (PerCom); 2015. p. 155–62. https://doi.org/10.1109/PERCOM.2015.7146522.
65. Hermsen S, Frost JH, Robinson E, Higgs S, Mars M, Hermans RCJ. Evaluation of a smart fork to decelerate eating rate. J Acad Nutr Diet. 2016; https://doi.org/10.1016/j.jand.2015.11.004.
66. Huang Q, Yang Z, Zhang Q. Smart-U: smart utensils know what you eat. In: IEEE INFOCOM 2018–IEEE Conference on Computer Communications; 2018. p. 1439–47. https://doi.org/10.1109/INFOCOM.2018.8486266.
67. Mertes G, Christiaensen G, Hallez H, Verslype S, Chen W, Vanrumste B. Measuring weight and location of individual bites using a sensor augmented smart plate. In: 40th Annual International Conference of the IEEE Engineering in Medicine and Biology Society (EMBC); 2018. p. 5558–61. https://doi.org/10.1109/EMBC.2018.8513547.
68. Burke R, Felfernig A, Göker MH. Recommender systems: an overview. AI Mag. 2011;32(3):13–8. https://doi.org/10.1609/aimag.v32i3.2361.
69. Shah K, Salunke A, Dongare S, Antala K. Recommender systems: an overview of different approaches to recommendations. In: International Conference on Innovations in Information, Embedded and Communication Systems (ICIIECS); 2017. p. 1–4. https://doi.org/10.1109/ICIIECS.2017.8276172.
70. Mika S. Challenges for nutrition recommender systems. In: Proceedings of the 2nd Workshop on Context Awareness in Retrieval and recommendation; 2011. p. 786.
71. Tran TN, Atas M, Felfernig A, Stettinger M. An overview of recommender systems in the healthy food domain. J Intell Inf Syst. 2017;50:501–26. https://doi.org/10.1007/s10844-017-0469-0.
72. Konstan JA, Riedl J. Recommender systems: from algorithms to user experience. User Model User-Adap Inter. 2012;22(1–2):101–23. https://doi.org/10.1007/s11257-011-9112-x.
73. Korpusik M, Glass J. Spoken language understanding for a nutrition dialogue system. IEEE Trans Audio Speech Lang Process. 2017;25:1450–61. https://doi.org/10.1109/TASLP.2017.2694699.
74. Casas J, Mugellini E, Khaled OA. Food diary coaching Chatbot. In: In Proceedings of the 2018 ACM International Joint Conference and 2018 International Symposium on Pervasive and Ubiquitous Computing and Wearable Computers (UbiComp '18); 2018. p. 1676–80. https://doi.org/10.1145/3267305.3274191.

75. Meso P, Jain R. Agile software development: adaptive systems principles and best practices. Inf Syst Manag. 2006;23(3):19–30. https://doi.org/10.1201/1078.10580530/46108.23.3.20060601/93704.3.
76. Duff O, Walsh D, Furlong B, O'Connor N, Moran K, Woods C. Behavior change techniques in physical activity eHealth interventions for people with cardiovascular disease: systematic review. J Med Internet Res. 2017;19(8):e281. https://doi.org/10.2196/jmir.7782.
77. Schembre S, Liao Y, Robertson M, Dunton G, Kerr J, Haffey M, et al. Just-in-time feedback in diet and physical activity interventions: systematic review and practical design framework. J Med Internet Res. 2018;20(3):e106. https://doi.org/10.2196/jmir.8701.
78. Sawesi S, Rashrash M, Phalakornkule K, Carpenter JS, Jones JF. The impact of information technology on patient engagement and health behavior change: a systematic review of the literature. JMIR Med Inform. 2016;4(1):e1. https://doi.org/10.2196/medinform.4514.
79. Greaves CJ, Sheppard KE, Abraham C, Hardemann W, Rode M, Evans PH, et al. Systematic review of reviews of intervention components associated with increased effectiveness in dietary and physical activity interventions. BMC Public Health. 2011;11:119. https://doi.org/10.1186/1471-2458-11-119.
80. Michie S, Abraham C, Whittington C, McAteer J, Gupta S. Effective techniques in healthy eating and physical activity interventions: a meta-regression. Health Psychol. 2009;28(6):690–701. https://doi.org/10.1037/a0016136.
81. Lara J, Evans EH, O'Brien N, Moynihan PJ, Meyer TD, Adamson AJ, et al. Association of behaviour change techniques with effectiveness of dietary interventions among adults of retirement age: a systematic review and meta-analysis of randomised controlled trials. BMC Med. 2014;12:177. https://doi.org/10.1186/s12916-014-0177-3.
82. Prochaska J, Velicer W. The Transtheoretical model of health behavior change. Am J Health Promot. 1997;12:38–48. https://doi.org/10.4278/0890-1171-12.1.38.
83. Klein M, Mogles N, van Wissen A. Why Won't you do What's good for you? Using intelligent support for behavior change. In: Salah AA, Lepri B, editors. Human behavior understanding. HBU 2011. *Lecture notes in computer science*, vol. 7065; 2011. https://doi.org/10.1007/978-3-642-25446-8_12.
84. Karoly P. Mechanisms of self-regulation: a systems view. Annu Rev Psychol. 1993;44(1):23–52. https://doi.org/10.1146/annurev.psych.44.1.23.
85. Swiss Society of Nutrition. Balanced Diet-Optimum Plate. http://www.sge-ssn.ch/fr/toi-et-moi/boire-et-manger/equilibre-alimentaire/assiette-optimale/. Accessed on 01 June 2020.
86. Cohen D, Farley TA. Eating as an automatic behavior. Prev Chronic Dis. 2008;5(1):A23.
87. Yau YH, Potenza MN. Stress and eating behaviors. Minerva Endocrinol. 2013;38(3):255–67.
88. Wac K. Quality of life technologies. In: Gellman M, editor. Encyclopedia of behavioral medicine. New York, NY: Springer; 2020. https://doi.org/10.1007/978-1-4614-6439-6_102013-1.

Chapter 3
Monitoring Technologies for Quantifying Medication Adherence

Murtadha Aldeer, Mehdi Javanmard, Jorge Ortiz, and Richard Martin

Abbreviation

AAL	Ambient Assistive Living
AHT	Assistive Health Technology
CPS	Cyber-Physical Systems
IC	Integrated Circuit
IMUs	Inertial Measurement Units
IoT	Internet of Things
NFC	Near Field Communication
RF	Radio Frequency
RFID	Radio Frequency Identification
RSSI	Received Signal Strength Indicator
TO	Transmit Only
UHF	Ultra-High Frequency
WSNs	Wireless Sensor Networks

Introduction

Human lifespans will continue increasing as the average quality of life improves. Evidence of this can be seen in recent reports that highlight the significant increase in aging population, especially in developed countries [1–3]. As one would anticipate, the global population of people aged 60 years and older will grow by 250% in 2050 as compared to 2013 [4]. Likewise, as society ages, long-term healthcare expenditures are projected to increase [5]. In order to maintain a healthy aging

M. Aldeer (✉) · M. Javanmard · J. Ortiz · R. Martin
WINLAB, Rutgers The State University of New Jersey, North Brunswick, NJ, USA
e-mail: maldeer@winlab.rutgers.edu; mehdi.javanmard@rutgers.edu;
jorge.ortiz@rutgers.edu; rmartin@scarletmail.rutgers.edu

© The Author(s) 2022
K. Wac, S. Wulfovich (eds.), *Quantifying Quality of Life*, Health Informatics,
https://doi.org/10.1007/978-3-030-94212-0_3

population, the employment of Assistive Health Technology (AHT) increases [4]. Based on this, great efforts are being made towards achieving greater expectations of the quality in healthcare systems [3]. There is no doubt that rapid technological advances will revolutionize research in the twenty-first century in a number of disciplines; namely human health. New approaches to monitor human health, behavior, and activity will be enabled. Medication adherence is an important component of health and well-being, with voluminous studies showing the importance of adequate medication adherence [6, 7].

Achieving healthy aging is challenging and thus requires several important strategies. Undoubtedly, correct medication is one of these strategies that are mainly related to the individual's behavior. In addition, it is well-known that medications are the primary approach for treating most illnesses [8]. Hence, it requires the individual to take the medication as directed by the healthcare professional [9]. However, medication adherence remains a common issue within the healthcare sector, and especially among older adults. In fact, more than 50% of the older people are living with multiple chronic illnesses. Thus, routine monitoring and assessment of the individual's adherence is crucial to improve their health outcomes [10]. To be successful, this should be performed using accurate assessment methods. Current assessment methods of medication adherence have advantages as well as limitations.

With the aim of describing how the state-of-the-art technology on medication adherence monitoring can improve healthcare systems, we divide the present chapter into several sections based on the main monitoring or sensing technology used. We also compare the different medication adherence monitoring techniques and approaches related to accuracy, energy efficiency, and user's comfort. Given the importance of technology embodiment in medication adherence systems, this chapter addresses the need of researchers and investigators of healthcare monitoring in both the engineering and medical societies.

Background

Medication Adherence

Medication adherence can be defined as the extent to which a person-taking medication adheres to a self-administered protocol [11]. In other words, medication adherence refers to the medication-intake behavior of the patient conforming to an agreed medication regimen specified by the healthcare provider with respect to timing, dosage, and frequency [12, 13]. From another point of view, non-adherence refers to the failure of taking medication as prescribed, including in-consistency, missing doses, and failing to re-fill the medication. Nonetheless, studies showed that failure to meet the medication-intake regime can result in emergence of drug resistance, accelerated progression of disease, many irrevocable health complications [13, 14], and increased mortalities [15].

The benefits of adhering to medication regimens are many. However, for the patient, high adherence to prescribed medication leads to less health complications, more treatments' benefits, and potentially active drug effect in the case of completely treated infectious disease [12]. Another benefit is that medication adherence helps in minimizing drug wastage and reducing healthcare costs [16]. On the other side, poor medication adherence proven come with degradation in the health of the patient that may potentially lead to lower quality of life.

Medication Adherence Monitoring

Full adherence to medication is required as the drug can be effective only when it is taken in the proper dosage [12]. Nonetheless, maintaining strict medication adherence is required that deems maintaining administration timing, dosage quantity, and frequency [17]. A wealth of reports revealed that up to 50% of the patients either never fill their medication prescriptions or do not use the medication as prescribed to them in medication regimens [18]. Unfortunately, poor adherence is prevalent among populations with chronic illnesses [19], which leads to hospital admission. In the US alone, poor medication adherence results in more than 100,000 mortalities annually, as well as hundreds of billion dollars of healthcare spending every year [20, 21]. A number of approaches have been used for the aim of monitoring medication adherence because it has been shown that improving adherence to medical therapy would substantially lead to both health and economic benefits.

In general, two key factors should be considered when discussing medication adherence. The first factor is monitoring, which is alternatively referred to as assessment, quantification, measurement, or evaluation. Medication monitoring means using some methods for observing if the patient has taken the medication or not. Hence, the effectiveness of the monitoring method plays a central role. The second factor is intervention. Intervention refers to the means that can be used for improving adherence to medication or correcting it once erroneous or drift is detected. How- ever, the latter is more in the domain of the psychological and social sciences as it requires understanding the cultural, psychological and social factors that affect the patient's behavior [22], and thus it is out of the scope of this chapter.

Methods that have been utilized for measuring medication adherence so far can be broadly divided into two categories, direct and indirect [23]. Direct methods of measurement of adherence include direct observation of the patient while taking the medication, laboratory detection of the drug in the biologic fluid of the patient (i.e., blood or urine), laboratory detection of the presence of nontoxic markers added to the medication in the biologic fluid of the patient, and laboratory detection of the presence of biomarkers in the dried blood spots [24]. Meanwhile, the patient's self-reporting, pill-counting, assessing pharmacy refill rates, and using electronic medication event tracking systems are examples of indirect methods of measuring adherence. There is not a gold standard measurement system that fulfills the criteria for an optimal medication adherence monitoring. Each category comes with benefits and limitations. Direct measures are accurate, but they are invasive and expensive. In comparison, indirect methods are less expensive and provide good estimation of the medication adherence. However, these methods relay on the reliability of the user [25]. As such, these factors should be taken into consideration when selecting the adherence measurement methodology.

Why Technology-Based Solutions?

The development of Cyber-Physical Systems (CPS) that integrate computation and physical processes for healthcare, are advancing rapidly [26]. More recently, such systems included few sensing and monitoring devices associated with mobile devices such as smart pill bottles, smart watches, smart phones, and wearables. The combination of these smart monitoring devices with interventions that remind the patient in case a deviation is detected has proven to improve medication adherence [27, 28]. Compared to manual approaches, electronic-based approaches can reduce the cost and effort from the user's interest. In addition, the accuracy of adherence measure, which is of great importance from the healthcare provider's point of view can be enhanced when using electronic-based systems. Furthermore, as we live in the era of the Internet of Things (IoT) [29], where everything is connected to the Internet, a connected health paradigm is becoming a more dominant field [30]. One expectation of connected health is the automated capability of communicating the collected adherence measurements to the provider, and the feature of issuing reminder and alert messages based on the processed information [31]. Moreover, electronic measurement systems can be portable and thus provide timely and long-term monitoring without restricting the user's mobility. In spite of the fact that electronic-based modalities can outperform traditional ones, the majority of electronic-based approaches come with limitations that act as burdens on the users, as we will see in Sect. 3.5. In fact, some of them have not achieved much success due to these burdens [23]. Based on this, we conclude that there is no optimal electronic-based solution for medication adherence evaluation and, for that, much additional efforts will be required to realize accurate, low cost electronic adherence monitoring.

Related Work

In the past, a wide number of review studies that addressed the medication adherence problem have been created. However, most reviews studied the medication adherence from a clinical point of view along with interventions [6, 7, 11]. Moreover, only a few studies have presented the electronic-based interventions [18, 23, 25, 32, 33]. Little attention has been paid towards employing technology in medication adherence monitoring and enhancement as compared to the traditional modalities. These reviews have elucidated the role of technology-based solutions for medication adherence assessment, the potential benefits and limitations, but, no detailed discussion on the cyber-physical system, including system design, hardware development, and data analytic of these solutions were given.

A rare number of studies describe technology-based interventions for adherence monitoring and enhancement. For example, Park et al. [33] presented an overview of a number of electronic systems and methods of medication measurement. Other review articles have discussed the smartphones' applications, and tablet applications technology [25] for medication adherence that are in the form of automated reminder systems. In [34], some technological medication reminder approaches have been briefly described. It is worth mentioning that only a recent study by Rokni et al. [35] has reported some commercially available technology-based solutions. In addition, they provided a brief discussion of some clinical studies that involved electronic medication monitoring. It also discussed the challenges associated with medication monitoring technologies from data analytics, reliability, and scalability sides. It is obvious that these survey studies are limited in providing a detailed discussion of the technical sides of the different technology-based sensing or monitoring approaches for medication adherence.

The main objective of this chapter is to explore this topic further by taking account of other medication monitoring systems such as ingestible biosensors, and discussing the trade-offs of each technology in multiple dimensions.

A Review of Medication Adherence Monitoring Systems

Medication non-adherence is an extensively studied complex problem. The common conclusion of these studies is that several interventions are required to improve medication adherence [18, 36]. Nonetheless, technological interventions are believed to be supportive tools in improving adherence. This is due to the fact that they allow timely monitoring, and generate useful information about the patient's behavior for the healthcare provider. To date, a considerable number of systems have been proposed and developed that utilize monitoring and tracking techniques in various health-related projects, including medication adherence monitoring. In this section, we categorize and review the existing approaches on designing monitoring systems for medication adherence applications using emerging technologies.

Our review includes articles from journals, and conference papers and proceedings. We excluded articles classified as editorials, book reviews, white papers, or newspaper reports. While searching for papers, electronic databases including Google Scholar, IEEE Xplore, ACM Digital Library, Springer Link, MDPI, and Science Direct, were used. The descriptors we used were "medication adherence", or "medication intake", or "medication monitoring", or "medication compliance" in combination with at least one of others, including "technology", "sensor", "smartwatch", "wearable", "smart bottle", "pill bottle", "pillbox", "vision system", "Radio Frequency Identification (RFID)" and "Near Field Communication (NFC)". The search was inclusive of all years from 2004 through 2019.

Using primarily the full text and the abstracts, we selected articles discussing medication adherence monitoring technologies and excluded papers discussing intervention applications. The literature review approach used in this paper follows an iterative and incremental procedure [37], and hence found and included new studies about medication adherence monitoring technologies and approaches to the surveyed studies.

Table 3.1 provides a taxonomy of the approaches reviewed in this chapter. Table 3.2 summarizes the key properties of existing technology-based systems reviewed in this chapter.

Table 3.1 A taxonomy of the technology-based approaches for medication adherence monitoring

Reference	Category	Main Technology	Secondary Technology	Monitored Activities and/ or Subjects
Hayes et al., (2009) [9]	**Sensor systems**	Smart pill container	–	Lid opening
Aldeer et al., (2019) [38]	**Sensor systems**	Smart pill container	–	Lid opening and closure, bottle picking and flipping/ shaking
Lee and Dey, (2015) [39]	**Sensor systems**	Smart pill container	–	Lid opening and closure, box manipulation
Kalantraian et al., (2016) [40]	**Sensor systems**	Wearable sensors	Smart pill container	Pill bottle pick up and pill swallowing
Wu et al., (2015) [41]	**Sensor systems**	Wearable sensors	Ingestible biosensors	Pill swallowing
Kalantraian et al., (2015) [42]	**Sensor systems**	Wearable sensors	–	Pill bottle opening, pill removal, pill pouring into the secondary hands, water bottle handling
Hezarjaribi et al., (2016) [43]	**Sensor systems**	Wearable sensors	–	Hand-to-mouth motion
Wang et al., (2014) [44]	**Sensor systems**	Wearable sensors	–	Taking a pill, drinking water and wiping mouth
Chen et al., (2014) [45]	**Sensor systems**	Wearable sensors	–	Cap twisting and hand-to-mouth actions
Mondol et al., (2016) [46]	**Sensor systems**	Wearable sensors	–	User's response in the form of voice commands

Table 3.1 (continued)

Reference	Category	Main Technology	Secondary Technology	Monitored Activities and/ or Subjects
Hafezi et al., (2015) [47]	**Sensor systems**	Ingestible biosensors	–	Medication ingestion
Chai et al., (2016) [14]	**Sensor systems**	Ingestible biosensors	–	Medication ingestion
Agarawala et al., (2004) [48]	**Proximity-based systems**	RFID	–	Pill bottle pick up
Becker et al., (2009) [49]	**Proximity-based systems**	RFID	–	Pill removal
Morak et al., (2012) [50]	**Proximity-based systems**	NFC	–	Pill removal
Batz et al., (2005) [51]	**Proximity-based systems**	Computer vision	–	Pill bottle opening, hand over mouth motion, bottle closing
Valin et al., (2006) [52]	**Vision-based systems**	Computer vision	–	Pill bottle opening, pill picking, pill swallowing, bottle closing
Dauphin and Khanfir, (2011) [53]	**Vision-based systems**	Computer vision	–	Pill bottle picking, drinking a glass of water, putting glass back
Huynh et al., (2009) [54]	**Vision-based systems**	Computer vision	–	Tracking the face, the mouth, the hands, and the medication bottle
Sohn et al., (2015) [55]	**Vision-based systems**	Computer vision	–	Bottle weight
Li et al., (2014) [56]	**Fusion-based systems**	RFID	Sensor networks	Pill removal, hand motion
Hasanuzzaman et al., (2013) [57]	**Fusion-based systems**	RFID	Computer vision	Pill bottle removal, tracking hands and medication bottle
Suzuki and Nakauchi, (2011) [58]	**Fusion-based systems**	Computer vision	Sensor networks	Pill bottle removal, user behavior prediction
Abbey et al., (2012) [59]	**Fusion-based systems**	Smart pill container	Mobile application	Pill removal
Boonnuddar and Wuttidittachotti, (2017) [60]	**Fusion-based systems**	Smart pill container	Mobile application	Bottle weight

Table 3.2 Summary of main applications, strengths, and limitations of the current technologies used in medication adherence

Category		Main Application Differences	Strengths	Limitations
Sensor Systems	Smart Pill Container	Detects cap opening and bottle pick up	Possibility to allow mobility Non-invasive	System's life is constrained by the battery Detect medication taking activity with low accuracy
	Wearable Sensors	Detects motions related to cap twisting, hand-to-mouth, pouring pill into the hand, and pill swallowing	Possibility to detect medication intake activity with sign accuracy Relatively easy to use Allow mobility	User's comfort and social acceptance due to their possible invasiveness Require frequent battery charging or replacement
	Ingestible Biosensors	Detect pill ingestion	Possibility to detect concurrent pills ingestion Allow mobility	User's comfort and social acceptance System's lifetime is constrained by the battery Security issues due to their limited resources
Proximity-Based Systems		Detects medication presence or absence within the proximity of reader's antenna	Non-invasive	Need to be coupled with other monitoring or sensing techniques for verification
Vision-Based Systems		Detects medication presence or absence within the scope of the camera	Non-invasive	Need to be coupled with tech or sensing techniques for verification
Fusion-Based Systems		Try to verify the operation of monitoring the medication taking activity	Higher accuracy as compared to standalone technology	Resource consuming Do not usually support mobility

Sensor-Based Systems

Recent years have seen the size, cost, and energy consumption of small wireless sensors decrease by several orders of magnitude [61]. Indeed, today, low-power wireless sensors can be bought for an affordable price. In the context of human health, sensor systems allow us to collect data on daily activities in a free-living environment and possibly over long time periods, seamlessly [62]. One promising application in that field is the monitoring and assessment of subject for medication intake [63]. In fact, sensor-based approaches are the most widely used among other approaches these days for adherence monitoring. Utilizing sensor networks into medicine intake and adherence monitoring systems comes with features and benefits. The regularity in measurements, remote monitoring capability, and context awareness are a few examples [63]. In general, wireless sensors in this area of monitoring can be put into two main categories based on the form of deployment: fixed and wearable. Fixed sensors are tied to minimally mobile objects such as pillboxes or pill bottles, and home apparatuses. Meanwhile, wearable sensors are lightweight, have high data fidelity, and mobile devices that are attached to the user's body. In vivo or intra body communication and networking [64] is another emerging sensor-based communication and network technology within the IoT family, which is enabling a new set of healthcare applications.

In this part, we describe the recent work on medication adherence monitoring using different forms of wireless sensing.

Smart Pill Containers

Pillboxes and pill bottles equipped with sensors have been developed for monitoring the medication-taking activity. In this context, Hayes et al. [9] developed MedTracker(Fig. 3.1). It is one of the earliest approaches that uses a 7-day multi-compartment pillbox embedding plungers in each compartment. It was designed to detect the lids of boxes opening as the plungers would activate a switch inside the pillbox that then triggers the micro-controller. The system uses Bluetooth technology for wireless transmission of the data to a nearby computer. Data was transmitted over the Bluetooth link every two hours for the aim of prolonging the lifetime of system, which was using a 9 V battery. The system includes RAM for storing medication taking events when there is no connection with the base station. However, it is obvi- ous that the system is simple and is error prone as it considers any lid opening event as medication taking. Regardless of its simplicity, the system achieved a lifetime of eight weeks only, given it was powered from a considerably big battery.

For a project that intended observing daily living of elderly people, Lee and Dey [39] developed a pillbox similar to that reported in [9]. A 7-day compartment has been equipped with a Microcontroller (MCU), a ZigBee wireless module, an accelerometer, and a battery (Fig. 3.2). Data were transmitted to a nearby laptop for further processing. The aim of this system was for human-computer interaction studies.

Fig. 3.1 MedTracker prototype pre- sented in [9]

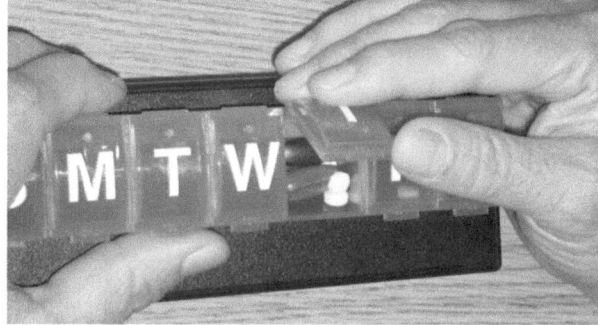

Fig. 3.2 The system
developed by Lee and
Dey [39]

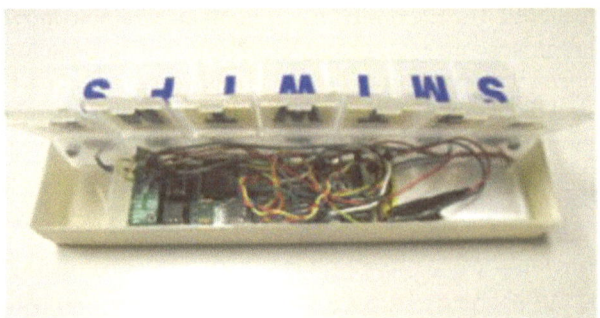

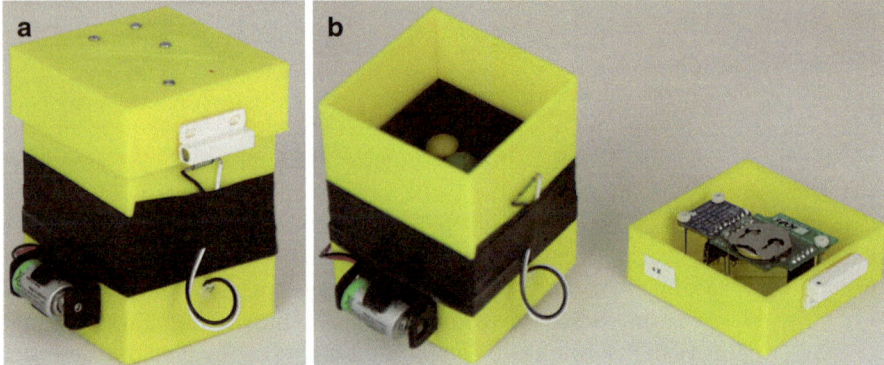

Fig. 3.3 The system prototype developed by Aldeer et al. [66]. (**a**) Pill bottle. (**b**) Bottle compartment and cap with the sensors shown

In another approach that was recently carried on by Aldeer et al. [65, 66], a smart pill bottle and a sensing framework for medication adherence monitoring have been proposed. As shown in Fig. 3.3, they built a 3D printed pill bottle equipped with a magnetic switch sensor, an accelerometer, and a load cell. Furthermore, the system uses PIP-Tag mote [67] as a platform for collecting the data from the employed sensors and then transmitting them wirelessly to a base station attached to a nearby computer.

Such an approach aims to eliminate the intervention and attachment of sensors to the human body, and by that it ensures user's comfort while maintaining accuracy by using the accelerometer sensor. However, the system does not ascertain if a pill is ingested or not by the user.

Wearable Sensors

In the recent years, Inertial Measurement Units (IMUs) have seen rapid achievements from both the cost and intelligence points of view [68]. IMUs usually consist of accelerometers, gyroscopes, and magnetometers, or a combination of these [69].

They have been widely used in healthcare applications by sensing motion and track- ing individuals [70]. Ultimately, the usage of motion sensors can help in revealing possible information about individual's health [66]. In this part, we present many wearable sensing systems and place them in two categories, depending on the place- ment location of the body, **neck-worn** and **wrist-worn**.

Neck-Worn Sensors: In one of the studies [40], the authors propose a wearable system for detecting user adherence up to the level of determining if the medication has been ingested. As shown in Fig. 3.4, they built a pendant-style necklace that includes a piezoelectric sensor, a Radio Frequency (RF) board, and battery. The piezoelectric sensor is used for sensing the mechanical stress resulting from skin motion during pill swallowing and generating voltage as a response. Major challenges associated with this approach pertain to user comfort and social acceptance [71] as the necklace needs to be worn by the patient and placed in contact with the skin during dose swallowing.

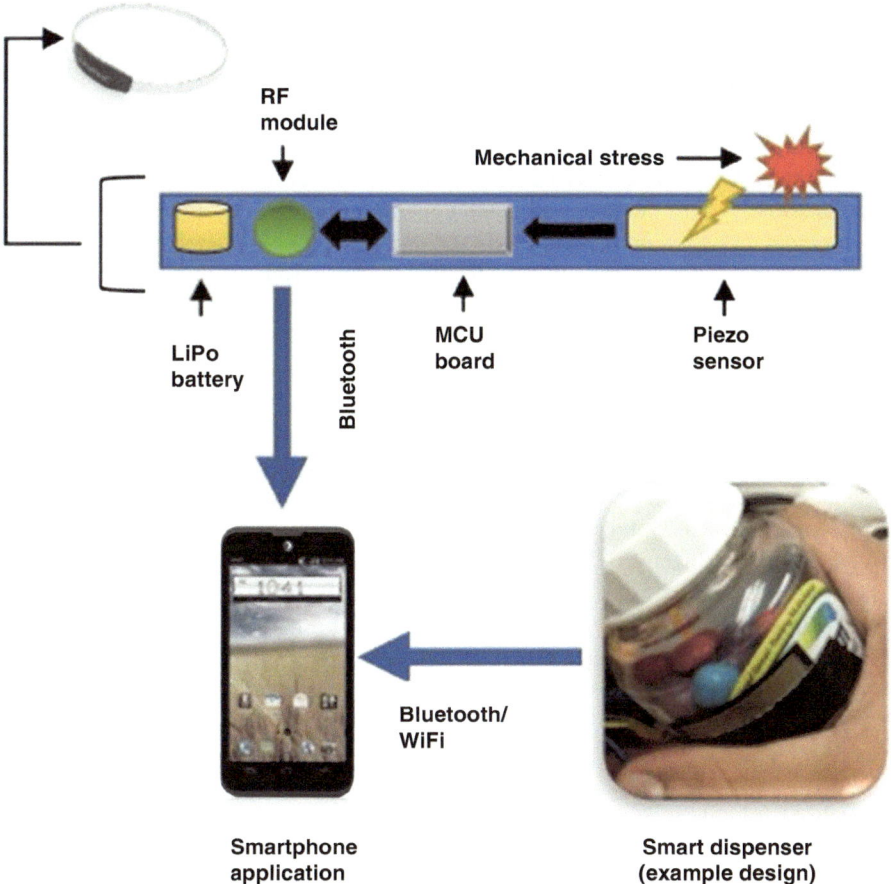

Fig. 3.4 The neck-worn system pre- sented in [40]

Fig. 3.5 The system developed by Wu et al. [41]

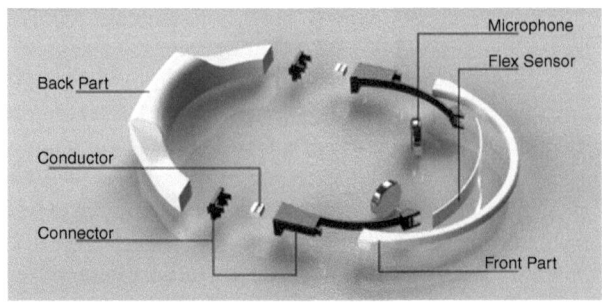

Another tool for assessing medication intake is using acoustic sensors in the form of neck wearables. Such an approach has been utilized for food intake monitoring applications [72]. Although this approach requires further research, it shows promise for being applicable to medication monitoring [73]. Only one prototype of this class of wearables was developed by Wu et al. [41]. The neckwear device contains microphones, a flex sensor, and an RFID reader (see Fig. 3.5). The microphones and the flex sensor are to be employed for sensing throat movement and chewing sound associated with medication swallowing activity. However, the study did not include any validation trials, thus making it difficult to make conclusions about the performance, social acceptance, and comfort of this approach.

Given the promise of acoustic sensing in food monitoring, it is highly likely that this technology will face the same challenges associated with other neck-worn sensors when applied in promoting medical compliance in older users [74].

Wrist-Worn Sensors: When reviewing sensor-based systems, one should not ignore personal sensors. Personal sensors are a class of wearables that can be used for fashion and tracking purposes, such as smartwatches [75]. Nonetheless, these wearables embed miniaturized and continuously progressing capabilities including Inertial Measurements Units (IMUs) (accelerometer, gyroscope, and magnetometer or a combination of these) [76, 77]. Thus, wearable and personal sensors have been recently used in many healthcare monitoring studies, including medication intake detection. The reason behind using IMUs in such systems is their ability to accurately recognize the intensity, direction, and angle of movements conjugated with medication intake activity in a 3D coordinate system [78]. Collecting such data will help in modeling the user's physical activity and then infer if it is associated with medication taking activity or not. In [43], accelerometer and gyroscope sensors embedded in a pair of smartwatches placed on both wrists of the user were used to sense and transmit readings associated with pill taking activity from 10 users. Using a decision tree classifier, the system was able to detect the wrist movement while taking medication with 78.3% accuracy using one smartwatch placed on either of the wrists. Moreover, the accuracy of the system was 86.2% when using two smartwatches for tracking the motion of both hands.

Wang et al. [44] used accelerometery data samples from wrist-watches and dynamic time warping technique to test if a sample belongs to either activities: taking a pill with water or drinking water and wiping mouth. Data from 25 individuals were used to classify the hand movement gestures associated with one of the previously mentioned activities. The system achieved 84.17% true positive rate. A further re- search study of Chen et al. featuring wearable sensors presents a system for detecting two actions "cap twisting" and "hand-to-mouth" from a triaxial accelerometer and a gyroscope [45]. Classification accuracies were 95% and 97.5% for cap twisting and hand-to-mouth actions, respectively.

Finally, termed MedRem, was presented in [46]. Unlike other approaches that used IMUs available on smartwatches, MedRem uses the speaker microphone on a smartwatch to provide reminders and track medication adherence via voice commands. When reminders are provided in the form of voice commands, it is expected that the user send a recording via the microphone sensor to confirm or postpone taking medication. The smartwatch then uses an android speech recognizer to analyze user's input and update a server. The capability of recognizing native and non-native English speakers' commands was 6.43% and 20.9% error rates.

Advantages of wearable sensors approaches include the ability of monitoring the user behavior in a free-living environment [72]. Another advantage is the accuracy of sensor-based systems. However, a main disadvantage that is pertained with wearable- based systems is the user acceptance and comfort, especially when considering old people [71]. This is due to the requirement that the sensor should be attached to the user for possibly a long time and recharged frequently, as wearables are usually powered by small batteries.

Ingestible Biosensors

The use of biosensors in connected health is in its infancy. However, with the introduction of In vivo communications, it can be expected that the biosensor technology will dramatically improve over time and increase in value to advancing healthcare delivery [64]. Ingestible devices are miniature capsule-looking devices that are digested and swallowed when taken through mouth like solid medications. These devices travel through the gastrointestinal tract and digestive system and collect data about specific physiological parameters [79]. One application of these devices can be for adherence monitoring, where data about drug consumption are collected and transmitted to a body-worn or nearby device for further post-processing [80].

Researchers from Proteus Digital Health, Inc. (Redwood City, CA, USA) have designed a micro biosensor that is intended to be integrated with pharmaceutical oral dose (pill or capsule) for evaluating medication ingestion [47, 81]. The sensor is built from an Integrated Circuit (IC) made of specific materials (including gold), with a food particle size. Upon contact with the gastric fluid, the ingestible sensor communicates with a wearable receiver worn by the patient and transmits a unique code. A mobile phone user interface can then identify the ingested medication based on the received code from the ingested biosensor. The designed device has been tested via multiple clinical studies. Furthermore, 412 subjects were involved in the clinical studies where they have performed more than 20,000 ingestions spanning 5656 days in total. The detection accuracy was more than 99%.

MyTMed is another system that is based on ingestible biosensors [14]. The central part of MyTMed is the digital capsule that can encapsulate oral medication. It is made of a standard gelatin pill capsule that includes a sesame seed size RFID tag. Upon ingestion by the patient, the gelatin capsule dissolves in the stomach and releases the medicine along with the RFID tag. The electro-chemical reaction between the tag's electrolytes in gastric acid forms a bio-galvanic battery that enables it to emit a unique code in the forms of packets to a body worn receiver. Eventually, the receiver utilizes short messaging service (SMS) to relay the packets to a cloud server that can be accessed by the caregiver. Based on a 10 participants trail study with 96 ingestion events, the system's detection accuracy was 87.3% [82].

Advantages of biosensor-based techniques include their ability to detect concurrent medication ingestion events with relatively high accuracy and no computational cost. However, as such systems require external receivers to be adhered to the individual's body, many users would object to wearing a banded device throughout the day and possibly for years (when considering people with chronic illnesses). Security and privacy are also an issue, with resource-constraint tags requiring low-energy and lightweight computing cryptographic tools [83].

Proximity Sensing

The visionary concept of IoT relays on some technologies, among which is the proximity detection [84]. Hence, objects usage in our daily life can be monitored by sensing their proximity to other things. Two important wireless communication technologies that are currently used for proximity detection and sensing are RFID [85] and NFC [86]. Overall, RFID and NFC are contactless short-range communication technologies that can be integrated in everyday life objects to sense the daily activities [87]. Here, we describe the RFID-based and NFC-based systems and their usefulness and shortcomings.

An early demonstration that applied RFID technology was designed by Agarawala et al. [48]. The system uses an RFID tag attached to a pill bottle that is placed on a platform embedding an RFID reader and LEDs (Fig. 3.6). The LEDs flash to notify the patient when it is time to take medication. Using this system, it is inferred that the medication is taken when the medication bottle is picked from the platform and it is not within the coverage radius of the RFID reader anymore. The caregiver can track the patient's adherence via an Ethernet connection with the platform. Another RFID- based system is SmartDrawer [49], Fig. 3.7. A drawer with an RFID reader that is capable of inventorying the pill bottles that are stored inside it as well as keeping a record of drug taking activities, is used. The pill bottles are equipped with RFID tags for identification and tracking. The system records the type of bottle and when it is removed from the drawer. In other words, it is assumed that the medication is taken when the bottle of that medicine is removed from the drawer and it is not within the scope of the RFID reader. Other short communications-based

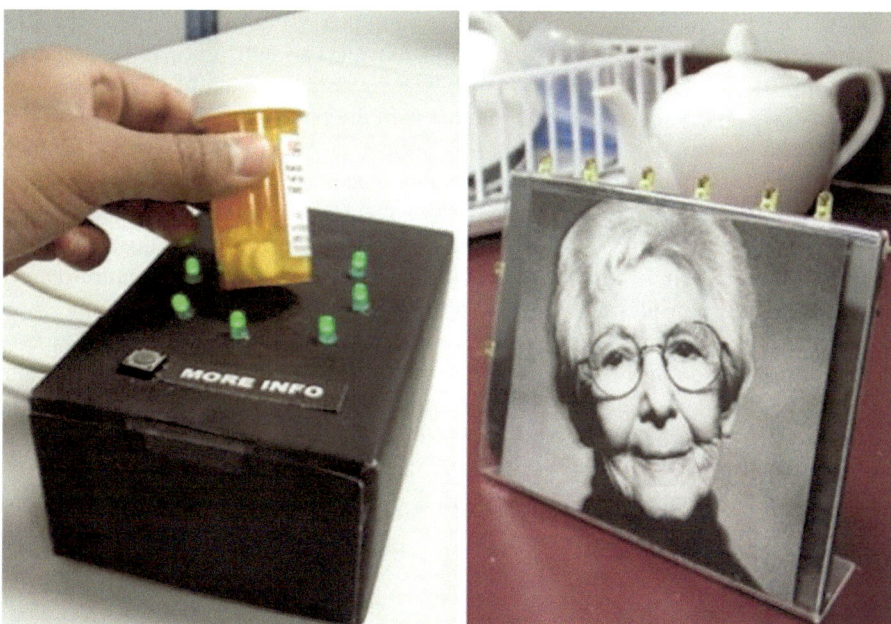

Fig. 3.6   The RFID-based system devel- oped by Agarawala et al. [48]

Fig. 3.7   SmartDrawer system developed by Becker et al. [49]

approaches designed a smart blister that is equipped with a μC along with the NFC technology available on mobile phones, to develop an adherence tele-monitoring system [50]. The idea is that the smart blister records the event of pill removal and reports this activity to a mobile phone that is in the proximity via NFC. The mobile phone then communicates this event to a remote server to be accessed by the caregiver that assesses the medication intake adherence.

Proximity sensing-based systems have advantages as well as limitations. The main advantage is the possibility of retrieving information such as dosage instructions that may include timing, frequency, and quantity. Such information can be helpful when considering elderly patients. Another advantage is the non-invasiveness, as sensing tags are usually attached to the pill containers. However, the main limitation of these systems is the requirement that the pill container being located within a short distance (several centimeters) of the vicinity of the main part of the system, which is the reader. Most importantly, there have been some studies that addressed possible harm to the fetus that are associated with the exposure to Ultra-High Frequency (UHF) RFID readers during pregnancy [88].

Vision-Based Systems

Recently, research in computer vision and image processing has attracted much attention, leading to the development of many algorithms for human activity representation and classification [89]. So far, vision-based systems have been the basis for a number of important healthcare applications. In the context of human activity recognition within smart environments or "Smart Homes" [75], where Ambient Assistive Living (AAL) technologies [2] exist; one choice for monitoring medication intake is to use vision modules for identifying and tracking inhabitants, motion, gestures, and subjects. In this section, we depict the current vision-based systems for medication intake monitoring and discuss their pros and cons.

In [51], a computer vision system was proposed for monitoring medication habits. The system uses one camera installed in the medication area, which may include a group of medication bottles (Fig. 3.8). The aim of this system was to track if the

Fig. 3.8 The vision-based system devel- oped by [51]

right medication is being taken by the user. In order for the system to work, it is required that only one user appears closely in the field of view of the camera during the medication taking session. Algorithms for skin color distinction have been used in order to distinguish between skin and non-skin colors. First, the systems extract all skin regions of the person in front of the camera. Then, this information is used for detecting hand/face occlusions and hand/hand occlusions. Researchers used four users in different environments to evaluate the system. Another computer vision system for monitoring medication intake was developed by Valin et al. [52]. The system considered multi-state scenarios including bottle opening, pill picking, pill swallowing, and bottle closing. It uses color classification algorithms for person detection and motion tracking by distinguishing the person's skin. In addition, colored bottles have been used for medication bottle detection. The recognition results were 90% classification accuracy for scenarios that differ from each other in the sequence of activities associated with medication taking.

The work in [53] focused on developing a technique for background suppression of videos captured by low resolution cameras. However, the technique was only tested with one participant and no accuracy measurements were reported. Furthermore, the system's accuracy may get affected for different colored clothes worn by the participants, as the experiments have been conducted with a participant wearing dark colors compared to the background. Another similar vision-based system developed by Huynh et al. [54] used a multi-level approach for detecting and tracking mobile objects during medication intake. The face, the mouth, the hands, a glass of water, and the medication bottle were tracked in this system. To achieve this, detection and tracking techniques for background subtraction, skin regions' segmentation, and using color information for bottle detection are used. The average success rate of activity recognition was 98% from a population of three subjects. In later work, the authors directly use two cameras for the aim of occlusion handling.

The literature also shows a monitoring system that consists of a digital scale and a camera that was presented in [55]. As illustrated in Fig. 3.9, a digital scale has been used such that it continuously measures and displays the medication bottle weight. The camera has been used to capture and send the scale's readings displayed

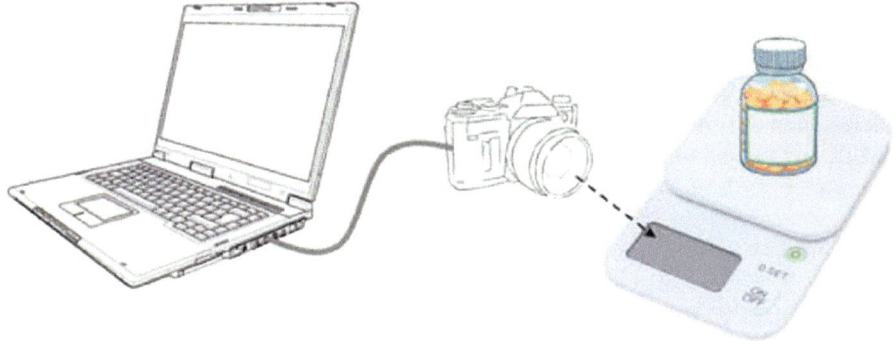

Fig. 3.9 The system developed by Sohn et al. [55]

on the screen to a nearby computer. Upon receiving the images, the computer then runs an image processing algorithm for processing the bottle's weight. From the bottle's weight decrease trend, the system can generate an alarm to remind the patient to take medication. It should be noted that although this work concentrates on vision analysis, it does not include any human subject tracking. It is obvious that such a system does not support mobility due to the fact that it requires the medication bottle to always be placed on the weight scale, and thus provides only a limited view. Although vision-based systems will play an important role in AAL environments, the main disadvantages of these approaches are their limitation in use and accuracy. In addition, these approaches may demand several resources, which can be expensive. Furthermore, as we progress further into the twenty-first century, users prefer fully mobile devices [90]. However, in contrast, vision-based approaches do not support mobility.

Finally, another limitation is that the user is required to be within the scope of the camera.

Fusion-Based Systems

It is seen from the studies we covered that each approach comes with drawbacks. As such, fusion-based systems have been developed that aim at blending advances available from multiple techniques for enhancing one or more technical drawback [72]. In this section, we subdivide fusion-based systems into several categories, based on the blend of techniques used.

Proximity-Sensor Systems

In [56], Li et al. have designed a system that was built with a cylindrically shaped 7-compartment pillbox, a wristband device, and a computer that all communicate with each other wirelessly. Fig. 3.10 shows the system. The pillbox is comprised of an Arduino MCU, a motor, a ZigBee transceiver, and an RFID reader. In addition, each compartment is embedded with a diode and a photo diode for detecting pill removal. The MCU controls the motor such that it rotates the compartment towards the user when it is time to take medication and when the RFID-based wristband is detected in the proximity of the pillbox. The wristband embeds an RFID tag, and an LED, and is used for collecting motion data associated with pill picking and taking.

Proximity-Visual Systems

A blend of RFID sensors and video camera has been used in [57] to characterize the medication taking activity in an in-home environment. In this work, medication bottles were equipped with RFID tags and stored in a medicine cabinet that embeds an RFID reader. The RFID technology is employed for identification purposes of

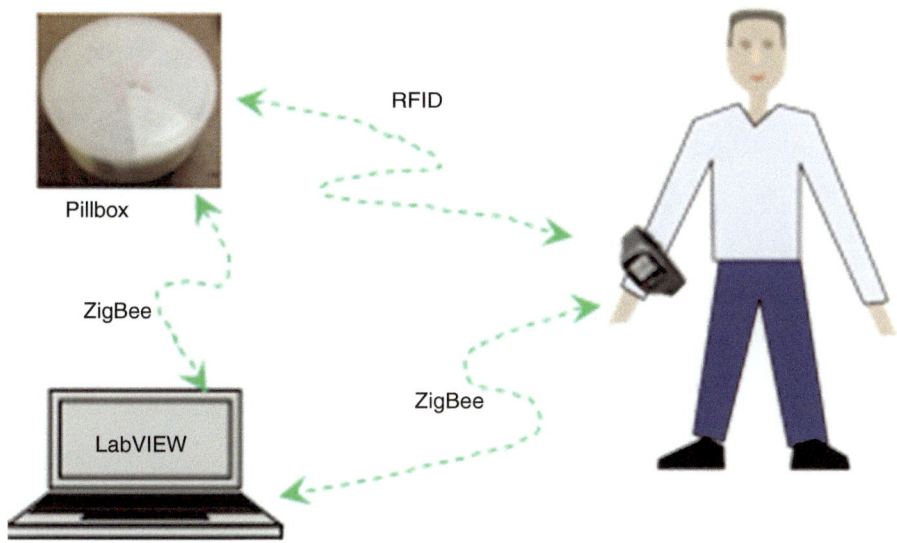

Fig. 3.10 The fusion-based system developed by [56]

the medication bottles placed in the cabinet. However, once a bottle is removed from the medication cabinet and it is out of the coverage of the reader's antenna, the identification process using RFID technology can not be achieved anymore. As such, the vision system is used such that it is activated once the medication bottle moves out of the range of the reader. The camera is used for tracking and verifying the occurrence of medication taking based on moving object detection and color model of the bottle.

Visual-Sensor Systems

Assistive living techniques have been used to track medication intake based on the patient's activity. One example is iMEC (Fig. 3.11), that has been developed by Suzuki and Nakauchi [58] for medicine timing and pill taking detection. Some home appliances (refrigerator, microwave oven, chair, and bed) have been attached with ubiquitous sensors for predicting the behavior of the patient. A medicine case equipped with a camera has been used for detecting pill removal. Eventually, the blend of data from these devices were used for confirming medication adherence.

Sensor-App Systems

Personal mobile device technology has witnessed a rapid progression in recent years. The services brought by mobile devices, such as the different means of communications and user applications, have enabled a host of possibilities. Thus, mobile applications' industry has been in race, including those for promoting healthcare of

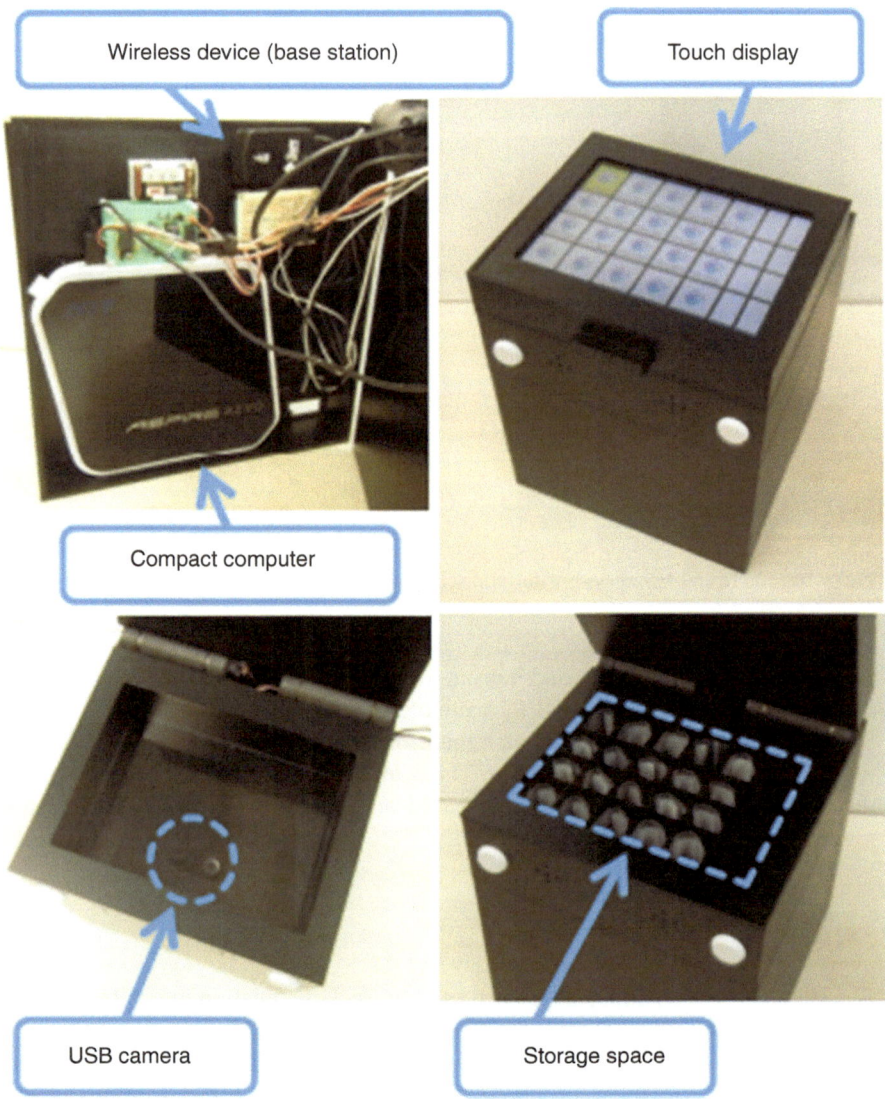

Fig. 3.11 iMEC system prototype devel- oped by Suzuki and Nakauchi [58]

older patients [25]. Specifically, many mobile and tablet-based applications have been developed in the form of automated reminder systems [91].

In this context, the sensor-app approach blends the use of sensor networks and mobile-app approaches for medication adherence tracking and monitoring. Abbey et al. [59] developed a pillbox containing multiple compartments with ambient light sensor fixed in each of them and a WiFi connection. Also, a mobile app has been developed that contains the medicine schedule. The pillbox and the mobile app are interconnected through an online data source. Hence, the mobile app generates alarms when it is the time of medication until the patient takes the medication from the pillbox or chooses to delay the action. In a recent study, Boonnuddar and Wuttidittachotti [60] proposed a pillbox-based system that uses the Arduino UNO WiFi and a load cell. Medication weight changes were reported to a server via the Internet. Also, a mobile application was developed that tracks the change in weight measurements and alerts the patient to take medication. The system was tested for 160 times of medication taking and the accuracy of the mobile application notification functionally was 96.88%.

Challenges and Future Trends

Technology is transforming healthcare as it brings new promises. Individuals rely on Quantified Self (QS) [92] technologies to collect multiple types data, such as sleep, location, mobility, and physical activity (including medication taking activity). However, still there are some technological challenges that need to be addressed in order for these systems to make a broader impact. As highlighted in Table 3.2, some weakening factors that may limit the adoption of such systems are the accuracy, energy consumption, and acceptability. However, there are other factors that are respectively related either directly or indirectly to these main factors such as lifetime, data fidelity, and user's comfort. Discussed below are these challenges and highlights on the trade-offs between them.

Challenges

System Accuracy and Data Fidelity

Achieving better healthcare requires accurate systems that capture the user's activity. This also applies to adherence monitoring systems. In general, accuracy is determined by the device being used for capturing the medication taking activity. Furthermore, the setting of medication taking can affect and limit the technology advances in use. For example, the system might operate at low-sampling rates as a trade-off for energy consumption minimization. However, this comes at the cost of lower data quality. Accuracy includes data quality, data precision, or data fidelity [67].

Data fidelity can be characterized by the sampling frequency, the sensor operation mode, and the duty cycling. Obtaining high accuracy data demands the system to be running at high-fidelity. However, high-fidelity systems deplete the battery energy at a fast rate, as their core should be set to run frequently for capturing the monitored event precisely. Thus, when engineering a tracking system, the energy consumption management should be considered carefully.

Energy Consumption and Lifetime

Monitoring systems can be battery-powered, for example, in the case of sensor networks and mobile device-based systems. This poses a challenge as the battery has limited energy budget [93]. From a system point of view, it is anticipated that a sufficient amount of electric current is being fed to the system to ensure its functionality. At the same time, from a user point of view, it is expected that the system lifetime lasts for as long as possible as application developers must either frequently replace batteries or use rechargeable batteries. This would likely be inadequate for user's acceptance and costly [94].

Even though only rare studies focused on the energy consumption of medication adherence monitoring systems, this is still central in this context as it can severely affect the performance and efficiency of the system [95]. This can be imagined by taking wearable systems powered by non-rechargeable batteries as an example. In general, the battery is a complex system that can behave unpredictably when affected by several factors and conditions, including the temperature and the applied load [96]. High-fidelity motion sensors are utilized within wearable devices for accurately sensing and quantifying the motion associated with medication taking activity. However, there is a trade-off between energy consumption and data fidelity. On the one hand, the sensor device should be operating continuously and sampling data frequently. On the other hand, even if temperature conditions are perfect, enabling the sensor(s) for frequent data sampling results in increasing the internal resistance of the battery and affecting its chemical and physical properties [97]. Operating the battery under such timing and intensity conditions will not enable it to provide voltage at a sufficient level that operates the connected device correctly, even with a considerable amount of unused charge being left. As a consequence of the experienced discharge behavior, the system's lifetime is directly affected. As such, wise battery usage is required [98]. Thus, techniques such as collaborative sensing to be employed for minimizing energy depletion in such systems. Once the energy consumption issue achieves notable progress, battery-powered systems such as wearable and portable systems can be used more widely in the area of adherence monitoring applications.

Acceptability and User's Comfort

The user's perception of a monitoring system has a great impact on its adoption and success. First, technological barriers such as battery energy consumption, mobility support, and others play a significant role as barriers to the wide acceptance of technology-based systems. Second, ethical challenges such as privacy and confidentiality also exist. Users are concerned about behaviors being monitored beyond medication taking and the potential of unintended users accessing the information collected [99]. In addition, users, and especially the older ones, tend to have social, physical, demographic, and cultural barriers towards using technology and, as a result, barring the user's acceptance of modern technology [41, 74].

Tampering, Authentication, and Active Non-Compliance

Two key challenges arise because users may try to actively deceive the system into thinking they are compliant when they are not. Tampering occurs when an unauthorized user receives the medication. The first challenge then becomes one of authentication—Is the person who is taking the medication who he claims to be? Tampering can arise for medications which can become addictive, such as opiates, where an addict or dealer has an incentive to fool the system. Authentication and authorization are analogous concepts in computer security—Is the person who they claim to be, and is this person authorized to take the medication? Although few projects have specifically tackled these security challenges, an array of wearables has investigated if a wearable is actually worn by the person it is supposed to [100]. A second set of approaches attempts to prevent unauthorized access with the use of physical barriers, such as locks on the pillboxes. A related set of approaches does not try to prevent unauthorized access, but rather take an auditing approach. For example, learning the wrist motions of different people can create an audit trail [101], which can then be used to identify tampering for later remediation.

The second challenge is observing active non-compliance, which is when a legitimate user actively deceives the system. Such behavior can occur when a user disagrees with a medical professional's treatment but appears to comply rather than challenge the professional's judgment. Active deception on the part of the user is more difficult to solve as the person using the system is legitimate but chooses not to consume the medication. A variety of approaches can be employed, such as video monitoring, but simple actions, such as placing medication in the mouth, faking a swallow, and then spitting it out later, will deceive most current technologies. The recent proposal of Quality of Life technologies [102] can help in monitoring different aspects of the individual's life. These can include social relationships and environment monitoring, that may impact the psychological and social factors and in result, patient's behavior. Creating monitoring systems that correctly identify active non-compliance remains an important research challenge.

Future Trends

It is clear from this review that most solutions have some sort of limitation. As such, the developed system may harness the advancements of a combination of technologies to achieve the ultimate goal. However, overcoming the challenges that were previously mentioned can be achieved as follows. To precisely monitor patient adherence, fine-grained sensors such as load cells, motion sensors for detecting and classifying gestures associated with hand-to-mouth movement, and sensors for cap opening and closure verification, are strong candidate technologies.

The integration of sensors that consume very little energy with limited fidelity along with sensors that report much higher fidelity of activity but also power-hungry on a single platform and decide what sensor and when to have it on, is an example of collaborative sensing that can be harnessed for prolonging the lifetime of a battery-powered system [67]. However, this requires sensor fusion algorithms that build a unified model based on different sensed and reported inputs—for example, Bayesian inference. In addition, since the wireless functionally in wireless-enabled systems constructs a bottleneck as it consumes a large portion from the battery energy, searching for low communication technologies is a must. An example of this can be the Transmit Only (TO) approach [67, 103] that can be employed rather than WiFi or Bluetooth. The TO technique is a single hop communication that does not demand handshaking or acknowledgment, and thus it minimizes the energy consumed for packet transmission to only a few tens of micro joules [67]. Finally, user's acceptability and comfort might be achieved by carefully designing a pill container that is low-energy consuming, smart, and wireless.

Conclusions

Medication non-adherence is a major problem in the healthcare sector. Poor medication adherence leads to healthcare resource wastage and sub-optimal treatment outcomes. As such, it has become an attractive research area for many researchers from multidisciplinary domains with the aim of developing new monitoring and interventions that can detect and correct medication taking regimens once they deviate. In this chapter, we have covered the technology-based techniques and systems for medication adherence monitoring. In addition, we put special stress on the advantages, disadvantages, and challenges associated with these approaches, but how those translate into changed operational and clinical outcomes requires more feedback and observations of both patients and clinical practitioners. From this review, we can conclude that work is still required to enhance technology-based systems that can overcome these challenges, especially the accuracy, user comfort, and battery consumption. In addition, assuring the whole workflow with minimal burden for the patients and health practitioners is still to be met.

References

1. Ortman JM, Velkoff VA, Hogan H. An aging nation: the older population in the United States. Suitland, MD, USA: United States Census Bureau; 2014. p. P25–P1140.
2. Iancu I, Iancu B. Elderly in the digital era. Theoretical perspectives on assistive Technologies. Technologies. 2017;5:60.
3. Koch S. Healthy ageing supported by technology–a cross-disciplinary research challenge. Inf Health Soc Care. 2010;35:81–91.
4. Garçon L, Khasnabis C, Walker L, Nakatani Y, Lapitan J, Borg J, Ross A, Velazquez Berumen A. Medical and assistive health technology: meeting the needs of aging populations. Gerontologist. 2016;56:S293–302.

5. Wamba SF, Anand A, Carter L. A literature review of RFID-enabled healthcare applications and issues. Int J Inf Manag. 2013;33:875–91.
6. Jimmy B, Jose J. Patient medication adherence: measures in daily practice. Oman Med J. 2011;26:155–9.
7. Lam WY, Fresco P. Medication adherence measures: an overview. Biomed Res Int. 2015;2015:217047. https://doi.org/10.1155/2015/217047.
8. Hutchins DS, Zeber JE, Roberts CS, Williams AF, Manias E, Peterson AM. Initial medication adherence–review and recommendations for good practices in outcomes research: An ISPOR medication adherence and persistence special interest group report. Value Health. 2015;18:690–9.
9. Hayes TL, Larimer N, Adami A, Kaye JA. Medication adherence in healthy elders: small cognitive changes make a big difference. J Aging Health. 2009;21:567–80.
10. Yap AF, Thirumoorthy T, Kwan YH. Systematic review of the barriers affecting medication adherence in older adults. Geriatr Gerontol Int. 2016;16:6993–7001.
11. Ho PM, Bryson CL, Rumsfeld JS. Medication adherence: its importance in cardiovascular outcomes. Circulation. 2009;709:3028–35.
12. van Heuckelum M, van den Ende CH, Houterman AE, Heemskerk CP, van Dulmen S, van den Bemt BJ. The effect of electronic monitoring feedback on medication adherence and clinical outcomes: a systematic review. PLoS One. 2017;12:e0185453. https://doi.org/10.1371/journal.pone.0185453.
13. Mrosek R, Dehling T, Sunyaev N. Taxonomy of health IT and medication adherence. Health Policy Technol. 2015;4:215–24. https://doi.org/10.1016/j.hlpt.2015.04.003.
14. Chai PR, Rosen RK, Boyer EW. Ingestible biosensors for real-time medical adherence monitoring: MyTMed. In: Proceedings of the 2016 49th Hawaii international conference on system sciences (HICSS), Koloa, HI, USA, 5–8; 2016. p. 3416–23.
15. Hugtenburg JG, Timmers L, Elders PJ, Vervloet M, van Dijk L. Definitions, variants, and causes of nonadherence with medication: a challenge for tailored interventions. Patient Prefer Adherence. 2013;7:675–82. https://doi.org/10.2147/PPA.S29549.
16. Connor J, Rafter N, Rodgers A. Do fixed-dose combination pills or unit-of-use packaging improve adherence? A systematic review. Bull World Health Org. 2004;82:935–9.
17. MacLaughlin EJ, Raehl CL, Treadway AK, Sterling TL, Zoller DP, Bond CA. Assessing medication adherence in the elderly. Drugs Aging. 2005;22:231–55.
18. Car J, Tan WS, Huang Z, Sloot P, Franklin BD. eHealth in the future of medi- cations management: personalisation, monitoring and adherence. BMC Med. 2017;15:73. https://doi.org/10.1186/s12916-017-0838-0.
19. Checchi KD, Huybrechts KF, Avorn J, Kesselheim AS. Electronic medication packaging devices and medication adherence: a systematic review. JAMA. 2014;312:1237–47. https://doi.org/10.1001/jama.2014.10059.
20. Balkrishnan R, Carroll CL, Camacho FT, Feldman SR. Electronic monitoring of medication adherence in skin disease: results of a pilot study. J Am Acad Dermatol. 2003;49:651–4. https://doi.org/10.1067/S0190-9622(03)00912-5.
21. Piette JD, Farris KB, Newman S, An L, Sussman J, Singh S. The potential impact of intelligent systems for mobile health self-management support: Monte Carlo simulations of text message support for medication adherence. Ann Behav Med. 2014;49:84–94. https://doi.org/10.1007/s12160-014-9634-7.
22. Easthall C, Barnett N. Using theory to explore the determinants of medication ad- herence; moving away from a one-size-fits-all approach. Pharmacy. 2017;5:50. https://doi.org/10.3390/pharmacy5030050.
23. Sachpazidis I, Sakas G. Medication intake assessment. In Proceedings of the 1st International Conference on PErvasive Technologies Related to Assistive Environments, Athens, Greece, 16–18 July 2008; p. 14.

24. Castillo-Mancilla J, Seifert S, Campbell K, Coleman S, McAllister K, Zheng JH, Hosek S. Emtricitabine-triphosphate in dried blood spots as a marker of recent dosing. Antimicrob Agents Chemother. 2016;60:6692–7. https://doi.org/10.1128/AAC.01017-16.
25. Dasgupta D, Chaudhry B, Koh E, Chawla NV. A survey of tablet applications for promoting successful aging in older adults. IEEE Access. 2016;4:9005–17. https://doi.org/10.1109/ACCESS.2016.2632818.
26. Stankovic JA. Research directions for cyber physical systems in wireless and mobile healthcare. ACM Trans Cyber-Phys Syst. 2017;1 https://doi.org/10.1145/2899006.
27. Liu X, Lewis JJ, Zhang H, Lu W, Zhang S, Zheng G, Liu M. Effectiveness of electronic reminders to improve medication adherence in tuberculosis patients: a clusterrandomised trial. PLoS Med. 2015;12:e1001876. https://doi.org/10.1371/journal.pmed.1001876.
28. Vervloet M, Linn AJ, van Weert JC, De Bakker DH, Bouvy ML, Van Dijk L. The effectiveness of interventions using electronic reminders to improve adherence to chronic medication: a systematic review of the literature. J Am Med Inf Assoc. 2012;19:696–704. https://doi.org/10.1136/amiajnl-2011-000748.
29. Gubbi J, Buyya R, Marusic S, Palaniswami M. Internet of things (IoT): a vision, architectural elements, and future directions. Future Gener Comput Syst. 2013;29:1645–60.
30. Hassanalieragh M, Page A, Soyata T, Sharma G, Aktas M, Mateos G, Andreescu S. Health monitoring and management using internet-of-things (IoT) sensing with cloud-based processing: opportunities and challenges. In: Proceedings of the 2015 IEEE International Conference on Services Computing (SCC), New York, NY, USA, 27 June–2 July 2015. p. 285–92.
31. Stegemann S, Baeyens JP, Cerreta F, Chanie E, Löfgren A, Maio M, Schreier G, Thesing-Bleck E. Adherence measurement systems and technology for medications in older patient populations. Eur Geriatr Med. 2012;3:254–60. https://doi.org/10.1016/j.eurger.2012.05.004.
32. Bosworth HB. How can innovative uses of technology be harnessed to improve medication adherence? Expert Rev Pharmacoecon Outcomes Res. 2012;12:133–5.
33. Park LG, Howie-Esquivel J, Dracup K. Electronic measurement of medication adherence. West J Nurs Res. 2015;37:28–49.
34. Mohammed HB, Ibrahim D, Cavus N. Mobile device based smart medication reminder for older people with disabilities. Qual Quant. 2018; https://doi.org/10.1007/s11135-018-0707-8.
35. Rokni SA, Ghasemzadeh H, Hezarjaribi N. Smart medication management, current technologies, and future directions. In: Handbook of research on healthcare administration and management. Hershey, PA, USA: IGI Global; 2017. p. 188–204.
36. Aldeer M, Martin RP. Medication adherence monitoring using modern technology. In: Proceedings of the IEEE 8th annual ubiquitous computing, electronics and Mobile communication conference (UEMCON), vol. 19–21. New York City, NY, USA; 2017. p. 491–7.
37. Lavallee M, Robillard PN, Mirsalari R. Performing systematic literature reviews with novices: An iterative approach. IEEE Trans Educ. 2014;57:175–81.
38. Aldeer M, Alaziz M, Ortiz J, Howard RE, Martin RP. A sensing-based framework for medication compliance monitoring. In: Proceedings of the 1st ACM International Workshop on Device-Free Human Sensing (DFHS'19), vol. 11. New York, NY, USA; November 2019. p. 52–6.
39. Lee ML, Dey AK. Sensor-based observations of daily living for aging in place. Pers Ubiquitous Comput. 2015;19:27–43. https://doi.org/10.1007/s00779-014-0810-3.
40. Kalantarian H, Motamed B, Alshurafa N, Sarrafzadeh M. A wearable sensor system for medication adherence prediction. Artif Intell Med. 2016;69:43–52. https://doi.org/10.1016/j.artmed.2016.03.004.
41. Wu X, Choi YM, Ghovanloo M. Design and fabricate neckwear to improve the elderly patients' medical compliance. In: Proceedings of the International Conference on Human Aspects of IT for the Aged Population, vol. 2–7. August. Los Angeles, CA, USA; 2015. p. 222–34.

42. Kalantarian H, Alshurafa N, Nemati E, Le T, Sarrafzadeh M. A smartwatch-based medication adherence system. In: Proceedings of the IEEE 12th International Conference on Wearable and Implantable Body Sensor Networks (BSN), vol. 9–12. June. Cambridge, MA, USA; 2015. p. 1–6.
43. Hezarjaribi N, Fallahzadeh R, Ghasemzadeh H. A machine learning approach for medication adherence monitoring using body-worn sensors. In: Design, Automation & Test in Europe Conference & Exhibition (DATE), Dresden, Germany, vol. 14–18. March; 2016. p. 842–5.
44. Wang R, Sitová Z, Jia X, He X, Abramson T, Gasti P, Farajidavar A. Automatic identification of solid-phase medication intake using wireless wearable accelerometers. In: Proceedings of the 36th Annual International Conference of the IEEE Engineering in Medicine and Biology Society, vol. 26–30. August. Chicago, IL, USA; 2014. p. 4168–71.
45. Chen C, Kehtarnavaz N, Jafari R. A medication adherence monitoring system for pill bottles based on a wearable inertial sensor. In: Proceedings of the 36th Annual International Conference of the IEEE Engineering in Medicine and Biology Society, vol. 26–30. August. Chicago, IL, USA; 2014. p. 4983–6.
46. Mondol MAS, Emi IA, Stankovic JA. MedRem: An interactive medication reminder and tracking system on wrist devices. In: Proceedings of the IEEE Wireless Health (WH), vol. 25–27. Bethesda, MD, USA; October 2016. p. 46–53.
47. Hafezi H, Robertson TL, Moon GD, Au-Yeung KY, Zdeblick MJ, Savage GM. An ingestible sensor for measuring medication adherence. IEEE Trans Biomed Eng. 2015;62:99–109. https://doi.org/10.1109/TBME.2014.2341272.
48. Agarawala A, Greenberg S, Ho G. The context-aware pill bottle and medication monitor. In: Proceedings of the Video Proceedings and Proceedings Supplement of the Sixth International Conference on Ubiquitous Computing (UBICOMP). Nottingham, UK; September 2004. p. 7–10.
49. Becker E, Metsis V, Arora R, Vinjumur J, Xu Y, Makedon F. SmartDrawer: RFID-based smart medicine drawer for assistive environments. In: Proceedings of the 2nd International Conference on PErvasive Technologies Related to Assistive Environments (PETRA '09). Corfu, Greece; June 2009. p. 9–13.
50. Morak J, Schwarz M, Hayn D, Schreier G. Feasibility of mHealth and near field communication technology based medication adherence monitoring. In: Proceedings of the Annual International Conference of the IEEE Engineering in Medicine and Biology Society. San Diego, CA, USA., 28 August–1 September; 2012. p. 272–5.
51. Batz D, Batz M, da Vitoria Lobo N, Shah M. A computer vision system for monitoring medication intake. In: Proceedings of the 2nd Canadian Conference on Computer and Robot Vision. Victoria, BC, Canada., 9–11 May; 2005. p. 362–9.
52. Valin M, Meunier J, St-Arnaud A, Rousseau J. Video surveillance of medication intake. In: Proceedings of the 28th Annual International Conference of the IEEE Engineering in Medicine and Biology Society (EMBS'06). New York, NY, USA., 30 August–3 September; 2006. p. 6396–9.
53. Dauphin G, Khanfir S. Background suppression with low-resolution camera in the context of medication intake monitoring. In: Proceedings of the 3rd European Workshop on Visual Information Processing (EUVIP), vol. 4–6. Paris, France; July 2011. p. 128–33.
54. Huynh HH, Meunier J, Sequeira J, Daniel M. Real time detection, tracking and recog- nition of medication intake. World Acad Sci Eng Technol. 2009;3:2801–8.
55. Sohn SY, Bae M, Lee DKR, Kim H. Alarm system for elder patients medication with IoT-enabled pill bottle. In: Proceedings of the International Conference on Information and Communication Technology Convergence (ICTC). Jeju, Korea., 28–30 October; 2015. p. 59–61.
56. Li J, Peplinski SJ, Nia SM, Farajidavar A. An interoperable pillbox system for smart medication adherence. In: Proceedings of the 36th Annual International Conference of the IEEE Engineering in Medicine and Biology Society, vol. 26–30. August. Chicago, IL, USA; 2014. p. 1386–9.

57. Hasanuzzaman FM, Yang X, Tian Y, Liu Q, Capezuti E. Monitoring activity of taking medicine by incorporating RFID and video analysis. Netw Model Anal Health Inf Bioinform. 2013;2:61–70. https://doi.org/10.1007/s13721-013-0025-y.
58. Suzuki T, Nakauchi Y. Intelligent medicine case for dosing monitoring and support. In: 2010 IEEE/RSJ International Conference on Intelligent Robots and Systems. Taipei; 2010. p. 3471–6.
59. Abbey B, Alipour A, Gilmour L, Camp C, Hofer C, Lederer R, Rasmussen G, Liu L, Nikolaidis I, Stroulia E, Sadowski C. A remotely programmable smart pillbox for enhancing medication adherence. In: Proceedings of the 25th International Symposium on Computer-Based Medical Systems (CBMS)., Rome, Italy, 20–22 June; 2012. p. 1–4.
60. Boonnuddar N, Wuttidittachotti P. Mobile application: patients' adherence to medicine intake schedules. In: Proceedings of the International Conference on Big Data and Internet of Thing, London, UK, 20–22 December; 2017. p. 237–41.
61. Polastre J, Szewczyk R, Culler D. Telos: enabling ultra-low power wireless research. In: Proceedings of the 4th International Symposium on Information Processing in Sensor Networks (IPSN), vol. 25–27. Los Angeles, CA, USA; April 2005. p. 364–9.
62. Aldeer MMN. A summary survey on recent applications of wireless sensor networks. In: Proceedings of the IEEE Student Conference on Research and Developement (SCOReD)., Putrajaya, Malaysia, 16–17 December; 2013. p. 485–90.
63. Alemdar H, Ersoy C. Wireless sensor networks for healthcare: a survey. Comput Netw. 2010;54:2688–710.
64. Shubair RM, Elayan H. In vivo wireless body communications: State-of-the-art and future directions. In: Proceedings of the Loughborough Antennas & Propagation Conference (LAPC)., Loughborough, UK, 2–3 November; 2015. p. 1–5.
65. Aldeer M, Martin RP, Howard RE. PillSense: designing a medication adherence monitoring system using Pill Bottle-Mounted Wireless Sensors. In: Proceedings of the IEEE International Conference on Communications Workshops (ICC Workshops), Kansas City, MO, USA, 20–24 May; 2018.
66. Aldeer M, Ortiz J, Howard RE, Martin RP. PatientSense: patient discrimination from in-bottle sensors data. In: Proceedings of the 16th EAI International Conference on Mobile and Ubiquitous Systems: Computing, Networking and Services (MobiQuitous '19). Houston, TX, USA, 12–14 November 2019; 2019. p. 143–52.
67. Aldeer MMN, Martin RP, Howard RE. Tackling the fidelity-energy trade-off in wire–less body sensor networks. In: Proceedings of the IEEE/ACM International Conference on Connected Health: Applications, Systems and Engineering Technologies (CHASE). Philadelphia, PA, USA., 17–19 July; 2017. p. 7–12.
68. Zeng H, Zhao Y. Sensing movement: microsensors for body motion measurement. Sensors. 2011;70:638–60.
69. Sprager S, Juric MB. Inertial sensor-based gait recognition: a review. Sensors. 2015;15:22089–127.
70. Büsching F, Kulau U, Gietzelt M, Wolf L. Comparison and validation of capacitive accelerometers for health care applications. Comput Methods Progr Biomed. 2012;696:79–88.
71. Kalantarian H, Alshurafa N, Sarrafzadeh M. A survey of diet monitoring technology. IEEE Pervasive Comput. 2017;16:57–65. https://doi.org/10.1109/MPRV.2017.1.
72. Vu T, Lin F, Alshurafa N, Xu W. Wearable food intake monitoring technologies: a comprehensive review. Computers. 2017;6:4.
73. Olubanjo T, Ghovanloo M. Real-time swallowing detection based on tracheal acoustics. In: Proceedings of the IEEE International Conference on Acoustics, Speech and Signal Processing (ICASSP)., Florence, Italy, 4–9 May; 2014. p. 4384–8.
74. Choi YM, Olubanjo T, Farajidavar A, Ghovanloo M. Potential barriers in adoption of a medication compliance neckwear by elderly population. In: Proceedings of the 35th Annual International Conference of the IEEE Engineering in Medicine and Biology Society (EMBC)., Osaka, Japan, 3–7 July; 2013. p. 4678–81.

75. Bennett J, Rokas O, Chen L. Healthcare in the smart home: a study of past, present and future. Sustainability. 2017;9:840. https://doi.org/10.3390/su9050840.
76. Chen H, Xue M, Mei Z, Bambang Oetomo S, Chen W. A review of wearable sensor Systems for Monitoring Body Movements of neonates. Sensors. 2016;16:2134. https://doi.org/10.3390/s16122134.
77. LeMoyne R, Mastroianni T. Wearable and wireless Systems for Healthcare I. Singapore: Springer; 2018.
78. Lai X, Liu Q, Wei X, Wang W, Zhou G, Han G. A survey of body sensor networks. Sensors. 2013;13:5406–47. https://doi.org/10.3390/s130505406.
79. Kiourti A, Nikita KS. A review of in-body biotelemetry devices: Implantables, ingestibles, and injectables. IEEE Trans Biomed Eng. 2017;64:1422–30. https://doi.org/10.1109/TBME.2017.2668612.
80. Kalantar-Zadeh K, Berean KJ, Ha N, Chrimes AF, Xu K, Grando D, Taylor KM. A human pilot trial of ingestible electronic capsules capable of sensing different gases in the gut. Nat Electron. 2018;1:79–87. https://doi.org/10.1038/s41928-017-0004-x.
81. Dua A, Weeks WA, Berstein A, Azevedo RG, Li R, Ward A. An in-vivo communication system for monitoring medication adherence. In: Proceedings of the Wireless Communications and Networking Conference (WCNC)., San Francisco, CA, USA, 19–22 March; 2017. p. 1–6.
82. Chai PR, Carreiro S, Innes BJ, Rosen RK, O'Cleirigh C, Mayer KH, Boyer EW. Digital pills to measure opioid ingestion patterns in emergency department patients with acute fracture pain: a pilot study. J Med Internet Res. 2017;19:e19. https://doi.org/10.2196/jmir.7050.
83. Trappe W, Howard R, Moore RS. Low-energy security: limits and opportunities in the internet of things. IEEE Secur Priv. 2015;13:14–21. https://doi.org/10.1109/MSP.2015.7.
84. Bolić M, Rostamian M, Djurić PM. Proximity detection with RFID in the internet of things. In: Proceedings of the 48th Asilomar Conference on Signals, Systems and Computers, vol. 2–5. Pacific Grove, CA, USA; November 2014. p. 770–14.
85. Roberts CM. Radio frequency identification (RFID). Comput Secur. 2006;25:18–26. https://doi.org/10.1016/j.cose.2005.12.003.
86. Coskun V, Ozdenizci B, Ok K. The survey on near field communication. Sensors. 2015;15:13348–405. https://doi.org/10.3390/s150613348.
87. Coskun V, Ozdenizci B, Ok K. A survey on near field communication (NFC) technology. Wirel Pers Commun. 2013;71:2259–94. https://doi.org/10.1007/s11277-012-0935-5.
88. Fiocchi S, Parazzini M, Liorni I, Samaras T, Ravazzani P. Temperature increase in the fetus exposed to UHF RFID readers. IEEE Trans Biomed Eng. 2014;61:2011–9. https://doi.org/10.1109/TBME.2014.2312023.
89. Zhang S, Wei Z, Nie J, Huang L, Wang S, Li Z. A review on human activity recognition using vision-based method. J Healthc Eng. 2017;2017 https://doi.org/10.1155/2017/3090343.
90. Alexander SM, Nerminathan A, Harrison A, Phelps M, Scott KM. Prejudices and perceptions: patient acceptance of mobile technology use in health care. Intern Med J. 2015;45:1179–81. https://doi.org/10.1111/imj.12899.
91. Silva JM, Mouttham A, El Saddik A. UbiMeds: a mobile application to improve accessi- bility and support medication adherence. In: Proceedings of the 1st ACM SIGMM international workshop on Media studies and implementations that help improving access to disabled users, vol. 23. Beijing, China; October 2009. p. 71–8.
92. Wac K. From quantified self to quality of life. In: Rivas H, Wac K, editors. Digital health. Cham: Springer; 2018. p. 83–108.
93. Rukpakavong W, Guan L, Phillips I. Dynamic node lifetime estimation for wireless sensor networks. IEEE Sensors J. 2014;14:1370–9. https://doi.org/10.1109/JSEN.2013.2295303.
94. Jovanov E, Milenkovic A, Otto C, de Groen PC. A wireless body area network of intel- ligent motion sensors for computer assisted physical rehabilitation. J Neuro Eng Rehabilit. 2005;2 https://doi.org/10.1186/1743-0003-2-6.

95. Yuan D, Kanhere SS, Hollick M. Instrumenting wireless sensor networks–a survey on the metrics that matter. Pervasive Mob Comput. 2017;37:45–62. https://doi.org/10.1016/j.pmcj.2016.10.001.
96. Feeney LM, Hartung R, Rohner C, Kulau U, Wolf L, Gunningberg P. Towards realistic lifetime estimation in battery-powered IoT devices. In: Proceedings of the 15th ACM Conference on Embedded Network Sensor Systems (SenSys '17)., Delft, The Netherlands, 6–8 November; 2017.
97. Feeney LM, Rohner C, Gunningberg P, Lindgren A, Andersson L. How do the dynamics of battery discharge affect sensor lifetime? In: Proceedings of the 11th Annual Conference on Wireless On-demand Network Systems and Services (WONS)., Obergurgl, Austria, 2–4 April; 2014. p. 49–56.
98. Feeney LM, Andersson L, Lindgren A, Starborg S, Tidblad AA. Using batteries wisely. In: Proceedings of the 10th ACM Conference on Embedded Network Sensor Systems (SenSys '12), vol. 6–9. Toronto, Canada; November 2012. p. 349–50.
99. Campbell JI, Eyal N, Musiimenta A, Haberer JE. Ethical questions in medical electronic adherence monitoring. J Gen Intern Med. 2016;31:338–42. https://doi.org/10.1007/s11606-015-3502-4.
100. Lewis A, Li Y, Xie M. Real time motion-based authentication for smartwatch. In: Proceedings of the IEEE Conference on Communications and Network Security (CNS)., Philadelphia, PA, USA, 17–19 October; 2016. p. 380–1.
101. Nixon KW, Chen Y, Mao ZH, Li K. User classification and authentication for mobile device based on gesture recognition. In: Robinson EP, editor. Network science and cybersecurity. New York, NY, USA: Springer; 2014. p. 125–35.
102. Wac K. Quality of life technologies. In: Gellman M, editor. Encyclopedia of behavioral medicine. New York, NY: Springer; 2020.
103. Aldeer M, Al Hilli A, Howard RE, Martin RP. Lifetime optimization of transmit- only sensor networks with adaptive Mobile cluster heads. In: Proceedings of the 2019 on Wireless of the Students, by the Students, and for the Students Workshop, vol. 21. Los Cabos, Mexico; October 2019. p. 3–5.

Chapter 4
Quantifying Energy and Fatigue: Classification and Assessment of Energy and Fatigue Using Subjective, Objective, and Mixed Methods towards Health and Quality of Life

Natalie Leah Solomon and Vlad Manea

Introduction

There are many ways to conceptualize "Energy" and "Fatigue" in the context of the WHO Quality of Life domain [1]. Energy and fatigue may be interrelated but may also be considered orthogonal. Low energy can be characterized by fatigue, lack of motivation, and lack of interest, while states of excessive energy can reach pathological levels that include disrupted sleep, restlessness and agitation, or even mania [2]. Although lacking energy can be burdensome and uncomfortable, it is simultaneously an adaptive symptom that is perceived as a need to rest or slow down [3]. Given that energy is a valuable resource, efficient spending and conservation of energy may result in the greatest chances of vitality and even survival [4]. Curiously, one tends to think of energy as a resource that is depleted and then replenished with rest. It is also frequently observed that using energy may be synonymous with generating energy (e.g. one may feel more replenished after engaging in an activity than they do following rest). This curious paradox highlights the potential value of classifying and enhancing our understanding of energy. Fatigue is both a normative experience as well as associated with many chronic illnesses and psychiatric disorders. Fatigue can be characterized by subjective feelings of "tiredness" and "lack of energy" [5] and can serve as a signal to prevent strain, damage, and injury [6].

In this chapter, energy refers to the strength and vitality required for sustained physical or mental activity. Lack of energy or fatigue is used to describe the subjective sensation (*perceived fatigue*) as well as the objective and quantifiable

---

N. L. Solomon (✉)
Stanford University School of Medicine, Stanford, CA, USA

V. Manea
Quality of Life Technologies Lab, Department of Computer Science, University of Copenhagen, Copenhagen, Denmark
e-mail: manea@di.ku.dk

© The Author(s) 2022
K. Wac, S. Wulfovich (eds.), *Quantifying Quality of Life*, Health Informatics,
https://doi.org/10.1007/978-3-030-94212-0_4

change in performance (*fatigability*) [7]. Fatigue can be classified as pathological or non-pathological. Pathological fatigue can be described as an overwhelming sense of tiredness at rest, exhaustion with activity, lack of energy that precludes daily tasks, or loss of vigour [7]. In healthy adults, non-pathological fatigue is predictable and does not interfere with usual daily activities. Non-pathological fatigue is typically brought about by prolonged exertion and diminishes with rest [8]. In addition to pathological and non-pathological fatigue, fatigue may also be subdivided as either physical or mental (cognitive/psychiatric) and further subdivided as primary (neurological) or secondary (non-neurological) [7–10]. Furthermore, performance refers to an individual's functioning in their daily environment while capacity refers to the maximal or optimized level of functioning.

Preliminary studies were conducted on energy and fatigue during the First World War when researchers investigated the impact of fatigue on efficiency and productivity of the industrial workforce [11]. This "occupational fatigue" continues to be a focus of research attention, especially in vocations and occupations in which fatigue carries serious implications. Traditionally, energy and fatigue have been assessed using qualitative, self-reported outcomes [12] and can be obtained from a number of validated scales [13, 14]. Most clinical fatigue studies use self-report measures that can broadly be classified as measuring perceptions of fatigue [8]. Despite the numerous scales that measure fatigue, there is no agreed-upon standard of which to compare subjective reports of fatigue [15, 16].

The use of technology to monitor and manage energy and fatigue has been investigated in order to help healthy individuals continue to live healthily [3, 6, 17], assist individuals with health issues [18–20], and address vocational or occupational fatigue to improve personal and workplace safety [21–23]. The monitoring of energy and fatigue helps individuals adapt their effort in recreational (e.g., amateur sport, exercise) and occupational (e.g., drivers, pilots, police officers, professional athletes, shift workers) settings to prevent negative effects (e.g., burnout, exhaustion, accidents, injury) and maintain quality of life [24, 25]. Further, energy and fatigue research is needed to examine their connection to underlying or potential health conditions as well as interventional studies to validate the operationalization of energy and fatigue monitoring in daily life.

In this chapter, we will classify energy and fatigue and present their measurement. The chapter is structured as methods of our work, classification of energy and fatigue (pathological as well as non-pathological), measurement and assessment of energy and fatigue, discussion of results, and conclusive remarks.

Methods

We conducted a scoping review of the existing literature between 2010 and 2020 in Google Scholar on the technology-enabled assessment of energy and fatigue. Search terms related to energy and fatigue (e.g., "fatigability", "tiredness") were coupled with terms pertaining to each of the following domains: (1) the population under study (e.g., "athlete", "driver"), (2) the health outcomes (e.g., "circulation",

Table 4.1 Domains of energy and fatigue literature review

Domain	Inclusion	Rationale	Search terms (selective)
Energy / fatigue	Mandatory	Energy and fatigue are often proxied by synonyms or antonyms.	Energy, fatigue, fatigability, tiredness, vitality
Population	Optional	Papers assessing energy and fatigue in healthy individuals often focus on a specific segment of the general population. For instance, two areas of focus are athletics and occupational fatigue.	Athlete, driver, performance, pilot, police, shift, sport, worker, employee
Health outcomes	Optional	Health outcomes are often delineated by specific elements of human physiology or pathology: Organs, systems, processes, and diseases. In addition, such elements can be further delineated by the population segment under study.	Cancer, cardiovascular, circulation, dementia, heart, kidney, mental, pulmonary, respiration
Measurement	Optional	Methodological measurements using technology can be described by the procedure, device, sensor, process, or result.	Accelerometer, app, application, camera, band, ecological momentary assessment, performance, capacity, electrocardiogram, electrooculogram, experience sampling method, Fitbit, galvanic, mobile, sensor, smart band, smartphone, smartwatch, vision, watch, wearable

"dementia", "heart"), and (3) the measurement (e.g., "accelerometer", "electrocardiogram", "wearable"). One example search phrase was "galvanic energy fatigue tiredness vitality". We also reviewed the relevant references of the identified literature. Table 4.1 reviews the domains, search terms (selective), and the rationale for choosing the domain.

Results

We found 40 reviews on energy and fatigue pertaining to the domains and 60 studies assessing fatigue by using technology. The search results included in this review either (1) reviewed energy and fatigue assessment for a specific population and/or health outcome, (2) provided evidence for the use of measurement to monitor or manage energy or fatigue, or (3) discussed human factors of technology towards monitoring energy and fatigue. The taxonomy of fatigue resulting from our literature review is depicted in Fig. 4.1.

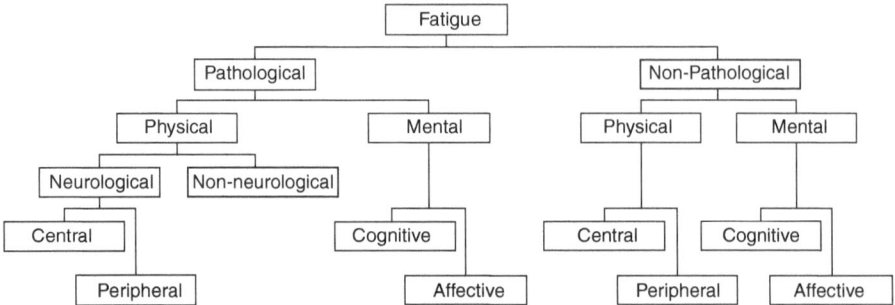

Fig. 4.1 Taxonomy of fatigue with pathological and non-pathological types [7–10, 26]

Energy and Fatigue Classification

Pathological Fatigue

Pathological fatigue is prolonged or chronic (>6 months), can be highly debilitating, and is much less common than non-pathological fatigue [27]. Pathological fatigue may be best understood as an amplified sense of normal (non-pathological) fatigue that can be induced by changes in one or more variables regulating work output [9]. For instance, a healthy individual may experience fatigue during or after exercising, but the same individual may perceive even more fatigue when exercising during an infectious disease [7]. Diseased individuals describe fatigue as an overwhelming sense of tiredness at rest, exhaustion with activity, loss of vigour, or lack of energy that precludes daily tasks, inertia or lack of endurance [28]. Pathological fatigue may be classified as physical or mental and is associated with multiple illnesses.

Physical Fatigue

Pathological physical fatigue includes neurological and non-neurological fatigue.

Neurological Fatigue

Neurological fatigue suggests that the physical expression of fatigue is mediated by central and peripheral mechanisms [27]. Therefore, neurological fatigue may be further classified as central or peripheral [9].

Central fatigue is generated at sites proximal to the peripheral nerves and referred to as a progressive decline in the ability to activate muscles voluntarily [29]. Central fatigue is due to impaired muscle performance that arises from the central nervous system [28]. A feeling of constant exhaustion is a characteristic of central fatigue [9]. Pathological central fatigue is found in Multiple Sclerosis, Traumatic Brain Injury, Parkinson's Disease, and many others.

Mechanisms of peripheral fatigue are usually attributable to a neuronal or muscular origin. Peripheral fatigue results from a lack of response in the neuromuscular system after central stimulation [27]. Peripheral fatigue is characterized by the failure to sustain the force of muscle contraction [9]. Pathological peripheral fatigue is found in neuromuscular disorder, rhabdomyolysis, muscle ischemia, restless legs syndrome and more.

In many of the previously mentioned health conditions, physical inactivity is a contributing factor to the increased fatigue of the patient [30]. Deconditioning, as a result of restricted physical activity, results in large decreases in muscle mass and strength, as well as increased fatigue due to changes in muscle metabolism [31, 32]. Physical fatigue is also increasingly observed as a secondary outcome in many diseases and health conditions during the performance of everyday activities [32].

Non-Neurological Fatigue

The exact mechanism of how non-neurological disease causes fatigue is not fully understood [7]. However, there are indications that peripheral proinflammatory cytokines signal the central nervous system to initiate fatigue [33]. A common non-neurological cause of temporary fatigue is an infection or the common cold. Non-neurological causes of chronic fatigue include infectious diseases (human immunodeficiency virus, mononucleosis, Borreliosis, and chronic pancreatitis), hematologic disease (anaemia and hemochromatosis), dehydration, immunological disease (celiac disease), rheumatological disease, cardiac disease (heart failure and cardiomyopathy), endocrinologic disorder (diabetes, Addison's disease, hypopituitarism, and hypothyroidism), renal disease (insufficiency and dialysis), lung disease (chronic obstructive lung disease and asthma), malnutrition (poor diet, irritable bowel disease, eating disorders and hypoproteinemia), liver disease, chronic pain, chronic fatigue syndrome, fibromyalgia, malignancy (cancer, sarcoma, lymphoma, and leukaemia), Gulf War disease, poisoning, mineral or vitamin deficiencies, drugs, or irradiation [7].

Drugs and medications may also be a cause of non-neurological fatigue. The drugs that cause fatigue include alcohol, antihistamines, benzodiazepines, antispasmodics, antiepileptic drugs, neuroleptics, and narcotics [7].

Mental Fatigue

Mental fatigue in the pathological domain includes cognitive and affective (psychological/psychiatric) fatigue. Cognitive fatigue has been studied in the context of MS [34], cancer [35], TBI [36], HIV [37], and other diseases. Affective fatigue is influenced by psychological factors (attitude, motivation, will, endurance, flexibility, inertia, persistence, concentration, and alertness) as well as psychiatric factors (depression, mania, psychosis, and addiction) [28]. Individuals with chronic fatigue report poorer mental health than their non-chronic fatigue counterparts [38].

Non-pathological Fatigue

In contrast to pathological fatigue, non-pathological fatigue is short term and remits with rest. Non-pathological fatigue is sometimes referred to as physiological fatigue in the scientific literature. Non-pathological fatigue alerts the individual to opportunity costs of current activities, as well as of the attraction of alternative activities [39]. Fatigue in healthy individuals is a universal experience and a natural occurrence after physical or mental efforts, usually relieved by rest. Research has examined biological explanations for pathological versus non-pathological fatigue [40], as well as self-report scales to distinguish fatigue associated disease from fatigue associated with healthy controls [41]. It has been reported that 55% of healthy individuals identified a physical sensation of fatigue and 24% identified a mental sensation of fatigue [26].

Physical Fatigue

From a physical perspective, fatigue is described as the inability of the muscles to maintain the required level of strength during exercise activities [42, 43]. It can also be characterized as an exercise-induced reduction in muscle's capability to generate force. There is no single cause of physical fatigue [44] and physical fatigue includes both central and peripheral fatigue.

Central fatigue designates a decrease in voluntary activation of the muscle, whereas, peripheral fatigue indicates a decrease in the contractile strength of the muscle fibres and changes in the mechanisms underlying the transmission of muscle action potentials [45]. Central and peripheral fatigue is a common experience during sport and exercise activities.

The impact of physical fatigue on cognitive performance depends both on the intensity and the duration of the exercise [46, 47]. Prolonged physical exercise leading to dehydration and physical fatigue is associated with a reduction in cognitive performance [48].

Mental Fatigue

Mental fatigue includes cognitive and affective fatigue and is an unfocused mental state, characterized by distraction, frustration, or discomfort. Mental fatigue is a psychobiological state caused by prolonged periods of demanding cognitive activity and characterized by subjective feelings of "tiredness" and "lack energy" [4].

In terms of cognitive activities, mental fatigue may be defined as the perception of feeling cognitively fatigued after performing demanding cognitive activities that involve concentration, attention, endurance, or alertness [49]. In the cognitive domain, fatigability can be measured as a decline in the reaction time, a decline in accuracy on continuous performance tasks, or a probe task that is given before and immediately after a fatiguing cognitive task [50, 51]. This cognitive fatigue is associated with problems completing tests, particularly where there is a requirement to sustain high levels of effort over time [39]. The effects of mental fatigue on cognitive performance [4, 51–53], and the skilled performance of drivers [54] and air pilots [55], have been investigated. Mental fatigue also limits physical performance [56] through perceived exertion [5]. Similarly, mental fatigue, following the performance of cognitive tasks, impairs emotion regulation [57].

Affective fatigue is characterized by low mood, tiredness, weariness, and lethargy [39]. It has been reported that 21% of healthy individuals identified an affective sensation of fatigue [26]. Non-pathological affective fatigue includes self-regulatory fatigue, empathy fatigue, and other fatigue associated with emotional depletion (burnout).

Factors Influencing Fatigue

Pathological and non-pathological fatigue is influenced by numerous factors, such as age, gender, physical condition, diet, latency to last meal, mental status, psychological conditions, personality type, life experience, and the health status of the individual [7]. Most studies found more fatigue in women than in men [38, 58–62]. Inconsistent findings have been reported regarding age and fatigue [38, 58, 62, 63]. Additionally, a high level of formal education has been associated with a lower prevalence of fatigue [61, 64, 65].

Sleepiness and fatigue are distinct and interrelated. Sleepiness refers to an increased propensity to fall asleep [66], while fatigue refers to tiredness resulting from exertion or illness. Fatigue may be regarded as a motivational drive to rest [67] and non-pathological fatigue will usually remit with rest. Sleepiness is related to circadian and homeostatic influences and remits after sleep [68], but not after rest.

Energy and Fatigue Measures

Fatigue perception is frequently measured by self-report scales, while fatigability is frequently assessed by performance, capacity, and technology-reported measures [69]. Subjective measures include scales and prompts for assessment while objective measures include performance and capacity tasks (physical and cognitive), physiological measurements (cardiac, ocular, neural), and markers (biological and behavioural).

Subjective Measures

Fatigue perception is frequently measured by application of patient-reported outcomes (PRO) [12] through validated scales prompted for assessment. These scales may be administered momentarily, daily, monthly etc. through paper, web, or smartphone.

Scale Instruments

Scales for self-reporting may be unidimensional, evaluating a single property, or multidimensional, evaluating multiple properties [49]. These instruments address different aspects of fatigue and energy and some address more than one aspect. No single measure of fatigue adequately captures the complexity of the phenomenon [15]. Researchers have pointed out that "in developing fatigue scales, there is a "catch 22″ situation: before a concept can be measured, it must be defined, and before a definition can be agreed upon, there must exist an instrument for assessing phenomenology. There is, unfortunately, no "gold standard" for fatigue, nor is there ever likely to be" [13]. Table 4.2 in this section depicts several scale instruments routinely used to measure energy and fatigue. The majority of these energy/fatigue self-report scales were designed for pathologic populations, but have been applied to non-pathologic populations as well.

Considerations in choosing a particular scale include recall period, unidimensionality or multidimensionality, scale structure and length, and suitable population. Scales differ in their scope, some measuring severity only, and others duration and impact on a range of functions [14]. Fatigue measures have been evaluated for the number of symptoms assessed, dimensions of fatigue explored, the time frame of the assessment, scale, method, the population on which the scale was developed, and psychometric properties [13, 14].

Some applications of these scales are illustrated below. SF-36 and PROMIS have been used in traditional studies assessing fatigue in the general population [14, 81]. FQ, FSS, and MAF have been employed to assess workplace-related fatigue [82, 83]. POMS has been used to assess fatigue in bus drivers [84] and sport athletes [24]. Scales were also used in traditional studies to assess energy and fatigue in individuals with a plethora of diseases, e.g., cancer [85, 86], cardiovascular disease [87, 88], chronic obstructive pulmonary disease [89], diabetes [90], fibromyalgia [91], hearing loss [16], inflammatory bowel disease [92–94], lupus [95], major depressive disorder [96], multiple sclerosis [97–99], psoriasis [100], pulmonary arterial hypertension [101], renal disease [102], rheumatic disease [103, 104], sleep apnea [105], stroke [106, 107], and traumatic brain injury [10].

Smartphone collection of self-reported energy and fatigue data has been utilized in the context of multiple sclerosis [108], cancer-related fatigue [109], and bipolar

Table 4.2 Scale instruments routinely used to measure energy and fatigue

Instrument	Recall period	Measures	Administration	Usage
Fatigue assessment scale (FAS) [70]	Usually ("refer to how you usually feel")	Unidimensional; fatigue severity	10 items, 5-level Likert scale	Pathologic and non-pathological (developed for chronic fatigue)
Functional Assessment of Chronic Illness Therapy (FACIT-F) Fatigue Subscale [71]	Past week	Unidimensional; general fatigue	13 items, 5-level Likert scale	Pathologic (people with various chronic illnesses, including cancer)
Fatigue impact scale (FIS) [72]	Past month, present time	Multidimensional; physical, cognitive, and psychosocial functioning, total fatigue	40 items, 5-level Likert scale	Pathologic (developed for infectious disease patients)
Fatigue questionnaire / fatigue scale (FQ / FS) [73]	Past month	Multidimensional; physical, mental, total, substantial, transient, and chronic fatigue	11 items, 4-level Likert scale	Pathologic and non-pathological (developed for use in hospital and community populations)
The fatigue severity scale (FSS) [74]	Past week	Unidimensional; fatigue severity	9 items, 7-level Likert scale	Pathologic and non-pathological (developed for patients with multiple sclerosis or systemic lupus erythematosus)
Multidimensional assessment of fatigue (MAF) [75]	Past week	Multidimensional; degree, severity, distress, and impact of fatigue	16 items, 4–10-level Likert scales	Pathologic and non-pathological (developed for patients with rheumatoid arthritis)
The multidimensional fatigue inventory (MFI) [76]	Lately ("refer to how you have been feeling lately")	Multidimensional; physical, mental, and general fatigue; reduced activity and motivation	20 items, 7-level Likert scale	Pathologic and non-pathological (used in chronically unwell and well populations)

(continued)

Table 4.2 (continued)

Instrument	Recall period	Measures	Administration	Usage
Medical outcomes study short form (SF-36) energy and fatigue subscale [77]	Past month	Multidimensional; physical, cognitive, social, and emotional functioning	4 items, 3–6-level Likert scales and yes/no	Pathologic and non-pathological (developed to measure the health status of individuals living in the community)
Patient-reported outcomes measurement information system (PROMIS), fatigue short form or computerized adaptive test [78]	Past week	Multidimensional; physical, mental, general, emotional, total, substantial, transient, chronic fatigue; reduced activity and motivation; physical, cognitive, psychosocial, social, emotional functioning; energy	Up to 95 items, 5-level Likert scale	Pathologic and non-pathological (can reliably estimate fatigue reported by the U.S. general population)
Profile of mood states (POMS), fatigue and vigour subscales [79]	Past week, present time	Multidimensional; physical and mental fatigue; energy	65 items, 5-level Likert scale	Non-pathological (adult version and adolescent version)
Visual analog scale to evaluate fatigue severity (VAS-F) [80]	Present time: "Right now"	Bidimensional; energy and fatigue	18 items, visual analogue	Pathologic and non-pathological (validated with adults aged 18–55 years)

disorder [110]. Smartphone data collection often incorporates validated scales. For example, a mobile phone application to collect data on self-reported fatigue for multiple sclerosis [108] incorporated PROMIS. Researchers concluded that a phone application incorporating PROMIS may be useful to provide estimates of fatigue to facilitate clinical monitoring of fatigue for clinic settings.

Momentary Assessments

The Ecological Momentary Assessment (EMA) is a technique that elicits a repeated, real-time measurement of behaviours or experiences as they occur in the naturalistic setting of an individual's daily life. This method was originally developed to perform *in situ* data collection for behavioural medicine [111]. The Experience Sampling Method (ESM) aims to assess participant thoughts, behaviours, and feelings during daily life by collecting self-reports, triggered at various moments during the day [112]. The two terms (EMA and ESM) are used interchangeably, and in practice, they are measured using the same methods [113].

Traditional studies employing EMA/ESM assessed fatigue and fatigability in segments of the general population. For instance, this method has been applied to demographic groups, work settings, and disease populations. Specifically, the relationship between women's passion for physical activity and vitality was examined using SF-36 scale [114]. Researchers have also employed POMS scale to examine occupational energy management strategies by hourly diary questions in academic workers [115]. A separate study examined the effects of breaks on regaining vitality in the workplace using an activation–deactivation adjective checklist [116]. Additionally, EMA/ESM assessment of energy/fatigue has been applied to disease populations including osteoarthritis ([117] researchers used SF-36 scale), kidney disease ([45] researchers used Daytime Insomnia Symptom Scale), and cancer ([118] researchers used a single-item fatigue intensity scale; [20] researchers used 10-point Likert scale for current fatigue).

Mobile-administered EMA/ESM has been applied to the management of diseases. For cancer and its treatment, fatigue is one of the most common and distressing side effects. Cancer-related fatigue causes disruption in all aspects of Quality of Life and may be a risk factor for reduced survival [119]. A mobile phone-based, symptom management system can assist in the management of chemotherapy-related toxicity in patients with breast, lung and colorectal cancer [109]. This system prompts patients to complete an electronic symptom questionnaire on their mobile phone twice a day. A systematic review of mobile apps for bipolar disorder [110] identified thirty-five symptom monitoring apps aiming at assisting users with symptom tracking.

Objective Fatigue and Energy Measures

Fatigability is primarily measured by quantifying the decline in one or more aspects of performance during the continuous performance of a prolonged task or comparing performance before and immediately after a prolonged performance of a separate fatigue-inducing task [8]. In pathological cases, individuals may experience fatigue even in activities of daily living [120]. When objectively measuring fatigue, it is important to indicate the domain examined and the task used to induce fatigability.

Fatigue-related decrements in task performance can be measured by following two common approaches. Ackerman [121] provides a classification of procedures for cognitive fatigue, and we argue that these same approaches pertain to physical fatigue as well. The indirect approach consists of the assessment of cognitive ability before and after a prolonged period of time during which effort may vary. The direct approach consists of the continuous measurement of fatigue during the difficult task. The benefits of the first method are that all participants can complete the same task, while the variation lies in the difference between ex-ante and ex-post fatigue among individuals. This method does not quantify the performance decrease as a continuous function of time. Conversely, the second method can monitor fatigue accumulation, but the tasks may vary. One example is vigilance tasks, where participants are required to maintain attention for target events while ignoring other stimuli [122, 123].

There is a distinction between capacity (describing a person's ability to execute a task in a standardized, optimized, or controlled environment), capability (describing what a person can do in their daily environment), and performance (describing what a person actually does in their daily environment) [124]. Capacity is the composite of all the physical and mental capacities that an individual can draw on and performance is what individuals do in their current environment [125]. It is beyond the scope of this chapter to classify past studies as capturing capacity, capability, performance, or a combination.

Physical Assessment

The monitoring of fatigue and energy has been examined as an approach to maintain health, assist in disease management, and improve performance, productivity, and safety. A plethora of methods have been employed in order to monitor fatigue and energy: performance-reported outcomes (PerfRO) [12] for physical and cognitive fatigability, and tech-reported outcomes (TechRO) [12] from physiological processes (cardiac, ocular, neural) and markers (biologic, behavioural).

Fatigability is usually quantified as a decline in peak force (torque), power (velocity of muscle contraction), speed, fatigue index (force change over time), sense of effort, perception of effort, or accuracy of performance after performing a task, which requires physical effort [7]. Characteristics of tasks include exercise type, intensity, load, tested muscle, and physical environment [28].

The first dimension of physical performance fatigue is "physical capacity" (i.e., maximum performance). The two most common indicators of physical capacity are (1) the aerobic capacity and (2) the power output capacity. Measures of aerobic capacity include the maximal oxygen volume (VO_2-max). Measures of power output include the peak power output. Momentary exercises leading to the assessment of these measures include aerobic and resistance training [98]. Example exercises routinely used, e.g., in professional sports players include various jump protocols, including squat and countermovement jumps, which can lead to indirect assessments of fatigability [24]. Direct measures of fatigue include a joint range of motion or flexibility of appendages such as the knee, hip, groin, and other joints during the exercise.

The second measured dimension of fatigue is "muscular strength." Studies measuring muscular strength included momentary resistance training of various types (weight machines, free weights, resistance bands, cycling ergometers) and other strength training (specialized locomotor training, cycling, aquatics) [98], muscular oxygen consumption (mVO_2), or electromyography (EMG).

The third dimension of fatigue is "mobility," which is more commonly measured in cases of pathological fatigue. Mobility measures include the momentary 6-Minute Timed Walk (6MTW) [126], the Timed 25-Foot Walk [127], and the Timed Up & Go [128].

Exercise-specific hardware used for such exercises include treadmills, weight machines, free weights, and resistance bands. Technology-enhanced exercises include the robotic-assisted treadmill and functional electrical stimulation-assisted cycling [98], and transcranial magnetic stimulation [7]. Figure 4.2 depicts an example of hardware used to measure physical performance.

Studies assessing non-pathological physical performance as a proxy for physical fatigue involved segments of the general population, e.g., physical fatigue in young adults using POMS, trail-making test on an iPad and mVO_2 [129], physical fatigue during a sit-to-stand physical test by using EMG and accelerometer (Samsung) in the lab [130], or PhysioLab, a physiological computing toolbox measuring multiple signals (ECG, EMG, and EDA) to study cardiorespiratory fitness in elderly populations [131], all momentary.

Other observational studies assessed physical fatigue in a pathological context with individuals with health conditions or diseases; assessments include the effects of caloric restriction on cardiorespiratory fitness and fatigue in older adults with obesity by using graded exercise tests measuring VO_2-max [132], the differences in motor fatigue between patients with stroke and patients with multiple sclerosis by using self-reported SF-36 and 6MWT [133], physical fatigue in lumbar disc herniation by using EMG [134].

Continuous monitoring studies assessed the effects of disease on fatigue, e.g., a rehabilitation program on aerobic fitness, cancer-related fatigue, and quality of life using subjective MFI and objective energy expenditure armbands (SenseWear) [135], or the fatigue monitoring system (FAMOS) which can monitor physiological parameters from multiple sclerosis patients and controls, all pathological.

Fig. 4.2 Hardware for physical performance example (ergometer). © Laufband Ergometer by Robowalk licensed under CC-BY-SA-4.0

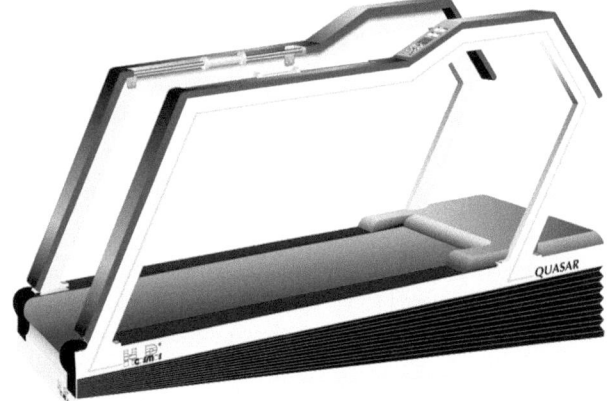

Cognitive Assessment

In the cognitive domain, fatigue leads to the degradation of cognitive performance [122], as reflected by degradations in verbal, visual, short, and long-term memory, processing speed, primary and divided attention, verbal fluency, motor speed, reading speed, visual scanning, orientation, calculation, success rate, and other measures.

Cognitive assessments were measured by using numerous momentary measures, which collectively assess the above degradations. Table 4.3 reviews several task-based tests yielding cognitive performance-reported outcomes [12].

Cognitive performance studies included fatigue assessment in non-pathological segments of the general population, e.g., alertness, vitality, and sleepiness by using Psychomotor Vigilance Task (PVT) and other tasks in different lighting settings [146], occupational fatigue, e.g., in healthcare and medical staff by using the rate of error [147], or airline pilots on the flight deck by using PVT [148].

Cognitive performance studies also included fatigue assessment in pathological settings, e.g., the relationships between health-related Quality of Life, fatigue, and exercise capacity in coronary artery disease individuals using MFI and a bicycle ergometer test [149].

Technology-driven studies include assessments of mental fatigue in a non-pathological context by performing tasks with a computer, e.g., keyboard and mouse interaction patterns [150] recovery from work exhaustion by use of Twitter [151], or in a pathological context. For example, those living with an acquired brain injury often have issues with cognitive fatigue due to factors resulting from the injury. Studies have shown fatigue to be one of the most disabling symptoms, regardless of the severity of brain injury [152–154]. Researchers presented a smartphone application for the evaluation of cognitive fatigue, which can be used daily to track cognitive performance in order to assess the influence of fatigue [155]. Researchers concluded that the presented smartphone application for the evaluation of cognitive fatigue could be utilized in everyday life.

Cardiac Physiology

Cardiac activity measures used to assess fatigue include the resting heart rate (HR), exercise heart rate (HRex), heart rate variability (HRV), and the heart rate recovery (HRR). The heart rate may increase or decrease in response to a variety of factors including physical and mental effort, distress, and anxiety that are potentially associated with fatigue [16]. Elevated HRV was observed during strenuous tasks in individuals with chronic fatigue [156] and healthy individuals of young age while performing a task [157]. HRR may serve as a marker of acute training-load alteration, however recent studies showed inconclusive results [24]. A more detailed measure of heart activity is the electrocardiogram (ECG), an electrophysiological method, which records the electric signals of the heart and from which the HR can be derived. Figure 4.3 depicts an electrocardiogram with an electrocardiograph and electrodes placed on the human body.

Table 4.3 Tasks and measures of cognitive performance

Task	Measures	Administration	Usage
Mini-mental state examination (MMSE) [136]	Orientation, short-term memory registration, attention, calculation, recall, language, and task reproduction	16 complex items: Qualitative and quantitative questions	Elders, potentially pathologic
Trail making test (TMT) [137]	Visual search, scanning, processing speed, mental flexibility, and executive functions.	Two items: The participant connects circles denoted by numbers and letters in ascending order	Non-pathological
Selective reminding test (SRT) [138]	Verbal memory	One item: The participant recalls as many as possible of 12 dictated unrelated words	Non-pathological
Spatial recall test (SPART) [139, 140]	Visuospatial learning, the susceptibility of such learning to proactive and retroactive interference, and the ability to recall visuospatial information following a period of delay	One item: The participant recalls as many as possible of 10 checkers on a 36-checkers square board	Pathologic, multiple sclerosis
Symbol digits modalities test (SDMT) [141]	Presence of organic cerebral dysfunction leading to neurological impairment	One item: The participant has 90 seconds to pair specific numbers with given geometric figures.	Pathologic, cerebral dysfunction
Paced auditory serial addition test (PASAT) [142]	Rate of information processing after recovering from trauma	Multiple items: The participant hears a series of digits, one every 3 seconds, and reports the sum of the last two digits.	Pathologic, multiple sclerosis
Word list generation (WLG) [143]	Neuropsychological measures of verbal fluency	One item: The participant generates words from a restricted category (e.g., starting with S or denoting animals) in 60 seconds.	Pathologic, dementia, multiple sclerosis
Rey auditory verbal learning test (RAVLT) [69]	Recent memory, verbal learning, susceptibility to interference, and retention of information after a certain period of time during which other activities are performed	Multiple items: 15 nouns read aloud each second for 5 consecutive trials followed by participant recall.	Non-pathological
Simple reaction time task (SRTT) [144]	Relationships between the deceleration of heart rate observed to anticipate both aversive and non-aversive stimuli, and several aspects of the somatic-motor activity.	One item: A square is shown on screen at different intervals. The participant selects a button to react to seeing the square.	Non-pathological

(continued)

Table 4.3 (continued)

Task	Measures	Administration	Usage
Psychomotor vigilance task (PVT) [23]	Impact of loss of sleep sustained wakefulness, and/or time of day on neurobehavioral performance	Multiple items: Ranging up to 10 minutes, similar to the SRTT.	Non-pathological
Brief Repeatable Battery of Neuropsychological Tests [145]	Selective short-term memory, spatial recall, symbol digit modalities, paced auditory serial addition, and word list generation; first used for multiple sclerosis	Multiple tests: Selective reminding test (SRT), spatial recall test (SPART), symbol digits modalities test (SDMT), paced auditory serial addition test (PASAT), delayed recall of the SRT, delayed recall of the SPART, and word list generation (WLG).	Pathologic, multiple sclerosis

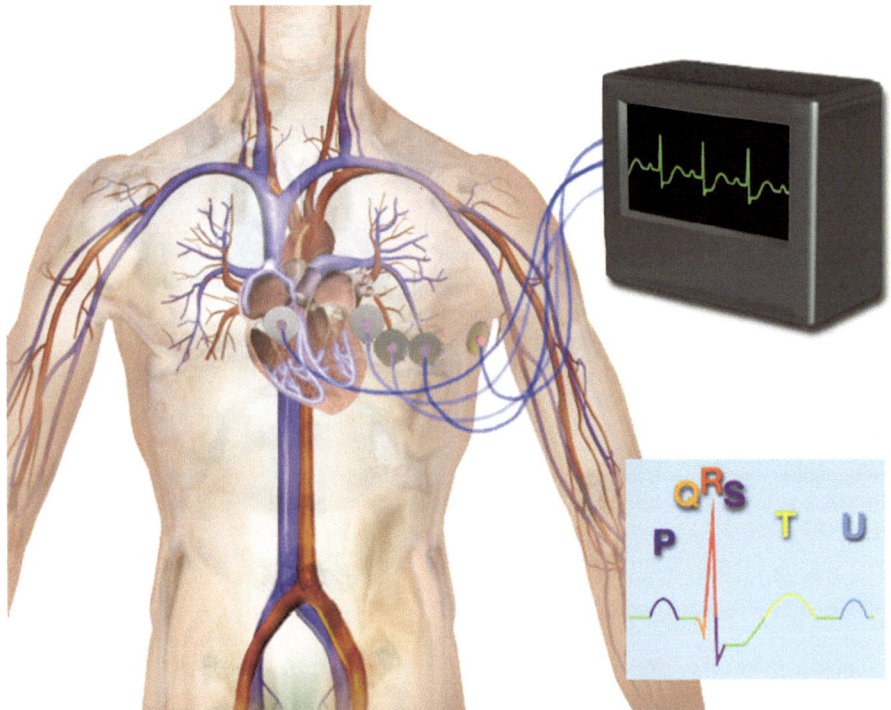

Fig. 4.3 Electrocardiogram. © Blausen Electrocardiogram by BruceBlaus licensed under CC-BY-3.0 [158]

Studies using the ECG to assess fatigue in a non-pathological, occupational context include airline crew [159], surgeons [160] or 3D TV watchers [161]. In these studies, the ECG was measured with electrode-based devices before and after the tiring task (i.e., via an indirect measurement approach). HRV pre- and post-task was used as a measure for fatigue in work settings, e.g., emergency and pre-hospital doctors [162].

Measurements of cardiac physiology have been performed during daily life (i.e., via a direct measurement approach) also in a non-pathological setting. A large body of research focused on assessing cardiac activity in healthcare and driving professionals. Medical interns were given Holter recorders throughout the day, measuring HR and HRV, in conjunction with resting ECG to assess fatigue [163]. Surgeon HRV (using EEG) was assessed in robot-assisted versus conventional cholecystectomy [164]. Drivers were assessed while driving, through an ECG device mounted on the steering wheel [165]. Another study assessed the impact of electroacupuncture on fatigue and Quality of Life using subjective SF-36 and objective HRV using ECG (SphygmoCor) [166]. A method aimed at estimating the perception of physical fatigue by predicting heart rate through smartphones has been proposed by estimating the oxygen consumption, using a smartphone acceleration and location (via accelerometer and GPS, respectively) [3]. The study yielded an adequate detection of fatigue when individuals performed daily-life activities under naturalistic conditions.

Ocular Physiology

Keeping the eye closed or having fixed changes in pupil diameter have been observed in a state of fatigue [167] due to monotony or sleep deprivation. Ocular physiology measures used for assessing fatigue include the spontaneous eye blink [168], pupil diameter [169], oscillations in pupil diameter (fatigue waves) [170, 171]. Another method used to detect fatigue is the electrooculogram (EOG), an electrophysiological method, which measures the resting electrical potential between the cornea and Bruch's membrane.

Studies using ocular physiology measures were primarily done to assess fatigue in non-pathological, occupational settings, e.g., in the military detecting sleep deprivation-induced fatigue by saccade peak velocity in the Navy using questionnaires (on PDAs), actigraphy (Actiwatch), and EOG (Natus, then Embla) during a saccade task [172] or assessing fatigue in the Air Force through saccadic velocity using software (Eyelink) in a dark room before and after a long flight [22]. For driver drowsiness, studies assessed fatigue by EOG using a device mounted next to the eyes for brief periods [21].

Smartphones have been utilized and applied to drivers as well. Researchers have presented an app, which uses information from both front and back cameras and others embedded sensors on the phone to detect and alert drivers to dangerous driving conditions inside and outside the car [173]. Researchers used computer vision and machine learning algorithms on the phone to monitor and detect whether the driver is tired or distracted using the front camera while at the same time tracking road conditions using the back camera. The front camera pipeline tracks the driver's head pose and direction as well as eyes and blinking rate as a means to infer drowsiness and distraction. Specifically, researchers used blink detection algorithms to detect periods of micro-sleep, fatigue and drowsiness. A more recent study improved EOG by mounting the device on the forehead to increase the duration of comfortable measurement [174].

Neural Physiology

Neural electrophysiological measures used to assess fatigue include the electroencephalogram (EEG), the evoked response potential (ERP), the Error Related Negativity, and lateralized readiness potential [16]. Magnetic resonance imaging (MRI) was also used to identify factors of fatigue [175]. This type of objective measure focuses on cognitive performance, described in a preceding section, by requiring the participants to conduct a task while monitoring takes place.

Studies have assessed neural physiology of fatigue in a non-pathological context by using EEG or ERP in the general population [176, 177], as well as EEG on occupational fatigue, e.g., drivers [178, 179], and surgeons while conducting a demanding task. For surgeons, Kahol [180] studied the impact of fatigue in surgical residents, which used a demanding task and measurement by EEG using a B-Alert device while Guru [181] assessed cognitive performance during robot-assisted surgery by EEG using a B-Alert device. Other studies which used electrophysiological measures in conjunction with other methods are elaborated on in the objective measures section of mix methods.

Biologic Markers

Fatigue-related biologic markers were studied in the pathological context of chronic disease: plasma glucose, associated with variations in transient physical and mental energy, effort, and fatigue with variable degrees of success [182, 183]; cortisol, an indirect marker of fatigue through stress level and energy expenditure associated with fatigue [184]; salivary alpha-amylase (sAA) associated with surrogate markers of nervous system activity [185] and task engagement/disengagement [186], with variable degrees of success; and melatonin following circadian patterns and disrupted in individuals with chronic disease and recurrent fatigue [187], used for sleep-related fatigue. In elite athletes, creatine kinase (CK), C-reactive protein (CRP), uric acid, testosterone, salivary immunoglobulin (S-IgA) were used as indirect markers of fatigue in the recovery period following intense physical activity. Biologic systems involved in the regulation of motor activity are intricately linked with sleep, feeding behaviour, energy, and mood [188].

Behavioural Markers

Common behavioural markers utilized to assess fatigue include sleep and physical activity. These markers can be assessed by research-grade devices and consumer devices alike, with various degrees of validated accuracy, wear comfort, and presence in the research lab for the procedure. Figures 4.4 and 4.5 depict research and consumer wearable devices, respectively. As opposed to the momentary measures above, the behavioural markers can also be monitored continuously (with very high frequency, e.g., seconds or milliseconds) and longitudinally (for an extended duration, e.g., weeks to years) in time.

Sleep can be assessed using polysomnography and actigraphy. Polysomnography (PSG) [189] is an electrophysiological sleep study, which assesses brain waves (EEG), oxygen levels in the blood, heart rate (ECG), eye movements (EOG), and muscle and skeletal muscle activation and movements (EMG), breathing functions, respiratory airflow, respiratory effort, and pulse oximetry (SpO_2). Polysomnography quantifies sleep duration, interruptions, stages (e.g., light, deep, rapid eye movement (REM)) and waking states (e.g., awake, asleep). Actigraphy [190] is a noninvasive electrophysiological method that assesses movement and is used to monitor humans at rest or during various types of physical activity. Examples of

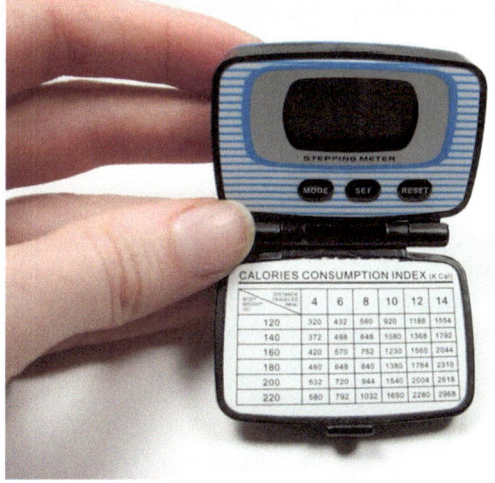

Fig. 4.4 Research-grade wearable (pedometer). © Pedometer Take-Apart by 11enore licensed under CC BY 2.0

Fig. 4.5 Consumer wearable (smartwatch). © Moto 360 Smartwatch by chrisf608 licensed under CC BY 2.0

research-grade wearable actigraph devices are ActiWatch[1] and ActiGraph.[2] The actigraph can be worn on the wrist or ankle during daily life, for several weeks. The actigraph allows for the continuous collection of data due to its non-invasive nature, however, widespread and longitudinal use is limited by its specific purpose of researching physical activity with limited considerations to the user experience and price.

More recent consumer wearable monitors, in the form of wristbands, smartwatches, sleep mattresses, or finger rings from manufacturers such as Fitbit,[3] Oura,[4] and Withings[5] [191] monitor sleep continuously by using a combination of movement, measured by a triaxial accelerometer, and HR/HRV, measured by photoplethysmography (PPG), non-invasive optical measurement of the volumetric variability of blood in the vessels under the skin. Consumer wearables can also measure behavioural markers pertaining to physical activity, e.g., duration, intensity

---

[1] https://www.usa.philips.com/healthcare/product/HC1046964/actiwatch-spectrum-activity-monitor

[2] https://www.actigraphcorp.com

[3] https://fitbit.com

[4] https://ouraring.com

[5] https://withings.com

(classified as, e.g., sedentary, low, moderate, and vigorous), type (using activity class recognition), effort (in metabolic equivalent of tasks (METs)), distance, elevation, step count, workouts, and other measures derived from the continuous multivariate data obtained from triaxial accelerometer and gyroscope sensors inside the device.

Studies assessing non-pathologic fatigue, sleep, and physical activity have been performed in segments of the general population and for several occupations, usually by combining subjective and objective measurements. In segments of the general population, Ellingson [17] studied the influence of active and sedentary behaviours on perceived energy and fatigue in women by using subjective POMS and SF-36 and objective physical activity by an accelerometer (Actigraph). For occupations, Rizzo [192] assessed the role of fatigue and sleepiness in drivers with obstructive sleep apnea by using subjective SF-36 and objective PSG. De Araújo Fernandes Jr. [193] quantified the impact of shift work on train drivers by using PVT and actigraphy (Actiwatch). Fernandes-Junior [193] assessed sleep, fatigue, and Quality of Life in night shift workers using subjective scale and actigraphy (Actiwatch). Towards the pathologic type of fatigue, Campbell [194] assessed fatigue and sleep in individuals having unexplained chronic fatigue by using subjective scales and objective PSG; Maher [195] quantified the relationships between fatigue, physical activity, and socio-demographic characteristics in children and adolescents with physical disabilities by using objective physical activity measurement using an accelerometer (Actigraph).

Numerous other studies have assessed pathologic fatigue in the context of a specific disease using PSG or actigraphy. Attarian [196], Kaynak [197], Veauthier [198], and Kaminska [199] studied relationships between sleep and fatigue in multiple sclerosis patients. Keefer (2006) and Shitrit [200] assessed sleep and fatigue in inflammatory bowel disease. Merikangas [188] used a combination of EMA and actigraphy to assess energy, mood, and activity in individuals with depressive disorders. Sun [201] assessed the relationships between daytime napping and fatigue and Quality of Life in cancer individuals by using subjective scale and objective sleep quality (Actigraph). Ancoli-Israel [202] assessed sleep, fatigue, and circadian activity in women with breast cancer by using subjective scale and objective circadian rhythms using actigraphy (Actiwatch). Holliday [203] assessed fatigue and sleep quality in prostate cancer patients by using a subjective scale of Quality of Life and actigraphy (Actiwatch). Cambras [204] studied circadian rhythm in patients of encephalomyelitis using actigraphy (ActTrust). Nicklas [205] assessed physical activity behaviours (using accelerometers) and fatigue (using SF-36) in adults of middle and old age with chronic inflammations. Nilsson [206] studied intensity levels of physical activity and fatigue in cancer patients by using an accelerometer (SenseWear). Vancampfort [207] studied the relationships between cardiorespiratory fitness and increased quality of life in people with bipolar disorder using, among others, the subjective SF-36 and an armband (SenseWear) for objective physical activity and sedentary behaviour measurement. Sheshadri [208] assessed the relationship between intensity levels of physical activity and fatigue in patients on dialysis by using step count from a pedometer (Accusplit).

More recent studies used wearables to assess wearable-measured sleep and physical activity in a pathologic context. Qazi [209] studied fatigue in patients with inflammatory bowel disease by using a Fitbit Charge HR. Sofia et al. [210] used the same wearable to associate sleep fragmentation with individuals having clinically active disease. Abbott [211] conducted an intervention study for physical activity in case of cancer-related fatigue patients by using activity trackers (undisclosed brand) without reporting measurements but reporting that the activity tracker was deemed helpful.

Mixed Methods

In our literature review, we identified numerous studies which combined two or more objective measures of fatigue. These studies focused on either cognitive or physical fatigue in the general population or specific occupations, or physical fatigue in specific segments of the population.

For non-pathologic cognitive fatigue in the general population, Zhang [174] estimated mental fatigue based on EEG (Neuroscan) and HRV from ECG while performing an arithmetic task using a personal computer, Ren [212] studied various degrees of mental fatigue by using multiple types of measurements: EEG, ECG as well as galvanic skin response (GSR), Smith [213] quantified the effects on cognitive tasks on mental fatigue indicators, using PVT and other two tasks and assessing fatigue through subjective VAS and objective HRV from EEG, and Brown [214] studied the effects of mental fatigue on exercise intentions and behaviour using cognitive and then physical exercises by using a cycle ergometer.

In the area of non-pathological physical fatigue, Kanitz [215] assessed the impact on eurythmy therapy on fatigue by using subjective MFI and objective HRV by ECG. For occupational fatigue, Smolders [216] studied the alertness during office hours induced by higher luminosity by using subjective measures, task performance (PVT, letter substitution test), and heart rate measures (ECG), Oriyama [217] studied fatigue in shift nurses by measuring objective HRV from ECG, and subjective EMA using VAS, and Singh [218] assessed the technical performance of surgeons when using robotic surgery where the task was a suture under time pressure, measured with a subjective surgical task scale and objective HR, and objective functional near-infrared spectroscopy (fNIRS).

In the area of pathological fatigue, Dishman [219] studied the effects of cycling exercise on fatigue among young adults who report persistent fatigue using incremental exercise test on an electronically braked, computer-driven cycle ergometer (Lode), and providing subjective POMS and objective HR (Polar), VO_2-max and expired gas (Parvo Medics), and EEG (Electrical Geodesics).

Table 4.4 Fatigue measurements and spectrums of characteristics from the literature review

Measurement	Subjective		Objective		
Location	Both office and daily life	Daily life	Office	Both office and daily life	Daily life
Reporting	Self-reported		Perf-reported	Tech-reported	
Administration	Scales	Prompts, e.g., EMA	Task hardware and devices	Research devices	Consumer devices
Validated	Yes	Partial	Yes	Yes	Partial
Quantifiable	No	No	Yes	Yes	Yes
Frequent	No	Yes	No	No	Yes
Continuous	No	No	Yes	Yes	Yes
Judgment-free	No	No	Partial	Yes	Yes
Mood-free	No	No	Yes	Yes	Yes
Memory-free	No	Partial	Yes	Yes	Yes
Owned	Partial	Partial	No	No	Yes
Contextual	No	Yes	No	Partial	Yes

Property Spectrums of Energy and Fatigue Measures

The findings from our literature review classify the energy and fatigue measurements by type (subjective and objective), location (clinician's office, daily life, or both/mixed), source (self-, performance/capacity-, and technology-reported, using the taxonomy by Mayo [12]), and administration (scales, prompts, tasks, and devices). In Table 4.4, we place each such measurement on spectrums for the following properties:

1. **Validated**: fatigue outcome reliability assessed by statistical analysis on the target population and scientific publication.
2. **Quantifiable**: fatigue outcomes interval or ratio at a minute or higher precision.
3. **Frequent**: often repeated administrations with one day or less between administrations.
4. **Continuous**: fatigue proxy variable measured on a time series with a minute or higher granularity.
5. **Judgment-free**: bias-free from the perception of judgment from the administrator; tasks and research devices allow some refraining.
6. **Mood-free**: bias-free from the voluntary or involuntary perception of self.
7. **Memory-free**: bias-free from the remembrance of the past; prompts allow for long-term memory loss.
8. **Owned**: whether the participant owns the device; scales and prompts are marked as partial in case they are delivered to a device owned by the participant.
9. **Contextual**: collected from settings daily life; research devices can be borrowed to the participant for a short time to wear in daily life context.

Discussion

Key Findings

Fatigue or lack of energy is a universal symptom experienced by those suffering from different medical and psychological illnesses as well as by healthy individuals in the general population. Overall, fatigue is a ubiquitous and multifaceted symptom that is challenging to define and measure. Fatigue may be classified as pathological or non-pathological, physical or mental, and can be measured subjectively or objectively.

Different approaches have been employed in order to measure energy and fatigue including scales, prompts, physical measures, cognitive measures, physiological markers, biological markers, behavioural markers, and mixed methods. Some measurement methods assess the effects of fatigue (e.g. performance decrements), some attempt to identify the source of fatigue (e.g. muscle dysfunction), while others adopt a behavioural perspective (e.g. decreased physical activity or prolonged sleep). Some methods focus on capacity while others assess performance. These varied methods each contain advantages and disadvantages in terms of traditional validation, access to continuous data, and ecological validity.

Subjective instruments instantiating self-reported outcomes [12] suffer from inherent shortcomings, in particular, they are infrequent and subjective. Furthermore, self-report by recall has an intrinsic problem: due to biases, such as mood states or sleepiness, individuals are not able to accurately recall past experience, particularly experiences that are frequent, mundane, and/or irregular [220]. In addition, the potential discrepancy between how one feels and how one thinks one should feel contributes to lack of ecological validity in self-reports of fatigue and requires further research [15]. Incorporating a real-time collection of fatigue data in naturalistic settings may reduce problems associated with retrospective recall of events, summarization of events, and artificial contexts or settings [118].

Objective measures obtained by tech-reported outcomes can be collected continuously from individuals in the context of daily life. To this end, both academia and industry are increasing their efforts to develop technological solutions, such as sensors which can measure, models which can assess, and artefacts which can manage energy and fatigue. Recent technological methods to monitor and manage energy and fatigue include sensors, smartphones and their applications, and research- and consumer-grade wearables. Technology-based monitoring of energy and fatigue could assist in the initial diagnosis and the early detection of diseases could enable one to monitor post-treatment evolution and could help assess the risk of certain medications on patients [3]. Furthermore, technology-based monitoring of energy and fatigue could assist healthy individuals in enhancing work performance, conserving and managing energy levels, and maintaining health.

Energy and fatigue are of great importance to diseased individuals. The connection between pathological fatigue and disease is well established in the literature. Fatigue frequently foreshadows conditions like multiple sclerosis [221], cancer [28], and HIV infection [222], among other diseases. Furthermore, fatigue, as well as increased energy, has been identified as a core symptom of mental health disorders including depressive disorders and bipolar disorders. Current literature on energy and fatigue is biased towards pathological, rather than healthy, populations. In addition, it is possible that the existence of healthy fatigue is a barrier to full comprehension of the impact of pathological fatigue, as pathological fatigue is more extreme and different. This highlights the importance of further research on both non pathological and pathological fatigue.

In addition to the comprehensive literature examining fatigue and disease, the monitoring of energy and fatigue has also been highlighted for specific vocational and occupational populations, such as professional athletes [24], police [25], and drivers [165]. The literature aims to gain an understanding of health, safety, occupational functioning, burnout, performance, and capacity. More efforts could be put toward studying healthy general populations, as in addition to affecting an individual's quality of life, fatigue impacts the economy because of the connection to productivity and illness.

Insights into the classification and measurement of energy and fatigue may also be applied broadly to the general population as mobile monitoring technology allows the assessment of these homeostatic systems in real-time [188]. Quality of Life Technologies (QoLT) refers to technologies for assessment or improvement of the individual's quality of life [223]. Optimal measurement of energy and fatigue would be moved out of the lab and into the real world, continuous rather than infrequent, and based on accurate, validated, yet minimally intrusive measures and devices. Future research could establish traditional validity for the continuous, daily life, measurement of energy and fatigue.

Assessing energy and fatigue could also contribute to the quantified self. The quantified self (QS) is any individual engaged in the self-tracking of any kind of biological, physical, behavioural, or environmental information. QS promotes a proactive stance toward obtaining information and acting on it [224]. One of the earliest recorded examples of quantified self-tracking is that of Sanctorius of Padua, who studied energy expenditure by tracking his food intake, weight, and elimination for 30 years in the sixteenth century [225]. State of the art energy and fatigue assessment could contribute meaningfully to the quantified self.

Limitations

A limitation of the current chapter stems from the pathological bias in the field. Namely, because the existing literature is biased toward pathological fatigue, we built the non-pathological (also referred to as physiological) classification system arm based on existing pathological models. This limitation is also related to our literature search strategy. Our method of reviewing the literature was based on a scoping review approach rather than a structured systematic review. We did not exclude studies based on methodologies used or populations studied.

Subjective measures discussed in this chapter contain limitations including being infrequent, involving recalls, and potential to be influenced by mood states, memory, and expectations. Wearable measurements also contain limitations related to the population that uses wearables. Specifically, device owners are more likely to be young individuals with disposable incomes who already lead healthy lifestyles and want to quantify their progress [226]. Future work should ensure that wearable data is representative and note this bias in current wearable data.

An additional limitation of the field is that there is not yet a validated calibration between objective measures and the concept of energy and fatigue. Therefore, much of our discussion is speculative. A major impediment in the understanding of fatigue and energy lies in the fact that for over 100 years, research has shown little relationship between self-report and actual, objective measurements of fatigue [167]. There are several definitions of energy and fatigue and these have not been conclusively associated with objective measures. This doesn't invalidate subjective or objective measures of fatigue but rather indicates that they may be describing something that is more complicated and cannot be whittled down to a single biological measure. Therefore, both subjective experience and objective measurements are being considered in the context of energy and fatigue, as they are important indicators for health and quality of life. Future research could aim to bridge the gap between subjective and objective measures by accounting for multiple variables and conducting calibration studies.

Opportunities

Energy and fatigue is a Quality of Life facet in which the successful assessment, exclusively through Quality of Life Technologies [223], has promising likelihood. The mass adoption of miniaturized devices in daily life (with large scale and diversity in personal and contextual characteristics of the data), the availability of relevant predictors of energy and fatigue in large scale data, and the presence of platforms that facilitate participation in research at scale contribute to the feasibility of the operationalization of this facet.

Currently, research is progressing in assessing pathological and non-pathological energy and fatigue by using subjective, objective, and mixed methods. Miniaturized devices, such as smartphones and wearables, increasingly accurately monitor daily life behaviours (e.g., physical activity and sleep), sense signals (e.g., heart rate, momentary electrocardiogram, etc.) and administer prompts (e.g., validated scales, items, and tasks). As the line between consumer health wearables and medical devices continues to blur, it is possible for a single wearable device to monitor a range of medical risk factors [227]. Adoption of wearables is increasing; 21% of Americans own a wearable [228], there are more than 200 models of wearables[6] and the market is expected to continue to increase by 2022 [229] towards available objective behavioural data at scale. Open health platforms are being employed to facilitate scalable participation and manage subjective, objective, and mixed data [230].

Co-calibrations of (1) subjective validated scales of energy and fatigue and (2) objective measures of daily life behaviours may rigorously validate objective measures of energy and fatigue and meet the aim of assessing energy and fatigue using QoLT. For example, a study aiming to co-calibrate subjective scales and objective

---

[6] https://www.inkin.com/wearables/

behaviours for occupational fatigue may collect multiple behavioural markers passively and continuously (e.g., physical activity, sleep, heart rate) from tens to hundreds of drivers for several months to years, during driving *and* daily living, and regularly administering validated energy and fatigue scales such that their recall periods cover the duration. Such a study may observe trends of fatigue longitudinally in time. Within a smaller sample size, a purely statistical approach would allow for the assessment of validity (e.g., by correlating the corresponding subjective and objective measures) and reliability (e.g., by measuring the same person's fatigue in similar days of week, months, or seasons) of the objective measure. Within a larger sample, a predictive approach would learn the subjective measures of energy and fatigue by using the objective measures of behaviours. These approaches can iteratively reduce the number of scale items. One step further, continuous behaviour monitoring during daily life facilitates the trigger of momentary assessments upon changes in objective behaviours that associate with changes in energy and fatigue. Such an approach may increase the accuracy of the co-calibration. Furthermore, alternative statistical or predictive risk scenarios can maintain energy (*"if you continue working at this pace, you will likely not get tired"*), prevent fatigue (*"if you continue working at this pace, you will likely accumulate occupational fatigue in two weeks"*), and compensate for the losses induced by fatigue (*"consider taking a break of one week to restore your productivity from three months ago"*).

Initially, co-calibrations may suffer from lower accuracy (e.g., revealing only basic trends and associations) or limited extent (e.g., applying for specific scale items, collecting limited objective behaviours, applying for limited energy and fatigue types) as the measured objective measures or available sample may not explain the energy and fatigue directly. In such cases, a directed graph of co-calibrations with additional Quality of Life facets (e.g., stress, health outcomes), using additional objective measures, may need to be constructed to represent the relationships accurately such that energy and fatigue are explained through a series of directed co-calibration paths originating exclusively from objective measures, essentially assessing energy and fatigue through QoLT exclusively.

A successful energy and fatigue assessment using QoLT would contribute to the *"Internet of everything"* 50-year vision of a digital future where *"internet use will be nearly as pervasive and necessary as oxygen"* [231]. Specifically, such an assessment would contribute to three of Stansberry's five hopeful visions of 2069. The first vision, *living longer and feeling better* where *"internet-enabled technology will help people live longer and healthier lives; scientific advances will continue to blur the line between human and machine"* [231] will be enabled by quantifying the relationships between energy, fatigue, behaviours, health, and Quality of Life outcomes. The second vision, *less work, more leisure* where *"artificial intelligence tools will take over repetitive, unsafe and physically taxing labour, leaving humans with more time for leisure"* [231] will be enabled through (short-term) the transition to increasingly passive reported outcomes that reduce the burden of participation in research and (longer-term) statistical and predictive optimization of physical and mental effort allocation for the occupations where energy and fatigue are prevalent.

The third vision, *individualized experiences* where *"digital life will be tailored to each user"* [230] will be enabled by interventions leveraging large scale data, accurate models, and alternative personalized scenarios addressing fatigue prevention, before management, and before compensation.

Conclusive Remarks

Energy and fatigue impact physical, cognitive, emotional, social, and occupational functioning and carry important implications for an individual's health and overall Quality of Life. Lacking energy carries consequences for an individual's routine functioning. Everyday activities, including work performance and self-care activities, can be impeded or even curtailed. Energy is required to sustain life and efficient spending of energy results in overall vitality. Paradoxically, one tends to think of energy of a resource that is depleted and then restored with rest, while at the same time, many observe that using energy generates additional energy. This curious paradox highlights the importance of future research and clarity. In addition, the classification of energy and fatigue is critical as it is possible that the existence of non pathological fatigue inhibits true appreciation of the impact of pathological fatigue.

The contributions of this chapter include a semi-structured literature review on energy and fatigue assessment and its potential within Quality of Life Technologies, a taxonomy of the field of energy and fatigue, and the identification of a research validation gap between subjective and objective measures of energy and fatigue. We foresee the necessity to conduct studies of increasing size in order to co-calibrate the subjective and objective measures towards the integration of exclusively objective measures in research and clinical practice.

The measurement of energy and fatigue has been complicated by difficulties in definition and assessment. We conclude that optimal classification and measurement of energy and fatigue would occur in the real world, continuously and in real-time, while being ecologically valid and informing the design of interventions aimed at maintaining energy and monitoring fatigue towards positive outcomes of health and Quality of Life.

References

1. "WHO | WHOQOL: Measuring quality of life," WHO. https://www.who.int/healthinfo/survey/whoqol-qualityoflife/en/index4.html. Accessed Jun. 24, 2020.
2. Stahl SM. The psychopharmacology of energy and fatigue. J Clin Psychiatry. 2002, Jan;63(1):7–8. https://doi.org/10.4088/jcp.v63n0102.
3. Hernández N, Favela J. Assessing the perception of physical fatigue using Mobile sensing. In: Wister M, Pancardo P, Acosta F, Hernández JA, editors. Intelligent data sensing and processing for health and Well-being applications. Academic; 2018. p. 161–73.
4. Boksem MAS, Tops M. Mental fatigue: costs and benefits. Brain Res Rev. 2008, Nov;59(1):125–39. https://doi.org/10.1016/j.brainresrev.2008.07.001.

5. Marcora SM, Staiano W, Manning V. Mental fatigue impairs physical performance in humans. J Appl Physiol. 2009, Mar;106(3):857–64. https://doi.org/10.1152/japplphysiol.91324.2008.

6. Al-Mulla MR, Sepulveda F, Colley M. A review of non-invasive techniques to detect and predict localised muscle fatigue. Sensors. 2011;11(4):4. https://doi.org/10.3390/s110403545.

7. Finsterer J, Mahjoub SZ. Fatigue in healthy and diseased individuals. Am J Hosp Palliat Med. 2014, Aug;31(5):562–75. https://doi.org/10.1177/1049909113494748.

8. Kluger BM, Krupp LB, Enoka RM. Fatigue and fatigability in neurologic illnesses. Neurology. 2013, Jan;80(4):409–16. https://doi.org/10.1212/WNL.0b013e31827f07be.

9. Chaudhuri A, Behan PO. Fatigue in neurological disorders. Lancet. 2004;363(9413):978–88. https://doi.org/10.1016/S0140-6736(04)15794-2.

10. Mollayeva T, Kendzerska T, Mollayeva S, Shapiro CM, Colantonio A, Cassidy JD. A systematic review of fatigue in patients with traumatic brain injury: the course, predictors and consequences. Neurosci Biobehav Rev. 2014, Nov;47:684–716. https://doi.org/10.1016/j.neubiorev.2014.10.024.

11. Ream E, Richardson A. Fatigue: a concept analysis. Int J Nurs Stud. 1996, Oct;33(5):519–29. https://doi.org/10.1016/0020-7489(96)00004-1.

12. Mayo NE, Figueiredo S, Ahmed S, Bartlett SJ. Montreal accord on patient-reported outcomes (PROs) use series–paper 2: terminology proposed to measure what matters in health. J Clin Epidemiol. 2017, Sep;89:119–24. https://doi.org/10.1016/j.jclinepi.2017.04.013.

13. Dittner AJ, Wessely SC, Brown RG. The assessment of fatigue: a practical guide for clinicians and researchers. J Psychosom Res. 2004, Feb;56(2):157–70. https://doi.org/10.1016/S0022-3999(03)00371-4.

14. Whitehead L. The measurement of fatigue in chronic illness: a systematic review of unidimensional and multidimensional fatigue measures. J Pain Symptom Manag. 2009, Jan;37(1):107–28. https://doi.org/10.1016/j.jpainsymman.2007.08.019.

15. Aaronson LS, et al. Defining and measuring fatigue. Image J Nurs Sch. 1999;31(1):45–50. https://doi.org/10.1111/j.1547-5069.1999.tb00420.x.

16. Hornsby BWY, Naylor G, Bess FH. A taxonomy of fatigue concepts and their relation to hearing loss. Ear Hear. 2016;37(Suppl 1):136S–44S. https://doi.org/10.1097/AUD.0000000000000289.

17. Ellingson L, Ae K, Vack N, Cook D. Active and sedentary behaviors influence feelings of energy and fatigue in women. Med Sci Sports Exerc. 2014, Jan;46(1):192–200. https://doi.org/10.1249/mss.0b013e3182a036ab.

18. Abdel-Kader K, et al. Ecological momentary assessment of fatigue, sleepiness, and exhaustion in ESKD. BMC Nephrol. 2014, Feb;15(1):29. https://doi.org/10.1186/1471-2369-15-29.

19. Cochran A, Belman-Wells L, McInnis M. Engagement strategies for self-monitoring symptoms of bipolar disorder with Mobile and wearable technology: protocol for a randomized controlled trial. JMIR Res Protoc. 2018;7(5):e130. https://doi.org/10.2196/resprot.9899.

20. Curran SL, Beacham AO, Andrykowski MA. Ecological momentary assessment of fatigue following breast cancer treatment. J Behav Med. 2004, Oct;27(5):425–44. https://doi.org/10.1023/B:JOBM.0000047608.03692.0c.

21. Chieh TC, Mohd M, Mustafa AH, Hendi SF, Majlis BY. Development of vehicle driver drowsiness detection system using electrooculogram (EOG). In: 2005 1st International Conference on Computers, Communications, Signal Processing with Special Track on Biomedical Engineering; 2005, Nov. p. 165–8. https://doi.org/10.1109/CCSP.2005.4977181.

22. Diaz-Piedra C, Rieiro H, Suárez J, Rios-Tejada F, Catena A, Stasi LLD. Fatigue in the military: towards a fatigue detection test based on the saccadic velocity. Physiol Meas. 2016, Aug;37(9):N62–75. https://doi.org/10.1088/0967-3334/37/9/N62.

23. Dinges DF, Powell JW. Microcomputer analyses of performance on a portable, simple visual RT task during sustained operations. Behav Res Methods Instrum Comput. 1985, Nov;17(6):652–5. https://doi.org/10.3758/BF03200977.

24. Thorpe RT, Atkinson G, Drust B, Gregson W. Monitoring fatigue status in elite team-sport athletes: implications for practice. Int J Sports Physiol Perform. Apr, 2017;12(s2):S2-27–34. https://doi.org/10.1123/ijspp.2016-0434.

25. Vila, Bryan, "Tired cops: the importance of managing police fatigue," Police Exec Res Forum, 2000.
26. Glaus A. Fatigue in patients with cancer: analysis and assessment. Springer Science & Business Media; 2012.
27. Jason LA, Evans M, Brown M, Porter N. What is fatigue? Pathological and nonpathological fatigue. PM&R. 2010, May;2(5):327–31. https://doi.org/10.1016/j.pmrj.2010.03.028.
28. Davis MP, Khoshknabi D, Yue GH. Management of fatigue in cancer patients. Curr Pain Headache Rep. 2006;10(4):260–9. https://doi.org/10.1007/s11916-006-0030-2.
29. Tanaka M, Watanabe Y. Supraspinal regulation of physical fatigue. Neurosci Biobehav Rev. 2012, Jan;36(1):727–34. https://doi.org/10.1016/j.neubiorev.2011.10.004.
30. Bogdanis GCP. Effects of physical activity and inactivity on muscle fatigue. Front Physiol. 2012;3 https://doi.org/10.3389/fphys.2012.00142.
31. Bloomfield SA. Changes in musculoskeletal structure and function with prolonged bed rest. Med Sci Sports Exerc. 1997;29(2):197–206.
32. Rimmer JH, Schiller W, Chen M-D. Effects of disability-associated low energy expenditure deconditioning syndrome. Exerc Sport Sci Rev. 2012, Jan;40(1):22–9. https://doi.org/10.1097/JES.0b013e31823b8b82.
33. Dantzer R, et al. Identification and treatment of symptoms associated with inflammation in medically ill patients. Psychoneuroendocrinology. 2008, Jan;33(1):18–29. https://doi.org/10.1016/j.psyneuen.2007.10.008.
34. Linnhoff S, Fiene M, Heinze H-J, Zaehle T. Cognitive fatigue in multiple sclerosis: An objective approach to diagnosis and treatment by transcranial electrical stimulation. Brain Sci. 2019, May;9(5):5. https://doi.org/10.3390/brainsci9050100.
35. Valentine AD, Meyers CA. Cognitive and mood disturbance as causes and symptoms of fatigue in cancer patients. Cancer. 2001;92(S6):1694–8. https://doi.org/10.1002/1097-0142(20010915)92:6+<1694::AID-CNCR1499>3.0.CO;2-S.
36. Kohl AD, Wylie GR, Genova HM, Hillary FG, DeLuca J. The neural correlates of cognitive fatigue in traumatic brain injury using functional MRI. Brain Inj. 2009, Jan;23(5):420–32. https://doi.org/10.1080/02699050902788519.
37. Byun E, Gay CL, Lee KA. Sleep, fatigue, and problems with cognitive function in adults living with HIV. J Assoc Nurses AIDS Care JANAC. 2016;27(1):5–16. https://doi.org/10.1016/j.jana.2015.10.002.
38. Wong WS, Fielding R. Prevalence of chronic fatigue among Chinese adults in Hong Kong: a population-based study. J Affect Disord. 2010, Dec;127(1):248–56. https://doi.org/10.1016/j.jad.2010.04.029.
39. Hockey B, Hockey R. The psychology of fatigue: work, effort and control. Cambridge University Press; 2013.
40. Light AR, White AT, Hughen RW, Light KC. Moderate exercise increases expression for sensory, adrenergic and immune genes in chronic fatigue syndrome patients, but not in normal subjects. J Pain Off J Am Pain Soc. 2009, Oct;10(10):1099–112. https://doi.org/10.1016/j.jpain.2009.06.003.
41. Schwartz JE, Jandorf L, Krupp LB. The measurement of fatigue: a new instrument. J Psychosom Res. 1993, Oct;37(7):753–62. https://doi.org/10.1016/0022-3999(93)90104-N.
42. Edwards RH. Human muscle function and fatigue. Ciba Found Symp. 1981;82:1–8.
43. Friedman JH, et al. Fatigue in Parkinson's disease: a review. Mov Disord. 2007;22(3):297–308. https://doi.org/10.1002/mds.21240.
44. Cairns SP, Knicker AJ, Thompson MW, Sjøgaard G. Evaluation of models used to study neuromuscular fatigue. Exerc Sport Sci Rev. 2005, Jan;33(1):9–16.
45. Abd-Elfattah HM, Abdelazeim FH, Elshennawy S. Physical and cognitive consequences of fatigue: a review. J Adv Res. 2015, May;6(3):351–8. https://doi.org/10.1016/j.jare.2015.01.011.
46. Kamijo K, Nishihira Y, Higashiura T, Kuroiwa K. The interactive effect of exercise intensity and task difficulty on human cognitive processing. Int J Psychophysiol. 2007, Aug;65(2):114–21. https://doi.org/10.1016/j.ijpsycho.2007.04.001.

47. Tomporowski PD. Effects of acute bouts of exercise on cognition. Acta Psychol. 2003, Mar;112(3):297–324. https://doi.org/10.1016/S0001-6918(02)00134-8.
48. Cian C, Barraud PA, Melin B, Raphel C. Effects of fluid ingestion on cognitive function after heat stress or exercise-induced dehydration. Int J Psychophysiol. 2001, Nov;42(3):243–51. https://doi.org/10.1016/S0167-8760(01)00142-8.
49. Falup-Pecurariu C. Fatigue assessment of Parkinson's disease patient in clinic: specific versus holistic. J Neural Transm. 2013, Apr;120(4):577–81. https://doi.org/10.1007/s00702-013-0969-1.
50. Persson J, Welsh KM, Jonides J, Reuter-Lorenz PA. Cognitive fatigue of executive processes: interaction between interference resolution tasks. Neuropsychologia. 2007, Apr;45(7):1571–9. https://doi.org/10.1016/j.neuropsychologia.2006.12.007.
51. van der Linden D, Eling P. Mental fatigue disturbs local processing more than global processing. Psychol Res. 2006, Sep;70(5):395–402. https://doi.org/10.1007/s00426-005-0228-7.
52. Lorist MM. Impact of top-down control during mental fatigue. Brain Res. 2008, Sep;1232:113–23. https://doi.org/10.1016/j.brainres.2008.07.053.
53. Cook DB, O'Connor PJ, Lange G, Steffener J. Functional neuroimaging correlates of mental fatigue induced by cognition among chronic fatigue syndrome patients and controls. NeuroImage. 2007,May;36(1):108–22. https://doi.org/10.1016/j.neuroimage.2007.02.033.
54. Lal SKL, Craig A. A critical review of the psychophysiology of driver fatigue. Biol Psychol. 2001,Feb;55(3):173–94. https://doi.org/10.1016/S0301-0511(00)00085-5.
55. Goode JH. Are pilots at risk of accidents due to fatigue? J Saf Res. 2003, Aug;34(3):309–13. https://doi.org/10.1016/S0022-4375(03)00033-1.
56. Van Cutsem J, Marcora S, De Pauw K, Bailey S, Meeusen R, Roelands B. The effects of mental fatigue on physical performance: a systematic review. Sports Med. 2017, Aug;47(8):1569–88. https://doi.org/10.1007/s40279-016-0672-0.
57. Grillon C, Quispe-Escudero D, Mathur A, Ernst M. Mental fatigue impairs emotion regulation. Emot Wash DC. 2015, Jun;15(3):383–9. https://doi.org/10.1037/emo0000058.
58. Chen MK. The epidemiology of self-perceived fatigue among adults. Prev Med. 1986, Jan;15(1):74–81. https://doi.org/10.1016/0091-7435(86)90037-X.
59. Hardy GE, Shapiro DA, Borrill CS. Fatigue in the workforce of national health service trusts: levels of symptomatology and links with minor psychiatric disorder, demographic, occupational and work role factors. J Psychosom Res. 1997, Jul;43(1):83–92. https://doi.org/10.1016/S0022-3999(97)00019-6.
60. Hickie IB, Hooker AW, Bennett BK, Hadzi-Pavlovic D, Wilson AJ, Lloyd AR. Fatigue in selected primary care settings: sociodemographic and psychiatric correlates. Med J Aust. 1996;164(10):585–8. https://doi.org/10.5694/j.1326-5377.1996.tb122199.x.
61. Loge JH, Ekeberg Ø, Kaasa S. Fatigue in the general norwegian population: normative data and associations. J Psychosom Res. 1998, Jul;45(1):53–65. https://doi.org/10.1016/S0022-3999(97)00291-2.
62. Pawlikowska T, Chalder T, Hirsch SR, Wallace P, Wright DJM, Wessely SC. Population based study of fatigue and psychological distress. BMJ. 1994, Mar;308(6931):763–6. https://doi.org/10.1136/bmj.308.6931.763.
63. David A, et al. Tired, weak, or in need of rest: fatigue among general practice attenders. BMJ. 1990, Nov;301(6762):1199–202.
64. Bultmann U, Kant I, Kasl SV. Fatigue and psychological distress in the working population psychometrics, prevalence, and correlates. U Bu. 2002:8.
65. Van Mens-Verhulst J, Bensing J. Distinguishing between chronic and nonchronic fatigue, the role of gender and age. Soc Sci Med. 1998, Sep;47(5):621–34. https://doi.org/10.1016/S0277-9536(98)00116-6.
66. Dement WC, Carskadon MA. Current perspectives on daytime sleepiness: the issues. Sleep. 1982, Sep;5(suppl_2):S56–66. https://doi.org/10.1093/sleep/5.S2.S56.
67. Shahid A, Shen J, Shapiro CM. Measurements of sleepiness and fatigue. J Psychosom Res. 2010, Jul;69(1):81–9. https://doi.org/10.1016/j.jpsychores.2010.04.001.

68. Dijk D, Czeisler C. Contribution of the circadian pacemaker and the sleep homeostat to sleep propensity, sleep structure, electroencephalographic slow waves, and sleep spindle activity in humans. J Neurosci. 1995, May;15(5):3526–38. https://doi.org/10.1523/JNEUROSCI.15-05-03526.1995.

69. Rey A. "L'examen clinique en psychologie." 1958.

70. Michielsen HJ, De Vries J, Van Heck GL. Psychometric qualities of a brief self-rated fatigue measure: the fatigue assessment scale. J Psychosom Res. 2003, Apr;54(4):345–52. https://doi.org/10.1016/S0022-3999(02)00392-6.

71. Webster K, Odom L, Peterman A, Lent L, Cella D. The functional assessment of chronic illness therapy (FACIT) measurement system: validation of version 4 of the core questionnaire. Qual Life Res. 1999;

72. Fisk JD, Ritvo PG, Ross L, Haase DA, Marrie TJ, Schlech WF. Measuring the functional impact of fatigue: initial validation of the fatigue impact scale. Clin Infect Dis. 1994, Jan;18(Suppl_1):S79–83. https://doi.org/10.1093/clinids/18.Supplement_1.S79.

73. Chalder T, et al. Development of a fatigue scale. J Psychosom Res. 1993, Feb;37(2):147–53. https://doi.org/10.1016/0022-3999(93)90081-P.

74. Krupp LB, LaRocca NG, Muir-Nash J, Steinberg AD. The fatigue severity scale: application to patients with multiple sclerosis and systemic lupus erythematosus. Arch Neurol. 1989, Oct;46(10):1121–3. https://doi.org/10.1001/archneur.1989.00520460115022.

75. Tack BB. "Dimensions and correlates of fatigue in older adults with rheumatoid arthritis.," Accessed: Jun. 24, 2020. [Online]. Available: https://elibrary.ru/item.asp?id=5833962.

76. Smets EMA, Garssen B, Bonke B, De Haes JCJM. The multidimensional fatigue inventory (MFI) psychometric qualities of an instrument to assess fatigue. J Psychosom Res. 1995, Apr;39(3):315–25. https://doi.org/10.1016/0022-3999(94)00125-O.

77. Ware JE Jr, Sherbourne CD. The MOS 36-item short-form health survey (SF-36): I. conceptual framework and item selection. Med Care. 1992:473–83.

78. Lai J-S, et al. How item banks and their application can influence measurement practice in rehabilitation medicine: a PROMIS fatigue item Bank example. Arch Phys Med Rehabil. 2011, Oct;92(10):S20–7. https://doi.org/10.1016/j.apmr.2010.08.033.

79. McNair D, Lorr M, Droppleman L. POMS manual for the profile of mood states. Educ Ind Test Serv. 1971;

80. Lee KA, Hicks G, Nino-Murcia G. Validity and reliability of a scale to assess fatigue. Psychiatry Res. 1991, Mar;36(3):291–8. https://doi.org/10.1016/0165-1781(91)90027-M.

81. Junghaenel DU, Christodoulou C, Lai J-S, Stone AA. Demographic correlates of fatigue in the US general population: results from the patient-reported outcomes measurement information system (PROMIS) initiative. J Psychosom Res. 2011, Sep;71(3):117–23. https://doi.org/10.1016/j.jpsychores.2011.04.007.

82. Alemohammad ZB, Sadeghniiat-Haghighi K. Risk of fatigue at work. In: Sharafkhaneh A, Hirshkowitz M, editors. Fatigue management: principles and practices for improving workplace safety. New York, NY: Springer; 2018. p. 181–91.

83. Sharafkhaneh A, Hirshkowitz M, editors. Fatigue management: principles and practices for improving workplace safety. New York, NY: Springer New York; 2018.

84. Xianglong S, Hu Z, Shumin F, Zhenning L. Bus drivers' mood states and reaction abilities at high temperatures. Transp. Res. Part F Traffic Psychol. Behav. 2018, Nov;59:436–44. https://doi.org/10.1016/j.trf.2018.09.022.

85. Mitchell SA. Cancer-related fatigue: state of the science. PM&R. 2010, May;2(5):364–83. https://doi.org/10.1016/j.pmrj.2010.03.024.

86. Pearson EJM, Morris ME, di Stefano M, McKinstry CE. Interventions for cancer-related fatigue: a scoping review. Eur J Cancer Care (Engl). 2018, Jan;27(1):e12516. https://doi.org/10.1111/ecc.12516.

87. Cohen R, Bavishi C, Haider S, Thankachen J, Rozanski A. Meta-analysis of relation of vital exhaustion to cardiovascular disease events. Am J Cardiol. 2017, Apr;119(8):1211–6. https://doi.org/10.1016/j.amjcard.2017.01.009.

88. Nagy A, et al. Association of Exercise Capacity with physical functionality and various aspects of fatigue in patients with coronary artery disease. Behav Med. 2018, Jan;44(1):28–35. https://doi.org/10.1080/08964289.2016.1189395.
89. Lewko A, Bidgood P, Jewell A, Garrod R. A comprehensive literature review of COPD-related fatigue. Curr Respir Med Rev. 2012, Nov;8(5):370–82. https://doi.org/10.2174/157339812803832476.
90. Jensen Ø, Bernklev T, Jelsness-Jørgensen L-P. Fatigue in type 1 diabetes: a systematic review of observational studies. Diabetes Res Clin Pract. 2017, Jan;123:63–74. https://doi.org/10.1016/j.diabres.2016.11.002.
91. Vincent A, Benzo RP, Whipple MO, McAllister SJ, Erwin PJ, Saligan LN. Beyond pain in fibromyalgia: insights into the symptom of fatigue. Arthritis Res Ther. 2013, Nov;15(6):221. https://doi.org/10.1186/ar4395.
92. Czuber-Dochan W, Ream E, Norton C. Review article: description and management of fatigue in inflammatory bowel disease. Aliment Pharmacol Ther. 2013;37(5):505–16. https://doi.org/10.1111/apt.12205.
93. Han CJ, Yang GS. Fatigue in irritable bowel syndrome: a systematic review and meta-analysis of pooled frequency and severity of fatigue. Asian Nurs Res. 2016, Mar;10(1):1–10. https://doi.org/10.1016/j.anr.2016.01.003.
94. Nocerino A, Nguyen A, Agrawal M, Mone A, Lakhani K, Swaminath A. Fatigue in inflammatory bowel diseases: etiologies and management. Adv Ther. 2020, Jan;37(1):97–112. https://doi.org/10.1007/s12325-019-01151-w.
95. Moldovan I, et al. Pain and depression predict self-reported fatigue/energy in lupus. Lupus. 2013, Jun;22(7):684–9. https://doi.org/10.1177/0961203313486948.
96. Ghanean H, Ceniti AK, Kennedy SH. Fatigue in patients with major depressive disorder: prevalence, burden and pharmacological approaches to management. CNS Drugs. 2018, Jan;32(1):65–74. https://doi.org/10.1007/s40263-018-0490-z.
97. Charvet L, Serafin D, Krupp LB. Fatigue in multiple sclerosis. Fatigue Biomed Health Behav. 2014, Jan;2(1):3–13. https://doi.org/10.1080/21641846.2013.843812.
98. Latimer-Cheung AE, et al. Effects of exercise training on fitness, mobility, fatigue, and health-related quality of life among adults with multiple sclerosis: a systematic review to inform guideline development. Arch Phys Med Rehabil. 2013, Sep;94(9):1800–1828.e3. https://doi.org/10.1016/j.apmr.2013.04.020.
99. Newland P, Starkweather A, Sorenson M. Central fatigue in multiple sclerosis: a review of the literature. J Spinal Cord Med. 2016, Jul;39(4):386–99. https://doi.org/10.1080/10790268.2016.1168587.
100. Skoie IM, et al. Fatigue in psoriasis: a controlled study. Br J Dermatol. 2017;177(2):505–12.
101. Weinstein AA, et al. Effect of aerobic exercise training on fatigue and physical activity in patients with pulmonary arterial hypertension. Respir Med. 2013, May;107(5):778–84. https://doi.org/10.1016/j.rmed.2013.02.006.
102. Artom M, Moss-Morris R, Caskey F, Chilcot J. Fatigue in advanced kidney disease. Kidney Int. 2014, Sep;86(3):497–505. https://doi.org/10.1038/ki.2014.86.
103. Stebbings S, Treharne GJ. Fatigue in rheumatic disease: an overview. Int J Clin Rheumatol. 2010, Aug;5(4):487–502. https://doi.org/10.2217/ijr.10.30.
104. do Espírito Santo RC, et al. Neuromuscular fatigue is weakly associated with perception of fatigue and function in patients with rheumatoid arthritis. Rheumatol Int. 2018, Mar;38(3):415–23. https://doi.org/10.1007/s00296-017-3894-z.
105. Bailes S, et al. Fatigue: the forgotten symptom of sleep apnea. J Psychosom Res. 2011, Apr;70(4):346–54. https://doi.org/10.1016/j.jpsychores.2010.09.009.
106. Aarnes R, Stubberud J, Lerdal A. A literature review of factors associated with fatigue after stroke and a proposal for a framework for clinical utility. Neuropsychol Rehabil. 2019, Mar;0(0):1–28. https://doi.org/10.1080/09602011.2019.1589530.
107. Duncan F, Wu S, Mead GE. Frequency and natural history of fatigue after stroke: a systematic review of longitudinal studies. J Psychosom Res. 2012, Jul;73(1):18–27. https://doi.org/10.1016/j.jpsychores.2012.04.001.

108. Newland P, Oliver B, Newland JM, Thomas FP. Testing feasibility of a Mobile application to monitor fatigue in people with multiple sclerosis. J Neurosci Nurs. 2019, Dec;51(6):331–4. https://doi.org/10.1097/JNN.0000000000000479.

109. Maguire R, McCann L, Miller M, Kearney N. Nurse's perceptions and experiences of using of a mobile-phone-based advanced symptom management system (ASyMS©) to monitor and manage chemotherapy-related toxicity. Eur J Oncol Nurs. 2008, Sep;12(4):380–6. https://doi.org/10.1016/j.ejon.2008.04.007.

110. Nicholas J, Larsen ME, Proudfoot J, Christensen H. Mobile apps for bipolar disorder: a systematic review of features and content quality. J Med Internet Res. 2015;17(8):e198. https://doi.org/10.2196/jmir.4581.

111. Shiffman S, Stone AA, Hufford MR. Ecological momentary assessment. Annu Rev Clin Psychol. 2008;4(1):1–32. https://doi.org/10.1146/annurev.clinpsy.3.022806.091415.

112. Larson R, Csikszentmihalyi M. The experience sampling method. In: Flow and the foundations of positive psychology. Dordrecht: Springer Netherlands; 2014. p. 21–34.

113. van Berkel N, Ferreira D, Kostakos V. The experience sampling method on Mobile devices. ACM Comput Surv. 2018, Jan;50(6):1–40. https://doi.org/10.1145/3123988.

114. Guérin E, Fortier MS, Williams T. 'I just NEED to move…': examining women's passion for physical activity and its relationship with daily affect and vitality. Psychol Well- Theory Res Pract. 2013, Nov;3(1):4. https://doi.org/10.1186/2211-1522-3-4.

115. Zacher H, Brailsford HA, Parker SL. Micro-breaks matter: a diary study on the effects of energy management strategies on occupational Well-being. J Vocat Behav. 2014, Dec;85(3):287–97. https://doi.org/10.1016/j.jvb.2014.08.005.

116. Rhee H, Kim S. Effects of breaks on regaining vitality at work: An empirical comparison of 'conventional' and 'smart phone' breaks. Comput Hum Behav. 2016, Apr;57:160–7. https://doi.org/10.1016/j.chb.2015.11.056.

117. Murphy SL, Smith DM. Ecological measurement of fatigue and fatigability in older adults with osteoarthritis. J Gerontol Ser A. 2010, Feb;65A(2):184–9. https://doi.org/10.1093/gerona/glp137.

118. Hacker ED, Ferrans CE. Ecological momentary assessment of fatigue in patients receiving intensive cancer therapy. J Pain Symptom Manag. 2007, Mar;33(3):267–75. https://doi.org/10.1016/j.jpainsymman.2006.08.007.

119. Bower JE. Cancer-related fatigue: mechanisms, risk factors, and treatments. Nat Rev Clin Oncol. 2014, Oct;11(10):597–609. https://doi.org/10.1038/nrclinonc.2014.127.

120. van der Linden D, Frese M, Meijman TF. Mental fatigue and the control of cognitive processes: effects on perseveration and planning. Acta Psychol. 2003, May;113(1):45–65. https://doi.org/10.1016/S0001-6918(02)00150-6.

121. Ackerman PL. 100 years without resting. In: Cognitive fatigue: multidisciplinary perspectives on current research and future applications. Washington, DC, US: American Psychological Association; 2011. p. 11–43.

122. Lieberman HR. Cognitive methods for assessing mental energy. Nutr Neurosci. 2007, Oct;10(5–6):229–42. https://doi.org/10.1080/10284150701722273.

123. Basner M, Dinges DF. Maximizing sensitivity of the psychomotor vigilance test (PVT) to sleep loss. Sleep. 2011, May;34(5):581–91. https://doi.org/10.1093/sleep/34.5.581.

124. Holsbeeke L, Ketelaar M, Schoemaker MM, Gorter JW. Capacity, capability, and performance: different constructs or three of a kind? Arch Phys Med Rehabil. 2009, May;90(5):849–55. https://doi.org/10.1016/j.apmr.2008.11.015.

125. World Health Organization, "World report on ageing and health.," World Health Organ., 2015.

126. Enright PL. The six-minute walk test. Respir Care. 2003;48(8):3.

127. Rudick R, et al. Recommendations from the national multiple sclerosis society clinical outcomes assessment task force. Ann Neurol. 1997;42(3):379–82. https://doi.org/10.1002/ana.410420318.

128. Podsiadlo D, Richardson S. The timed 'up & go': a test of basic functional mobility for frail elderly persons. J Am Geriatr Soc. 1991;39(2):142–8. https://doi.org/10.1111/j.1532-5415.1991.tb01616.x.

129. Boolani A, O'Connor PJ, Reid J, Ma S, Mondal S. Predictors of feelings of energy differ from predictors of fatigue. Fatigue Biomed. Health Behav. 2019, Jan;7(1):12–28. https://doi.org/1 0.1080/21641846.2018.1558733.

130. Roldán Jiménez C, Bennett P, Ortiz García A, Cuesta Vargas AI. Fatigue detection during sit-to-stand test based on surface electromyography and acceleration: a case study. Sensors. 2019, Jan;19(19):19. https://doi.org/10.3390/s19194202.

131. Muñoz JE, Gouveia ER, Cameirão MS, Badia SB. PhysioLab–a multivariate physiological computing toolbox for ECG, EMG and EDA signals: a case of study of cardiorespiratory fitness assessment in the elderly population. Multimed Tools Appl. 2018, May;77(9):11521–46. https://doi.org/10.1007/s11042-017-5069-z.

132. Nicklas BJ, et al. Effects of caloric restriction on cardiorespiratory fitness, fatigue, and disability responses to aerobic exercise in older adults with obesity: a randomized controlled trial. J Gerontol A Biol Sci Med Sci. 2019, Jun;74(7):1084–90. https://doi.org/10.1093/gerona/gly159.

133. Sehle A, Vieten M, Mündermann A, Dettmers C. Difference in motor fatigue between patients with stroke and patients with multiple sclerosis: a pilot study. Front Neurol. 2014;5 https://doi.org/10.3389/fneur.2014.00279.

134. Dedering Å, Gnospelius Å, Elfving B. Reliability of measurements of endurance time, electromyographic fatigue and recovery, and associations to activity limitations, in patients with lumbar disc herniation. Physiother Res Int. 2010;15(4):189–98. https://doi.org/10.1002/pri.457.

135. Schmitt J, Lindner N, Reuss-Borst M, Holmberg H-C, Sperlich B. A 3-week multimodal intervention involving high-intensity interval training in female cancer survivors: a randomized controlled trial. Physiol Rep. 2016;4(3):e12693. https://doi.org/10.14814/phy2.12693.

136. Folstein MF, Robins LN, Helzer JE. The mini-mental state examination. Arch Gen Psychiatry. 1983, Jul;40(7):812. https://doi.org/10.1001/archpsyc.1983.01790060110016.

137. U. S. Army, "Army individual test battery," Man. Dir. Scoring, 1944.

138. Hannay HJ, Levin HS. Selective reminding test: An examination of the equivalence of four forms. J Clin Exp Neuropsychol. 1985, Jun;7(3):251–63. https://doi.org/10.1080/01688638508401258.

139. Barbizet J, Cany E. Clinical and psychometrical study of a patient with memory disturbances. Int J Neurol. 1967;7(1):44–54.

140. Rao SM, Hammeke TA, McQuillen MP, Khatri BO, Lloyd D. Memory disturbance in chronic progressive multiple sclerosis. Arch Neurol. 1984, Jun;41(6):625–31. https://doi.org/10.1001/archneur.1984.04210080033010.

141. A. Smith, Symbol digit modalities test. Western Psychological Services Los Angeles, 1973.

142. Gronwall DMA. Paced auditory serial-addition task: a measure of recovery from concussion. Percept Mot Skills. 1977, Apr;44(2):367–73. https://doi.org/10.2466/pms.1977.44.2.367.

143. Barr A, Brandt J. Word-list generation deficits in dementia. J Clin Exp Neuropsychol. 1996, Dec;18(6):810–22. https://doi.org/10.1080/01688639608408304.

144. Obrist PA, Webb RA, Sutterer JR. Heart rate and somatic changes during aversive conditioning and a simple reaction time task. Psychophysiology. 1969;5(6):696–723.

145. Bever CT Jr, Grattan L, Panitch HS, Johnson KP. The brief repeatable battery of neuropsychological tests for multiple sclerosis: a preliminary serial study. Mult Scler J. 1995;1(3):165–9.

146. Smolders KC, de Kort YA. Bright light and mental fatigue: effects on alertness, vitality, performance and physiological arousal. J Environ Psychol. 2014;39:77–91.

147. Westbrook JI, Raban MZ, Walter SR, Douglas H. Task errors by emergency physicians are associated with interruptions, multitasking, fatigue and working memory capacity: a prospective, direct observation study. BMJ Qual Saf. 2018;27(8):655–63.

148. Hartzler BM. Fatigue on the flight deck: the consequences of sleep loss and the benefits of napping. Accid Anal Prev. 2014;62:309–18.

149. Staniute M, Bunevicius A, Brozaitiene J, Bunevicius R. Relationship of health-related quality of life with fatigue and exercise capacity in patients with coronary artery disease. Eur J Cardiovasc Nurs. 2014, Aug;13(4):338–44. https://doi.org/10.1177/1474515113496942.

150. Pimenta A, Carneiro D, Novais P, Neves J. Monitoring mental fatigue through the analysis of keyboard and mouse interaction patterns. In: International Conference on Hybrid Artificial Intelligence Systems; 2013. p. 222–31.
151. Foti K, Xanthopoulou D, Papagiannidis S, Kafetsios K. The role of tweet-related emotion on the exhaustion–recovery from work relationship. In: Conference on e-Business, e-Services and e-Society; 2019. p. 380–91.
152. Dikmen S, Machamer J, Temkin N, McLean A. Neuropsychological recovery in patients with moderate to severe head injury: 2 year follow-up. J Clin Exp Neuropsychol. 1990;12(4):507–19.
153. Ponsford JL, Olver JH, Curran C. A profile of outcome: 2 years after traumatic brain injury. Brain Inj. 1995;9(1):1–10.
154. Van Zomeren AH, Van den Burg W. Residual complaints of patients two years after severe head injury. J Neurol Neurosurg Psychiatry. 1985;48(1):21–8.
155. Price E, Moore G, Galway L, Linden M. Towards a mobile assistive technology for monitoring and assessing cognitive fatigue in individuals with acquired brain injury. In: 2015 IEEE International Conference on Computer and Information Technology; Ubiquitous Computing and Communications; Dependable, Autonomic and Secure Computing; Pervasive Intelligence and Computing; 2015. p. 1487–91.
156. Tran Y, Wijesuriya N, Tarvainen M, Karjalainen P, Craig A. The relationship between spectral changes in heart rate variability and fatigue. J Psychophysiol. 2009;23(3):143–51.
157. Segerstrom SC, Nes LS. Heart rate variability reflects self-regulatory strength, effort, and fatigue. Psychol Sci. 2007;18(3):275–81.
158. Blausen.com staff. Medical gallery of Blausen medical 2014. WikiJournal of Medicine. 2014;1(2)
159. Samel A, Wegmann H-M, Vejvoda M. Aircrew fatigue in long-haul operations. Accid Anal Prev. 1997;29(4):439–52.
160. Goldman LI, McDonough MT, Rosemond GP. Stresses affecting surgical performance and learning: I. correlation of heart rate, electrocardiogram, and operation simultaneously recorded on videotapes. J Surg Res. 1972;12(2):83–6.
161. Chen C, Li K, Wu Q, Wang H, Qian Z, Sudlow G. EEG-based detection and evaluation of fatigue caused by watching 3DTV. Displays. 2013;34(2):81–8.
162. Holdsworth L, Evens T. Is heart rate variability a useful marker of stress and fatigue in emergency and pre-hospital clinicians?–a systematic review. Resuscitation. 2017;118:e95.
163. Stamler JS, Goldman ME, Gomes J, Matza D, Horowitz SF. The effect of stress and fatigue on cardiac rhythm in medical interns. J Electrocardiol. 1992;25(4):333–8.
164. Heemskerk J, et al. Relax, it's just laparoscopy! A prospective randomized trial on heart rate variability of the surgeon in robot-assisted versus conventional laparoscopic cholecystectomy. Dig Surg. 2014;31(3):225–32.
165. Jung S-J, Shin H-S, Chung W-Y. Driver fatigue and drowsiness monitoring system with embedded electrocardiogram sensor on steering wheel. IET Intell Transp Syst. 2014;8(1):43–50.
166. Díaz-Toral LG, Banderas-Dorantes TR, Rivas-Vilchis JF. Impact of electroacupuncture treatment on quality of life and heart rate variability in fibromyalgia patients. J Evid-Based Complement Altern Med. 2017;22(2):216–22.
167. DeLuca J. "19 Fatigue: its definition, its study, and its future," Fatigue Window Brain, p. 319, 2005.
168. Stern JA, Boyer D, Schroeder D. Blink rate: a possible measure of fatigue. Hum Factors. 1994;36(2):285–97.
169. Sirois S, Brisson J. Pupillometry. Wiley Interdiscip Rev Cogn Sci. 2014;5(6):679–92.
170. Eggert T, et al. The pupillographic sleepiness test in adults: effect of age, gender, and time of day on pupillometric variables. Am J Hum Biol. 2012;24(6):820–8.
171. Lowenstein O, Feinberg R, Loewenfeld IE. Pupillary movements during acute and chronic fatigue: a new test for the objective evaluation of tiredness. Invest Ophthalmol Vis Sci. 1963;2(2):138–57.

172. Hirvonen K, Puttonen S, Gould K, Korpela J, Koefoed VF, Müller K. Improving the saccade peak velocity measurement for detecting fatigue. J Neurosci Methods. 2010;187(2):199–206.
173. You C-W, et al. CarSafe: a driver safety app that detects dangerous driving behavior using dual-cameras on smartphones. In: Proceedings of the 2012 ACM Conference on Ubiquitous Computing; 2012. p. 671–2.
174. Zhang Y-F, Gao X-Y, Zhu J-Y, Zheng W-L, Lu B-L. A novel approach to driving fatigue detection using forehead EOG. In: 2015 7th International IEEE/EMBS Conference on Neural Engineering (NER); 2015. p. 707–10.
175. Morris G, Maes M. Myalgic encephalomyelitis/chronic fatigue syndrome and encephalomyelitis disseminata/multiple sclerosis show remarkable levels of similarity in phenomenology and neuroimmune characteristics. BMC Med. 2013;11(1):1–23.
176. Lorist MM, Bezdan E, ten Caat M, Span MM, Roerdink JB, Maurits NM. The influence of mental fatigue and motivation on neural network dynamics; an EEG coherence study. Brain Res. 2009;1270:95–106.
177. Sheng Y, Huang K, Zou J, Wang L, Wei P. Exploring the relationship between neural mechanism and detection in mental fatigue by genetic algorithm and hierarchical clustering. In: 2019 IEEE 7th International Conference on Bioinformatics and Computational Biology (ICBCB); 2019. p. 120–4.
178. Craig A, Tran Y, Wijesuriya N, Nguyen H. Regional brain wave activity changes associated with fatigue. Psychophysiology. 2012;49(4):574–82.
179. Zhao C, Zheng C, Zhao M, Tu Y, Liu J. Multivariate autoregressive models and kernel learning algorithms for classifying driving mental fatigue based on electroencephalographic. Expert Syst Appl. 2011;38(3):1859–65.
180. Kahol K, Smith M, Brandenberger J, Ashby A, Ferrara JJ. Impact of fatigue on neurophysiologic measures of surgical residents. J Am Coll Surg. 2011;213(1):29–34.
181. Guru KA, Shafiei SB, Khan A, Hussein AA, Sharif M, Esfahani ET. Understanding cognitive performance during robot-assisted surgery. Urology. 2015;86(4):751–7.
182. Gold AE, MacLeod KM, Deary IJ, Frier BM. Hypoglycemia-induced cognitive dysfunction in diabetes mellitus: effect of hypoglycemia unawareness. Physiol Behav. 1995;58(3):501–11.
183. Newsholme EA, Blomstrand E, Ekblom B. Physical and mental fatigue: metabolic mechanisms and importance of plasma amino acids. Br Med Bull. 1992;48(3):477–95.
184. Olsson EM, Roth WT, Melin L. Psychophysiological characteristics of women suffering from stress-related fatigue. Stress Health J Int Soc Investig Stress. 2010;26(2):113–26.
185. Kuebler U, et al. Norepinephrine infusion with and without alpha-adrenergic blockade by phentolamine increases salivary alpha amylase in healthy men. Psychoneuroèndocrinology. 2014;49:290–8.
186. Hopstaken JF, van der Linden D, Bakker AB, Kompier MA. The window of my eyes: task disengagement and mental fatigue covary with pupil dynamics. Biol Psychol. 2015;110:100–6.
187. Van Heukelom RO, Prins JB, Smits MG, Bleijenberg G. Influence of melatonin on fatigue severity in patients with chronic fatigue syndrome and late melatonin secretion. Eur J Neurol. 2006;13(1):55–60.
188. Merikangas KR, et al. Real-time mobile monitoring of the dynamic associations among motor activity, energy, mood, and sleep in adults with bipolar disorder. JAMA Psychiatry. 2019;76(2):190–8.
189. Bloch KE. Polysomnography: a systematic review. Technol Health Care. 1997;5(4):285–305.
190. Sadeh A, Hauri PJ, Kripke DF, Lavie P. The role of actigraphy in the evaluation of sleep disorders. Sleep. 1995;18(4):288–302.
191. Wac K. From quantified self to quality of life. In: Digital health. Springer; 2018. p. 83–108.
192. Rizzo D, et al. The role of fatigue and sleepiness in drivers with obstructive sleep apnea. Transp Res Part F Traffic Psychol Behav. 2019;62:796–804.
193. Fernandes-Junior SA, Ruiz FS, Antonietti LS, Tufik S, Túlio de Mello M. Sleep, fatigue and quality of life: a comparative analysis among night shift workers with and without children. PLoS One. 2016;11(7):e0158580.

194. Campbell R, Tobback E, Delesie L, Vogelaers D, Mariman A, Vansteenkiste M. Basic psychological need experiences, fatigue, and sleep in individuals with unexplained chronic fatigue. Stress Health. 2017;33(5):645–55.
195. Maher C, et al. Fatigue is a major issue for children and adolescents with physical disabilities. Dev Med Child Neurol. 2015;57(8):742–7.
196. Attarian HP, Brown KM, Duntley SP, Carter JD, Cross AH. The relationship of sleep disturbances and fatigue in multiple sclerosis. Arch Neurol. 2004, Apr;61(4):525–8. https://doi.org/10.1001/archneur.61.4.525.
197. Kaynak H, et al. Fatigue and sleep disturbance in multiple sclerosis. Eur J Neurol. 2006;13(12):1333–9.
198. Veauthier C, et al. Fatigue in multiple sclerosis is closely related to sleep disorders: a polysomnographic cross-sectional study. Mult Scler J. 2011;17(5):613–22.
199. Kaminska M, et al. Obstructive sleep apnea is associated with fatigue in multiple sclerosis. Mult Scler J. 2012;18(8):1159–69.
200. Shitrit AB-G, et al. Sleep disturbances can be prospectively observed in patients with an inactive inflammatory bowel disease. Dig Dis Sci. 2018;63(11):2992–7.
201. Sun J-L, Lin C-C. Relationships among daytime napping and fatigue, sleep quality, and quality of life in cancer patients. Cancer Nurs. 2016;39(5):383–92.
202. Ancoli-Israel S, et al. Sleep, fatigue, depression, and circadian activity rhythms in women with breast cancer before and after treatment: a 1-year longitudinal study. Support Care Cancer. 2014, Sep;22(9):2535–45. https://doi.org/10.1007/s00520-014-2204-5.
203. Holliday EB, Dieckmann NF, McDonald TL, Hung AY, Thomas CR Jr, Wood LJ. Relationship between fatigue, sleep quality and inflammatory cytokines during external beam radiation therapy for prostate cancer: a prospective study. Radiother Oncol. 2016;118(1):105–11.
204. Cambras T, Castro-Marrero J, Zaragoza MC, Díez-Noguera A, Alegre J. Circadian rhythm abnormalities and autonomic dysfunction in patients with chronic fatigue syndrome/Myalgic encephalomyelitis. PLoS One. 2018;13(6):e0198106.
205. Nicklas BJ, Beavers DP, Mihalko SL, Miller GD, Loeser RF, Messier SP. Relationship of objectively-measured habitual physical activity to chronic inflammation and fatigue in middle-aged and older adults. J Gerontol Ser Biomed Sci Med Sci. 2016;71(11):1437–43.
206. Nilsson M, Arving C, Thormodsen I, Assmus J, Berntsen S, Nordin K. Moderate-to-vigorous intensity physical activity is associated with modified fatigue during and after cancer treatment. Support Care Cancer. 2019:1–8.
207. Vancampfort D, et al. Higher cardio-respiratory fitness is associated with increased mental and physical quality of life in people with bipolar disorder: a controlled pilot study. Psychiatry Res. 2017;256:219–24.
208. Sheshadri A, Kittiskulnam P, Johansen KL. Higher physical activity is associated with less fatigue and insomnia among patients on hemodialysis. Kidney Int Rep. 2019;4(2):285–92.
209. Qazi T, Farraye FA. Sleep and inflammatory bowel disease: an important bi-directional relationship. Inflamm Bowel Dis. 2019;25(5):843–52.
210. Sofia MA, Andersen M, Rubin DT. Sleep dysfunction and inflammatory bowel disease. In: Sleep effect on gastrointestinal health and disease: translational opportunities for promoting health and optimizing disease management. Nova Science Publishers, Inc; 2018. p. 177–99.
211. Abbott L. Energy through motion\copyright: An activity intervention for cancer-related fatigue in an ambulatory infusion center. Clin J Oncol Nurs. 2017;21(5):618.
212. Ren D, An Y, Li Z. The discriminative model of mental fatigue based on comprehensive parameter analysis. In: 2019 Prognostics and System Health Management Conference (PHM-Qingdao); 2019. p. 1–6.
213. Smith MR, Chai R, Nguyen HT, Marcora SM, Coutts AJ. Comparing the effects of three cognitive tasks on indicators of mental fatigue. J Psychol. 2019;153(8):759–83.
214. Brown DM, Bray SR. Effects of mental fatigue on exercise intentions and behavior. Ann Behav Med. 2019;53(5):405–14.

215. Kanitz JL, et al. The impact of eurythmy therapy on fatigue in healthy adults–a controlled trial. Eur J Integr Med. 2012;4(3):e289–97.
216. Smolders KC, De Kort YA, Cluitmans PJM. A higher illuminance induces alertness even during office hours: findings on subjective measures, task performance and heart rate measures. Physiol Behav. 2012;107(1):7–16.
217. Oriyama S, Miyakoshi Y, Kobayashi T. Effects of two 15-min naps on the subjective sleepiness, fatigue and heart rate variability of night shift nurses. Ind Health. 2013;
218. Singh H, et al. Robotic surgery improves technical performance and enhances prefrontal activation during high temporal demand. Ann Biomed Eng. 2018;46(10):1621–36.
219. Dishman RK, Thom NJ, Puetz TW, O'Connor PJ, Clementz BA. Effects of cycling exercise on vigor, fatigue, and electroencephalographic activity among young adults who report persistent fatigue. Psychophysiology. 2010;47(6):1066–74.
220. Zeng Z, Pantic M, Roisman GI, Huang TS. A survey of affect recognition methods: audio, visual, and spontaneous expressions. IEEE Trans Pattern Anal Mach Intell. 2008;31(1):39–58.
221. Braley TJ, Chervin RD. Fatigue in multiple sclerosis: mechanisms, evaluation, and treatment. Sleep. 2010;33(8):1061–7.
222. Adinolfi A. Assessment and treatment of HIV-related fatigue. J Assoc Nurses AIDS Care. 2001, Sep;12:29–34. https://doi.org/10.1016/S1055-3290(06)60155-6.
223. Wac K. Quality of life technologies. NY USA: Encycl. Behav. Med. Springer N. Y; 2020.
224. Swan M. The quantified self: fundamental disruption in big data science and biological discovery. Big Data. 2013;1(2):85–99.
225. Neuringer A. Self-experimentation: a call for change. Behaviorism. 1981;9(1):79–94.
226. Jones S. "Health & Fitness Wearables," 2014. https://www.juniperresearch.com/research-store/devices-technology/digital-health-wearables-research-report. Accessed Jun. 19, 2020.
227. Piwek L, Ellis DA, Andrews S, Joinson A. The rise of consumer health wearables: promises and barriers. PLoS Med. 2016;13(2):e1001953.
228. Vogels EA. About one-in-five Americans use a smart watch or fitness tracker. Wash. DC Pew Res. Cent; 2019.
229. Ubrani J, Llamas R, Shirer M. "IDC forecasts sustained double-digit growth for wearable devices led by steady adoption of smart-watches," International Data Corporation, 2020. https://www.idc.com/getdoc.jsp?containerId=p. Accessed Jun. 15, 2020.
230. Estrada-Galiñanes V, Wac K. "Collecting, exploring and sharing personal data: why, how and where," Data Sci., no. Preprint, pp. 1–28.
231. Stansberry K, Anderson J, Rainie L. Experts optimistic about the next 50 years of digital life. Pew Research Center: Internet, Science & Tech; 2019. https://www.pewresearch.org/internet/2019/10/28/experts-optimistic-about-the-next-50-years-of-digital-life/   (accessed Jun. 25, 2020)

Chapter 5
Quantifying Mobility in Quality of Life

Nancy E. Mayo and Kedar K. V. Mate

What Is Mobility?

According to Merriam-Webster's Dictionary mobility is the "ability or capacity to move".[1]

In the scientific community mobility has been defined "*as the ability to move oneself (either independently or by using assistive devices or transportation) within environments that expand from one's home to the neighborhood and to regions beyond*" [1]. The life-space within a person can move has also been recognized as ranging from the person's room, home, outdoors, and neighborhood to the service community of shops, banks, healthcare facilities etc., surrounding area within person's own country, and the world. This mobility is recognized to be constrained or influenced by financial, environmental, and psychosocial conditions as well as physical and cognitive capabilities. Gender, culture, and the person's life-experience also affect mobility. In order to conceptualize mobility more coherently, Webber et al. [1] proposed a framework that links factors relevant to walking, wheeling, driving, and taking alternate forms of transportation within different life-spaces.

In the context of global and public health, mobility has been defined within the World Health Organization's International Classification of Function, Disability and Health (ICF) [2] according to the components of: changing and maintaining body position (d410-d429); carrying, moving and handling objects (d430-d449); walking and moving (d450-d469);moving around using transportation (d470-d489). Table 5.1 lists these components that are based on the person doing the movement.

---

[1] https://www.merriam-webster.com/dictionary/mobility

N. E. Mayo (✉) · K. K. V. Mate
Center for Outcomes Research and Evaluation, Research Institute of McGill University
Health Center, Montreal, Canada
e-mail: nancy.mayo@mcgill.ca; kedar.mate@mail.mcgill.ca

© The Author(s) 2022
K. Wac, S. Wulfovich (eds.), *Quantifying Quality of Life*, Health Informatics,
https://doi.org/10.1007/978-3-030-94212-0_5

Table 5.1 Components of
Mobility according to the ICF

Mobility component [ICF code]
Changing and maintaining body position
[d410] changing basic body position
[d415 maintaining a body position
[d420] transferring oneself
Carrying, moving and handling objects
[d430] lifting and carrying objects
[d435 moving objects with lower extremities
[d440] Fine hand use
[d445 hand and arm use
Walking and moving
[d450] walking
[d455] moving around
[d460] moving around in different locations
[d465] moving around using equipment

What is immediately obvious from the definition using the ICF is, that mobility is a necessary (but not sufficient) capacity required for many other activities such as basic activities of daily living, more complex activities required for maintain self and living space, work, and recreation and leisure including sports. These downstream activities that depend on some degree of mobility are themselves important contributor to QOL [3].

The ICF definition of mobility also includes two other very important qualifying constructs: capacity, what the person can do usually in a test situation; and performance, what the person actually does [4]. Capacity and performance constructs have important implications for measuring mobility. In the ICF, capacity refers to the person's ability to execute a task in a standard environment. This tends to refer to clinical testing. Whereas Webber et al. [1] also include capacity as a factor in their mobility model but refers to having the biological capability such as having sufficient joint mobility or strength to make mobility possible, areas that fall under the body structure and function component of the ICF.

In the ICF context, performance is poorly described. In another view, Loechte et al. [5] considered mobility in relationship to movement away from the home and other parameters related to what the person is doing. Action range is quantified as to how far a person moves outside their house. Distance is a parameter that can be measured independently from the home reference point such as distance covered per unit time such as per minutes (Six Minute Walk Test (6MWT) [6], per hour, per day. Other mobility parameters are duration (how long someone is mobile), pace (how many steps per minute, e.g. 100 steps per minute), and frequency of mobility events (how many 10 minute bouts of walking). There are important mobility parameters that are related to physical activity guidelines [7]. Mobility can also be characterized qualitatively, without measurement units such as time-of-day, alone or accompanied, and places. For example, places people move to more than 5 minutes can be captured and qualified by location (parks, malls, cafes, museums), activity (socializing, exercising), environment (noisy), or either usual or novel location. As an example, consider two different people.

Mark leaves the house at the same time every day, rides his bicycle 4 km to work taking the same route, and comes home. On the weekend he goes to the gym and does family-related errands and activities.

Eleanor, works from home and regularly exercises by walking, biking, or swimming at her community pool; she sometimes calls up a friend to go with her. These activities are done at different times during the day depending on her work schedule. Sometimes she works in cafes and makes a point of going to different ones. On the weekend, she likes to explore different parts of the city or take small trips out of the city. These are done alone or sometimes with a friend.

In each case, Mark and Eleanor would be classified with the same mobility capacity and performance indicators on the ICF, and they would have the same "action distance" values, but their mobility is realized through very different patterns. The richness of this mobility variety is not easy to capture without technology. Mark may accurately self-report because of the routine nature of his mobility but Eleanor would not be able to provide an average mobility rating for the past week or past month owing to her mobility variation.

While capacity ICF indicators are necessary in the context of a person with a health conditions, they are not sufficient for tracking mobility in healthy populations where performance in the real world is the relevant QOL indicator.

What Is QOL?

The Dictionary of Quality of Life and Health Outcomes Measurement [8] has this to say about QoL:

> QOL is a term often used erroneously to refer to health-related quality of life or health status, but is broader than just health and includes components of material comforts, health and personal safety, relationships, learning, creative expression, opportunity to help and encourage others, participation in public affairs, socializing, and leisure [9]. The World Health Organization (WHO) has defined quality of life as *"individuals' perception of their position in life in the context of the culture in which they live and in relation to their goals, expectations, standards and concerns"*. In the context of health research, quality of life goes beyond a description of health status, but rather is a reflection of the way that people perceive and react to their health status and to other, nonmedical aspects of their lives. According to Aristotle, quality of life would be *the best kind of life, the happiest life* [10].

Clearly measuring this happiest of lives is a challenge. QoL is not the same as health as health is only one of many QoL components [9]. Three approaches have been taken to measure QoL. *QoL profiles* are derived from measures that are made up of multi-item domains that produce domain scores and a total score. The WHOQOL-100 [11] (100 items; six domains: physical health, psychological, level of independence, social relationships, environment, spirituality, plus two single indicators of general health and global QoL) and WHOQOL-Bref [12] (26 items; four domains: physical health, psychological, social relationships, environment and the same two single indicators), are examples of profile QoL measures where the domains are physical health (including mobility), psychological health, social relationships and the environment. The CASP-19 is a quality-of-life measure comprising four domains, control, autonomy, pleasure and self-realization [13]. QoL has also been measured using *health indices* such as the EQ-5D from the EUROQOL Group [14] or the Health Utilities Index [15] but these measures do not cover many, if any, domains beyond health [16]. The advantage of indices is that they comprise multiple dimensions (usually with one item per each dimension) and a single score is derived by weights related to how much a typical citizen is willing to trade off years of life for these dimensions health. The third way in which QoL is measured is by a single item rated on an ordinal scale (e.g. Excellent, Very good, Good, Fair, Poor) or a visual analogue scale (from worst, 0, to best, 10).Only the person can assess their QoL using these methods and, other than the single-item method, only periodic assessments are possible [17].

Instead of relying on people to periodically sample and report on their QoL, Wac [18] proposes a new way of measuring QOL, using technologies. Quality of Life Technologies (QoLT) refers any technologies that can be used for assessment or improvement of the individual's QoL. QoLT will be the way of the future owing to the increasing availability of miniaturized computing, storage, and communication capacity that are now embedded within various personal devices and made available through smartphones and wearables.

How much can we infer about QoL by quantifying domains related to QoL? This chapter addresses this novel way of thinking about QoL. Here we address one of these quantifiable domains, namely individual's (body) mobility, a construct that lends itself well to being quantified by harnessing the power of these existing and emerging technologies.

Is Mobility Important for QoL?

Now that mobility has been described in terms of capacity and performance, and QoL has been defined as "the happiest kind of life", we discuss if mobility is important to QoL and, if so, what aspects of mobility are important. The answer to this question needs to consider who is being asked and how it is asked. For mostly healthy members of the general population, mobility, particularly walking, is the most important of five key HRQL items. A well known and widely used measure of HRQL is the EQ-5D from the EUROQOL [14] group. The weight, in terms of degree of detraction from perfect health, that members of the general population (from the United States) put on having no mobility as represented by being unable to walk about, is -0.558 (on a scale from 0 to 1; where '0' is the worst possible health state and '1' is perfect health). In contrast, the detracting effect of extreme problems in self-care (-0.471), usual activities (-0.374), pain (-0.537) and mood (-0.450). This translates to be willing to trade off nearly 6 years of life in order to live with no problems with walking about. Research on the effect of mobility limitations on risk of death shows that people, again from the general population of the United States, who walk very slowly have a risk of death 1.89 times higher (95% CI: 1.46–2.46), in comparison to the fastest walkers Liu [19].

In the EQ-5D classification system, the walking component of mobility is one of five HRQL areas clearly underlining its importance. One way of knowing if mobility is important to QoL is to identify how often and how comprehensively mobility is included in recognized QoL or health-related QoL measures. Mayo et al. [16] found that in generic QoL or HRQL measures, mobility was represented in 1/8 items in the HUI [20]; 10/36 items in the SF-36 (1/12 in SF-12) [21, 22]; 2/35 items or 1/8 dimensions of the AQOL [23], 1/26 items in the WHOQOL-BREF [12], and 9/71 items of the QWB [24].

Mobility is clearly important for people with health conditions. Almost every health condition can have an effect on mobility, permanent or transient. There are ICF core sets for some 30 health conditions and all except mental health conditions included one or more aspects of mobility.[2, 3] Mobility is more important to QoL once it is limited. People tend to take mobility for granted until the limitations set in, but when asked how they would imaging their life without mobility, they imagine it poorer than with other health challenges. This is one of the reasons why, when valuing health for the purposes of allocation of scarce resources or evaluation of cost-effectiveness of medical interventions, those with the health condition under consideration are not asked to provide a valuation from their point of view as this valuation is considered to be too influenced by their current health state [25–27].

The importance of mobility to QoL has been investigated extensively using many different methods, quantitative and qualitative, and in many different health conditions.

What Do People Say?

One of the best ways of answering the question about how mobility relates to QoL is to ask people directly. This is made possible through a synthesis of the qualitative literature or when people are asked open-ended questions.

A synthesis of 11 qualitative studies on QoL after hip fracture [28] identified mobility as a key contributor. Limited mobility affected this population's opportunities to make free choices about their activities and social interactions, impacted on independence, and was a threat to preservation of self-image. Thirteen qualitative studies from people with leg ulcers also confirmed the importance of mobility to QoL [29].

---

[2] https://www.merriam-webster.com/dictionary/mobility
[3] https://www.icf-research-branch.org/icf-core-sets

In a systematic review of 20 papers using an individualized approach to identifying areas important for QoL in people with cancer, Aburub et al. [30] found that in 11 or the 20 papers, mobility/physical activity was listed a one of the top 10 areas. In a review of the 10 most important areas of QoL across four health conditions, Mayo et al. [31] reported that for people with stroke, mobility was the number one area of importance; for people with Multiple Sclerosis mobility was number 6, and for people with cancer and HIV mobility was also selected as important to QoL.

What Do the Data Say about the QoL of the Body

The quantitative investigation of the importance of mobility to QoL is challenging because, as conceptualized using the ICF model [2], mobility limitations are caused by impairments of body structure and function and also act to limit other important activities and restrict participation in key personal, family, and societal roles that have a more direct influence on QoL. Figure 5.1 show these theoretical influences by combining the ICF model with the Wilson-Cleary model [32]. The ICF focuses on the observable manifestations of disability, while the Wilson-Cleary model goes beyond these to consider the effects on health perception and QOL, considered by the ICF model to reflect satisfaction with the observable manifestations and akin to well-being.

Having established that mobility is definitely important for QOL, other questions arise. Is mobility more, less, or as important as other health, social, and environmental domains? What are the best methods of partitioning out the role of mobility in quality of life? If we were to use mobility to quantify QOL, how much will we under or overestimate QOL and in whom and under what circumstances?

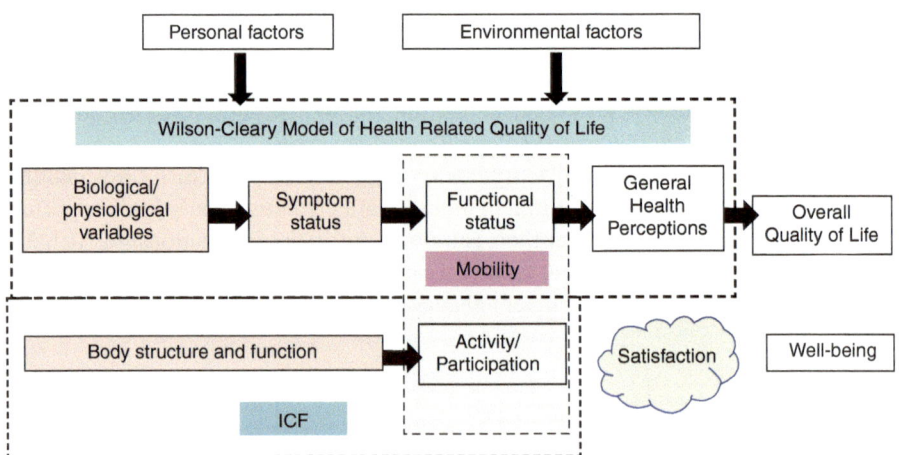

Fig. 5.1 Integration of the Wilson-Cleary Model and the ICF showing the place of mobility

In order to sort out the relative importance of mobility and other factors to QOL it is necessary to have a strong theoretical model linking both the capacity and performance aspects of mobility to QOL. One such model is the Wilson-Cleary model [32] that has been shown in Fig. 5.1. The Wilson-Cleary model shows the links between biological and physiological measures taken on the body, symptoms reported by the person (pain, fatigue, mood), function (what the person can do physically and mentally), health perception (how the person actually feels) and QOL and also shows that these links are affected by factors related to the person (age and sex but also beyond these to include lifestyle and preferences), and their environment.

This complexity requires that mobility be considered in a multi-factorial framework and Structural Equation Modeling (SEM) [33] is an ideal statistical method for carrying out a fair assessment of the impact of mobility. However, it is important when using SEM that the outcome is a QOL measure and not a composite measure that includes the constructs under investigation. For example, the WHOQOL-Bref includes a domain for physical health and this includes items about mobility. A selection of 18 relevant papers from a structured review are summarized in Table 5.2.

Table 5.2 Studies Using Structural Equation Modeling to Estimate the Importance of Mobility to QOL and Related Constructs

Study # / First Author	Country	Population	Number	N Model Variables	Mobility Rank
1. Li [34]	China	Older persons	4245	3	1
Outcome: Life satisfaction	**Other model variables**: Duration of disability, social engagement			**Mobility variable:** Difficulty with running, walking, climbing stairs, bending, reaching, and lifting objects	
2. Shahrbanian [35]	Canada	Multiple sclerosis	188	4	1
Outcome: Participation	**Other model variables**: Fatigue, pain, mood			**Mobility variable:** Physical function	
3. Alonso [36]	22 countries: 5 low and lower-middle income; 5 upper-middleincome; 11 high income.	General population	51,344	9	1
Outcome: Overall physical and mental health (0–100 VAS)	**Other model variables**: Chronic conditions (9 mental, 10 physical), domains of WHO-DAS (cognition, self-care, getting along, family burden, stigma, life activities, participation) along with age, sex, employment, country			**Mobility variable:** WHO disability assessment schedule 2.0 (WHODAS) mobility domain: Standing 30 minutes, rising from chair, moving around home, leaving home, walking long distance.	

(continued)

Table 5.2 (continued)

Study # / First Author	Country	Population	Number	N Model Variables	Mobility Rank
4. Lampinen [37]	Finland	Older adults	663	5	1
Outcome: Mental Well-being (revised Beck depression inventory)	**Other model variables**: Physical activity, leisure activity, age, chronic illness,			**Mobility variable:** Difficulty with climbing stairs and walking 2 km without stopping.	
5. Bentley [38]	USA	Older persons	677	2	PCS 2, MCS 1
Outcome: PCS, MCS	**Other model variables**: Functional status			**Mobility variable:** Life-space mobility	
6. Huang [39]	Taiwan	Osteoporosis	161	10	PCS 1, MCS 3
Outcome: PCS, MCS	**Other model variables**: Disease characteristics (duration, pain level, chronic diseases, fracture experiences, and ADL), social support dimensions, and PQoL and MQoL of osteoporosis patients with (age, marital status, school years, income, and exercise habits).			**Mobility variable:** ADL, exercise habits	
7. Lee [40]	Korea	Parkinson's	217	6	Tests: 6; motor signs: 1
Outcome: PDQL	**Other model variables**: Age, disease-related factors (motor signs, disease duration), quality of sleep, pain, and depression			**Mobility variable:** Grip strength, balance, functional reach motor signs: Speech, facial expression, tremor, rigidity, finger tapping, rapid alternating movements	
8. Tannenbaum [41]	Canada	Older persons	2311	5	2
Outcome: Latent SRH (SF-12, EQ-VAS)	Life style latent (perception health living, exercise, nutrition), health conditions/ polypharmacy, social health, mental health (SF-12 MCS)			**Mobility variable:** SF-12 (PCS)	
9. Kalpinski [42]	USA	Traumatic brain injury	312	6	2
Outcome: Life satisfaction, SRH	**Other model variables**: Injury severity, FIM cognition, FIM Independence, occupational activity, social engagement			**Mobility variable:** Ability to and frequency of freely moving around residence and community including using transportation	
10. Shim [43]	Korea	Rheumatic disease	360	5	2

Table 5.2 (continued)

Study # / First Author	Country	Population	Number	N Model Variables	Mobility Rank
Outcome: WHOQOL-BREF	**Other model variables:** Pain, pain catastrophising, depression, fear avoidance beliefs			**Mobility variable:** Health assessment questionnaire (HAQ): Degree of difficulty with dressing, rising, eating, walking, maintaining hygiene, reaching, gripping, and other common activities)	
11. Barclay [44]	Canada	Stroke	227	4	2
Outcome: SRH (EQ-VAS)	**Other model variables:** Indoor and outdoor mobility, depression			**Mobility variable:** Gait speed	
12. Bouchard [45]	Canada	Multiple sclerosis	189	9	2,4,5,6,9
Outcome: Illness intrusiveness	**Other model variables:** Fatigue, depression, pain, health perception			**Mobility variable:** Balance [2], 6MWT, Physical function, Power	
13. Perruccio [46]	Canada	Join replacement	449	3	3
Outcome: SRH	**Other model variables:** Two latents: Mental health (anxiety, depression); social health (participation, transportation)			**Mobility variable:** Physical health latent: Pain on activity, ADL, physically demanding activities, fatigue	
14. Aree-Ue [47]	Thailand	Osteoarthritis	200	4	3
Outcome: OA knee and hip quality of life (OAKHQOL)	**Other model variables:** Pain, fatigue, depression			**Mobility variable:** Timed-up-and-Go test	
15. Mayo [48]	Canada	Stroke	533	4	4
Outcome: SRH	**Other model variables:** Latent variables for: Biological variables, symptoms, function, health perception, personal and environmental factors following Wilson-Cleary model			**Mobility variable:** Physical function	
16. Soh [49]	Australia	Parkinson's	210	12	4,7
Outcome: PDQ-39	**Other model variables:** Age, co-morbidities, disease duration, disease severity, fall history, sex, social support, motor and non-motor impairment (UPDRS), self-care limitations (UPDRS-II)			**Mobility variable:** Motor impairments, (UPDRS), timed-up-and-Go test;	
17. Biederman [50]	Netherlands	Older persons	193	5	5
Outcome: CASP-19	**Other model variables:** Social functioning (partner, loneliness, social network, social support), depressive symptoms, self-efficacy, socioeconomic status			**Mobility variable:** Leg strength, aerobic endurance, dynamic balance	

(continued)

Table 5.2 (continued)

Study # / First Author	Country	Population	Number	N Model Variables	Mobility Rank
18. Mayo [51]	Canada	Older HIV men	707	18	7

Outcome:	**Other model variables**: Latent variables for:	**Mobility variable:**
Single QOL item	Biological variables, symptoms, function, health perception, personal and environmental factors following Wilson-Cleary model	Physical function

ADL Activities of Daily Living; *MCS* Mental Component Summary (SF-36 / RAND-36); *OA* Osteoarthritis; *PCS* Physical Component Summary (SF-36 / RAND-36); *PDQ(L)* Parkinson's Disease Quality of Life; Physical Function: Subscale of 10 self-report items from SF-36/RAND-36 (limitation in vigorous and moderate activities, stairs, walking short and long distances, bending, lifting, and self-care); *SRH* Self-rated Health; *UPDRS* United Parkinson's Disability Rating Scale; *WHO* World Health Organization

A feature of these papers is that the impact of mobility is evaluated in the context of other variables including personal factors, environmental factors, symptoms, and other activities. However, a limitation of these studies is there is no universally accepted measure of QoL and the outcomes modeled covered domains that are part of, but not, QoL such as physical health (often the PCS from the SF-36), general health, mental health, participation, illness intrusiveness, or generic or condition specific profile measures of HRQL. Two constructs closer to QoL were life-satisfaction and well being. Only two of the 18 studies measured of QoL, one with a QoL measure, CASP-19 (Study #16, and one with a single item (Study # 17 from Table 5.2). The studies are ordered according to the rank of mobility in explain QoL outcomes.

In these studies, mobility was measured in two ways, through self-reported limitations or difficulties with mobility related activities and through tests of physical capacity. For the latter, these measures included impairments affecting mobility such as strength, balance, and aerobic capacity and performance tests of mobility such as gait speed, Timed-up-and-Go, and 6MWT.

The first seven studies had at least one component of the mobility variable as the most important. Of these seven studies, four were from general population samples or older persons (Study # 1,3,4,5) and three were from clinical populations (Study # 2,6,7). Five studies had mobility as the second most important variable (Study # 8–12) and these were all of clinical populations.

The two studies (Study # 17,18) where mobility variables were not related to QoL were the same two studies that used actual QoL measures rather than measures of related constructs. These measures relate to QoL of the person. Study #17 used the CASP-19 which measures the extent to which the older person feels control, autonomy, pleasure and self-realization. Study #18 used a single QoL item in men living with HIV, a condition that does not affect mobility primarily. The outcomes in many of the other studies included aspects of the body's QoL for which mobility would have a stronger influence.

This new distinction between QoL of the person and *QoL of the body* is important because health care targets the body. Monitoring the body's QoL could be a valuable way for people to know about their body and to reduce the impact of illness and life's stresses on the body. Monitoring the body's QoL could also be an effective health care surveillance strategy particularly if it can be done unobtrusively and in real-time.

QoL of the person is best reflected through global QoL measures including those of life satisfaction, whereas QoL of the body is reflected in outcomes related to in health, participation, and illness' intrusiveness, by, for example impairing mobility. The importance of this distinction could be profound, as the person, with assistance for a growing portfolio of technologies, is better placed to monitor the QoL of their body in real time and react or adjust accordingly. Also, it is recognized that QoL of the person has a profound effect on the body through stress reaction or, in general, behaviors aimed at managing negative emotions [52], that the person could be made aware of through monitoring QoL of the body.

Using Mobility to Quantify QoL

If we were to use mobility to quantify QoL, how much will we under or overestimate QoL and in whom and under what circumstances? While the WHO provides a definition of QoL, in 1978, Flanagan [9] identified 15 components of QoL which are listed in Fig. 5.2.

Based on the literature [36, 53, 54] and the clinical and research experience of the authors, an estimated 1/3 of QoL of the person would be explained by mobility, with helping others and active recreation most affected and learning, understanding self, and creativity the least affected (see Fig. 5.3).

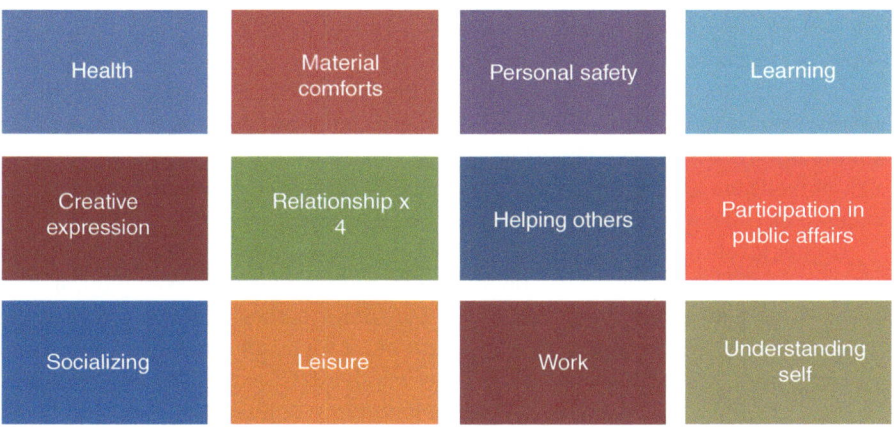

Fig. 5.2 Flanagan's (1978) Components of QOL

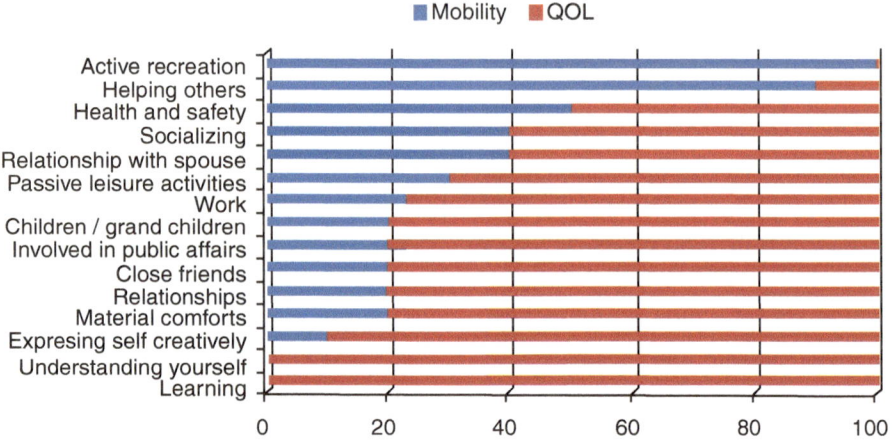

Fig. 5.3 Amount of QOL explained by mobility

What Mobility Parameters Can Be Monitored Via Technologies?

Table 5.2 provides a list of mobility parameters that people have linked to QoL. Starting at the most global is life-space mobility [55], which relates to how far from home a person moves and has been shown to be associated with social support, capacity to drive, and gait speed [56]. One can tell a lot about a person through this one measure, which can be monitored through Global Positioning System (GPS) technology (or some combination of Cell-IDs [57]) that comes with most smart phones or smart wearables. Second on the list, and also very telling about the body's QoL is the amount of vigorous activities carried out which can be monitored using heart rate and physical accelerations. Increased heart rate without movement would indicate stress (physical or emotional), fear, or cardiac pathology. Amount of time spent in activity would be another very important indicator of the body's QoL and this is easily monitored using simple wearable devices that monitor accelerations. A common metric is sedentary time [58], which is known to be detrimental to health. Wearable accelerometers also provide information on speed of movement such as step cadence and duration of activity bouts. Canadian Physical Activity Guideline [7] for adults advocate a minimum of 150 minutes of moderate activity accumulated over a week in bouts of 10 minutes; walking at a cadence of 100 steps a minute would meet this recommendation [59].

Higher levels of physical activity such as is achieved through climbing stairs can also be tracked on many smart phone/watch applications. For example a 4 hour game of golf on a hilly golf course that the senior author (NM) regularly walks accumulates 20,000 steps and the equivalent of 38 flights of stairs.

Gait speed while considered the sixth vital sign [60] is actually less important than cadence and more difficult to track as it requires a measure of distance not just stepping frequency. Gait speed is easier to measure clinically where a fixed course can be walked and timed, raising its importance as a clinical indicator because of ease of measurement. The importance of gait speed for safety cannot be disputed as people who cannot walk a speeds of greater than 0.5 m/sec (i.e., in terms of their capacity) are at risk of falls and benchmarks of >0.7 m/sec and > 1.0 m/sec indicate safety risk when crossing 2- and 4-lane streets [61]. However, the frequency at which people walk at different cadence bands a more relevant indicator of how the body is doing in everyday life. Tudor-Locke [59] showed that despite having capacity to walk at a health promoting pace when tested clinically, it is rare for the North American senior to do this in the real world for more than a few minutes a day. Mate et al. [62] showed the same was true for people with Multiple Sclerosis in that they do not reproduce what they are capable of doing on a clinical test in their real world environment, except of course if they lack capacity.

Another clinical test that shows importance for QoL is the Timed-Up-and-Go (TUG) test [63]. This test is indicative of mobility as it requires standing up from a chair, walking 3 meters, turning around, and return to sit back down on the chair. Again, this test is easily done in the clinic, as it requires only a standard chair and 3 meters of walking space. In the real world, the number of transitions from sitting to standing can be captured, as it is another metric available on standard accelerometers.

These mobility measures relate to activity but it is also possible to track motor impairments that lead to mobility limitations such as slowness of movements that can result from stiff joints, resting and intention tremors, poor posture, balance, and poor gait quality. Monitoring these impairments would require different technologies distributed to different parts of the body, but all are possible. There are apps for tremor and balance that require the person hold a smart phone. Posture, balance and gait quality can be measured using inertial devices attached to the back [64] or to the shoe [65–68]. A selection of wearable devices for mobility and health are shown below (Table 5.3).

Table 5.3 Examples of wearable technologies that can monitor the QoL of the body

	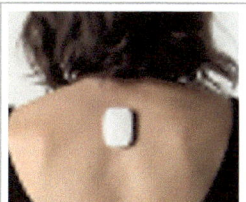
Garmin activity tracker	UPRIGHT Posture trainer

Wearable oximeter (AARC)	Gait analyser
	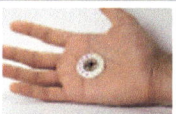
Heel2Toe™ PhysioBiometrics Inc.	Sweat monitor

How do these measurements relate to QoL of the body? Research show that slowness of movement, poor and irregular gait, poor posture, lack of activity can indicate states of pain, fatigue, low mood, apathy, or anxiety [69–73].

Mapping the Future

We are at the cusp of changing the way we think about monitoring and remediating [74] and technology is poised to empower people to take charge of their own QoL, including, or starting from, their body's QoL. Given the emergence of today's and tomorrow's personalized and miniaturized technologies, the future of monitoring the QoL of the body is inevitable [18]. We envision that the individuals will wear an accurate, well designed smartwatch that monitors activity, heart rate, oxygen saturation, and tremor, amongst other variables. For people with specific health challenges, their posture, stability, and gait quality would be also monitored by small unobtrusive sticky devices placed on their spine, shoulder, ankle or shoe. They also would receive updates on how they are doing, as their devices are continually connected to their smartphone. They will be able to see how their body reacts to internal and external stimulae and learn how to respond to signs of threat [18]. They will also be able to program these devices to provide engaging and effective feedback for optimal performance, literally stamping in good mobility habits, and influencing positively their QoL in the long term.

References

1. Webber SC, Porter MM, Menec VH. Mobility in older adults: a comprehensive framework. Gerontologist. 2010 Aug;50(4):443–50.
2. WHO. World Health Organization. International classification of functioning, disability and health. 2nd ed. Geneva; 2001.
3. Mayo NE, Wood-Dauphinee S, Cote R, Durcan L, Carlton J. Activity, participation, and quality of life six months post-stroke. Arch Phys Med Rehab. 2002;83:1035–42.
4. Mate KK, Kuspinar A, Ahmed S, Mayo NE. Comparison between common performance-based tests and self-reports of physical function in people with multiple sclerosis: does sex or gender matter? Arch Phys Med Rehabil. 2019 May;100(5):865–73.
5. Loechte N, Von BT, Haux R. Mobility parameters in health applications–a literature review. Stud Health Technol Inform. 2014;205:523–7.
6. Crapo RO, Casaburi R, Coates AL, Enright PL, MacIntyre NR, Mckay RT, et al. ATS statement: guidelines for the six-minute walk test. Am J Respir Crit Care Med. 2002 Jul 1;166(1):111–7.
7. Canadian Physical Activity Guideline for Older Adults. http://www.csep.ca/CMFiles/Guidelines/CSEP_PAGuidelines_older-adults_en.pdf. 2015.
8. Mayo NE. Dictionary of quality of life and health outcomes measurement. 1st ed. International Society for Quality of Life; 2015.

9. Flanagan JC. A research approach to improving our quality of life. Am Psychol. 1978;33:138–47.
10. McKeon R. Aristotle. Nicomachean Ethics. Introduction to Aristotle. New York: Modern Library; 1947.
11. The WHOQOL Group. The World Health Organization quality of life assessment (WHOQOL): position paper from the World Health Organization. Soc Sci Med. 1995 Nov;41(10):1403–9.
12. Development of the World Health Organization WHOQOL-BREF quality of life assessment. The WHOQOL group. Psychol Med. 1998 May;28(3):551–8.
13. Sim J, Bartlam B, Bernard M. The CASP-19 as a measure of quality of life in old age: evaluation of its use in a retirement community. Qual Life Res. 2011 Sep;20(7):997–1004.
14. EuroQol Group. EQ-5D. http://www.euroqol.org/. 2016.
15. Feeny D, Torrance G, Furlong W. Health utilities index. In: Spilker B, editor. Quality of life and Pharmacoeconmics in clinical trials. 2nd ed. Philadelphia: Lippincott-Raven; 1996. p. 239–52.
16. Mayo NE, Moriello C, Asano M, van der Spuy S, Finch L. The extent to which common health-related quality of life indices capture constructs beyond symptoms and function. Qual Life Res. 2011 Jun;20(5):621–7.
17. Wac K. Quality of life technologies. In: Gellman M, editor. Encyclopedia of behavioral medicine. New York: Springer; 2020.
18. Wac K. From quantified self to quality of life. Digital health. Springer; 2008. p. 83–108. https://link.springer.com/chapter/10.1007/978-3-319-61446-5_7
19. Liu B, Hu X, Zhang Q, Fan Y, Li J, Zou R, et al. Usual walking speed and all-cause mortality risk in older people: a systematic review and meta-analysis. Gait Posture. 2016 Feb;44:172–7.
20. Feeny D, Furlong W, Boyle M, Torrance GW. Multi-attribute health status classification systems–health utilities index. Pharmaco Economics. 1995;7(6):490–502.
21. Ware J Jr, Kosinski M, Keller SD. A 12-item short-form health survey: construction of scales and preliminary tests of reliability and validity. Med Care. 1996;34(3):220–33.
22. Ware JE Jr, Sherbourne CD. The MOS 36-item short-form health survey (SF-36): I. conceptual framework and item selection. Med Care. 1992;30(6):473–81.
23. Richardson J, Iezzi A, Khan MA, Maxwell A. Validity and reliability of the assessment of quality of life (AQoL)-8D multi-attribute utility instrument. Patient. 2014;7(1):85–96.
24. Kaplan RM. The quality of Well-being index. Quality of life and Pharmacoeconomics in clinical trials. 2nd ed. Philadelphia: Lippincott-Raven; 1996.
25. McPherson K, Myers J, Taylor WJ, McNaughton HK, Weatherall M. Self-valuation and societal valuations of health state differ with disease severity in chronic and disabling conditions. Med Care. 2004 Nov;42(11):1143–51.
26. Insinga RP, Fryback DG. Understanding differences between self-ratings and population ratings for health in the EuroQOL. Qual Life Res. 2003 Sep;12(6):611–9.
27. Brazier J, Akehurst R, Brennan A, Dolan P, Claxton K, McCabe C, et al. Should patients have a greater role in valuing health states? Appl Health Econ Health Policy. 2005;4(4):201–8.
28. Ehlers MM, Nielsen CV, Bjerrum MB. Experiences of older adults after hip fracture: an integrative review. Rehabil Nurs. 2018 Sep;43(5):255–66.
29. Phillips P, Lumley E, Duncan R, Aber A, Woods HB, Jones GL, et al. A systematic review of qualitative research into people's experiences of living with venous leg ulcers. J Adv Nurs. 2018 Mar;74(3):550–63.
30. Aburub AS, Mayo NE. A review of the application, feasibility, and the psychometric properties of the individualized measures in cancer. Qual Life Res. 2017 May;26(5):1091–104.
31. Mayo NE, Aburub A, Brouillette MJ, Kuspinar A, Moriello C, Rodriguez AM, et al. In support of an individualized approach to assessing quality of life: comparison between patient generated index and standardized measures across four health conditions. Qual Life Res. 2017 Mar;26(3):601–9.

32. Wilson IB, Cleary PD. Linking clinical variables with health-related quality of life. A conceptual model of patient outcomes. J Am Med Assoc. 1995 Jan 4;273(1):59–65.
33. Kline RB. Principles and practice of structural equation modeling. 1st ed. Guilford Press; 1998.
34. Li L, Loo BP. Mobility impairment, social engagement, and life satisfaction among the older population in China: a structural equation modeling analysis. Qual Life Res. 2017 May;26(5):1273–82.
35. Shahrbanian S, Duquette P, Ahmed S, Mayo NE. Pain acts through fatigue to affect participation in individuals with multiple sclerosis. Qual Life Res. 2016 Feb;25(2):477–91.
36. Alonso J, Vilagut G, Adroher ND, Chatterji S, He Y, Andrade LH, et al. Disability mediates the impact of common conditions on perceived health. PLoS One. 2013;8(6):e65858.
37. Lampinen P, Heikkinen RL, Kauppinen M, Heikkinen E. Activity as a predictor of mental Well-being among older adults. Aging Ment Health. 2006 Sep;10(5):454–66.
38. Bentley JP, Brown CJ, McGwin G Jr, Sawyer P, Allman RM, Roth DL. Functional status, life-space mobility, and quality of life: a longitudinal mediation analysis. Qual Life Res. 2013 Sep;22(7):1621–32.
39. Huang CY, Liao LC, Tong KM, Lai HL, Chen WK, Chen CI, et al. Mediating effects on health-related quality of life in adults with osteoporosis: a structural equation modeling. Osteoporos Int. 2015 Mar;26(3):875–83.
40. Lee J, Choi M, Jung D, Sohn YH, Hong J. A structural model of health-related quality of life in Parkinson's disease patients. West J Nurs Res. 2015 Aug;37(8):1062–80.
41. Tannenbaum C, Ahmed S, Mayo N. What drives older women's perceptions of health-related quality of life? Qual Life Res. 2007 May;16(4):593–605.
42. Kalpinski RJ, Williamson ML, Elliott TR, Berry JW, Underhill AT, Fine PR. Modeling the prospective relationships of impairment, injury severity, and participation to quality of life following traumatic brain injury. Biomed Res Int. 2013;2013:102570.
43. Shim EJ, Hahm BJ, Go DJ, Lee KM, Noh HL, Park SH, et al. Modeling quality of life in patients with rheumatic diseases: the role of pain catastrophizing, fear-avoidance beliefs, physical disability, and depression. Disabil Rehabil. 2018 Jun;40(13):1509–16.
44. Barclay R, Ripat J, Mayo N. Factors describing community ambulation after stroke: a mixed-methods study. Clin Rehabil. 2015 Aug;29(29):509–21.
45. Bouchard V, Duquette P, Mayo NE. Path to illness intrusiveness: what symptoms impact the life of people living with multiple sclerosis? Arch Phys Med Rehabil. 2017 Apr 15;98(7):1357–65.
46. Perruccio AV, Davis AM, Hogg-Johnson S, Badley EM. Importance of self-rated health and mental Well-being in predicting health outcomes following total joint replacement surgery for osteoarthritis. Arthritis Care Res (Hoboken). 2011 Jul;63(7):973–81.
47. Aree-Ue S, Kongsombun U, Roopsawang I, Youngcharoen P. Path model of factors influencing health-related quality of life among older people with knee osteoarthritis. Nurs Health Sci. 2019 Sep;21(3):345–51.
48. Mayo NE, Scott SC, Bayley M, Cheung A, Garland J, Jutai J, et al. Modeling health-related quality of life in people recovering from stroke. Qual Life Res. 2015 Jan;1:41–53.
49. Soh SE, McGinley JL, Watts JJ, Iansek R, Murphy AT, Menz HB, et al. Determinants of health-related quality of life in people with Parkinson's disease: a path analysis. Qual Life Res. 2013 Sep;22(7):1543–53.
50. Bielderman A, de Greef MH, Krijnen WP, van der Schans CP. Relationship between socio-economic status and quality of life in older adults: a path analysis. Qual Life Res. 2015 Jul;24(7):1697–705.
51. Mayo NE, Brouillette MJ, Scott SC, Harris M, Smaill F, Smith G, et al. Relationships between cognition, function, and quality of life among HIV+ Canadian men. Qual Life Res. 2019 Sep 9;
52. Nanni MG, Caruso R, Sabato S, Grassi L. Demoralization and embitterment. Psychol Trauma. 2018 Jan;10(1):14–21.
53. Collins JJ, Baase CM, Sharda CE, Ozminkowski RJ, Nicholson S, Billotti GM, et al. The assessment of chronic health conditions on work performance, absence, and total economic impact for employers. J Occup Environ Med. 2005 Jun;47(6):547–57.

54. Diehr PH, Thielke SM, Newman AB, Hirsch C, Tracy R. Decline in health for older adults: five-year change in 13 key measures of standardized health. J Gerontol A Biol Sci Med Sci. 2013 Sep;68(9):1059–67.
55. Phillips J, Dal GE, Ritchie C, Abernethy AP, Currow DC. A population-based cross-sectional study that defined normative population data for the life-space mobility assessment-composite score. J Pain Symptom Manag. 2015 May;49(5):885–93.
56. Kuspinar A, Verschoor CP, Beauchamp MK, Dushoff J, Ma J, Amster E, et al. Modifiable factors related to life-space mobility in community-dwelling older adults: results from the Canadian longitudinal study on aging. BMC Geriatr. 2020 Jan 31;20(1):35.
57. Fanourakis M, Wac K. ReNLocAn anchor-free localizaiton algorithm for indirect ranging. 6th International Symposium on A World of Wireless, Mobile and Multimedia Networks (WoWMoM); 2015 Jun 15; Institute of Electrical and Electronics Engineers (IEEE}; 2015 p. 1–19.
58. Benatti FB, Ried-Larsen M. The effects of breaking up prolonged sitting time: a review of experimental studies. Med Sci Sports Exerc. 2015 Oct;47(10):2053–61.
59. Tudor-Locke C, Camhi SM, Leonardi C, Johnson WD, Katzmarzyk PT, Earnest CP, et al. Patterns of adult stepping cadence in the 2005-2006 NHANES. Prev Med. 2011 Sep;53(3):178–81.
60. Fritz S, Lusardi M. White paper: "walking speed: the sixth vital sign". J Geriatr Phys Ther. 2009;32(2):46–9.
61. Hornyak V, Van Swearingen JM, Brach JS. Measurement of gait speed. Topics in Geriatric Rehabilitation. 2012;28(1):27–32.
62. Mate K, Mayo NE. Clinically assessed walking capacity does not always reflect real world walking performance: an example from people with multiple sclerosis. International Journal of MS Care. 2020;22(3):147–54.
63. Podsiadlo D, Richardson S. The timed "up & Go": a test of basic functional mobility for frail elderly persons. J Am Geriatr Soc. 1991;39:142–8.
64. Collett J, Esser P, Khalil H, Busse M, Quinn L, DeBono K, et al. Insights into gait disorders: walking variability using phase plot analysis, Huntington's disease. Gait Posture. 2014 Sep;40(4):694–700.
65. Vadnerkar A, Figueiredo S, Mayo NE, Kearney RE. Design and validation of a biofeedback device to improve heel-to-toe gait in seniors. IEEE J Biomed Health Inform. 2018 Jan;22(1):140–6.
66. Mate KK, Abou-Sharkh A, Morais JA, Mayo NE. Real-time auditory feedback-induced adaptation to walking among seniors using the Heel2Toe sensor: proof-of-concept study. JMIR Rehabil Assist Technol. 2019 Dec 11;6(2):e13889.
67. Mate KKV, Abou-Sharkh A, Morais JA, Mayo NE. Putting the best foot forward: relationships between indicators of step quality and cadence in three gait vulnerable populations. NeuroRehabilitation. 2019;44(2):295–301.
68. Carvalho LP, Mate KKV, Cinar E, Abou-Sharkh A, Lafontaine A, Mayo NE. A new approach toward gait training in patients with Parkinson's disease. Gait Posture. 2020;81(9):14–20.
69. Mansoubi M, Weedon BD, Esser P, Mayo N, Fazel M, Wade W, et al. Cognitive performance, quality and quantity of movement reflect psychological symptoms in adolescents. J Sports Sci Med. 2020 Jun;19(2):364–73.
70. Esser P, Collett J, Maynard K, Steins D, Hillier A, Buckingham J, et al. Single sensor gait analysis to detect diabetic peripheral neuropathy: a proof of principle study. Diabetes Metab J. 2018 Feb;42(1):82–6.
71. Harvey RH, Peper E, Mason L, Joy M. Effect of posture feedback training on health. Appl Psychophysiol Biofeedback. 2020 Jun;45(2):59–65.
72. Serrano-Checa R, Hita-Contreras F, Jimenez-Garcia JD, Achalandabaso-Ochoa A, Aibar-Almazan A, Martinez-Amat A. Sleep quality, anxiety, and depression are associated with fall risk factors in older women. Int J Environ Res Public Health. 2020 Jun;5:17(11).

73. Michie S, van Stralen MM, West R. The behaviour change wheel: a new method for characterising and designing behaviour change interventions. Implement Sci. 2011 Apr;23(6):42.
74. Holloway C, Dawes H. Disrupting the world of disability: the next generation of assistive technologies and rehabilitation practices. Healthc Technol Lett. 2016 Dec;3(4):254–6.

Chapter 6
iSenseYourPain: Ubiquitous Chronic Pain Evaluation through Behavior-Change Analysis

Matteo Ciman

Introduction

In 1979, the International Association for the Study of Pain defined pain as "an unpleasant sensory and emotional experience associated with actual or potential tissue damage or described in terms of such damage" [1]. Pain is a subjective experience corresponding to an unpleasant situation that may be both physical and psychological. Acute and chronic pain are considered two distinct medical conditions. Acute pain is provoked by a specific injury or disease, generally lasts no longer than 6 months, and goes away when the underlying cause is gone [2]. Chronic pain, on the other hand, is a long-term condition. When associated with a specific injury or disease, it is considered a disease itself because it outlasts the normal healing time. It may even arise from a psychological state with no biological cause and with no recognizable endpoint [2]. In patients with chronic pain, overall quality of life is diminished [3], and for some patients, this pain can persist for one's entire life. Aspects of quality of life that are usually influenced by chronic pain include, but are not limited to, sleep, cognitive and brain function, mood, mental health, and even sexual function [3–6]. Moreover, the amount one's quality of life decreases is strongly correlated with the severity of chronic pain experienced [3], so that higher levels of pain lead to more significant reductions in quality of life.

The impact of chronic pain on quality of life calls for a reliable and continuous approach to pain monitoring and evaluation in order to provide the best possible support to patients. The aim of such an approach should be to assess patients' pain throughout their lives in a timely and accurate manner without the use of self-reporting and to help patients cope with their situations so that they can avoid a considerable decrease in their quality of life. For this reason, this chapter presents *iSenseYourPain*,

M. Ciman (✉)
Geneva School of Economics and Management, Center for Informatics, Quality of Life Technologies Lab, University of Geneva, Geneva, Switzerland

Department of Mathematics, University of Padua, Padua, Italy

© The Author(s) 2022   137
K. Wac, S. Wulfovich (eds.), *Quantifying Quality of Life*, Health Informatics,
https://doi.org/10.1007/978-3-030-94212-0_6

a system design that aims to gather continuous data about patients' behavior in order to understand how their pain experiences influence their lives. For example, the system seeks to determine if changes in sleep occur by measuring the number of hours and times of day when one sleeps, along with data such as the number of steps taken each day and the amount of time spent outside. The data is collected using ubiquitous devices and personal sensors integrated into patients' everyday lives. This way, patients do not need to carry intrusive, special medical devices to collect the information. By facilitating real-time assessment, the *iSenseYourPain* system is primarily designed to determine when increases in a patient's pain occur while measuring daily life activities and recognizing patterns of activity and pain experiences. In addition, the system aims to provide feedback to help patients manage pain exacerbations when needed with the help of their physician or relatives.

This chapter is structured as follows. Section 2 discussed the pain and its assessment via self-report methods, while Sect. 3, its assessment via devices. Section 4 presents the *iSenseYourPain* design n choices and Sect. 5 concludes the chapter.

Self-Reported Pain and Chronic Pain Assessment

Chronic pain requires frequent patient follow-up focusing on how pain is triggered [3], how it fluctuates with different behavioral patterns, patients' medication regimens and daily life contexts, and the ways pain influences overall health and life quality. Currently, the evaluation of chronic pain is generally based on the use of paper questionnaires and scales for which patients provide self-reports at predefined time intervals responding to a set of questions concerning their symptoms in the previous days or months. The most commonly used scales are described in what follows.

Boonstra et al. [7] have developed a scale model to evaluate pain based on the **Numeric Rating Scale**. A common challenge with this type of scale, however, is defining the different cut-off points for the mild, moderate, and severe pain categories into which the scale is divided. The **Visual Analog Scale** (VAS) is another common method for quantifying the severity of pain. It is a continuous outcome measure consisting of a scale 100 mm in length ranging from 0 to 100 with low- and high-end points corresponding to no pain and the worst pain. The VAS is easy to administer and has been validated for both adults and older children. It has also proven to be a reliable and valid technique to measure acute pain in emergency departments [8]. Meanwhile, the **Brief Pain Inventory** is commonly used to measure a patient's pain intensity and how much this pain influences their ability to live their everyday life. It consists of two different categories, namely pain intensity and pain interference [9]. The **Medical Outcomes Study Pain Measures** is another questionnaire that evaluates pain according to intensity, frequency, duration, and its impact on behavior and mood [10]. The *Oswestry Low Back Pain Disability Questionnaire* and the **Back Pain Classification Scale** are tools used by researchers and disability evaluators to evaluate low back functional disability or psychological disturbance [11, 12]. The **Pain and Distress Scale**, another frequently used tool, is a measurement of mood and behavior that may be associated with acute pain. It does not describe the severity of patients' pain itself but rather the physical and emotional reactions that can be attributed to limitations caused by pain in daily activities, including increased anxiety, depression, and decreased alertness [13]. In pediatric populations, **facial expression drawings** or "faces scales" are a popular method of assessing pain severity. A variety of faces scales exist, each of which uses a series of facial expressions to illustrate a spectrum of pain intensity. Faces scales are ordinal outcome measures consisting of a limited number of categorical responses ordered in a specific pattern. Although the optimum design of the facial expressions is frequently debated, the literature suggests that face-based rating scales are the preferred method of pain reporting among children. The **Wong-Baker FACES Scale** in particular has been implemented in multiple pediatric settings for pain assessment [14]. The *McGill Pain Questionnaire (MPQ)* is used to evaluate and monitor the pain over time, even to determine the effectiveness of any intervention [15]. The *Pain Perception Profile* (PPP) uses four different points of view to describe in the pain experience of each patient [16].

Finally, the **Chronic Pain Grade Questionnaire** is a multidimensional measure that assesses two dimensions of overall chronic pain severity: pain intensity and pain-related disability [17].

Table 6.1 provides an overview of the different pain scales, including the items to be filled out by the patient in each scale, the recall period, and the number of

Table 6.1 Recap of different scales used to measure pain in patients

Scale name	Number of items	Type of pain Recall period	Number of output levels
The visual analog scale (VAS) [8]	1	Acute and chronic pain Now	10
The Brief Pain Inventory [9]	11	Acute and chronic pain Now, recent days and past weeks	11
The Medical Outcomes Study Pain Measures [10]	12	Chronic pain Past four weeks	12
The Oswestry Low Back Pain Disability Questionnaire [11]	10	Chronic pain	6
The Back Pain Classification Scale [12]	103	Acute and chronic pain	Checklist – 1
The Pain and Distress Scale [13]	20	Acute pain Recent days	4
The McGill pain questionnaire (MPQ) [15]	20	Chronic pain Now	1
Pain perception profile (PPP) [16]	37	Acute pain Now	4
Chronic Pain Grade Questionnaire [17]	7	Chronic pain	10

output levels provided by the scale based on the type of pain (acute or chronic) being assessed.

One of the aims of the assessment system that is proposed in this chapter is to realize a shift from a paper-based self-reporting approach to an automatic and ubiquitous one. Self-reporting requires time and cognitive effort to carry out and cannot be performed frequently, especially to the degree required for constant monitoring. The latter approach is based on the use of various ubiquitous devices that can constantly assess pain-related behavior changes. The *iSenseYourPain* system, which is presented in Sect. 4, is pervasive, ubiquitous, and non-invasive, requiring no input from the patient beyond an initial calibration phase and no equipment besides the technologies and devices already used in one's everyday-life, thus increasing patient acceptance. Before the presentation of the system, however, Section 3 provides a brief overview of how smart devices have been used for pain detection until now in order to contextualize the development of *iSenseYourPain.*

Pain Detection Using Smart Devices

In the last decade, there has been an exponential increase in the number of mobile and ubiquitous devices in use, while a considerable number of self-monitoring applications have been introduced that aim to assess and improve individuals' overall quality of life. Several studies have pointed out that, for these solutions to be effective, they should be easy to use, customizable, and adaptable to the routine and lifestyle of each person, including their location, social interactions, and healthcare needs, while providing timely and personalized suggestions [18].

Given that chronic pain can have a severely detrimental impact on quality of life, several systems and platforms have been developed to provide support to patients suffering from it. One such platform is that of online peer-support forums. Several studies have been conducted that analyze their interactive use among patients with similar symptoms and diseases to promote the exchange of information and personal experience, provide distraction, and facilitate social or peer support. Meta-analysis of several trials in which online forums were leveraged in patients' care showed that patients who used such forums experienced a significant reduction in pain and anxiety, loneliness, and withdrawn behavior, as well as a greater willingness to return for treatment [19–21].

Besides online forums, the use of personal device systems is becoming increasingly common in the context of chronic pain. Kristjánsdóttir et al. [22] have developed a smartphone intervention system for women with widespread chronic pain. The intervention involves one face-to-face session between the patient and a nurse and four weeks of written communication between the patient and their therapist through the device. In Kristjánsdóttir et al.'s study of the system, participants filled daily smartphone diary entries to support their awareness and reflect on pain-related thoughts, feelings, and activities. The registered diaries were made available to a therapist who provided personalized written feedback to the patient based on cognitive-behavioral therapy principles. The results suggest that a smartphone-delivered intervention with diaries and personalized feedback can reduce "catastrophizing" and prevent increases in functional impairment and symptom levels in women with widespread chronic pain following inpatient rehabilitation.

Meanwhile, multiple applications have been developed that focus on the management and assessment of chronic pain. *Painometer* [23] is an app that helps users assess pain intensity, including four different pain scales (including a faces scale) that can be used by patients to report pain experiences to their physician. Its main purpose is to encourage patients to report their pain and make the act of reporting more acceptable by offering a simpler and more accurate means of communication between patient and physician (see Fig. 6.1). The *iCanCope with Pain* [24] program is an integrated web and smartphone application for children and adolescents suffering from chronic pain. The goal of the application is to address the self-management needs of adolescents with chronic pain by improving access to disease information and symptom-management strategies while providing functionality for the self-monitoring of symptoms, the setting of personalized goals, pain coping skills training, chronic pain education, and peer-based social support.

Other studies in pain management have investigated the implementation of less ubiquitous or pervasive methods such as the use of external sensors such as electrocardiogram (ECG) or electroencephalogram (EEG) (as shown in Fig. 6.2). These devices have been used to collect data for biological values such as levels of oxygen saturation in the blood (SPO$_2$), body temperature, heart rate (HR), heart rate

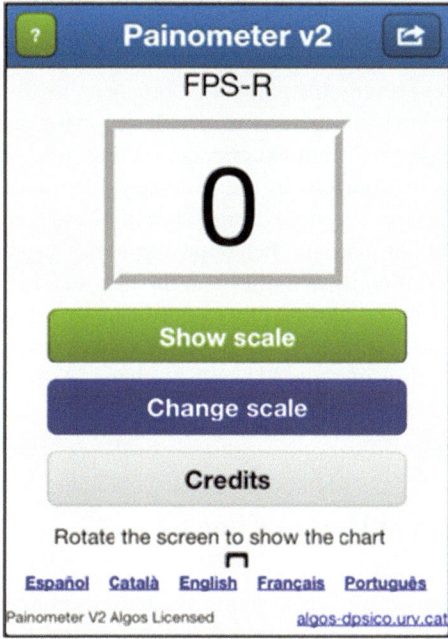
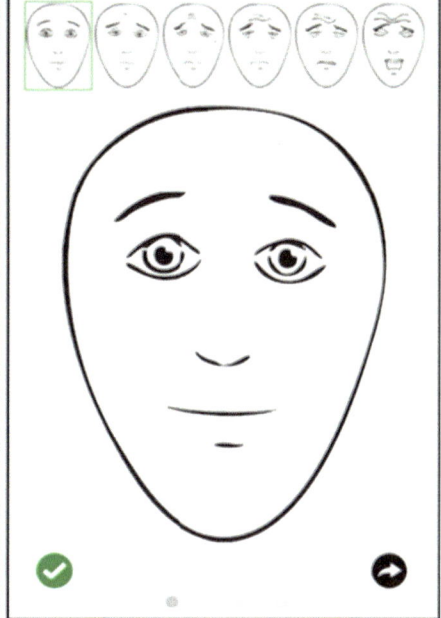

Fig. 6.1 Painometer App

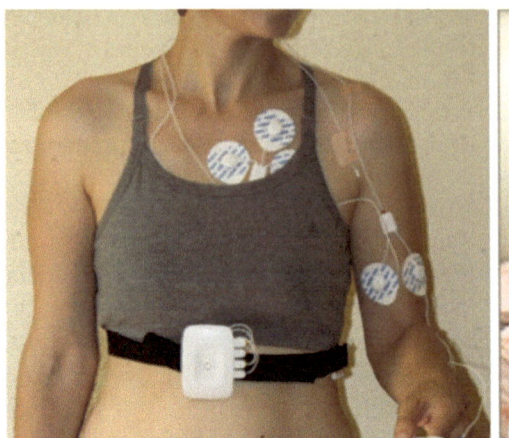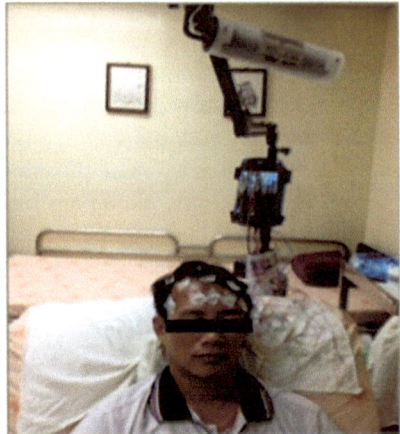

Fig. 6.2 Example of external sensor-based systems for pain assessment [25, 26]

variability (HRV), and galvanic skin response (GSR) [26, 27], as well as to perform simple electrocardiograms (ECGs) [25, 28] to detect and predict migraines in patients. Despite their effectiveness, it is clear that such approaches are far from easy to adopt for routine evaluation and monitoring of individuals' behavior metrics. Such external sensors are generally more invasive than personal wearable sensors, which in most cases are capable of performing the same measurements. For example, the Holter ECG monitor that is used to measure HR (shown in Fig. 6.2) is much more intrusive than a simple smartwatch, despite the fact that smartwatches are capable of monitoring HR continually with nearly the same accuracy.

As this section has demonstrated, current research and medical practices involving the use of smart devices for pain detection are either based on self-reporting and thus do not assess real-time changes in patients' pain experience, or they rely on external devices that are likely to affect or interfere in the everyday life of the patient. For these reasons, this chapter proposes a system designed to unobtrusively collect accurate and timely information about patients' behavior and identify correlations between changes in patients' everyday life activities and their experiences of pain. In contrast to assessments that are based on self-reporting, the system also aims to provide immediate evaluations and, eventually, to support patients before chronic pain dramatically impacts the quality of their life and the lives of those around them.

iSenseYourPain: System Criteria and Design Choices

In this section, we outline the proposed *iSenseYourPain* system's main requirements and components, beginning with a discussion of previous research concerning smart devices that informs the system's overall approach. As indicated earlier in the chapter, the system's main purpose is to use everyday devices to collect different types of data, analyze the data, and determine if a patient is experiencing a higher level of pain than normal and in which circumstances.

The design of the system and its components is based on the results of previous research that highlighted how behavior, and the interactions between a user and several daily-life smart objects, are influenced by emotional states and moods [29, 30]. Pain is not an emotion or feeling but a physical and/or mental state that may alter an individual's usual behavior. Such changes in behavior can be measured using several common devices, such as smartphones and other personal ubiquitous devices. The acquired data may have varying levels of granularity and encompass diverse modalities, ranging from the number of steps taken during the day to more complex measurements such as sleep quality or HR variability. Ultimately, the method of collecting data from everyday ubiquitous devices to assess the behaviors and behavior changes associated with pain meets the criteria that were initially set for the system: namely, the realization of automatic and minimally intrusive assessments that leverage patients' daily life environments.

The design of the *iSenseYourPain* system is presented in Fig. 6.3 below. The design includes four main components:

- A sensing component, which consists of a set of sensors and ubiquitous devices capable of collecting data about patients' lives that are embedded in the patient's environment (Fig. 6.3, left side);
- A self-report component occasionally used by the patient to self-report pain levels (Fig. 6.3, left side);

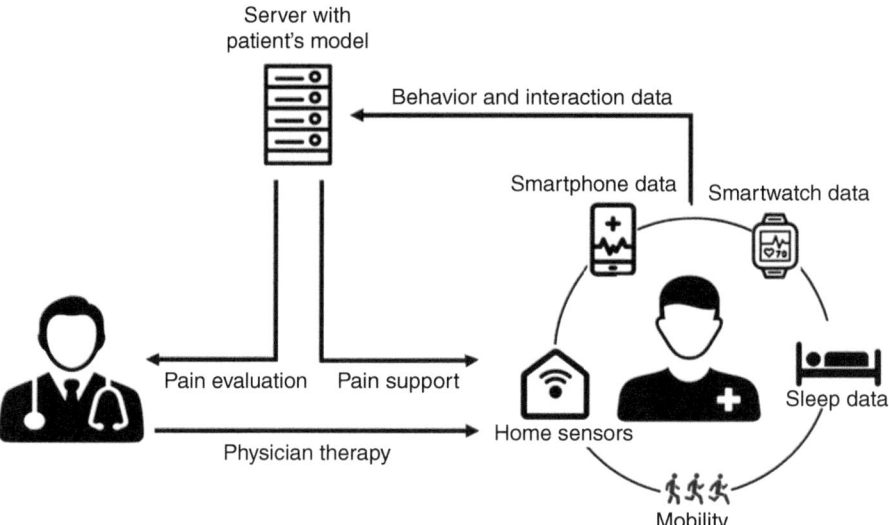

Fig. 6.3 iSenseYourPain system high-level design

- The core analytics component, which is hosted at a dedicated secure server and which analyzes the collected data to model and evaluate individuals' behaviors and correlate them with their pain levels (Fig. 6.3, middle);
- A visualization component for the physician and/or the individual's relatives that summarizes the patient's pain experiences and aspects of their behavior and indicates potential areas of focus for additional interventions or usual care (Fig. 6.3, right side).

Data Sources and Collection

The system uses the data it collects to qualitatively and quantitatively describe the patient's behavior and define objective metrics in order to identify patterns linked to the patient's experienced pain levels. Table 6.2 below indicates how data corresponding to the different aspects of daily life potentially influenced by chronic pain can be sampled and collected within the system. In addition to data collected automatically through devices, a questionnaire based on the Chronic Pain Based Questionnaire (CPBQ) [17] is used occasionally to collect subjective pain experience data that is later correlated with data collected by the devices.

Table 6.2 Aspects of daily life influenced by pain and corresponding data types and rates

Aspects of daily life	Type of data sources	Sampling rate	Other factors
General activity	Smartphone, hybrid home sensors	Hourly, daily	Highly variable between days, seasons, weeks
Mood	Smartphone, hybrid home sensors	Weekly, monthly	Highly subjective, additional data collected via self-reports
Walking ability	Smartphone, wearables	Daily	Strongly influenced by pain level
Work activity	Smartphone interactions, wearables/sensors	Daily, weekly	High variability according to the subject, additional data collected via self-reports
Relations with other people	Smartphone, hybrid home sensors	Daily, weekly, monthly	Number of interactions and time spent vary highly according to subject
Sleep	Smartphone, wearables	Daily	Influenced by various factors: Food, alcohol, physical activity, etc.
Enjoyment of life	Smartphone interactions, wearables/sensors	Daily	Highly subjective, additional data collected via self-reports
Perceived pain levels (CPBQ [17])	Structured questionnaire	Varies, frequency progressively decreases	Subjective

There are two important aspects of the design of the data collection methods: modularity and ubiquity. The collection methods are ubiquitous because all data sources are embedded within everyday life devices such as smartphones, wearables, the home environment, TVs, and appliances. This design choice aims to reduce the system's intrusiveness, leading to higher levels of acceptance from the patient while lowering the influence of the system itself on everyday behaviors. On the other hand, the system's modularity is essential to making it as scalable as possible and adaptable to the specific set of sensors or devices available to and used by each patient. As is explained in Sect. 4.2 below, each patient is characterized by a unique model corresponding to their everyday behaviors and pain-related behavior changes. These models are continually developed and evaluated based on the available sensors and devices and the data they collect.

As indicated above, patients are also occasionally asked to self-report pain levels by filling in a brief questionnaire on their smartphone in a way that leverages the Experience Sampling Method (ESM [31]). This self-reporting is carried out frequently at the beginning of the system use period, as it is important for creating and calibrating the patient's personalized model. The estimated duration of this initial calibration phase is one to three weeks, depending on the regularity of the patients' usual behaviors (such as sleep and activities) and the extent to which pain affects these activities. Once the model is pre-trained, the frequency with which patients submit self-reports via ESM will decrease, as the behavior–pain model will require only occasional smaller adjustments. In addition to the behaviors incorporated into the patient-specific model, there are some unusual behavioral indicators that may be strongly connected with pain experiences, such as the patient's staying a full day indoors or spending unnaturally long periods in bed. These behaviors are treated as "red flags" indicating a pain experience regardless of the patient's personalized model. When such unusual behaviors are detected by the system's technologies, the patient will be prompted to self-report so that the system can develop its pain model more accurately.

Data Analysis and Modeling

The core analytics component of *iSenseYourPain* is hosted at a secure server and focuses on providing analytics of the features derived from the system's various data sources. A model specifically derived for each patient is used to correlate behavior with the patient's pain levels. Table 6.3 provides examples of the features that can be derived from different data sources.

As every patient changes their behavior in different ways in response to pain experiences, the proposed system aims to identify the variations from normal behavior that are associated with pain for each particular patient. For example, sleep sensors can be used to model when a patients' quality of sleep decreases based on the number of hours the patient is sleeping, the number of times they wake up during the night, and the duration of time they spend awake. Similarly, patients

Table 6.3 Examples of possible features derived from different data sources

Type of data	Source	Features
Smartphone interactions	Smartphone	Use time, screen interactions (touch, placement, strength of touch), time spent with different applications
Home life	Hybrid home sensors (for activity, light, noise, etc.)	Time spent inside/outside home, time spent in different rooms, types of activities performed at home
Sleep data	Wearables Smartphone	Sleep duration, timing of sleep during 24 h period, quality of sleep (especially sleep interruptions), time of different sleep phases (potentially inaccurate) Smartphone use around sleep and wake-up time
Physical activity	Wearables smartphone	Number of steps, time spent walking or inactive, timing of activities during 24 h period
Heart data	Wearables	Average HR, resting HR, time spent in different HR zones, HRV
Social interactions	Smartphone	Number of calls and messages sent/received (including social media usage), people met during the day (indoor/outdoor meetings, types of locations visited)

experiencing pain may take fewer steps outside their house than they normally do while constantly moving inside the house and changing their position (e.g., sitting or standing) or moving from room to room, all of which are behaviors that can be modeled by the system. Patients may also have different ranges of HR or HRV and smartphone habits. As each patient reacts to pain in different ways, the *iSenseYour-Pain* system uses the data collected during the initial training phase to understand each patient's behavior and develop a model that corresponds to their specific behavioral patterns.

Conclusions

Chronic pain is considered a permanent medical condition. It may be due to a specific disease or injury, or it can arise from a psychological state, both of which may lead to chronic pain conditions with no predictable endpoint [2] and an overall decrease in quality of life. In patients with chronic pain, the behavioral changes associated with pain experiences tend to vary. Currently, these changes are generally assessed via self-reported measures that are infrequent, subjective, and memory based.

In this chapter, we have proposed the *iSenseYourPain* system design as a means to constantly evaluate the pain experienced by patients through assessment of observed behavioral patterns. *iSenseYourPain* collects data about patients' everyday behaviors and models relevant aspects of their daily life. Based on a specific model developed for each patient, it then assesses and predicts patients' pain levels based on their behaviors. The system can inform a patient's physician about increased pain levels or inform their relatives about their potential care needs.

The *iSenseYourPain* system entails a significant contribution to the development of quality of life technologies [32] and the potential use of everyday technologies to quantify different aspects of individuals' lives [33]. Overall, implementing systems such as *iSenseYourPain* may facilitate the achievement of better life quality for patients and for those around them.

Future studies will be conducted that focus on implementing the system, both on the side of data collection and on that of the development of the patient model. Moreover, we plan to investigate which aspects of patients' daily lives that can be measured with personal and ubiquitous devices are most representative of behavioral change associated with pain experiences in specific types of patients.

References

1. Merskey H. Pain terms: a list with definitions and notes on usage. Recommended by the IASP subcommittee on taxonomy. Pain. 1979;6:249–52.
2. Grichnik KP, Ferrante FM. The difference between acute and chronic pain. Mount Sinai J Med. 1991;58(3):217–20.
3. Smith BH, Elliott AM, Chambers WA, Smith WC, Hannaford PC, Penny K. The impact of chronic pain in the community. Fam Pract. 2001;18(3):292–9.
4. Breivik H, Collett B, Ventafridda V, Cohen R, Gallacher D. Survey of chronic pain in Europe: prevalence, impact on daily life, and treatment. Eur J Pain. 2006;10(4):287.
5. McCarberg B, Nicholson B, Todd K, Palmer T, Penles L. The impact of pain on quality of life and the unmet needs of pain management: results from pain sufferers and physicians participating in an internet survey. Am J Ther. 2008;15
6. Sheu R, Lussier D, Rosenblum A, Fong C, Portenoy J, Joseph H, Portenoy RK. Prevalence and characteristics of chronic pain in patients admitted to an outpatient drug and alcohol treatment program. Pain Med. 2008;9(7):911–7.
7. Boonstra AM, Stewart RE, Köke AJA, Oosterwijk RFA, Swaan JL, Schreurs KMG, Schiphorst Preuper HR. Cut-off points for mild, moderate, and severe pain on the numeric rating scale for pain in patients with chronic musculoskeletal pain: variability and influence of sex and catastrophizing. Front Psychol. 2016;7:1466.
8. Bijur PE, Silver W, Gallagher EJ. Reliability of the visual analog scale for measurement of acute pain. Acad Emerg Med. 2001;8(12):1153–7.
9. Cleeland, C., Ryan, K.: Pain assessment: global use of the brief pain inventory. Annals, Academy of Medicine, Singapore 1994.
10. Stewart AL, Ware JE. Measuring functioning and Well-being: the medical outcomes study approach. duke university Press; 1992.
11. Fairbank J, Couper J, Davies J, O'brien J, et al. The Oswestry low back pain disability questionnaire. Physiotherapy. 1980;66(8):271–3.
12. Leavitt F, Garron DC. Validity of a back pain classification scale for detecting psychological disturbance as measured by the MMPI. J Clin Psychol. 1980;36(1):186–9.
13. Zung WW. A self-rating pain and distress scale. Psychosomatics. 1983;24(10):887–94.
14. Garra G, Singer AJ, Taira BR, Chohan J, Cardoz H, Chisena E, Thode HC Jr. Validation of the Wong-baker FACES pain rating scale in pediatric emergency department patients. Acad Emerg Med. 2010;17(1):50–4.
15. Melzack, R., Katz, J.: The McGill pain questionnaire: appraisal and current status. 2001.

16. Tursky B, Jamner LD, Friedman R. The pain perception profile: a psychophysical approach to the assessment of pain report. Behav Ther. 1982;13(4):376–94.
17. VonKorff M, Ormel J, Keefe FJ, Dworkin SF. Grading the severity of chronic pain. Pain. 1992;50(2):133–49.
18. Wulfovich S, Fiordelli M, Rivas H, Concepcion W, Wac K. "I must try harder": design implications for Mobile apps and wearables contributing to self-efficacy of patients with chronic conditions. Front Psychol. 2019;10:2388.
19. Bender JL, Radhakrishnan A, Diorio C, Englesakis M, Jadad AR. Can pain be managed through the internet? A systematic review of randomized controlled trials. Pain. 2011;152
20. Holden G, Bearison DJ, Rode DC, Fishman-Kapiloff M, Rosenberg G, Onghena P. Pediatric pain and anxiety: a meta-analysis of outcomes for a behavioral telehealth intervention. Res Soc Work Pract. 2003;13(6):693–704.
21. Holden G, Bearison DJ, Rode DC, Kapiloff MF, Rosenberg G, Rosenzweig J. The impact of a computer network on pediatric pain and anxiety: a randomized controlled clinical trial. Soc Work Health Care. 2002;36(2):21–33.
22. Kristjánsdóttir ÓB, Fors EA, Eide E, Finset A, Stensrud TL, vanDulmen S, Wigers SH, Eide H. A smartphone-based intervention with diaries and therapist-feedback to reduce catastrophizing and increase functioning in women with chronic widespread pain: randomized controlled trial. J Med Internet Res. 2013;15
23. De la Vega R, Roset R, Castarlenas E, Sánchez-Rodríguez E, Solé E, Miró J. Devel- opment and testing of painometer: a smartphone app to assess pain intensity. J Pain. 2014;15(10):1001–7.
24. Stinson JN, Lalloo C, Harris L, Isaac L, Campbell F, Brown S, Ruskin D, Gordon A, Galonski M, Pink LR, et al. iCanCope with pain: user-centred design of a web-and mobile-based self-management program for youth with chronic pain based on identified health care needs. Pain Res Manag. 2014;19(5):257–65.
25. Cao ZH, Ko LW, Lai KL, Huang SB, Wang SJ, Lin CT. Classification of migraine stages based on resting-state EEG power. In: 2015 international joint conference on neural networks (IJCNN). IEEE; 2015. p. 1–5.
26. Pagán J, DeOrbe MI, Gago A, Sobrado M, Risco-Martín JL, Mora JV, Moya JM, Ayala JL. Robust and accurate modeling approaches for migraine per-patient prediction from ambulatory data. Sensors. 2015;15(7):15419–42.
27. Pagán J, Risco-Martín JL, Moya JM, Ayala JL. Grammatical evolutionary techniques for prompt migraine prediction. In: Proceedings of the genetic and evolutionary computation conference 2016; 2016. p. 973–80.
28. Cao Z, Lin CT, Chuang CH, Lai KL, Yang AC, Fuh JL, Wang SJ. Resting-state EEG power and coherence vary between migraine phases. J Headache Pain. 2016;17(1):102.
29. Ciman M, Wac K. Individuals' stress assessment using human-smartphone interaction analysis. IEEE Trans Affect Comput. 2018;9(1):51–65.
30. Ciman M, Wac K, Gaggi O. iSenseStress: assessing stress through human-smartphone interaction analysis. In: 2015 9th International Conference on Pervasive Computing Technologies for Healthcare (PervasiveHealth); 2015. p. 84–91.
31. Csikszentmihalyi M, Larson R. Validity and reliability of the experience sampling method; 2014. p. 35–54.
32. Wac K. Quality of life technologies. In: Gellman M. (eds) encyclopedia of behavioral medicine. Springer, New York, NY. USA.
33. Wac, K.: From quantified self to qualify of life. In Digital health (pp. 83-108). Springer.

Chapter 7
Technologies for Quantifying Sleep: Improved Quality of Life or Overwhelming Gadgets?

Sirinthip Roomkham, Bernd Ploderer, Simon Smith, and Dimitri Perrin

The Importance of Sleep

Sleep is, broadly, a recuperative and restorative process. As behavior, it is characterized by reduced activity, withdrawal, quiescence, and reduced responsiveness. Put simply, humans typically tend to move to a designated sleep space (such as a bed), lie down, close their eyes and lie still. Sleep is also a very complex neurophysiological process, of which only some features can be readily observed or measured [1], with associated complex changes in physiology, including reduced respiration and heart rate, changes in heart-rate variability, cyclical changes in muscle tone, amongst others.

Humans are regarded as diurnal (day active) animals, and as adults they tend to achieve their major sleep episode during the night-time hours. This timing of sleep is largely governed by circadian processes, and particularly by the evening expression of the hormone melatonin, together with a 'homeostatic' increase in the likelihood of sleep, the longer a person is awake. These two processes ideally function to promote regular, consistent, and sufficient sleep. However, the achievement of good sleep also depends on opportunity and environment. In the contemporary context, sleep is often disrupted by choice or by externalities. These can include preferences

S. Roomkham · B. Ploderer
School of Computer Science, Queensland University of Technology (QUT),
Brisbane, Australia

S. Smith
Institute for Social Science Research (ISSR), The University of Queensland,
Brisbane, Australia

D. Perrin (✉)
School of Computer Science, Queensland University of Technology (QUT),
Brisbane, Australia

Centre for Data Science, Queensland University of Technology (QUT), Brisbane, Australia
e-mail: dimitri.perrin@qut.edu.au

© The Author(s) 2022
K. Wac, S. Wulfovich (eds.), *Quantifying Quality of Life*, Health Informatics,
https://doi.org/10.1007/978-3-030-94212-0_7

around work, study, or leisure hours (including engagement in social media, gaming and other device use), education demands, the presence of illness and pain, stress and worry, environmental factors including noise and extremes of temperature, natural and other disasters, and the suppression of melatonin expression by artificial light.

While there is robust debate around the idea of an 'epidemic of sleeplessness' [2], many people are not getting sufficient sleep. Current evidence-based recommendations for sleep duration [3] suggest that 7 or more hours of sleep is required by most adults, and that more than 9 hours might be needed by young adults. In contrast, the US Centres for Disease Control report that 30–40% of people habitually achieve fewer than 7 hours of sleep [4], defined as short sleep duration, with almost 70% of high school students sleeping fewer than 8 hours [5]. Although normal sleep duration can vary between individuals, sleep restricted to less than 6 hours per night has been associated with significant impairments in cognitive performance, vigilance, and affect [6].

The effects of sleep loss can be seen very acutely. For example, sleep restricted by just a few hours can be observed in objective performance the next day [7, 8]. However, sleep loss can also accumulate over nights, so that an hour less sleep each night over a week may be equivalent in effect to a full night of sleep deprivation. The impact of chronic partial sleep restriction may not be observed immediately, but may be seen many years later through increased health and mental health problems [9]. This possibility has been raised by data showing strong links between involvement in shift work and later long-term health consequences. This means that sleep needs to be understood across timescales ranging from a single night to decades, and the nature, level, and intensity of measurement needs to reflect those timescales.

The two most common sleep disorders are insomnia and Obstructive Sleep Apnoea (OSA). It is possible that as many as a billion adults globally have Obstructive Sleep Apnoea, with a prevalence approaching 50% in some countries [10, 11]. The prevalence of insomnia may range from 6–10% by strict clinical definitions, to over 30% for poor or unsatisfactory sleep [12]. The direct (medical and industrial) and indirect (social) costs associated with these sleep disorders, and with other forms of poor or disrupted sleep, are very high, e.g. over $680 Billion across just 5 OECD nations [13]. The understanding, identification, and treatment of sleep disorders have seen very rapid growth over the past decade. While current treatment approaches to these disorders are regarded as both effective and cost effective [14, 15], this growth still represents a very high additional cost. Due to the high global burden of OSA and insomnia, health-care systems will face major cost and logistic challenges, and must adopt more effective and efficient diagnostic and management strategies so that the negative health impacts can be minimized. There is a specific and growing need for broader public health, individual health, early intervention, and prevention approaches to sleep health.

Sleep is clearly not a unitary state. Instead, it has distinct dimensions including duration, quality, timing and regularity, each of which can have corresponding objective and subjective meaning. Further, sleep matters for daytime function so that constructs such as daytime alertness, clarity, speed, energy and satisfaction are also very important. This multidimensionality means that approaches to the measurement of sleep by devices must also be multidimensional, or be understood to be constrained to one or more dimensions, in order to understand sleep meaningfully. Another implication is that a measure optimized for measurement along one dimension may not function well to measure another. Rather than asking '*how well does this device measure sleep*', we might ask '*what is it about sleep that this device measures*'. Sleep science is a relatively new field, and the parallel rise in personal miniaturized sensing technology suggests a great opportunity for increased measurement, and increased range and mode of measurement, to inform new understanding of this complex state.

Individuals are increasingly interested in their own sleep, mirroring increasing public interest and understanding of the role of sleep in overall health and wellbeing alongside nutrition and exercise. Quality of Life Technologies (QoLT) provide immediate feedback through smart devices that people can use in their daily life to assess and enhance their health and well-being [16]. Most sleep tracking devices appear to have achieved the fundamental aims of QoLT, for instance, in that they claim to enable individuals to quantify their sleep through apps or wearable devices. However, important questions remain. Are these devices addressing a need that cannot be met by existing methods? Can they provide an objective and reliable assessment of sleep? Are there potential pitfalls in the use of such devices? Our Chapter aims to cover all these questions.

Why Are Existing Methods Not Adequate?

Sleep disorders are defined by recognized clinical criteria such as the International Classification of Sleep Disorders, ICSD-3 [17], the International Classification of Diseases, ICD-11 [18], and the Diagnostic and Statistical Manual of Mental Disorders, DSM-V [19]. Meeting these criteria generally requires an overnight sleep study using polysomnography (PSG) to confirm the clinical diagnosis of a specialist sleep physician (e.g. for suspected OSA), or may require careful clinical diagnosis by a psychologist without reference to PSG (in the case of insomnia).

Polysomnography (PSG) is regarded as the gold standard for diagnosis of specific sleep disorders (shown in Fig. 7.1). PSG objectively quantifies sleep-related indices from a combination of electroencephalogram (EEG, brain states), electrocardiogram (ECG, cardiac states), electromyogram (EMG, muscle tension), electrooculogram (EOG, eye movements in sleep) and indices related to sleep disorders such as nasal airway pressure and flow, respiratory effort, infra-red video, sound recording, and blood oxygen saturation. Individuals typically spend one or more nights in a hospital or clinic sleep laboratory, where they are monitored throughout the night, although there is increasing uptake of home-based and limited channel studies [20].

While PSG remains the gold standard for assessment of specific sleep disorders and is an evolved medical technology, it has certain features that limit its use. One or two nights of assessment does not allow for longer-term tracking of change, and does not assess the habitual sleep of individuals in their natural sleep environments

Fig. 7.1 A polysomnography test (image courtesy of Prof. David Lovell)

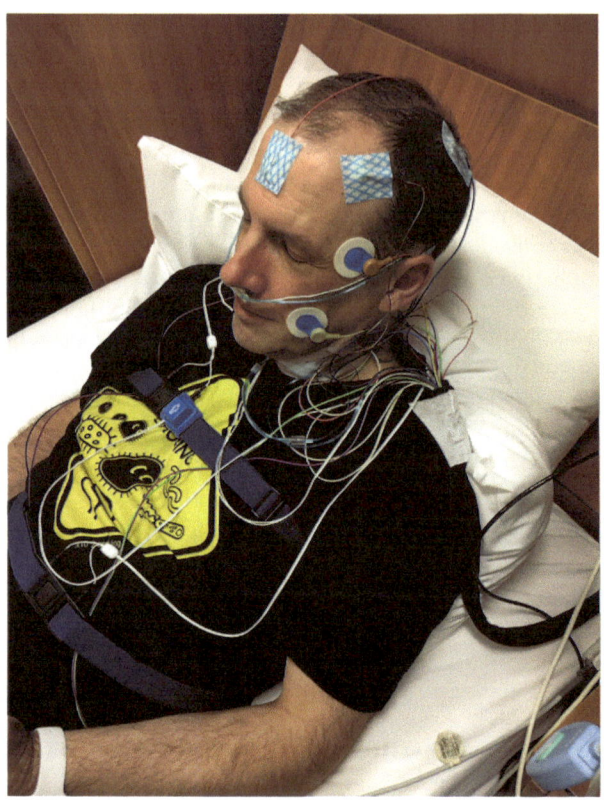

and schedules. PSG requires extensive resources including equipment, staff, and software. In addition, it is not always easy for someone to undertake a sleep study when needed as waiting times for access can be considerable [21]. The interpretation of sleep data is another challenge. Variability in the use of established rule sets for visual scoring of the PSG results in unreliability in key PSG outcome indices (such as the Apnea-Hypopnea Index, AHI) both within individual scorers and across laboratories [22], even when a single rule set is adopted [23]. For this reason, automated analysis of studies using Machine Learning have been proposed as a way to reduce scoring variability and error. While these analytical models increase the consistency of scoring, and perform well against expert visual scoring, no single model has achieved widespread acceptance. The promise of neural network models and more recent Artificial Intelligence models has been recognized [24, 25] as a way to increase efficiency and accuracy in sleep medicine, but also to provide a deeper understanding of sleep and circadian biology.

A second key method used to understand sleep is through elicitation of self-reported experience. This approach is required to capture the qualitative or subjective dimensions of sleep [26]. In some cases, this experience cannot be measured objectively (e.g. satisfaction with sleep), or is necessary to understand predictors or consequences of poor sleep (such as increased alcohol use, or poor work performance). This can be done through structured 'sleep diaries' or time use surveys [27], or through clinical and non-clinical rating scales and other measures. These measures can be used to complement objective measurements from PSG, or may stand-alone for use in other settings such as epidemiological surveys.

Subjective assessment using instruments such as the Pittsburgh Sleep Quality Index [28] is seen as relatively inexpensive and quick, and for those reasons is often used as an initial screen for sleep diagnosis [29]. However, the use of paper-and-pencil measures in particular can lead to high transcription and scoring costs, significant user effort, and simple measurement error. There is typically a requirement for careful development of subjective measures along sound psychometric principles, and a need for high quality normative studies (e.g. stratified by age, gender, socioeconomics or other factors) to inform meaningful interpretation. This work has been done in a number of recent initiatives, e.g. Yu et al. [30], but these measures are not in widespread use in sleep medicine.

A final common method for sleep and circadian rhythm measurement is actigraphy. This is typically in the form of a wrist-worn 'movement watch' based on accelerometry or other movement capture methods and which can be used to measure sleep over extended periods in a naturalistic environment including the person's home, work, or other settings. Generation of these devices have been widely used in the past two decades for sleep assessment [31].

The American Academy of Sleep Medicine (AASM) provides a guideline to establish clinical practice recommendations for using actigraphy in sleep medicine. The purpose of using actigraphy is not to replace the gold standard of sleep measurement but rather to assist in deriving helpful metrics for sleep disorder assessment and treatment [32]. Actigraphy devices are typically continuously worn for 24-hours a day for several days to months. Sleep parameters are extracted from the movement data via specific sleep detection algorithms [33], and other derivatives such as circadian rhythm parameters and estimates of activity level can also be generated. By using actigraphy, clinicians are able to obtain unique information about sleep in a person's natural or habitual sleep environment. This can be particularly important when sleep schedules are dictated by work demands (e.g. shift work), circadian phase shifts (e.g. international travel) or other circumstances where variation over time is predicted. Although actigraphy has been well studied and has been validated against PSG in specific populations, there are essential limitations to the concordance between the two approaches. The devices used in research and clinical practice tend to prioritize reliability, standardization, long battery life, and capacity for re-use over other considerations that might be more important for consumer-grade devices.

Methods and Tools for Objective and Quantitative Assessment of Daily Sleep

Consumers now have access to a number of technologies to objectively and quantitatively assess their daily sleep in their home environment. Below we review mobile applications, wearable devices (smartwatches, rings, headbands), as well as co-called 'nearables' that are near the individual, embedded in the sleep environment (e.g. in mattresses). Broadly speaking, mobile applications are more widely available to many consumers. Wearables and nearables, on the other hand, are more expensive; yet potentially provide a holistic view of our health and well-being along metrics such as daily activities, heart rate, and sleep. Wearables typically also contain a wider variety of sensor technology such as Electrocardiogram (ECG), Electroencephalogram (EEG) and Photoplethysmogram (PPG), which can lead to more accurate data.

Mobile applications are readily available at the consumer's fingertips. Consumers can manually record their sleep via sleep diaries or use mobile-embedded sensors to track their sleep automatically. Sleep logs and diaries allow consumers to record their bedtime, number of awakenings, activities before going to sleep and after awakening. There are two main approaches to keep track of sleep diaries: paper and electronic sleep diaries. Tonetti et al. [34] suggested that both methods achieved similar results, whereas electronic sleep diaries provide more benefits over paper diaries in terms of reducing time for data entry and automatically recording the time when the diary is logged. Choe et al. [35] emphasized that electronic sleep diaries are more beneficial to individuals when the diary application is quick to use, engaging, and encourages self-reflection.

Automatic sleep tracking apps require minimal or no data entry. They are easy to use and inexpensive for self-tracking. However, their accuracy is often neglected, and their sleep results are often over claimed. Most sleep apps are not clinically validated against laboratory PSG, the gold standard of sleep tracking, because PSG is labor intensive, expensive and difficult to access. Sleep Cycle, a popular sleep-tracking app, monitors a sleep-wake stage through motion or sound sensors, and provides a summary of sleep quality. Fino et al. [36] showed that the Sleep Cycle app failed to show adequate reliability when compared against PSG for sleep-wake detection. Despite the lack of reliability, Sleep Cycle app has continuously been used widely due to word of mouth (e.g. Editors' Choice on the Apple App Store in October, 2020), and its potential benefits. Robbins et al. [37] report the use of Sleep Cycle to understand sleep duration and quality for four-year trends in general population. The results reveal helpful information. However, there is still some concern about its reliability and validity of sleep outcomes. Similar to Sleep Cycle, Sleep Time app shows poor correlation with PSG in terms of sleep parameters (e.g., sleep duration and number of awakenings) and sleep-wake stages [38]. Overall, further studies are needed to assess applications' utility and examine how much trust that we could potentially improve so that we are more confident using it.

Popular smartwatches like the Apple Watch, Samsung Gear and Fitbit become part of our daily living and are potentially helpful to remind us to do more exercise or to sleep more. Current studies [39, 40] investigate the accuracy of using the Apple Watch to determine sleep-wake stages and their results yielded reliable performance compared with a clinical-grade device and PSG. Roomkham's study suggested that using an Apple Watch to monitor sleep-wake stages could be an add-on to traditional actigraphy as well as a way to study the broader population. Both studies have transparent methods to evaluate the reliability of smartwatches, which increases the confidence in using such devices or adopting their techniques for tracking sleep in medicine, research, and self-tracking. There is clear potential for smartwatches to monitor not only sleep itself but also sleep quality [41].

Smart rings are becoming less intrusive and more attractive for consumers. Smart rings usually contain similar sensors to smartwatches, such as motion, heart rate, pulse rate, and body temperature. A recent investigation of healthy adolescents shows promising sleep outcomes beyond the sleep-wake stage into the whole sleep cycle [42]. However, the ŌURA ring used in this study employs a proprietary sleep algorithm to derive sleep stages, and most studies can only validate the overall sleep outcomes against PSG.

A variety of headbands exist that track sleep through EEG sensors around the head. At the time of writing, the Dreem headband is a popular choice amongst consumers and researchers. The Dreem headband is able to measure brain activity (EEG), breathing, and movement. The results of heart rate, breathing frequency and respiratory rate variability are reliable which resulted in providing precision sleep stages and sleep parameters [43].

Nearables [44] are another type of consumer-grade device. They are placed near the sleep environment, e.g. under the mattress or on the bedside table. An example of nearables is the Beddit Sleep tracker, which measures heart rate through ballisto-cardiography (BCG) derived from pressure sensors under the mattress. Tuominen et al. [45] reported that the accuracy of the sleep tracking is low and that consumers should be careful interpreting these results. Similar with other types of consumer-grade devices, total sleep time was overestimated, and wake after sleep onset under-estimated. These results were also based on healthy participants only, and the reliability of such devices for users with obstructive sleep apnea, insomnia, and other types of sleep disorders is unclear. Another example is the DoppleSleep [46], a contactless sleep system that is placed next to the bed and uses radar signals to gather movement, heart rate, and breathing data. It performed well for sleep-wake classification (~90% recall), and to a lesser extent on sleep staging (~80% recall for REM vs. Non-REM). The S+ system by ResMed is another nearable, which relies on ultra-low power radiofrequency waves to monitor the user's movements and breathing. Schade et al. [47] evaluated it against PSG and actigraphy and reported showed good sleep detection accuracy (~93%) and lower wake detection accuracy (69–73%), as for most devices. It is worth noting that wake detection was better than actigraphy (at 48%), thanks to a higher accuracy in detecting wake before sleep onset.

Overall, it is important that convenience does not come at the expense of the accuracy and usefulness of the information that can be derived from consumer-grade devices [49]. Recent work from Depner et al. [48] identified important metrics for sleep using wearable devices, and circadian metrics that may influence sleep (such as a level of exercise, which can be captured passively through the wearable devices). Tables 7.1 and 7.2 capture important metrics that are essential for validation against the gold standard.

Another crucial consideration is that when the sleep algorithm is provided by the device manufacturer, it is often a proprietary black box. One consequence is that data consistency is at the mercy of a software update over which the users often have no control. This can be a problem, especially for long-term tracking. The device may also suddenly disappear from the market. Users, in particular in a research context, should favor devices that provide access to the raw data and allow the deployment of open algorithms, and tracking methods that have been validated for a broad range of users (age, healthy vs. sleep disorders, etc.).

Benefits, Limitations, and Potential Pitfalls

There are many benefits for consumers from having relatively easy access to sleep data from apps, wearables and nearables. A major benefit is a better awareness of sleep. People often purchase devices for other purposes, e.g. to track steps, and sleep data comes as a bonus [50, 51]. Once people have access to sleep data, they are curious about what the data may tell them about themselves, and whether they can see any trends. Based on such data, sleep may become more of a priority because they become more conscious of how much sleep they actually get, as well as how much sleep they ought to get for a healthy night of sleep. Having an awareness of sleep and tracking sleep often goes hand in hand with goal setting, e.g. aiming for at least 7 hours of sleep. Many apps provide virtual rewards for getting enough sleep that can increase motivation [52]. However, the main challenge here is that (unlike with the number of steps walked) people cannot voluntarily control how much time they spend in a particular sleep stage. Another benefit is the ability to explore links between sleep and other quality of life data, e.g. to identify if you sleep more on days when you exercise [50]. Exploring such links in the data is particularly useful

Table 7.1 Sleep. What a Wearable Should/Can Measure (Adapted from [48]). Legend: a green tick (✓) represents metrics a wearable can measure; an orange tilde (~) represents metrics that only some devices can measure; a red exclamation mark (!) indicates proprietary methods

Sleep Duration	Sleep Quality !	EEG	Physiology	Respiration
Time in bed ✓	Sleep Onset Latency ✓	Sleep Staging ✓!	Heart Rate (HR) ~	Blood Oxygen Levels (SPO2)
Total Sleep Time (TST) (Bed time, wake time) ✓	Wake After Sleep Onset (WASO) ✓	Sleep Spindles	Heart Rate Variability (HRV) ~	Respiratory rate (RR) ~ & effort
	REM Latency ✓	Sleep Wave Activity	Blood Pressure (BP)	Nasal Pressure
	Sleep Efficiency ✓	Slow Oscillations	Body Position	Air Flow
	Periodic Limb Movement		Skin Conductivity (GSR) ~	Snoring
				Apnea Hypopnea Index (AHI)

Table 7.2 Circadian Physiology. What a Wearable Should/Can Measure (Adapted from [48]). Legend: a green tick (✓) represents metrics a wearable can measure; an orange tilde (~) represents metrics that only some devices can measure; a red exclamation mark (!) indicates proprietary methods

Behavioural	Physiological	Environmental (nearables ~)
Sleep onset time ✓	Heart Rate (HR) ✓	24h light exposure pattern (intensity and wavelength)
Wake time ✓	Blood Pressure (BP) ✓	Ambient temperature (also: noise, humidity, air quality) ~
Timing of physical activity ✓	Skin Temperature (Temp) ~	Geographical location ~
Timing of energy intake: food, caffeine, alcohol ✓	Core body temperature (CBT) ✓	EMA/self-assessments of environmental aspects (bed partner, kids, pets)
Timing of supplement and mediation intake	Skin Conductivity (GSR) ~	
EMA/self-assessments of physical/mental state aspects (pain, mood, etc.) ~		

for people with sleep disorders and other chronic conditions, who seek to identify reasons for their sleep problems and other health conditions [53]. Going one step further, some people benefit from experimenting based on sleep and other quality of life data. For example, members of the Quantified Self community—a worldwide community of self-tracking enthusiasts—report self-experiments with sleep. These include experiments to improve sleep, e.g. by changing their sleep patterns, environmental aspects like reducing the noise and light in their bedroom, or sleep health strategies such as reducing their caffeine consumption. Conversely, sleep data also allows for experiments on the impact of sleep on other lifestyle factors, e.g. if they feel more alert at work and in everyday life [54].

Studies of consumers tracking their own sleep highlight also several challenges and pitfalls. In collecting and organizing data, consumers face the challenge of working with potentially inaccurate data, e.g. because accelerometer data alone cannot reliably determine insights into sleep stages (without EEG data), and accelerometer may also inadvertently respond to the movements of partners in the same bed [55]. Consumer devices that contain EEG sensors potentially provide more accurate data, but the downside is that they are often uncomfortable to wear and impede sleep, and as a result are not worn for longer periods of time [56]. Consumers also report that they lack information on triggers (e.g. why they went to bed late) and contextual information (e.g. their stress and wellbeing) that may help to explain sleep data [54]. Even when consumers track such information through other means, they may not have the skills and tools to confidently relate data from different devices and sources to establish relationships, cause and effect [50].

The interpretation of sleep data also poses several challenges to consumers. People often report a lack of time to revisit and interpret data [57]. Sleep data can be overwhelming or even stressful when the desired outcomes do not materialize [58].

Particular information about sleep stages can be difficult to interpret, because we do not consciously experience sleep stages, nor is the relationship between objective sleep stages and subjective sleep quality clear to consumers [53]. A related challenge is that sleep data and what might objectively constitute quality sleep (e.g. more than 8 hours of sleep), may not relate to how consumers experience sleep or how refreshed they feel when they wake up [56]. There appears to be a disconnect between the various scientific metrics for quantifying sleep (e.g. sleep efficiency and sleep stages) and the way people experience and understand their sleep, which is more aligned with sleep duration and how satisfied or alert they feel as a result [59]. Unfortunately, this disconnect is not helped by the fact that most consumer devices are black boxes that conceal how data is collected and processed, which is partly a result of the complexity of sensor hardware and algorithms used [56], but partly also a deliberate choice of companies to protect their intellectual property [60].

A final challenge for sleep-trackers is that seeing a problem in the data is not the same as finding an opportunity to change it [61]. If our fitness-tracking device tells us that we have not achieved our 10,000 steps goal, then we can at least in principle remedy this by going for an extended walk. On the other hand, if the same device tells us that we only get 5 hours of sleep or not enough deep sleep, then the opportunity to change this is less clear [50, 62]. We can change the sleep environment (e.g. block out light with curtains) and we can also improve our sleep hygiene (e.g. reduce the consumption of caffeinated products) to increase the chances of getting more sleep, but we cannot voluntarily control sleep itself [53]. Furthermore, in making adjustments to our environment and sleep hygiene we often face external constraints, e.g. we may not be able to adjust the start time for school or work to better suit our own sleep needs, nor do we have control over autonomous conditions like our mood, stress, and hormone cycle [50].

To help mitigate these challenges, sleep-tracking at home is often conducted in collaboration with others. Family members, and especially bed partners, can play a vital role in supporting sleep tracking. Parents and older siblings often need to set up technology, interpret data, encourage healthy sleep hygiene, manage medical appointments, etc. [63]. Health professionals continue to play a crucial role to interpret the data, ask questions to reflect on sleep, and to provide advice on what actions to take [64].

Coming back to our original question, we have shown that personal sleep tracking technologies can improve quality of life by increasing awareness of sleep, and by offering opportunities to explore connections with other quality of life data and potential sleep problems. However, we have also shown that collecting data, making sense of it to take action can be difficult and perhaps even overwhelming, because it requires expertise on sleep and tracking technology. We hope that this chapter will provide a useful introduction to these issues to get the most out of quantifiable sleep data. However, we also add a note of caution: quantification is important, but sleep cannot yet be reduced to a collection of measures. As we have discussed above, there is a mismatch between scientific sleep metrics and our subjective experience of sleep. Quantifying sleep also invites people to search for signs of sleep problems, instead of promoting good sleep health [59]. No matter how elaborate, measures will always only provide a partial picture of sleep. A complete picture requires complementary information from subjective assessments, a broad perspective (overall health, but also inputs and interactions with other Quality of Life indicators), and an understanding of the varying needs and concerns of different individuals over a life course.

References

1. Dement WC. The stanford sleep book. 2006.
2. Marshall NS, Lallukka T. Sleep pirates-are we really living through a sleep deprivation epidemic and what's stealing our sleep? Eur J Pub Health. 2018;28(3):394–5.
3. Watson NF, Badr MS, Belenky G, Bliwise DL, Buxton OM, Buysse D, Dinges DF, Gangwisch J, Grandner MA, Kushida C, Malhotra RK, Martin JL, Patel SR, Quan SF, Tasali E. Recommended amount of sleep for a healthy adult: a joint consensus statement of the American Academy of sleep medicine and Sleep Research Society. Sleep. 2015;38(6):843–4.
4. Shockey TM, Wheaton AG. Short sleep duration by occupation group–29 states, 2013–2014. Morb Mortal Wkly Rep. 2017;66:207–13.
5. Wheaton AG, Jones SE, Cooper AC, Croft JB. Short sleep duration among middle school and high school students–United States, 2015. Morb Mortal Wkly Rep. 2018;67:85–90.
6. Hirshkowitz M, Whiton K, Albert SM, Alessi C, Bruni O, DonCarlos L, Hazen N, Herman J, Katz ES, Kheirandish-Gozal L, Neubauer DN. National Sleep Foundation's sleep time duration recommendations: methodology and results summary. Sleep Health. 2015;1(1):40–3.
7. Belenky G, Wesensten NJ, Thorne DR, Thomas ML, Sing HC, Redmond DP, et al. Patterns of performance degradation and restoration during sleep restriction and subsequent recovery: a sleep dose-response study. J Sleep Res. 2003;12(1):1–12.
8. Rossa KR, Smith SS, Allan AC, Sullivan KA. The effects of sleep restriction on executive inhibitory control and affect in young adults. J Adolesc Health. 2014;55(2):287–92.
9. Kecklund G, Axelsson J. Health consequences of shift work and insufficient sleep. BMJ. 2016;355:i5210.
10. Benjafield AV, Ayas NT, Eastwood PR, Heinzer R, Ip MS, Morrell MJ, Nunez CM, Patel SR, Penzel T, Pépin JL, Peppard PE. Estimation of the global prevalence and burden of obstructive sleep apnoea: a literature-based analysis. Lancet Respir Med. 2019;7(8):687–98.
11. Lyons MM, Bhatt NY, Pack AI, Magalang UJ. Global burden of sleep-disordered breathing and its implications. Respirology. 2020;25:690–702.
12. Morin CM, Jarrin DC. Epidemiology of insomnia: prevalence, course, risk factors, and public health burden. Sleep Med Clin. 2013;8(3):281–97.
13. Hafner M, Stepanek M, Taylor J, Troxel WM, Van Stolk C. Why sleep matters–the economic costs of insufficient sleep: a cross-country comparative analysis. Rand health quarterly. 2017;6(4)
14. Reynolds SA, Ebben MR. The cost of insomnia and the benefit of increased access to evidence-based treatment: cognitive behavioral therapy for insomnia. Sleep Med Clin. 2017;12(1):39–46.
15. Streatfeild J, Hillman D, Adams R, Mitchell S, Pezzullo L. Cost-effectiveness of continuous positive airway pressure therapy for obstructive sleep apnea: health care system and societal perspectives. Sleep. 2019;42(12):zsz181.
16. Wac K. Quality of life technologies. In: Gellman M, editor. Encyclopedia of behavioral medicine. New York, NY: Springer; 2020. https://doi.org/10.1007/978-1-4614-6439-6_102013-1.
17. American Academy of Sleep Medicine. International classification of sleep disorders. 3rd ed. Darien, IL: American Academy of Sleep Medicine; 2014.

18. World Health Organization International classification of diseases for mortality and morbidity statistics (11th revision). 2018. Retrieved from https://icd.who.int/browse11/l-m/en

19. American Psychiatric Association. Diagnostic and statistical manual of mental disorders. 5th ed. Arlington, VA: American Psychiatric Publishing; 2013.

20. Corral J, Sánchez-Quiroga MÁ, Carmona-Bernal C, Sánchez-Armengol Á, de la Torre AS, Durán-Cantolla J, Egea CJ, Salord N, Monasterio C, Terán J, Alonso-Alvarez ML. Conventional polysomnography is not necessary for the management of most patients with suspected obstructive sleep apnea. Noninferiority, randomized controlled trial. Am J Respir Crit Care Med. 2017;196(9):1181–90.

21. Rotenberg BW, George CF, Sullivan KM, Wong E. Wait times for sleep apnea care in Ontario: a multidisciplinary assessment. Can Respir J. 2010;17

22. Collop NA. Scoring variability between polysomnography technologists in different sleep laboratories. Sleep Med. 2002;3(1):43–7.

23. Ruehland WR, O'Donoghue FJ, Pierce RJ, Thornton AT, Singh P, Copland JM, Stevens B, Rochford PD. The 2007 AASM recommendations for EEG electrode placement in polysomnography: impact on sleep and cortical arousal scoring. Sleep. 2011;34(1):73–81.

24. Goldstein CA, Berry RB, Kent DT, Kristo DA, Seixas AA, Redline S, Westover MB. Artificial intelligence in sleep medicine: background and implications for clinicians. J Clin Sleep Med. 2020;16(4):609–18.

25. Stephansen JB, Olesen AN, Olsen M, Ambati A, Leary EB, Moore HE, et al. Neural network analysis of sleep stages enables efficient diagnosis of narcolepsy. Nat Commun. 2018;9(1):1–15.

26. Shahid A, Wilkinson K, Marcu S, Shapiro CM, editors. STOP, THAT and one hundred other sleep scales. Springer Science & Business Media; 2012.

27. Carney CE, Buysse DJ, Ancoli-Israel S, Edinger JD, Krystal AD, Lichstein KL, Morin CM. The consensus sleep diary: standardizing prospective sleep self-monitoring. Sleep. 2012;35(2):287–302.

28. Buysse DJ, Reynolds CF, Monk TH, Berman SR, Kupfer DJ. The Pittsburgh sleep quality index: a new instrument for psychiatric practice and research. Psychiatry Res. 1989;28(2):193–213.

29. Ibáñez V, Silva J, Cauli O. A survey on sleep assessment methods. PeerJ. 2018;6:e4849.

30. Yu L, Buysse DJ, Germain A, Moul DE, Stover A, Dodds NE, Johnston KL, Pilkonis PA. Development of short forms from the PROMIS™ sleep disturbance and sleep-related impairment item banks. Behav Sleep Med. 2012;10(1):6–24.

31. Ancoli-Israel S, Cole R, Alessi C, et al. The role of actigraphy in the study of sleep and circadian rhythms. American Academy of sleep medicine review paper. Sleep. 2003;26(3):342–92.

32. Smith MT, McCrae CS, Cheung J, Martin JL, Harrod CG, Heald JL, Carden KA. Use of actigraphy for the evaluation of sleep disorders and circadian rhythm sleep-wake disorders: an American Academy of sleep medicine clinical practice guideline. J Clin Sleep Med. 2018;14(7):1231–7.

33. Martin JL, Hakim AD. Wrist actigraphy. Chest. 2011;139(6):1514–27.

34. Tonetti L, Mingozzi R, Natale V. Comparison between paper and electronic sleep diary. Biol Rhythm Res. 2016;47(5):743–53.

35. Choe EK, Consolvo S, Watson NF, Kientz JA. Opportunities for computing technologies to support healthy sleep behaviors. In: Proceedings of the SIGCHI conference on human factors in computing systems; 2011. p. 3053–62.

36. Fino E, Plazzi G, Filardi M, Marzocchi M, Pizza F, Vandi S, Mazzetti M. (not so) smart sleep tracking through the phone: findings from a polysomnography study testing the reliability of four sleep applications. J Sleep Res. 2020;29(1):e12935.

37. Robbins R, Affouf M, Seixas A, Beaugris L, Avirappattu G, Girardian J-L. Four-year trends in sleep duration and quality: a longitudinal study using data from a commercially available sleep tracker. J Med Internet Res. 2020;22(2):e14735.

38. Bhat S, Ferraris A, Gupta D, Mozafarian M, DeBari VA, Gushway-Henry N, Gowda SP, Polos PG, Rubinstein M, Seidu H, Chokroverty S. Is there a clinical role for smartphone sleep apps?

Comparison of sleep cycle detection by a smartphone application to polysomnography. J Clin Sleep Med. 2015;11(7):709–15.

39. Roomkham S, Hittle M, Cheung J, Lovell D, Mignot E, Perrin D. Sleep monitoring with the apple watch: comparison to a clinically validated actigraph. F1000Research. 2019;8(754):754.

40. Walch O, Huang Y, Forger D, Goldstein C. Sleep stage prediction with raw acceleration and photoplethysmography heart rate data derived from a consumer wearable device. Sleep. 2019;42(12):zsz180.

41. Alfeo AL, Barsocchi P, Cimino MG, La Rosa D, Palumbo F, Vaglini G. Sleep behavior assessment via smartwatch and stigmergic receptive fields. Pers Ubiquit Comput. 2018;22(2): 227–43.

42. de Zambotti M, Rosas L, Colrain IM, Baker FC. The sleep of the ring: comparison of the ŌURA sleep tracker against polysomnography. Behav Sleep Med. 2019;17(2):124–36.

43. Arnal PJ, Thorey V, Ballard ME, Hernandez AB, Guillot A, Jourde H, et al. The Dreem headband as an alternative to polysomnography for EEG signal acquisition and sleep staging. Sleep. 2019;43(11):662734.

44. Bianchi MT. Sleep devices: wearables and nearables, informational and interventional, consumer and clinical. Metabolism. 2018;84:99–108.

45. Tuominen J, Peltola K, Saaresranta T, Valli K. Sleep parameter assessment accuracy of a consumer home sleep monitoring ballistocardiograph beddit sleep tracker: a validation study. J Clin Sleep Med. 2019;15(3):483–7.

46. Rahman T, Adams AT, Ravichandran RV, Zhang M, Patel SN, Kientz JA, Choudhury T. Dopplesleep: a contactless unobtrusive sleep sensing system using short-range doppler radar. In: Proceedings of the 2015 ACM international joint conference on pervasive and ubiquitous computing; 2015. p. 39–50.

47. Schade MM, Bauer CE, Murray BR, Gahan L, Doheny EP, Kilroy H, et al. Sleep validity of a non-contact bedside movement and respiration-sensing device. J Clin Sleep Med. 2019;15(7):1051–61.

48. Depner CM, Cheng PC, Devine JK, Khosla S, de Zambotti M, Robillard R, et al. Wearable technologies for developing sleep and circadian biomarkers: a summary of workshop discussions. Sleep. 2020;43(2):zsz254.

49. Hunasikatti M. Non-contact sensors: need for optimum information is more important than convenience. J Clin Sleep Med. 2019;15(11):1707.

50. Liang Z, Ploderer B, Liu W, Nagata Y, Bailey J, Kulik L, Li Y. SleepExplorer: a visualization tool to make sense of correlations between personal sleep data and contextual factors. Pers Ubiquit Comput. 2016;20(6):985–1000.

51. Whooley, M., Ploderer, B., & Gray, K. On the integration of self-tracking data amongst quantified self. Proceedings of British HCI 2014. 2014. Retrieved from http://tinyurl.com/m3lj2c7

52. Rooksby J, Rost M, Morrison A, Chalmers M. Personal tracking as lived informatics. In: Proceedings of the SIGCHI conference on human factors in computing systems; 2014. p. 1163–72.

53. Ravichandran R, Sien SW, Patel SN, Kientz JA, Pina LR. Making sense of sleep sensors: how sleep sensing technologies support and undermine sleep health. In: Proceedings of the 2017 CHI conference on human factors in computing systems; 2017. p. 6864–75.

54. Choe EK, Lee NB, Lee B, Pratt W, Kientz JA. Understanding quantified-selfers' practices in collecting and exploring personal data. In: Proceedings of the SIGCHI conference on human factors in computing systems; 2014. p. 1143–52.

55. Roomkham S, Lovell D, Cheung J, Perrin D. Promises and challenges in the use of consumer-grade devices for sleep monitoring. IEEE Rev Biomed Eng. 2018;11:53–67.

56. Liang Z, Ploderer B. How does Fitbit measure brainwaves: a qualitative study into the credibility of sleep-tracking technologies. Proceedings of the ACM on Interactive, Mobile, Wearable and Ubiquitous Technologies. 2020;4(1):1–29.

57. Li I, Dey A, Forlizzi J. A stage-based model of personal informatics systems. In: Proceedings of the SIGCHI conference on human factors in computing systems; 2010. p. 557–66.

58. Lupton D. Data mattering and self-tracking: what can personal data do? Continuum-Journal of Media & Cultural Studies. 2020;34(1):1–13. https://doi.org/10.1080/10304312.2019.1691149
59. Buysse DJ. Sleep health: can we define it? Does it matter? Sleep. 2014;37(1):9–17.
60. Gillespie T. Designed to 'effectively frustrate': copyright, technology, and the agency of users. New Media Soc. 2006;8(4):651–69.
61. Munson SA. Rethinking assumptions in the design of health and wellness tracking tools. Interactions. 2017;25(1):62–5. https://doi.org/10.1145/3168738.
62. Liu W, Ploderer B, Hoang T. In bed with technology: challenges and opportunities for sleep tracking. In: Proceedings of the annual meeting of the Australian special interest Group for Computer Human Interaction; 2015. p. 142–51.
63. Pina LR, Sien S-W, Ward T, Yip JC, Munson SA, Fogarty J, Kientz JA. From personal informatics to family informatics: understanding family practices around health monitoring. Paper presented at the Proceedings of the 2017 ACM Conference on Computer Supported Cooperative Work and Social Computing, Portland, Oregon, USA; 2017. https://doi.org/10.1145/2998181.2998362.
64. Costa Figueiredo M, Chen Y. Patient-generated health data: dimensions, challenges, and open questions. Foundations and trends®. Human-Computer Interaction. 2020;13(3):165–297. https://doi.org/10.1561/1100000080.

Chapter 8
Improving Work Capacity and HRQoL: The Role of QoL Technologies

Joan Julia Branin

Introduction

Physical activity is closely linked with health and well-being; however, many Americans do not engage in regular exercise. Only one in five adults in the US meets the CDC physical activity guidelines of 150 min of aerobic activity and 2 days of muscle strengthening activity per week [1]. This trend of inactivity increases with age and can interfere with an individual's capacity to work. The consequences of these trends are that the average worker can no longer deliver a full day's effort in a physically, psychologically, and cognitively demanding job. Degenerative diseases associated with inactivity and obesity are epidemic. The benefits of physical activity and fitness extend beyond job performance and work capacity and include longer life and enhanced quality of life. Fit workers are more productive, are absent fewer days, and have a more positive attitude toward work and life in general [2].

Physical activity and fitness can be quantified by leveraging the available, affordable fitness technology. Fitness technology, including trackers and smartphone applications (apps), have become increasingly popular for measuring and encouraging physical activity in recent years. Such technology encompasses individual fitness trackers that can stand alone, a fitness tracker paired with a companion app, or an app that can be downloaded onto a smartphone without the need for an extra device. The fitness tracker market is currently thriving, with estimates of almost 1.5 billion dollars in revenue last year alone [3] and is expected to increase to a five-billion-dollar industry by 2019 [4]. A 2013 analysis revealed that there are over 41,000 health and fitness apps currently available to the public via iTunes (e.g., Map My Walk, Runkeeper, My Fitness Pal) [5] and over half of smartphone users report having downloaded such an app [6]. Other personalized QoL technologies exist to

J. J. Branin (✉)
Center for Health & Aging Research, Pasadena, CA, USA

K. Wac, S. Wulfovich (eds.), *Quantifying Quality of Life*, Health Informatics,
https://doi.org/10.1007/978-3-030-94212-0_8

assess and improve physical activity and work capacity such as heart rate monitors, sleep trackers, and smart scales, to name a few.

These physical activity and fitness trackers and apps are among the increasing number of Quality of Life Technologies (QoLT) for the assessment or improvement of an individual's QoL. QoLT leverage the increasing availability of miniaturized computing, storage, and communication sensor- and actuator-based, context-rich technologies that can be embedded within various personal devices, for example, smartphones and wearables. They rely on hardware technologies (i.e., devices or physical interaction elements) or software technologies (e.g., apps or web-based interfaces) or, most likely, a combination of both. QoLT on an increasing scale can be personalized to satisfy the intended needs of the user anywhere and anytime and can be used in a continuous and longitudinal yet minimally intrusive way, in the individual's daily life [7]. Overall personalized, and miniaturized computing QoL technologies have the potential to be change agents for an individual's physical work capacity and health- related quality of life and significantly impact public health, research, and policy.

This chapter addresses (1) the factors associated with variations in work capacity and quality of life; (2) the state-of-art of personalized, miniaturized computing QoL technologies for measuring and improving individual work capacity; (3) the use of activity trackers to quantify work capacity; and (4) strategies to enhance use of Web-based and non-Web-based tools and fitness technology for behavioral change, health management, and rehabilitation interventions for the self-management of work capacity and enhancement of health-related quality of life across the lifespan. This chapter concludes with guidelines for the monitoring and evaluating of digital health technologies and eHealth interventions and suggestions for future development of tools for the assessment and remediation of working capacity. The research question that guides this literature review is: How does the use of self-management QoL technologies affect work capacity and reported health-related quality of life?

Definition and Measurement of Work Capacity

According to the WHOQOL theoretical model, *working capacity* is the facet of physical health that examines a person's use of his or her energy for work. "Work" is defined as any major activity in which the person is engaged. Major activities might include paid work, unpaid work, voluntary community work, full-time study, care of children and household duties. This facet focuses on a person's ability to perform work, regardless of the type of work [8].

In the past, many research studies on physical activity and work capacity have relied on self-report instruments. However, such subjective measures of activity can be highly unreliable, stemming from memory, social acceptability and other biases, revealing both higher and lower estimates than an objective measurement [9, 10]. Since then, several standardized instruments have been developed to measure work capacity.

One of these standardized instruments, **The Work Ability Index (WAI)**, is an instrument consisting of seven items, which takes into consideration the demands of work and the worker's health status and resources and is used to assess the work ability of workers during health examinations and workplace surveys. The purpose of the WAI is to help define necessary actions to maintain and promote work ability. The validity and reliability of the WAI has been assessed in correlation analyses. The WAI and all its items have been shown to reliably predict work disability, retirement, and mortality [11]. More recently, the validity of WAI has been studied by Radkiewich [12] and WAI's test–retest reliability by de Zwart [13]. The WAI has become the standard for assessing work capacity in occupational health research and the daily practice of occupational health care and has been translated into 24 languages.

Another instrument, **The Work-ability Support Scale (WSS),** is a newer tool designed to assess vocational ability and support needs following onset of acquired disability to assist decision-making in vocational rehabilitation. It is designed to be used both for people who are working, or, as a planning tool for those considering returning to work. The tool has 16 items across three domains of work functioning: physical/environment, thinking and communicating, and social/behavioral. Scores ranges from 1 for constant support to 7 for independence. There are also an additional seven items related to contextual factors outside the workplace that could affect work functioning. Its scoring accuracy and rater reliability has been supported [14, 9].

Work capacity can also be assessed through a **Functional Capacity Evaluation (FCE)** which evaluates an individual's capacity to perform work activities related to his or her participation in employment. The FCE process is a set of tests, practices and observations highly specific for a type of job for which the individual is being assessed, as it compares the individual's health status, and body functions and structures to the demands of this job and the work environment. It can provide an accurate measurement of an individual's ability to perform critical work tasks. This can help to determine an individual's capability/ability to return to work or their employability. An FCE is performed on a one-on-one basis and can last up to 4 hours. A well-designed FCE should consist of a battery of standardized assessments that offer results in performance-based measures and demonstrates predictive value about the individual's return to work. The FCE report includes an overall physical demand level, a summary of job-specific physical abilities, a summary of performance consistency and overall voluntary effort, job match information, adaptations to enhance performance, and treatment recommendations, if requested [15].

Two self-reports have been utilized to assess the physical aspects of work capacity in cross-cultural settings. One instrument, **The International Physical Activity Questionnaire- Long (IPAQ-L)** is a 27-item questionnaire for use by either telephone or self-administered methods that can be used to obtain internationally comparable data on health–related physical activity in theses domains (domestic physical activity (PA), occupational PA, leisure-time PA, active transportation and sitting time) and intensities of PA (vigorous, moderate, and walking). The instrument has undergone extensive reliability and validity testing across 12 countries (14 sites). IPAQ has high reliability and moderate criteria validity in comparison with accelerometers. Good test-retest reliability for total PA, occupational PA, active transportation, and vigorous intensity activities was shown. The results suggest that these measures have acceptable measurement properties for use in many settings and in different languages and are suitable for national population-based prevalence studies of participation in physical activity [16, 17].

Another instrument, **The Global Physical Activity Questionnaire (GPAQ)** is a 16-item test developed by WHO as an improvement of the International Physical Activity Questionnaire (IPAQ) for use in cross-cultural settings. It collects information on physical activity participation in three settings or domains (activity at work, travel to and from places, and recreational activities) and sedentary behavior. It assesses work-related abilities such as able to "perform work involve vigorous-intensity activity that causes large increases in breathing or heart rate like [carrying or lifting heavy loads, digging or construction work] for at least 10 minutes continuously" and "time spent walking or bicycling for travel on a typical day." Studies have shown fair-to-moderate validity of the GPAQ in a self-administered format in German, French, and Italian [18].

A few questionnaires have been designed for wide-scale, population-based surveillance of occupational physical activity (PA) behaviors. The **Occupational Physical Activity Questionnaire (OPAQ)** is a seven-item survey that identifies the average time per week spent in occupational tasks, e.g., sitting or standing, walking, and heavy labor activities. The modifications made when designing OPAQ improved its reliability for persons with stable work patterns, but at the expense of poorer reliability for persons with more variable PA. OPAQ did not have superior validity to IPAQ. OPAQ showed moderate to high 2-week test–retest reliability and moderate criterion validity when compared with detailed occupational PA records. The validity of the OPAQ is similar to other established occupational PA questionnaires [19].

In addition, certain occupations have developed work capacity tests to assess the specific demands of their required work. An example, **The Work Capacity Test** (WCT), is a family of tests to determine firefighters' physical capabilities to perform the duties of wildland firefighting and to meet National Wildfire Coordinating Group (NWCG) standards for wildland firefighters. There are three levels of tests known as the "pack test" (arduous), "field test" (moderate), and "walk test" (light). The Arduous Pack Test is intentionally stressful as it tests the capacity of muscular strength and aerobic endurance of firefighters. Considerable effort has been spent on validating the test to the work demands of US wildland firefighters, for whom the test displays content validity; however, work is still needed to verify its reliability and criterion and construct validity [20].

Research Studies on Variations in Work Capacity and QoL

Studies have documented greater exercise and physical work capacity among people who are active compared with sedentary individuals [21, 22]. Individuals with self-rated good health are more likely to be employed [21]. Participants who had the capacity to work graded themselves as having both better health and HRQoL than those with a non-capacity to work [21].

Physical activity in older adults not only improves their physical work capacity but also reduces the risk of chronic diseases, such as cardiovascular disease, stroke, obesity, and hypertension; improves cognitive and mental health; lowers the chance of falls; and helps maintain a longer independent life. While a great deal of variation exists, significant declines in physical work capacity have been reported between the ages of 40 and 60 years due to decreases in aerobic and musculoskeletal capacity. These physical declines can lead to increases in work- related injuries and illness. Differences in habitual physical activity, among other factors, greatly influences the variability seen in individual physical work capacity and its components [23].

Chronic diseases have an enormous impact on the ability to work. Being able to work is particularly important for the quality of life of people with chronic diseases [24]. Among individuals diagnosed with chronic conditions such as multiple sclerosis (MS), a majority suffer with fatigue, which strongly influences their everyday life. Flensner [25] found that individuals with MS who had the capacity to work reported significantly less fatigue compared to those with no capacity to work. Additionally, the level of work capacity was significantly higher among those participants who were less sensitive to heat, while those who were sensitive to heat showed significantly more often a non-capacity to work. This study lends support to some existing evidence of the beneficial impact of good health on work ability in patients with MS.

The relationship between physical and functional capacity and quality of life among elderly people who have a chronic disease was demonstrated in a study by Oztürk. Oztürk [26]. found that there are differences among elderly female and male individuals with a chronic disease in terms of the number of chronic diseases, types of chronic disease, mobility level, functional status, and QoL. One difference, mobility level, is related to functional capacity and QoL particularly in females; the higher the mobility level, the higher the QoL in females Rehabilitation programs to improve physical and functional capability and participation in daily activities may improve quality of life.

Participation in regular physical activity has been associated with better cardio-metabolic indices [27], skeletal health [28], and cognitive and academic performance [29] in young people aged 7–18 years. Conversely, doing little or no physical activity in youth has been related to poor health outcomes and decreased quality of life in adulthood resulting in extended medical care and associated costs [30]. One study showed that the cost of MS in Sweden is about €600 million a year, one third of which are indirect costs associated with loss of production [31].

Reduced capacity to work is a major cause of high medical care costs and personal care costs in the later stages of a disease [30].

The State-of-Art of Personalized Computing QoL Technologies for Measuring Physical Activity

Types of Wearable Activity Trackers and Other Devices

In the past, many research studies on physical activity and work capacity have relied on self- report and self-report instruments to assess exercise behavior and capacity. However, such subjective measures of activity can be highly unreliable, revealing both higher and lower estimates than an objective measurement. People may over-estimate activity intensity or time spent to present a favorable impression. Self-reports often focus on discrete activities such as going for a run or working out at a gym. Objective measurement of physical activity can record activity that may be missed in a questionnaire or self-report. This may be especially relevant for older adults who are less likely to go to a gym on a regular basis [32]. Thus, there is an increased interest in using objective activity monitors and other available, affordable QoL technologies to quantify physical activity and work capacity.

Activity trackers refer to sensor-based personalized wearable devices that automatically track and monitor various indicators of physical activity, such as steps taken, stairs climbed, duration and quality of sleep, pulse or heart rate, and self-reported calories burned. Newer devices feature an electrodermal activity (EDA) sensor to help measure your stress level, an ECG sensor to assess your heart rhythm, an SpO2 sensor to measure the amount of oxygen in your blood, and a skin temperature sensor. Some include built-in GPS. Activity trackers synchronize this data with users' personal accounts, ensuring easy access from any device by the user. Activity trackers provide relatively unbiased data about basic physical activities and have the advantage of boosting physical activity through the integration of empirically tested behavioral change techniques such as goal setting, self-monitoring, social support, social comparison, feedback, and rewards [33], in contrast to antecedent technologies, such as pedometers. Self-monitoring and goal setting using activity trackers have been especially effective in promoting self-efficacy and physical activity in interventions to improve capacity for work [28]. Studies have shown that physical activity and fitness is increased using wearable activity trackers [21, 23].

Good fitness level may increase the physical capacity for work as well as enable one to operationalize the factors important for self-reported work capacity. For example, wearable activity trackers data may enable one to quantify the answer for the WAI factor of "work ability in relation to the demands of the job" or "estimated work impairment due to diseases." As for the WSS scale, fitness trackers may enable one to quantify the "physical and motor" or "mobility and access", "community mobility" and even "stamina and pacing" factors. Also, activity trackers may be useful in quantifying physical capabilities and capacity for certain work functions.

Fitness Trackers Fitness trackers are at the heart of the fitness technology movement and have broad appeal. Their main appeal is that they create a snapshot of an individual's physical fitness, which can empower individuals to make healthy changes that impact their ability to engage in everyday work and nonwork activities. According to 2020 PCMag [33], a few examples of trackers that do a little of everything are the *Fitbit Inspire HR* and *Fitbit Charge 3*. Easy-to-use and easy-to-read collected data, they are both excellent options for first-time users. If an individual is more interested in having a full-fledged smartwatch that includes fitness tracking, then the *AppleWatch Series 5* (which does a bit of everything except measure sleep) or *Samsung Galaxy Fit* are the best options. With a smartwatch, an individual gets apps and a lot more functionality, such as the ability to send text messages from one's watch. One major disadvantage is battery life. The best fitness trackers can last a week or more, but smartwatches usually need to be charged once a day. For runners, cyclists, and anyone else who is already invested in fitness and monitoring peak work capacity, the best option is a runner's watch that doubles as an all-day fitness tracker such as the *Forerunner 45*.

Heart Rate Monitors (HRM)   HRMs read an individual's pulse while working out or active and are usually either chest straps or watch-style devices. They are not meant for monitoring heart rate 24/7 for medical purposes; for that, one needs an HRM that is FDA- approved or the equivalent in another country. Polar's heart rate monitors, including the *Polar OH1*, have excellent tools for finding heart rate zones as well as explaining the activity's benefits for the heart and body. Some fitness trackers or running watches have a heart rate sensor built in and allow one to see their heart rate in real time. Some such as *JBL's Reflect Fit* are headphones, which take the pulse from the ear. HRMs may enable better quantification of stamina and pacing contributing to increased work capacity (WSS).

Smartphone Fitness Apps   Smartphone fitness and exercise apps provide users with data to quantify physical activities such as their "physical and motor" capacity and "stamina and pacing" with the additional advantage of immediate data aggregation and analysis. Users can then modify their behaviors or work activities accordingly. Additionally, dedicated smartphone- based apps may enable one to quantify the "sensory and perceptual skills" (WSS) required by the job.

For free workouts, *The Johnson & Johnson Official 7-Minute Workout* app has a variety of workouts for people of all fitness levels and of different lengths which are great for people who are just getting started with exercise and for frequent travelers to use in their hotel room.

Run-tracking apps, such as *Runkeeper* and *Strava*, use the phone (or a compatible watch or fitness tracker) to record a runner's pace, distance, mileage, and more. Apps such as *MapMyFitness* can track non-sport activities—for instance, shoveling snow, raking leaves, or walking briskly. *MyFitnessPal* app is one of the best apps for logging the foods one eats to count calories and get a nutritional breakdown. *Weight Watchers* and *Noom* include access to coaches through their apps as well as community aspects so the user will not be alone on your fitness journey. If one wants hardcore training by an MMA champion, *Touchfit: GSP* gives an individual a series of progressively harder videos by fighter George St-Pierre. Jillian.

Michaels has a similar app with video-based workouts and a diet plan. Then there are apps and sites that specialize in one type of class, such as ballet, barre, and yoga or where you get a personal trainer to work with you on a personal fitness plan. Additionally, these dedicated smartphone-based apps resources may assist in quantifying "work ability" (WAI) and "sensory and perceptual skills" (WSS) required by the job.

Sleep Trackers Sleep may influence one's stamina, pacing and physical capacity to work and thus is an important variable to assess. Sleep trackers can track the user's sleep so that an individual can learn more about his/her sleep patterns. Most fitness trackers now include sleep tracking. There are also smart mattresses with tracking technology built in. The most advanced, such as the *Sleep Number 36* makes adjustments during the night that are tailored to the user. Or one could consider a smart pillow that plays white noise and detects snoring, such as the *REM- Fit Zeeq Smart Pillow*.

Other Devices There are other aspects of daily life and its relation to potential work capacity that are worth tracking and operationalizing. Smart water-bottles track your water consumption and remind an individual to drink to minimize dehydration. Smart clothing gives feedback to runners about their form in real time, commenting on heel strikes, cadence, etc.

Some fitness enthusiasts believe that the secret to maximizing their fitness potential is in their blood. A service called *InsideTracker* will send a phlebotomist to one's home or office to collect a blood sample and send it to a lab. There is even a wearable *DNAband* that helps an individual choose groceries based on their DNA, but it is now only available in the UK.

Evidence for the Use and Effectiveness of Activity Trackers

Academic and industry research has shown that the use of activity trackers can increase physical activity through continuous monitoring of activity progress, motivational messages, social support, and many other empirically tested behavioral change techniques [34–36]. That, in turn, may influence the work capacity [2, 26].

Wearable activity trackers have been shown to be effective in measuring and increasing the types of physical activity (e.g., steps, distance, calories expended, heart rate) that lead to increase work capacity. Adults who started using wearable activity trackers have been shown to increase daily activity levels [21]. A 7-month study of 18 participants (aged 36 to 73 years) who were given a wearable tracker found that 16 participants continued to use it after 7 months. The benefits of use included weight loss, social connection, and increased activity awareness [37].

Participants aged 60 years and older who were given a tracker reduced waist circumference and increased step count during another 12-week study [35]. African American and Hispanic older female participants, who tested a newly developed tracking device in a 7-week study, increased their physical activity level, lost weight, and lowered blood pressure levels [21]. Activity trackers have been found to be more effective than their predecessors, where sedentary female older adults who used digital trackers significantly increased their physical activity compared with those who used pedometers [21]. A tracker that delivered prompts via a short message service has also been found effective in increasing moderate-to-vigorous physical activity among overweight and obese adults [38].

Activity trackers facilitate physical activity in both young and older adults and are particularly beneficial for older adults because of the protective power of physical activity against diseases associated with older ages [36]. Despite the evident benefits of activity trackers for older generations, digital care today is more available to younger populations, leaving older adults on the periphery of the industry [39]. As little as 7% of older adults owned an activity tracker in 2014 [40]. Although many adults are now aware of this technology and its increased popularity, this population still shows slow rates of adoption that depend on many factors, including activity tracker trial and price [41]. Almost 84% of older adults aged 65 years and older do not meet the aerobic and muscle-strengthening physical activity requirements [42], which makes activity trackers a particularly relevant technology for this age group. Physical activity recommendations for older adults tend to focus on moderate-intensity aerobic and muscle-strengthening activities such as walking, jogging, bicycle riding, yard work, and gardening [43]. Some of these activities are tracked by wearable technology.

Activity trackers have the advantage of boosting physical activity through the integration of empirically tested behavioral change techniques such as goal setting, self-monitoring, social support, social comparison, feedback, and rewards [36] in contrast to antecedent technologies, such as pedometers. Self-monitoring and goal setting have been especially effective in promoting self-efficacy and physical activity in functional capacity interventions. Although wearing a new piece of health technology is a novel activity for older adults, they appreciate the activity tracker's contribution to self-awareness and goal setting. Activity trackers provide older adults with relatively unbiased data about basic activities. In addition, older adults view activity trackers as helpful motivators in achieving walking goals and competing with themselves [44].

Another important advantage of activity tracker usage in the 65+ population is social connection. For a population that is characterized by social isolation and loneliness, technology such as activity trackers that addresses social connectivity needs is perceived as helpful in overcoming barriers to increase physical activity [45].

Although activity trackers can be helpful in increasing physical activity, this technology is not ideal. For example, in a study with 8 adults who were aged 75 years or older, 3 participants experienced technical problems with the activity trackers, preventing them from gathering any activity feedback. Participants reported that they could only get the activity tracker to work 78% of the time [46].

Thus, the full potential of quantifying and leveraging the available, affordable fitness technology in enhancing the capacity to work has only begun to be realized.

Each year additional new and upgraded wearable devices and new apps with expanded tracking features such as monitoring blood oxygen saturation levels, feedback on sleep quality and stamina levels, and anxiety and panic attack tracking are being introduced to enhance the ability of these devices to monitor and facilitate major improvements in physical work capacity.

Validity and Reliability of Fitness Trackers for Measuring Physical Activity

Many studies have tested the utility of fitness trackers for measuring physical activity, with varied results [46, 47]. For each fitness tracker, there are many models, with a variety of algorithms that provide activity estimates. It is important to note the brand, model, and body placement used to facilitate comparisons across studies. Early studies found that fitness trackers including the wrist-worn Fitbit and Fitbit Ultra and the waist-worn Fitbit One have acceptable reliability and validity comparable to research–standard devices in the lab [47]. Van Remoortel.

[47] conducted a systematic review to identify whether available activity monitors (AM) have been appropriately validated for use in assessing physical activity in healthy populations and in those with chronic diseases. The latter population walk more slowly than healthy individuals which is reflected in their six-minute walking distance. This review suggests that most monitors (and pedometers) are less accurate during slower walking speeds. Validation studies of activity monitors are highly heterogeneous which is partly explained by the type of activity monitor and the activity monitor outcome. Since activity monitors are less accurate at slow walking speeds and information about validated activity monitors in chronic disease populations is lacking, proper validation studies in these populations are needed prior to their inclusion in clinical trials.

From Activity Trackers to Quantified Work Capacity

With respect to use of the evidence of fitness data and work capacity, wearable activity trackers data may enable one to quantify the answers for self-reports and self-report instruments.

Good fitness level may increase the physical capacity for work as well as enable one to operationalize the factors important for self-reported work capacity. For example, the wearable activity trackers data may enable one to quantify the answer for the WAI factor of "work ability in relation to the demands of the job" or "estimated work impairment due to diseases". As for the WSS scale, the fitness trackers may enable to quantify the "physical and motor" or "mobility and access", "community mobility" and even "stamina and pacing" factors. In addition, they may be useful in quantifying physical capabilities and capacity for certain work functions.

Evaluating the match between the physical, mental, social, environmental, and the organizational demands of a person's work and his or her capacity to meet these demands are important in assessing work capacity. Measurement of workability requires consideration of a range of factors, including physical ability to perform tasks, ability to cope with the cognitive/communication demands of the job, and to function appropriately in the social and environmental context of the work. Wearable activity trackers can quantify many of these factors and are especially useful for individuals with physically demanding work such as a mail carrier/postman and a wildlands firefighter (see Table 8.1).

Strategies to Improve Work Capacity Using Behavioral Change Techniques and QoLT Interventions

Work capacity is the employee's ability to accomplish production goals without undue fatigue, and without becoming a hazard to oneself or coworkers. It is a complex composite of aerobic and muscular fitness, natural abilities, intelligence, skill, experience, acclimatization, nutrition, and, of course, motivation. Even the most highly motivated workers may fail if they lack the strength or endurance required by the job. For prolonged arduous work, fitness is the most important determinant of work capacity [2]. While a great deal of variation exists, an average decline of 20% in physical work capacity has been reported between the ages of 40 and 60 years, due to decreases in aerobic and musculoskeletal capacity. These declines can contribute to decreased work capacity, and consequential increases in work-related injuries and illness [23].

Variations in work capacity over time can be quantified and self-monitored using quality of life technologies. The increasing availability of miniaturized computing, storage, and communication sensor-and actuator-based, context-rich technologies that can be embedded within various personal devices, such as smartphones and wearables, offer the opportunity to increase physical activity and lessen some work capacity declines through the continuous measuring, monitoring, and modifying of lifestyle behaviors such as the type, duration, and intensity of exercise, nutrition, sleep quality, and water hydration [48–50].

Use of Digital Fitness Devices and Behavioral Change Management Techniques

One strategy to improve work capacity is by using personalized, wearable activity tracking and physical monitoring devices with behavioral change techniques (BCTs) and evidence-based behavioral interventions [51]. Some of the more common behavioral change techniques are goal setting, feedback, rewards, social support,

Table 8.1 Quantification of Work Capacity Using Wearable QoL Technologies

Scale and Question/ Factor on Standardized Instrument	Data Source to Quantify the Factor	Sampling Rate	Notes, Other Factors	Examples: Mail Carrier/ Postman and Wildlands Firefighter
General use				
WAI: "Work ability in relation to the demands of the job"	Fitness trackers: Steps, distance, ECG	Daily	Need benchmark to quantify "demands of the job"	Delivery of the mail to door either on foot with or without a mail buggy or motorized vehicle/ electric bicycle. Ability to lift heavy mail sacks.
WAI: "Estimated work impairment due to diseases"	Fitness trackers: Steps, distance, ECG, SpO2	Daily/ weekly depending on nature of illness	Need benchmark capacity to quantify "impairment"	Having a limit of max number of km/ mi a day due to illness.
WSS: "Physical and motor"	Fitness trackers: Steps, distance, ECG detailed motor skills via accelerometer	Weekly report	Need benchmark of motor skills required (e.g., lifting, dexterity, coordination, balance).	Ability to bike and/or drive long distances and stretches of time. Dexterity and ability to open different types of doors and manipulate a huge number of letters/ packages with high accuracy.
WSS: "Sensory and perceptual"	Smartphone: Use, tests, GPS, water hydration levels	Daily/ weekly	Need benchmark of visual, auditory, and perceptual skills required.	Recognition of where the post man is and delivery location. Mapping mail route using GPS Navigational system degree of water hydration.
WSS: "Mobility and access"	Fitness trackers: Steps, distance, smartphone: Indoor/outdoor, mobility type, duration, exercise modality	Daily	Need benchmark of mobility skills required.	Covering specified number of km/mi a day (biking+ walking).

Table 8.1 (continued)

Scale and Question/ Factor on Standardized Instrument	Data Source to Quantify the Factor	Sampling Rate	Notes, Other Factors	Examples: Mail Carrier/ Postman and Wildlands Firefighter
WSS: "Community mobility"	Smartphone: Indoor/outdoor, mobility type, duration, exercise modality	Weekly, as needed	Need definition of community and mobility job requirements.	Covering specified number of km/mi a day (biking+ walking), number of residences or businesses or geographic area a day
WSS: "Stamina and pacing"	Sleep HRM, HRV Distance per time (km/h or mph),	Weekly	Self-reported energy and fatigue levels.	Speed of biking or walking (app, tracker) may also use speech recognition for energy assessment.
FCE form "climb stairs, ladders" "walk on uneven ground"	Fitness trackers: Steps, distance, ECG, SpO2	As needed	Observation/ evaluation by independent third party. Standards vary for return-to-work by job category. Employment or union standards.	Climb stairs and steep driveways to deliver mail walk over uneven terrain or over objects to deliver mail.
IPAQ-L; "Vigorous activities" "moderate activities" and "walking". "As part of your work": Duration and number of times	Fitness trackers steps, distance, ECG, minutes cardio levels	Daily/ typical week	Self-report job- related physical activity levels in past 7 days. Need benchmark of stamina and motor skills requirements. Established by employer or union	Delivery of the mail to door either on foot with or without a mail buggy or motorized vehicle. Ability to lift heavy mail sacks

(continued)

Table 8.1 (continued)

Scale and Question/ Factor on Standardized Instrument	Data Source to Quantify the Factor	Sampling Rate	Notes, Other Factors	Examples: Mail Carrier/ Postman and Wildlands Firefighter
Global physical activity questionnaire (GPAQ). "Perform work that involves vigorous-intensity activity" "large increases in breathing or heart rate"" carrying or lifting heavy loads... for at least 10 minutes continuously" "time walking or bicycling for work Travel"	Fitness trackers, steps, distance, time, HRM, cardio levels oxygen saturation levels, SpO2, ECG	Daily/ typical week	Self-report job- related physical activity levels in past 7 days. Need benchmark of stamina and motor skills requirements.	Delivery of the mail to door either on foot with or without a mail buggy or motorized vehicle. Ability to carry heavy mail sacks.
Physical activity assessment tool (PAAT) «been regularly physically active for the last 1–5 months"	Fitness trackers, steps, distance, time, HRM, ECG	Monthly	Need benchmark to quantify demands of the work vs non work activities	Ability to deliver mail, walk distances.
OPAQ: "Hours walk at work" "hours sitting or standing during work" "hours perform heavy labor activities at work"	Fitness trackers, steps, distance, HRM, ECG, accelerometer,	Weekly	Need benchmark to quantify specific demands of the work	Ability to stand to deliver mail and walk distances; carry heavy mail sacks.
Specific use				
WCT: (job Specific test used by wildlands firefighters) WCT: "Pack test"	Fitness trackers: Steps, distance, water hydration, HRM, HRV, ECG, SpO2	Daily, as required	Pack test benchmark of 3-mile hike with a 45-pound pack over level terrain in 45 minutes, approximates an aerobic fitness score of 45--the established standard for wildland Firefighters.	Example: Wildlands firefighter Ability to perform arduous work over prolong periods of time.

Table 8.1 (continued)

Scale and Question/ Factor on Standardized Instrument	Data Source to Quantify the Factor	Sampling Rate	Notes, Other Factors	Examples: Mail Carrier/ Postman and Wildlands Firefighter
WCT" field test"	Fitness trackers: Steps, distance, HRM, ECG, SpO2	Daily, as required	Field test benchmark of 2- mile hike with a 25-pound pack in 30 minutes. The passing score, approximates an aerobic fitness score of 40.	Ability to perform moderately strenuous duties e.g., considerable walking over irregular ground, standing for long periods, lifting 25 to 50 pounds.
WCT "walk test"	Fitness trackers: Steps, distance, stress	Daily, as required	Walk test benchmark of 1 mile in 16 minutes with a load, Approx. An aerobic fitness score of 35.	Ability to meet emergencies and evacuate to a safety zone.

Note: *WAI* The Work Ability Index; *WSS* The Work-ability Support Scale; *FCE* Functional Capacity Evaluation; *IPAQ-L* International Physical Activity Questionnaire-Long Version; *GPAQ* Global Physical Activity Questionnaire; *OPAQ* Occupational Physical Activity Questionnaire; *WCT* Work Capacity Test

coaching, identifying barriers/problem solving, and action planning. Self-regulatory behavior change techniques such as goal setting, self- monitoring, and social support have been associated with greater increases in physical activity than those that did not [32]. The extent that fitness trackers can enhance physical capacity to work may be dependent on *goal setting* such as the daily step goal of 10,000 steps or 250 steps within an hour. *Feedback* and *rewards,* another set of behavior change techniques closely tied to goal setting, include the activity tracker reminders, text messages, and real-time alerts when the user has met a goal or has been sedentary for too long which have been reported to be effective in motivating and increasing physical activity and fitness [52, 53]. Social factors such as *social support or competition* have been shown to increase engagement, adherence, and completion in physical activity interventions beneficial to modifying work capacity [52, 54]. *Coaching* is often used in fitness technologies and interventions to motivate increases in physical activity through weekly information sessions that encouraged healthy behaviors and a weekly 30-min group walking session [52]. BCT techniques used with fitness technology can lead to increases in physical activity, endurance, muscle strength, and decreases in physical work fatigue [32].

However, certain BCT strategies may be more effective for different age groups in enhancing the physical aspects of work capacity using wearable personal activity devices. A critical analysis by Mercer [54] noted that self-regulation techniques present in wearable activity trackers such as goal setting, feedback, and social support are effective for younger adults, whereas older adults may benefit more from problem-solving, rewards for successful behavior, and modeling or demonstrating behavior.

Some of these behavioral change techniques are included in fitness technology. An analysis of seven commercially available fitness trackers (Jawbone UP24, Nike Fuelband, Polar Loop, Misfit Shine, Withings Pulse, Fitbit Zip, and Spark) revealed that most or all of these trackers included goal setting, feedback, rewards, self-monitoring, and social support [54]. Many fitness trackers include a "mobile coach" within their connected application or website for encouraging increases in physical activity, stamina, and endurance. More recently, Lyons [36] used BCTs to systematically analyze 13 wearable activity trackers and found that all of these trackers helped users to self-monitor behavior, obtain feedback on behavior, and add objects to the environment while also generally support users in goal setting and comparing their behavior with their goal. However, many health and fitness apps do not include BCT strategies thus their impact on the physical activity and the strength, and endurance aspects of work capacity may be limited [32].

Challenges Using Quality of Life Technologies in the Improvement of Work Capacity

Several challenges exist in using quality of life technologies in the improvement of work capacity. One challenge for the use of wearable activity trackers for BCT for physical activity and work capacity interventions is that there is no guarantee that the user will encounter a technique even if it is present on the device. Another challenge is that trackers are complex tools with multiple features and different users are likely to have different experiences. Similarly, a user who has a lower health or technology literacy may not explore the features as deeply as a health professional or expert technology user, regardless of age. Thus, the potential of wearable activity trackers to increase fitness behaviors through BCTs leading to improved capacity for physical activity and work can vary greatly.

Discussion and Recommendations

The Quantified Self Movement (QS) is a relatively recent trend wherein QS practitioners rely on the wealth of digital data originating from wearables, applications, and self-reports to enable them to assess diverse domains of daily life---physical state, (e.g., mobility, steps), psychological state (e.g., mood), social interactions (e.g., number of Facebook "likes") and environmental context they are in (e.g., pollution) which contribute to an individual's Quality of Life (QoL). The collected QS data enables an individual's state and behavioral patterns to be assessed through these different QoL domains, based on which individualized feedback can be provided, in turn enabling the improvement in the individual's state and QoL [55, 56].

Continuous monitoring is at the heart of the Quantified Self movement since it often gives a more accurate picture of human motion realities than short periods of laboratory testing and may provide a more accurate assessment of individual work capacity across different time periods and work-related activities. The long-term vision of QS activity is that of a systemic monitoring approach where an individual's continuous personal information climate provides real-time performance optimization suggestions. The availability of powerful, personalized, and wearable mobile devices facilitates the provision of ubiquitous computing applications that enable clinicians and employers to better assess and monitor work capacity in the workplace and everyday life and to develop more effective interventions to enhance the physical and emotional aspects of capacity to work than previously possible.

While the concept of the quantified self may have begun with self-tracking at the individual level, the term is quickly being extended to include "group data" and the idea of aggregated data from multiple quantified selves as self-trackers share and work collaboratively with their data. Using QS group device data could be helpful in quantifying potential healthcare cost reductions and savings and for verifying user behaviors in behavioral change management programs in the public and private sectors. QS group data has the potential to benefit society by going beyond data creation and information generation to meaning-making and action-taking strategies with applications for impacting work capacity on the population level.

However, more research on the effectiveness of QS monitoring with QoL technologies is needed. To date, relatively few high-quality studies have been conducted examining the correlation between physical activity and work capacity and the overall effectiveness of physical activity trackers and smartphone interventions. Future studies should describe these interventions, the device and app features, and the AI and algorithms used with adequate detail so results can be reproduced, and lessons learned to advance work capacity and QoLT research.

Future research should also be directed toward enhancing an understanding of the time course of intervention effects in enhancing one's capacity for work. Of particular interest is a better understanding of the timepoint at which peak effect size is reached, the timepoint at which user engagement decreases, and the factors that underpin these phenomena. The relatively short- term nature of positive effects suggest that additional efforts are required to design app features which help sustain user engagement with the app over time, perhaps through modules, unlockable content, gaming, and rewards. Sustaining user engagement is particularly important for smartphone-based interventions quantifying the physical aspects of work capacity due to the absence of human support and minimal supportive accountability.

Mobile technology used for assessing and monitoring work capacity should continue to emphasize ease of use, function, feedback, tailored information, ability to personalize design, and design-aesthetic as highly ranked engagement strategies. It will be useful for future app designs to incorporate long-term engagement strategies as increased exposure to the intervention can lead to larger, longer lasting effects in improving physical fitness and in turn, influencing work capacity.

Fitness technology should continue to include theoretically derived behavior change techniques that are useful for their intended population. Strategies such as goal setting, self- monitoring, feedback, rewards, social support, and coaching seem to be especially helpful in increasing activity and healthy behaviors that impact the work capacity of younger adults; while older adults may benefit more from problem-solving, rewards for successful behavior, and modeling or demonstrating behavior. The use of QS technologies as different types of affordances supporting the goal-oriented actions by individuals can, in turn, improve capacity to work and their QoL.

Regardless of the type of intervention, efforts should be made to address the barriers that keep inactive people, especially older adults and low-SES populations, from using wearable activity devices to better help them engage in regular physical activity resulting in decreased sedentary behaviors and increased work capacity especially work requiring physical fitness and endurance.

It is recommended that app and mobile device developers and behavior change experts collaborate when developing apps used to monitor physical activity, physical fitness, or the physical aspects of work capacity. By understanding app usage, guidelines can be developed to create apps based on health behavior research to better quantify and promote long-term physical activity and sustainable work capacity.

Lastly, the use of digital, mobile, and wireless QoL technologies and tools in an evidence-based, structured environment can lead to a fundamental transformation of the patient- professional relationship and employer-employee relationship into collaborative partnerships focused on quantifying and improving work capacity and enhancing the overall health-related quality of life of individuals.

References

1. CDC. Facts about physical activity, 2014. Available from https://www.cdc.gov/physicalactivity/data/. Accessed 10 February 2020.
2. Sharkey B, Gaskill S. Fitness and work capacity 2009 edition. NWCG PMS 304-2. Boise, ID: National Wildfire Coordinating Group, safety and health working team, National Interagency Fire Center. 98p.
3. Ridley D. Year-over-year wearables spending doubles, according to NPD. The NPD Group, Inc; 2016. Available from: https://www.npd.com/wps/portal/npd/us/news/press-releases/2016/year-over-year-wearables-spending-doubles-according-to-npd/?utm_source=twitter&utm_medium=social&utm_content=Oktopost-twitter-profile&utm_campaign=Oktopost-Press+Releases
4. Lamkin P. Fitness tracker market to top $5bn by 2019. Available from http://www.wareable.com/fitness-trackers/fitness-tracker-market-to-top-dollar-5-billion-by-2019-995.
5. Aitken M, Gauntlett C. Patient apps for improved healthcare. NJ: Parsippany; 2013. Available from: http://obroncology.com/imshealth/content/IIHI%20Apps%20report%20231013F_interactive.pdf
6. Krebs P, Duncan DT. Health app use among US mobile phone owners: a national survey. JMIR Mhealth Uhealth. 2015, Nov 4;3(4):e101. https://doi.org/10.2196/mhealth.4924.
7. Wac K. Quality of life technologies. In: Gellman M, editor. Encyclopedia of behavioral medicine. New York, NY: Springer; 2020. https://doi.org/10.1007/978-1-4614-6439-6_102013-1.

8. WHO Working Group. The World Health Organization quality of life assessment (WHOQOL): development and general psychometric properties. Soc Sci Med. 1998;46(12):1569–85. https://doi.org/10.1016/s0277-9536(98)00009-4.

9. Falck RS, McDonald SM, Beets MW, Brazendale K, Liu-Ambrose T. Measurement of physical activity in older adult interventions: a systematic review. Br J Sports Med. 2015;50 https://doi.org/10.1249/01.mss.0000477099.67521.f6.

10. Prince SA, Adamo KB, Hamel ME, Hardt J, Gorber SC, Tremblay M. A comparison of direct versus self-report measures for assessing physical activity in adults: a systematic review. Int J Beh Nutr Phys Act. 2008;5:1–24. https://doi.org/10.1186/1479-5868-5-56.

11. Ilmarinen J, Tuomi K. Past, present and future of work ability, people and work res reports, 2004;65. Finnish Institute of Occupational Health, Helsinki, 2004. p 1–25.

12. Radkiewicz P, Widerszal-Bazy M. Psychometric properties of work ability index in the light of comparative survey study. International congress series 1280. The Netherlands: Elsevier; 2005. p. 304–9.

13. de Zwart BC, Frings MH, van Duivenbooden JC. Test–retest reliability of the work ability index questionnaire. Occup Med Rehabil. 2002 June;52(4):177–81. https://doi.org/10.1093/occmed/52.4.177.

14. Turner-Stokes L, Fadyl J, Rose H, Williams H, Schlüter P, McPherson K. The work-ability support scale: evaluation of scoring accuracy and rater reliability. J Occup Rehabil. 2013;24 https://doi.org/10.1007/s10926-013-9486-1.

15. Soer R, van der Schans CP, Groothoff JW, Geertzen JH, Reneman MF. Towards consensus in operational definitions in functional capacity evaluation: a Delphi survey. J Occup Rehabil. 2008;18:389–400. https://doi.org/10.1007/s10926-008-9155-y.

16. Craig CL, Marshall A, Sjostrom M, Bauman A, Booth M, Ainsworth B. International physical activity questionnaire: 12-country reliability and validity. Med Sci Sports Exerc. 2003;35:1381–95. https://doi.org/10.1249/01.MSS.0000078924.61453.FB.

17. van Poppel MNM, Chinapaw MJM, Mokkink LB, van Mechelen W, Terwee CB. Physical activity questionnaires for adults: a systematic review of measurement properties. Sports Med. 2010;40:565–600. https://doi.org/10.2165/11531930-000000000-00000.

18. Cleland CL, Hunter RF, Kee F, Cupples ME, Sallis JF, Tully MA. Validity of the global physical activity questionnaire (GPAQ) in assessing levels and change in moderate-vigorous physical activity and sedentary behaviour. BMC Public Health. 2014;14:1255. https://doi.org/10.1186/1471-2458-14-1255.

19. Reis JP, Dubose KD, Ainsworth BE, Macera CA, Yore MM. Reliability and validity of the occupational physical activity questionnaire. Med Sci Sports Exerc. 2005;37(12):2075–83. https://doi.org/10.1249/01.mss.0000179103.20821.00.

20. Petersen A, Payne W, Phillips M, Netto K, Nichols AD. Validity and relevance of the pack hike wildland firefighter work capacity test: a review. Ergonomics. 2010;53(10):1276–85. https://doi.org/10.1080/00140139.2010.513451.

21. Kononova A, Li L, Kamp K, Bowen M, Rikard RV, Cotten S, Peng W. The use of wearable activity trackers among older adults: focus group study of tracker perceptions, motivators, and barriers in the maintenance stage of behavior change. JMIR Mhealth Uhealth. 2019;7(4):e9832. https://doi.org/10.2196/mhealth.9832.

22. Macera C, Hootman JM, Sniezek JE. Major public health benefits of physical activity. Arthritis Rheum. (Arthritis Care & Research). 2003;7(1):122–8. https://doi.org/10.1002/art.10907.

23. Kenny GP, Yardley JE, Martineau L, Jay O. Physical work capacity in older adults: implications for the aging worker. Am J Ind Med. 2008 Aug;51(8):610–25. https://doi.org/10.1002/ajim.20600.

24. Stanton AL, Revenson TA, Tennen H. Health psychology: psychological adjustment to chronic disease. Annu Rev Psychol. 2007;58:565–92. https://doi.org/10.1146/annurev.psych.58.110405.085615.

25. Flensner G, Landtblom AM, Söderhamn O, Ek AC. Work capacity and health-related quality of life among individuals with multiple sclerosis reduced by fatigue: a cross-sectional study. BMC Pub Health. 2013 Mar;15(13):224. https://doi.org/10.1186/1471-2458-13-224.

26. Öztürk A, Şimşek TT, Yümin ET, Sertel M, Yümin M. The relationship between physical, functional capacity and quality of life (QoL) among elderly people with a chronic disease. Arch Geront Geriatr. 2011 Dec-Nov;53(3):278–83. https://doi.org/10.1016/j.archger.2010.12.011.

27. Ekelund U, Luan J, Sherar LB, Esliger DW, Griew P, Cooper A. Moderate to vigorous physical activity and sedentary time and cardiometabolic risk factors in children and adolescents. JAMA. 2012;307(7):704–12. https://doi.org/10.1001/jama.2012.156.

28. Gracia-Marco L, Moreno LA, Ortega FB, Leon F, Sioen I, Kafatos A, et al. Levels of physical activity that predict optimal bone mass in adolescents: the HELENA study. Am J Prev Med. 2011;40(6):599–607. https://doi.org/10.1016/j.amepre.2011.03.001.

29. Kwak L, Kremers SP, Bergman P, Ruiz JR, Rizzo NS, Sjostrom M. Associations between physical activity, fitness, and academic achievement. J Pediatr. 2009;155(6):914–8. e911. https://doi.org/10.1016/j.jpeds.2009.06.019.

30. Lee IM, Shiroma EJ, Lobelo F, Puska P, Blair SN, Katzmarzyk PT. Effect of physical inactivity on major non-communicable diseases worldwide: an analysis of burden of disease and life expectancy. Lancet. 2012;380(9838):219–29. https://doi.org/10.1016/S0140-6736(12)61031-9.

31. Berg J, Lindgren P, Fredrikson S, Kobelt G. Costs and quality of life of multiple sclerosis in Sweden. Eur J Health Econ. 2006;7(Suppl 2):S75–85. https://doi.org/10.1007/s10198-006-0379-5.

32. Sullivan AN, Lachman ME. Behavior change with fitness technology in sedentary adults: a review of the evidence for increasing physical activity. Front Public Health. 2017;4:289. https://doi.org/10.3389/fpubh.2016.00289.

33. Duffy J. The ultimate guide to health and fitness tech in 2020. 2020. PCMag. 27 Jan 2020. Available from https://www.pcmag.com/picks/the-ultimate-guide-to-health-and-fitness-tech. Accessed 2 Feb 2020.

34. Cadmus-Bertram LA, Marcus BH, Patterson RE, Parker BA, Morey BL. Randomized trial of a Fitbit-based physical activity intervention for women. Am J Prev Med. 2015 Sep;49(3):414–8. https://doi.org/10.1016/j.amepre.2015.01.020.

35. O'Brien T, Troutman-Jordan M, Hathaway D, Armstrong S, Moore M. Acceptability of wristband activity trackers among community dwelling older adults. Geriatr Nurs. 2015 Apr;36(2 Suppl):S21–5. https://doi.org/10.1016/j.gerinurse.2015.02.019.

36. Lyons EJ, Lewis ZH, Mayrsohn BG, Rowland JL. Behavior change techniques implemented in electronic lifestyle activity monitors: a systematic content analysis. J Med Internet Res. 2014 Aug 15;16(8):e192. https://doi.org/10.2196/jmir.3469.

37. Randriambelonoro M, Chen Y, Pu P. Can fitness trackers help diabetic and obese users make and sustain lifestyle changes? Computer. 2017 Mar;50(3):20–9. https://doi.org/10.1109/MC.2017.92.

38. Wang JB, Cadmus-Bertram LA, Natarajan L, White MM, Madanat H, Nichols JF, Ayala GX, Pierce JP. Wearable sensor/device (Fitbit one) and SMS text-messaging prompts to increase physical activity in overweight and obese adults: a randomized controlled trial. Telemed J E Health. 2015 Oct;21(10):782–92. https://doi.org/10.1089/tmj.2014.0176.

39. Levine DM, Lipsitz SR, Linder JA. Trends in seniors' use of digital health technology in the United States, 2011-2014. J Am Med Assoc. 2016 Aug 02;316(5):538–40. https://doi.org/10.1001/jama.2016.9124.

40. Ledger D, McCaffrey D. Endeavour Partners. Cambridge, MA: Endeavour Partners; 2014. [2019-03-20]. Inside wearables: how the science of human behavior change offers the secret to long-term engagement. https://medium.com/@endeavourprtnrs/inside-wearable-how-the-science-of-human-behavior-change-offers-the-secret-to-long-term-engagement-a15b3c7d4cf3 webcite.

41. Puri A, Kim B, Nguyen O, Stolee P, Tung J, Lee J. User acceptance of wrist-worn activity trackers among community-dwelling older adults: mixed method study. JMIR Mhealth Uhealth. 2017 Nov 15;5(11):e173. https://doi.org/10.2196/mhealth.8211.

42. Center for Disease Control and Prevention. [2018-01-29]. Adult participation in aerobic and muscle-strengthening physical activities - United States, 2011 https://www.cdc.gov/mmwr/preview/mmwrhtml/mm6217a2.htm webcite.

43. Elsawy B, Higgins KE. Physical activity guidelines for older adults. Am Fam Physician. 2010;81(1):55–9. http://www.aafp.org/link_out?pmid=20052963. Accessed 2 Feb 2020

44. Mercer K, Giangregorio L, Schneider E, Chilana P, Li M, Grindrod K. Acceptance of commercially available wearable activity trackers among adults aged over 50 and with chronic illness: a mixed-methods evaluation. JMIR Mhealth Uhealth. 2016;4(1):e7. https://doi.org/10.2196/mhealth.4225. http://mhealth.jmir.org/2016/1/e7/

45. Newall NE, Menec VH. Loneliness and social isolation of older adults: why it is important to examine these social aspects together. J Soc Pers Relationships. 2017;36(3):925–39. https://doi.org/10.1177/0265407517749045.

46. Ehn M, Eriksson LC, Åkerberg N, Johansson AC. Activity monitors as support for older persons' physical activity in daily life: qualitative study of the users' experiences. JMIR Mhealth Uhealth. 2018;6(2):e34. https://doi.org/10.2196/mhealth.8345.

47. Van Remoortel H, Giavedon S, Raste Y, Burtin C, Louvaris Z, Gimeno-Santos E, Langer D, Glendenning A, Hopkinson NS, Vogiatzis I, Peterson BT, Wilson F, Mann B, Rabinovich R, Puhan MA, Troosters T. Validity of activity monitors in health and chronic disease: a systematic review. Int J Behav Nutrition Phys Act. 2012;9:84. https://doi.org/10.1186/1479-5868-9-84.

48. Coughlin SS, Whitehead M, Sheats JQ, Mastromonico J, Smith S. A review of smartphone applications for promoting physical activity. Jacobs J Comm Med. 2016;2(1):021.

49. Romeo A, Edney S, Plotnikoff R, Curtis R, Ryan J, Sanders I, Crozier A, Maher C. Can smartphone apps increase physical activity? Systematic review and meta-analysis. J Med Internet Res. 2019;21(3):e12053. https://doi.org/10.2196/12053.

50. Scheid JL, West SL. Opportunities of wearable technology to increase physical activity in individuals with chronic disease: an editorial. Int J Environ Res Public Health. 2019;16(17):3124. https://doi.org/10.3390/ijerph16173124.

51. Chia GLC, Anderson A, McLean LA. Behavior change techniques incorporated in fitness trackers: content analysis. JMIR Mhealth Uhealth. 2019;7(7):e12768. https://doi.org/10.2196/12768.

52. Conroy DE, Yang CH, Maher JP. Behavior change techniques in top-ranked mobile apps for physical activity. Am J Prev Med. 2014;46:649–52. https://doi.org/10.1016/j.amepre.2014.01.010.

53. Normand MP. Increasing physical activity through self-monitoring, goal setting, and feedback. Behav Interv. 2008;23:227–36. https://doi.org/10.1002/bin.267.

54. Mercer K, Li M, Giangregorio L, Burns C, Grindrod K. Behavior change techniques present in wearable activity trackers: a critical analysis. JMIR Mhealth Uhealth. 2016;4(2):e40. https://doi.org/10.2196/mhealth.4461. http://mhealth.jmir.org/2016/2/e40/

55. Wac K. From quantified self to quality of life. In: Rivas H, Wac K, editors. Digital health. Health informatics. Springer; 2018. https://doi.org/10.1007/978-3-319-61446-5_7.

56. Wac K, Fiordelli M, Gustarini M, Rivas H. Quality of life technologies: experiences from the field and key challenges. In. IEEE Internet Comput. 2015, July-Aug;19(4):28–35. https://doi.org/10.1109/MIC.2015.52.

Part III
Psychological Health

Chapter 9
Quantifying the Body: Body Image, Body Awareness and Self-Tracking Technologies

Arianna Boldi and Amon Rapp

Introduction

Among the plethora of terms used to refer to self-tracking, "Quantified Self" (QS) is commonly employed to describe a community of people that attempt to gain "self-knowledge through numbers", believing that tracking is an essential starting point to make a change in the direction of an "optimal self". In Quantified Selfers' perspective, precise measurements and accurate data interpretation should lead to better awareness and improved knowledge, informing their everyday decisions, shaping their future, and, eventually, their identity [1].

In QS rhetoric, technological devices can overcome the natural limits that people encounter when they seek to gain self-knowledge, like a poor sense of time, a limited, fallible memory, and cognitive biases that negatively affect the opportunities for collecting relevant information to make decisions. Exact numbers collected by technology, instead, are powerful as they are not subject to memory distortion and, most importantly, *"they hold secrets they can't afford to ignore, including answers to questions they (people) have not yet thought to ask"* [2]. This belief is entangled with the empiricist idea that "observation" can convey a neutral, objective and clear comprehension of phenomena: unlike language, which is ambiguous and multivalent, data speak for themselves [3].

The availability of wearable devices and ubiquitous technologies recently boosted the popularity of self-tracking technologies even outside the strict circle of Quantified Selfers, reaching the broader population [4–6]. QS rhetoric then seeped

A. Boldi (✉)
Psychology Department, University of Torino, Torino, Italy
e-mail: arianna.boldi@unito.it

A. Rapp
Computer Science Department, University of Torino, Torino, Italy
e-mail: amon.rapp@unito.it

into the everyday use of these technologies, which are now integrated into a variety of practices in domains as diverse as health, sport, wellness, and safeness.

Achieving "self"-knowledge is strongly emphasized within the QS discourse, but do these technologies really support the development of an integrated knowledge about the self [7]? Or do they fragment the image that we have of ourselves into a variety of unrelated patterns of data? In this perspective, the body and the representation that self-tracking technologies convey of it gain a central importance.

Self-trackers are involved in a complex process of knowledge development, but this cannot be achieved without knowing the body, as knowledge is always situated and embodied [8]. However, self-tracking devices seem to embrace an abstract and scattered conception of the body, based on unrelated numbers, graphs, and depictions. This representation appears to not integrate into a coherent image that takes into account the body complex nature made up of perceptions, proprioceptive sensations, and self-representations. This may turn into biases and distortions of how we look at our bodies, worsening, rather than improving, our self-knowledge [9].

In order to understand the ways through which the progressive "quantification" introduced by self-tracking technologies is affecting the body, we need to preliminary explore a series of theoretical constructs concerning the body, which appear to be addressed differently by literature pertaining to different disciplines (e.g., Human-Computer Interaction, psychology, sociology, neurology). This diversity may entail unclear definitions and theoretical overlaps that may cloud our understanding of the current changes produced by technology on our bodies. *How are the concepts of the body and the self conceived? What are their relations? What kind of relationship is there among body schema, body image, and body awareness?* These are some of the questions we address in the first part of the chapter. The second part, instead, illustrates how individuals' body image and awareness are affected by the usage of self-tracking technologies in the sports domain. It clarifies the concepts introduced above, by surfacing how athletes use wearable data to inform their sports practices and eventually develop an understanding of their body. It shows both the opportunities and the risks introduced by the quantification of the body, by highlighting that self-tracking technologies may either increase the understanding of the athlete's body, or turn it into a series of aseptic information, which may distance the athlete from her body sensations and excessively "rationalize" her sports activity. This part builds on the empirical data collected through 20 interviews conducted with amateur and elite athletes, which have been previously published in a TOCHI article [9].

The Self and the Body

Self-trackers are interested in achieving a better knowledge on themselves, which can be useful to enrich or change several aspects of their lives. At least in principle, this goal should be achieved by placing the body at the center of the knowledge development process. However, it appears that the body, in its materiality and multifaceted nature, is clouded in self-tracking practices.

A core concept in critical investigations upon self-tracking is that of "digital double", also known as "data twin", "data double" or "datafied self", which results from data assemblages [3, 10]. These investigations emphasize that, albeit we naturally have a body made up of sensations, contemporary technologies feed back a "screen body", an abstract object dematerialized in a variety of data points. The body becomes something to be observed from a distance, controlled and managed with the help of technology.

Contemporary medicine is certainly the field in which technology has produced the most visible shift in the way bodies are treated: in the clinical practice, the symptoms recounted by the patients are losing their relevance, in favor of the visual examination mediated by technology (X-ray, RM, TAC), which is in charge of finding the "truth" about the body [11]. Likewise, self-tracking devices collect data that are not immediately visible and display them to the user, generating a "virtual" version of the body, a repository of storable and processable data [12, 13]. As an emerging effect of the quantification of human body through biometric practices, bodies are turned into numbers [14]. Nonetheless, numbers are not the natural way through which we represent our bodies.

People have different ways to relate to their body and, through them, they interact with the world and build their own sense of the self. Body schemata, body image and body awareness are theoretical constructs that point to particular ways of representing the body. These body representations are built upon a set of sensations, which are the object of perception. Human beings, however, are not purely reactive agents and perception is not something that "simply happens to us" [15, 16] nor senses are passive receptors. Rather, cognitive, emotional and even cultural factors influence the perceptive process, even when we consider the most primary aspects of the body, such as the heartbeat, which are tracked and measured by self-tracking devices.

This entails that sensations and body processes cannot simply translated into objective numbers aimed at capturing the "immediate" nature of our body. Actually, our relationship with our body is mediate by our representations, and there is a considerable gap between body sensations and their subjective appraisal. The goal of the next paragraphs is to provide a greater understanding of the ways we have to mediate the relationship between the body and our selves.

Body Constructs

The scientific literature about body representations points out six main theoretical constructs that operationalize the way we relate to our bodies: "body schema", "body image", "body awareness", "interoception", "exteroception", and "proprioception". The first three constructs concern the representations people have of their own body, resulting from the integration of various signals (e.g., touch, hearing, sight) and their processing at different levels (e.g., cognitive, sociocultural). Instead, the latter refer to the perceptive processes concerning the state of the body in relation to endogenous and exogenous stimuli. More specifically, proprioception is defined as the awareness of the body position and of the movements of the body; exteroception refers to the perception of the body arising from exogenous stimuli; interoception is a multidimensional construct that concerns the perception of sensations connected with body internal processes, like organ functioning [17–19].

Even though all these constructs are equally important to understand body-self relation, in the following we focus on the constructs concerning body representations, as they are more tightly related with the issues arising from the use of self-tracking devices to monitor body parameters.

Body Schema, Body Image, and Self-Tracking

There is large consensus in psychological and philosophical literature over the existence of two distinct types of body representation: body schema and body image [20, 21].

The concept of "body schema" has been first introduced by Bonnier [22] and further defined, by Head and Holmes [23], as a representation, mostly unconscious, of the body's position in space. By contrast, body image is depicted as a more conscious and intentional representation of the body, or a set of beliefs about the body.

These ways of representing the body are essential for building the self: the internal stream of sensations, which makes a person feel the body as her own, has been recognized as central for developing a stable sense of identity. Contemporary neuropsychological research showed that both deficits and distortions of body schema and body image lead not only to a variety of deficits in bodily experiences, such as personal neglect, apraxia or autotopagnosia [21, 24], but also to more complex disorders, such as anorexia and bulimia nervosa [25]. Sometimes, distortions in body schema and body image may be intertwined in the same syndrome, without a clear separation between them [26]. Nevertheless, body schema and body image should be treated separately since they refer to different ways of representing the body, as we will see in the next paragraphs.

Body Schema

Body schema is a representation of the body's spatial properties, a constantly updated postural model, mainly unconscious. The first investigations on body schema were focused on the somatosensory capacities of our bodies and their relation with the self [22, 23]. Later authors confirmed that the sensorimotor capacities of our bodies are fundamental for the construction of the self, as they shape our pragmatic possibilities to interact with the environment [21]. The fact that we have a body that moves in certain ways drives the way we perceive and act and this, in turn, contributes to shape our self in relation to the world.

The relation among sensoriality, movement, the body and, ultimately, the self has been further addressed by both ecological [among others, 27] and enactive or sensorimotor theories [e.g., 28]. According to the latter, sensory systems are active systems that function as simulators of action. Then, perception is connected with the ability of the brain to anticipate action by using internal models, which simulate and somehow predict the interaction among the body, the environment and other entities [15]. In other words, to perceive is, essentially, to simulate through internal models [29]. In this line, recent works enriched the concept of body schema by conceptualizing it as an integrated internal model of the body, which represents and simulates the spatial properties of the body and its surroundings.

To summarize, body schema is important for the interaction of the person with the environment, and having a coherent and stable body schema is essential for developing an integrated sense of the self situated in the world. The construct of body schema, however interesting, allows us to see only one side of the problem, as it focuses on the spatial aspects of perceiving and representing the body. It leaves apart representations that involve more complex factors, such as beliefs, emotions, and values, also including socio-cultural norms. To account for all these elements, it is needed to introduce a more complex construct, namely, the concept of body image.

Body Image

Body image points to a more conceptual representation of the body, even though a univocal definition of its characteristics is difficult to achieve. It appears to be connected with body schema, as the experiment of the rubber hand shows [30]. In this experiment, the body image acts in a top-down manner upon the body schema [20], making the individual believe that she feels sensations on a rubber hand. Body image also seems to be entangled with the evaluations people make of various characteristics of their own bodies (like shape and size), as well as the emotions associated to those evaluations [31].

Body image, therefore, is a cognitive representation of the body, but is not an exact copy of the body as it appears from the outside (as the image that the body reflects in a mirror), nor of the functioning of the internal organs or the autonomous nervous system [32]. Rather, body image appears to be related to the narrative aspect of the self, which concerns the stories that we tell about ourselves [21].

The close relationship between body image and the self is particularly evident in people with a distorted body image. Dissatisfaction with weight and body shape has been associated with several psychological problems: in particular, it is considered a predictive factor for eating disorders [31]. Moreover, researchers found correlation between Identity Problems, according to DSM IV, and body image: for example, Vartanian [33] emphasizes that the body defines the self and having a problematic body image may lead to an equally disturbed sense of the self. As there may be multiple representations of the body [29], individuals may have multi-faceted self-definitions [34]. However, it is important that these facets are stable, coherent, and clear, since coherence is considered a protective factor with respect to bodily and identity disorders [33].

The complexity of body image construct is apparent if we examine how it is operationalized in questionnaires aimed at analyzing the body image. To assess body image more than 150 measures have been used [35]. Kling et al. [35] synthesized the psychometric properties of several self-report measures about body image: the revised Body Appreciation Scale (BAS) [36], the Body Esteem Scale for Adolescent and Adults (BEESA) [37], the Body Shape Questionnaire (BSQ) [38], the Centre for Appearance Research Valence Scale (CAR-VAL) [39], the Drive for Muscularity Scale (DMS) [40], the Weight and Shape Concerns Subscales of the Eating Disorders Examination Questionnaire (EDE-Q) [41], the Body Dissatisfaction subscale of the Eating Disorder Inventory-3 (EDI-3) [42] and, finally, the Appearance Evaluation subscale and the Body Areas Satisfaction Scale of the Multidimensional Body Relations Questionnaire (MBSRQ) [43].

This variety may depend on the multidimensionality of the construct, which led researches to develop different body image measures. Here, therefore, we can define body image as a multidimensional construct, which encompasses thoughts, attitudes, beliefs, emotions, and cultural values related to the body [44]. Body image, in fact, is also affected by cultural stereotypes associated e.g., to gender [45].

Self-Tracking

If we consider body schema and body image as representations that mediate our relation with the body, we can affirm that self-tracking technologies should account for this mediation. By collecting and feeding back data about our bodies, they do not simply transform our body processes into numbers, but also affect the ways we represent our bodies. Likewise, the ways we look at our bodies may impact on how we use self-tracking technologies. What role does body image play in self-tracking practices? What happens when people have the availability of a large amount of body data, which integrate (or do not integrate) into the images they have of their own bodies?

On the one hand, Edwards [45] showed that activity tracker use (i.e., Fitbit) may be affected by the image that users have of their bodies: dissatisfaction with body image does provide motivation for using a Fitbit and dissatisfied users look to improve their bodies in some way (N = 9; age range = 16–64; females = 5). On the other hand, the "schizophrenic phenomenon" can shed light on the issues that people are encountering when using self-tracking devices. For example, Hortensius et al. [46] pointed out the frustration and sense of fragility that trackers feel when they cannot link a measure (e.g., of their food intakes, or blood glucose levels) to their personal experience, or when the device prompts undesirable data (N = 28; age range = 40–76; females = 15). Numbers that are not coherent with the user's body images seem no to give her any cues for improving her self-understanding: rather, they can lead to a sensation of despair [47].

People use narratives to constitute the self [48]. Such narratives are commonly built in retrospection, upon reminiscences: however, current perceptions and mental events play an important role in this process. The self is not something that merely lives in the past, through its memories, or in the future, through mental simulations: it exists here and now, in the *hic and nunc*, and it constantly changes along with our internal perceptions, experiences, and actions [7]. The "self", therefore, is made up of a multitude and mutable representations, elicited by a flow of sensations and bodily actions that occur in the moment. The sense of coherence we experience about our self is due to more stable configurations that sediment over time, like body image, and to the narratives we tell us about ourselves.

Self-tracking technologies, to be effective, should then integrate into these aspects of the self, encouraging, rather than disrupting, a coherent image about the body and, consequently, a coherent narration about the self. In other words, people can effectively use self-tracking devices to build their identity provided that the "digital self" emerging from the data becomes integrated into the body representations and self-narratives they have constructed over time. More precisely, self-tracking technologies can develop self-knowledge, if they are able to support people in generating coherent images and stories about their body and their self [3, 48]. The integration of the data in a coherent self guarantees a stable sense of identity and serves as a protective factor for mental health.

In so doing, they should take into account the flux of mental and bodily events continuously affecting the "present self", especially those of which the person is aware. This leads to consider the notion of body awareness, which differs from both the concept of body schema and that of body image, albeit is strongly connected with both of them.

Body Awareness and Self-Tracking

The "body awareness" construct emerged across a wide range of health topics and has been described as "an innate tendency of our organism to self-organize and to feel the unity with oneself" [49]. It stems from the concepts of proprioception and interoception, but has a more nuanced meaning.

Body Awareness

Interoception, as we have seen, refers to the perception of sensations concerning the internal parts of the body, like heartbeat and respiration. Nevertheless, there is a distinction among the actual body-related events, their subjective perception [50] and the way each person evaluates her ability to accurately identify internal body states, which is a metacognitive skill [51]. In this perspective, body awareness is more than the simple focus on one's own body, as it requires recognizing the interplay between body states and the cognitive appraisal of those body states [52].

Therefore, body awareness may be considered as an interface between top-down and bottom-up information: on the one hand, there are visual, tactile, olfactory, gustative and proprioceptive stimuli; on the other hand, there is the cognitive-affective processing of those physiological perceptions [53].

Body awareness is a key element for affect regulation and for the sense of self [18, 54] and it strictly depends on mental processes, included attitudes, affects, beliefs, memories, and cultural imprints [54]. It seems, in fact, that those mental processes can modify the subjective experience of body parts and of the body in general.

Pylvänäinen and Lappalainen [55] highlight that body awareness and body image are strictly tied together. This is evident, for example, among depressed patients: it has been observed that depressed individuals having dissatisfaction with body image also lack mindful body awareness [55]. Likewise, patients with fibro-myalgia overestimate their body size due to the experience of pain in certain body areas: *"as pain increased, the patients described changes in the perception of their body size and its relationship with space: they felt their body becoming larger and as though space was shrinking"* [56 , p. 2]. We may say that body image refers to a more stable representation of the body, which has been developed over time, whereas body awareness accounts for the momentary conscious stimuli that con-tinuously affect our bodies. Both the representations involve cognitive, emotional and cultural aspects.

Self-Tracking

Considering body awareness when we investigate self-tracking practices may allow better understanding the impact of self-tracking devices on the body, as well as their potential positive and harmful consequences.

Sharon and Zandbergen [57] argued that self-trackers use their devices to have a more "active and watchful mind", helping them be aware of body sensations, actions, and habits that are commonly unperceivable. Self-tracking allows people to sense elements concerning their internal perception, like the time of the day, or to acquire new capabilities, like identifying the calories and the weight of a portion of food just by looking at it. *"In such examples, numerical data are not all the end-goal of tracking; they are more like an unsophisticated, intermediate stage towards more augmented senses."* [57 , p. 1700]. Here, self-tracking serves to raise bodily aware-ness, to learn to better feel the body through the data [3] and to improve the users' confidence in perceiving their own body. Research confirmed that augmenting per-ception of body stimuli through data could improve body awareness [49, 58], and this could have positive impacts on people's health [59].

However, paying more attention to body states by using self-tracking technolo-gies may not be beneficial for all the individuals and, in certain cases, it may elicit discomforting sensations [4, 27]. People are different, are situated in diverse con-texts, and have different reasons to collect data about their bodies: they may need to monitor very specific body aspects that may be crucial for their health, or to gain another perspective on their bodily sensations. However, they can use self-tracking technologies also to reinforce some maladaptive behaviors.

An excessive focus on the self and on the body can be linked to emotional distress, anxiety and depression disorders, as well as eating disorders and sexual dysfunctions in certain individuals [60]. For instance, continuous health feedback, prompted by tracking technology, may worsen anxiety and stress symptoms leading to preoccupation with one's health, especially in people with certain personality traits, such as neuroticism and anxiety sensitivity [4]. This entails that self-tracking technologies should account for the different predispositions that different individuals may have, depending on their personality traits, and even on previous psychological disorders.

Moreover, body quantification may give an excessive emphasis to numbers and data to the detriment of feelings and sensations, yielding a sense of disembodiment [11, 61]. A virtual self, made up of disembodied data, could alienate the individual from herself and from the others. Berardi [62] stressed that alienation describes the contemporary age characterized by the impossibility of enjoying the presence of the other, in the form of physical presence. With the word "derealization", he refers to the difficulty of the "animated body" in accessing the "animated body" of the others. Technologies have made remote interaction possible, that is interaction in the absence of the bodies [63]. Nevertheless, "in presence" social interaction is considered fundamental for the building of the self [64]. We need others' corporality to grasp the nonverbal cues that tell us their attitude toward us and, finally, to understand who we are: when the bodies are substituted with data, and communication is replaced by sharing information, the risk is that we form a more opaque image of ourselves. Users, especially adolescents, who compare their "virtual body" with that of other users on social media platforms are more exposed to several health-related psychological outcomes, like anxiety, depression or sleep problems [65]. Technology may further worsen symptoms of people who already have trouble with their body image, as in the case of patients suffering from bulimia and anorexia nervosa using weight-loss app [66].

In sum, the quantification of the body operated by technology and its subsequent dematerialization open both opportunities for and threats to the ways we relate to our bodies and our selves. The double-edged consequences of self-tracking on body representations will be further exemplified in the next paragraphs, where we report on the findings collected during a qualitative study conducted with amateur and elite athletes about the use of self-tracking in sport. The next Section summarizes parts of the findings reported in Rapp and Tirabeni [9], focusing on how personal data are affecting the way athletes relate to their bodies.

Self-Tracking and Sport

We interviewed 20 athletes to investigate the impact of self-tracking technologies on physical activity.

Method

We recruited 8 amateur athletes (A1-A8) and 12 elite athletes (E1-E12) (mean age = 31,7; SD = 6,5; females = 8) who have been using a self-tracking device for more than three months, asking them to recount their experience with it. All the recruited participants owned a smartphone and a wearable device aimed at capturing sports-related data. While elite athletes competed at least nationally during their career, amateur athletes exercised at least three times per week, spending five hours or more practicing. We included in our sample different sports, involving both endurance and non-endurance athletes. The sports addressed were cycling, swimming, triathlon, cross-country skiing, ski mountaineering, trekking, alpinism, free climbing, soccer, and sprint running. Almost all the athletes were educated and numerate. We aligned the size of the sample to other Human-Computer Interaction (HCI) studies with similar purposes and design, also following a data saturation criterion.

The interviews were semi-structured and lasted an average of 58 minutes (min = 40 min.; max = 70 min). Questions were addressed to explore athlete's attitude towards their discipline, use of personal data, and effects of use of technology on their sports experience. We allowed participants to explore topics not listed in the interview guide, and we prompted new questions when we needed to better understand their recounts. Interviews were audio recorded and transcribed. The analysis of the collected data followed standard open and axial coding techniques. Data were coded independently by two researchers who generated initial codes. Then, they reviewed the codes to assess their consistency. All the discrepancies were discussed and resolved.

In the following we outline how self-tracking technologies affect the ways the athletes represent their own body. Most of the reported quotes are extracted from Rapp & Tirabeni [9].

Findings

Self-tracking devices, at first glance, appear to have a positive impact on body awareness, especially for amateur athletes. A1, for instance, reports that such devices provide him with "*an awareness that you couldn't have before*". Amateurs agree that trackers can support the athlete in developing a greater awareness of her body, by prompting fixed measures to which compare those signals that are tied to a specific level of heart rate. Being in a certain hear rate zone, in fact, is a primary goal for athletes who want to achieve a certain standard of performance: "*if you're within the zone and you know how you feel, then you try to memorize it, and then when you do a race or a workout, and you're without the heartbeat* [tracker], *you try to understand in which zone you are, if you're in a medium that you can manage for the whole race*", says E4 [9].

The device not only can make the athlete directly aware of internal body processes that she is not able to identify by herself (i.e., the heart rate); but also can support the athlete in learning how to "read" her body, in order to detect such hidden processes. When the athlete feels certain body signals, she may not be able to retrace them to a specific heart rate zone. This is due to their high variance: they may differ depending on contextual factors, like the weather or the athlete's physical condition. The device allows the athlete to progressively bring those signals that are meaningful for her sports performance back to certain heart rate zones, thus "teaching" her how to become more aware of the internal processes of her own body.

The increased body awareness that self-tracking technologies may produce, however, is not exempt from side effects. In fact, the device, rather than being used as a tool for learning how to listen to the body, may become essential for the athlete's sports practice. The elite athletes emphasize that self-tracking devices may undermine the athlete's confidence in what they call "sensations", in favor of a complete reliance on the data provided by the device. Such sensations refer to the body and go beyond the signals of being in a certain heart rate zone. Actually, they point to fine-grained information about the body that allows the elite to tune her performance on the basis of the continuous changing context. It is a superior form of body awareness that elite athletes develop over the years, by carefully listening to their bodies: "*To use sensation means that I search some reference points in my body, the rhythm of the hair on the shoulders, how the foot hits the ground, if it's heavy, or more round... [...] It's even the sensation that I have at that moment. Some days when I don't want to push forward at all... and then I precisely hear the exertion of the legs, the sensation of being more or less light*" [9]. These body sensations are used to tune their sports performance, regulate their rhythms, understand their level of fatigue, and recognize when they are reaching their body limits.

Awareness of sensations, however, can be jeopardized by an excessive use of self-tracking devices, which, in turn, can worsen the sports performance. E11 highlights that "*it happens to see athletes, non-professional athletes, athletes of the next generation... you tell them 'run slow for an hour' and they're not capable of running slow because they don't have a reference, they don't have the watch* [the tracker] *that can tell them that they're running slow, they can't manage themselves*" [9]. Elite athletes use their device simply as a commodity, rather relying on their "superior form" of body awareness during races, when the technology is actually left apart.

The tracker, in fact, may also give information contrasting their current body representation, and this may produce anxiety and worries during important events. E10 says that "*So many times you feel good and you push forward, and maybe the heartbeat goes beyond the rate that you think you should keep, and maybe by looking at the watch you get frightened thinking that the rhythm that you're keeping is wrong, when maybe your body is actually adapting itself [...], you're managing everything all right, even if it's a little harder than what you had set in advance*" [9]. The discrepancy between the representation of the body prompted by the device and the representation owned by the athlete may thus be perceived as disturbing and counterproductive for the athlete's goals.

This situation, however, is slowly changing due to technological advancements. More fine-grained instruments are progressively allowing the measurement of "sensations" that were previously identifiable only by the athlete. In cycling, for instance, the power meter allows to capture the cyclist's legwork. *"It has been the real revolution of both workouts and competitions, for now only available for bikes [...] If you keep 395 you'll blow away at the last kilometer. If you see others that maybe begin the rise stronger than you, you don't care, you look at your instrument, you keep that power, because you know that you can keep it"*, says E8 [9]. Nonetheless, in the elites' eyes, "these advancements" are seen as a worsening of the overall sports experience. Despite the undeniable positive impacts of devices such as the power meter on the sports performance, such devices are slowly affecting the athlete's body image, shifting it from a "living body" to a "mechanical body". E7, for instance, emphasizes that *"the watch is a machine and measures your activity as if you were a machine, but the human body... there is a mental part and other mechanisms that the watch can't compute"* [9]. This points out that the complex nature of the body can hardly be turned into numbers without producing impacts on the body image.

Discussion

Findings of this study highlight that the body is so variable in its reactions to both endogenous (e.g. stress, fatigue) and exogenous (environmental) factors, that parameters, like the heart rate, which technology aims to capture and turn into univocal numbers, can hardly account for it. The body is made up of meaningful "sensations", and the body awareness that elite athletes develop is addressed to detect the richness of such sensations. The increased awareness about the heart rate (and other "objective parameters" pertaining to the body) produced by self-tracking technologies, therefore, may cloud the athlete's opportunities for recognizing body sensations. Actually, it may decrease the awareness of the whole body in its multifaceted variability. In other words, the greater awareness of body parameters induced by the data may mislead the athlete's "superior" body awareness, which is considered by the elites more reliable and fruitful when there is a lot at stake. Furthermore, it may prevent the development of such ability in the amateur athletes. A subsequent study substantially confirmed the insights coming from this research [67].

However, the development of more "precise" devices, capable of directly tying the body measures to the sports outcomes, seems to be progressively changing this landscape. The power meter, widely employed in professional cycling, anticipates a future when self-tracking devices could provide an efficient substitute, in terms of their instrumental value, of the sensations that are currently leading the elite's conduct in a large variety of sports. This, however, is seen as an impoverishment of the sports experience also by those who are using this kind of device. Moreover, elite athletes emphasize that these instruments might change their body image, transforming it in something that is shaped by the data collected by the device, to which the real body would need to adapt. This sort of "mechanical body" would regulate the rhythms and dynamics of the real body, constraining it to respond to the incoming data in a continuous and unavoidable feedback loop.

The concerns about the body arisen by our participants parallel those emphasized by authors who are starting to outline a critical discourse towards the assumptions embedded in QS culture. Lupton [11], for instance, highlights that trackers appear to extend the capacities of the body by supplying data that can then be used to display the body's limits and capabilities and allow users to employ these data to work on themselves. However, these technologies conceptualize the body not as a sensing body through which one can gain self-knowledge, but as a data generating device that has to be coupled with technology in order to be known [13]. In this perspective, the repository of the body knowledge shifts from the individual to the device. This also entails the individual's subservience to technology, since these "data-doubles" feed back information to the user in ways that are intended to encourage the user's body to act in certain ways.

What seems relevant, therefore, is that the benefits on body awareness that self-tracking instruments are bringing in the sports domain may be blurred not only by the reduction of the athlete's ability in becoming aware of her body sensations, but also by the athlete's loss of control over her own body, induced by an externalized body image that is imposed by the device.

Beyond the Athletes

As we have already noticed, the availability of a variety of commercial wearable devices that automatically collect personal body data has boosted the popularity of self-tracking outside the circle of specific populations, like quantified selfers, athletes, and people with a health condition, reaching the general public [68]. A variety of smartwatches, activity trackers, and smart clothes [69–71] are now available on the market, promising to collect data on body aspects as diverse as blood pressure, body movements, and respiration.

Nonetheless, we are currently far away from the possibility of providing such "general users" with complete and reliable representations of their body, which could be used to increase their body awareness and feed back an insightful image of their body. Trackers still exclusively focus on "objective" parameters and the representations that they give of the body is rarely meaningful for people that are not used to manage a large amount of quantitative data [72].

What kind of data and/or design techniques do we need for supporting users in increasing their body awareness and developing their body image, even helping them improve the perception they have of themselves? If we look at, for instance, Body Appreciation Scale 2 [36], which is a positive body image measure assessing individuals' acceptance of, favorable opinions toward, and respect for their bodies, we see the multidimensionality of body image construct and, consequently, the complexity of capturing and feeding back an image of the user's body that could really help her ameliorate the image of her body.

Scale items span from body acceptance and love (e.g., "I feel love for my body", "I feel good about my body"). inner positivity influencing outer demeanor (e.g., "My behavior reveals my positive attitude toward my body; for example, I walk holding my head high and smiling"), to appreciating the functionality of the body (e.g., "I feel that my body has at least some good qualities"), taking care of the body via healthy behaviors (e.g., "I respect my body," "I am attentive to my body's needs"), and internalization of media appearance ideals ("I feel like I am beautiful even if I am different from media images of attractive people (e.g., models, actresses/actors"). In order to take into account all these aspects of the body image construct, it is not possible to simply rely on the functionalities of current trackers, but we need to envision novel ways of collecting and visualizing data.

For instance, self-reporting appears essential to grasp the subjective meanings that people ascribe to their body, and thus capture body acceptance and love, as well as appreciation and taking care of the body. Future research, therefore, needs to explore novel ways for eliciting the self-reporting of body data, which may be burdensome by requiring a high degree of compliance [73]. Self-reporting may be complemented by content analysis of social media, especially with reference to the goal of understanding how media appearance ideals may affect the user's body image; or automated tracking focusing on "specific aspects" of body data, like those related to "body-harm" (e.g., lack of sleep, bad food, and lack of exercise), which could help to infer how people take care of their own body; or those connected with posture and face expressions/emotions (maybe collected when the user is looking at herself in front of a mirror), which may work toward a better understanding of the inner positivity toward the body.

Furthermore, finding novel ways for representing body data becomes essential if we want to give a meaningful body image of the user, also pushing her toward a more positive representation of her body. Adopting concrete representations of the body data collected by a wearable may support the projection into and the development of an emotional connection with them [74], possibly promoting the development of a more positive body image. Rapp et al. [6], for instance, proposed a visualization of the user's body merged with the personal data collected by a variety of self-tracking devices, as if the user were looking into a mirror. The visualization does not allow for a precise quantification of the user's data, but conveys a general impression about the user's body, engaging her, at the same time, in an immersive interaction.

All these lines of research, which could work toward making self-tracking data closer to people's needs, by providing them with more meaningful body representations also pushing them toward more positivity, are still in their infancy. Their exploration would lead to design novel Quality of Life Technologies (QoLT), which have been defined as any technologies for assessment or improvement of the individual's QoL [75]. In fact, allowing people to construct a more positive image of their own body, as well as to develop a greater body awareness, could ultimately increase their overall self-knowledge providing benefits to many different aspects of their everyday life. However, much more research is needed to find insightful depictions of body data and novel ways to unobtrusively collect them.

Conclusion

In this chapter we have highlighted the potential impacts of self-tracking technologies on the body. In order to assess such impacts, we stressed that we need to consider the ways we naturally relate to our bodies, namely through different body representations. On the basis of the examination of three theoretical constructs that refer to the different ways through which we represent our own body, we outlined how technology is affecting our body conceptions, particularly highlighting the potential negative outcomes that may stem from the usage of body data. In this line, examples coming from a study in the sports domain pointed out the main risks for the athlete's body opened up by tracking technology.

However, all the issues we pointed out may be counterbalanced by the great opportunities that self-tracking instruments seem to open. Esmonde [68] highlights that the boundaries of the body are not defined by the skin, as they extend to the outside world, the environment and technology. In this perspective, self-tracking devices could have the potentialities to expand the limits of our body and, eventually, of our self. To achieve this goal, however, researchers and practitioners should start rethinking not only the way technology is designed, but also the theoretical frame in which it is inserted, in order to account for the complex modalities through which we relate to our bodies and our selves.

References

1. Marcengo A, Rapp A. Visualization of human behavior data: the quantified self. In: Huang ML, Huang W, editors. Innovative approaches of data visualization and visual analytics. Hershey, PA: IGI Global; 2014. p. 236–65. https://doi.org/10.4018/978-1-4666-4309-3.ch012.
2. Wolf G. Know thyself: tracking every facet of life, from sleep to mood to Pain, 24/7/365. In: Wired Magazine. 2009. https://www.wired.com/2009/06/lbnp-knowthyself/. Accessed 25 Jan 2020.
3. Bode M, Kristensen DB. The digital doppelgänger within. A study on self-tracking and the quantified self movement. In: Canniford R, Bajde D, editors. Assembling consumption: researching actors, networks and markets. Oxon: Routledge; 2016. p. 119–35.
4. van Dijk ET, Westerink JH, Beute, F, IJsselsteijn, WA. In sync: The effect of physiology feedback on the match between heart rate and self-reported stress. 2015. BioMed research international, 2015. https://doi.org/10.1155/2015/134606.
5. Rapp A, Cena F. Personal informatics for everyday life: how users without prior self-tracking experience engage with personal data. Int J Human-Comp Stud. 2016;94:1–17. https://doi.org/10.1016/j.ijhcs.2016.05.006.
6. Rapp A, Marcengo A, Buriano L, Ruffo G, Lai M, Cena F. Designing a personal informatics system for users without experience in self-tracking: a case study. Behav Inform Technol. 2018;37:335–66. https://doi.org/10.1080/0144929X.2018.1436592.
7. Rapp A, Tirassa M. Know thyself: a theory of the self for personal informatics. Human-Computer Interaction. 2017;32:335–80. https://doi.org/10.1080/07370024.2017.1285704.
8. Ihde D. Bodies in technology, vol. 5. University of Minnesota Press; 2002.
9. Rapp A, Tirabeni L. Personal informatics for sport: meaning, body, and social relations in amateur and elite athletes. ACM Transactions on Computer-Human Interaction. 2018;25:1–30. https://doi.org/10.1145/3196829.

10. Lupton D. You are your data: self-tracking practices and concepts of data. In: Selke S, editor. Lifelogging. Springer VS: Wiesbaden; 2016. p. 61–79.
11. Lupton D. Quantifying the body: monitoring and measuring health in the age of mHealth technologies. Crit Public Health. 2013;23:393–403. https://doi.org/10.1080/0958159 6.2013.794931.
12. Ajana B. Governing through biometrics: the biopolitics of identity. Basingstoke: Palgrave Macmillan; 2013.
13. Schüll ND. Data for life: wearable technology and the design of self-care. BioSocieties. 2016;11:317–33. https://doi.org/10.1057/biosoc.2015.47.
14. Van der Ploeg I. The illegal body: "Eurodac" and the politics of biometric identification. Ethics Inf Technol. 1999;1:295–302. https://doi.org/10.1023/A:1010064613240.
15. Morasso P, Casadio M, Mohan V, Rea F, Zenzeri J. Revisiting the body-schema concept in the context of whole-body postural-focal dynamics. Front Hum Neurosci. 2015; https://doi.org/10.3389/fnhum.2015.00083.
16. Noë A. Action in perception. Cambridge: MIT Press; 2004.
17. Barrett LF, Quigley KS, Bliss-moreau E, Aronson KR. Interoceptive sensitivity and self-reports of emotional experience. 2004;87:684–697. https://doi.org/10.1037/0022-3514.87.5.684.
18. Cameron O. Visceral sensory neuroscience. In: Interoception. New York (NY): Oxford University Press; 2002.
19. Sherrington CS. The integrative action of the nervous system. New Haven, CT: Yale University Press; 1947.
20. Longo MR, Schüür F, Kammers MPM, Tsakiris M, Haggard P. Self-awareness and the body image. Acta Psychol. 2009;132:166–72. https://doi.org/10.1016/j.actpsy.2009.02.003.
21. Gallagher S. How the body shapes the mind. New York: Oxford University Press; 2005.
22. Bonnier P, Asomatognosia P, L'aschématie B. Rev Neurol. 1905;13:605–9. Epilepsy & Behavior. 2009;16:401-403. https://doi.org/10.1016/j.yebeh.2009.09.020.
23. Head H, Holmes G. Sensory disturbances from cerebral lesions. Brain. 1991;34:102–254. https://doi.org/10.1093/brain/34.2-3.102.
24. Schwoebel J, Coslett HB. Evidence for multiple, distinct representations of the human body. J Cogn Neurosci. 2005;17:543–53. https://doi.org/10.1162/0898929053467587.
25. Glashouwer KA, Van Der VRML, Adipatria F, Jong PJ, De Vocks S. The role of body image disturbance in the onset, maintenance, and relapse of anorexia nervosa: a systematic review. Clin Psychol Rev. 2019;74 https://doi.org/10.1016/j.cpr.2019.101771.
26. De Vignemont F, De. Body schema and body image–pros and cons. Neuropsychologia. 2010;48:669–80. https://doi.org/10.1016/j.neuropsychologia.2009.09.022.
27. Gibson, JJ. The senses considered as perceptual systems. 1966.
28. Maturana HR, Varela FJ. Autopoiesis and cognition: the realization of the living. Boston, MA: Reidel Pub. Co.; 1980.
29. Berthoz A. The vicarious brain, creator of worlds. Harvard University Press; 2017.
30. Tsakiris M, Haggard P. The rubber hand illusion revisited: Visuotactile integration and self-attribution. J Exp Psychol Hum Percept Perform. 2005;31:80–91. https://doi.org/10.1037/0096-1523.31.1.80.
31. Sabik NJ, Lupis SB, Geiger AM, Wolf JM. Are body perceptions and perceived appearance judgments by others linked to stress and depressive symptoms? J Appl Biobehav Res. 2019;24 https://doi.org/10.1111/jabr.12131.
32. Kamps CL, Berman SL. Body image and identity formation: the role of identity distress. Revista Latinoamericana de Psicologia. 2011;43:267–77. https://doi.org/10.14349/rlp.v43i2.739.
33. Vartanian LR. When the body defines the self: self-concept clarity, internalization, and body image. J Soc Clin Psychol. 2009;28:94–126. https://doi.org/10.1521/jscp.2009.28.1.94.
34. Linville PW. Self-complexity and affective extremity: Don't put all of your eggs in one cognitive basket. Soc Cogn. 1985;3:94–120.
35. Kling J, Kwakkenbos L, Diedrichs PC, Rumsey N, Frisén A, Brandão MP, Silva AG, Dooley B, Rodgers RF, Fitzgerald A. Systematic review of body image measures. Body Image. 2019;30:170–211. https://doi.org/10.1016/j.bodyim.2019.06.006.

36. Tylka TL, Wood-Barcalow NL. The body appreciation Scale-2: item refinement and psychometric evaluation. Body Image. 2015;12:53–67. https://doi.org/10.1016/j.bodyim.2014.09.006.
37. Mendelson BK, Mendelson MJ, White DR. Body-esteem scale for adolescents and adults. J Pers Assess. 2001;76:90–106. https://doi.org/10.1207/S15327752JPA76016.
38. Cooper PJ, Taylor MJ, Cooper Z, Fairburn CG. The development and validation of the body shape questionnaire. Int J Eat Disord. https://doi.org/10.1002/1098-108X.
39. Moss TP, Rosser BA. The moderated relationship of appearance valence on appearance self consciousness: development and testing of new measures of appearance schema components. PLoS One. 2012;7 https://doi.org/10.1371/journal.pone.0050605.
40. McCreary DR, Sasse DK. An exploration of the drive for muscularity in adolescent boys and girls. J Am Coll Health Assoc. 2000;48:297–304. https://doi.org/10.1080/07448480009596271.
41. Fairburn CG, Bèglin SJ. Assessment of eating disorders: interview or self-report questionnaire? Int J Eat Disord. 1994;16:363–70. https://doi.org/10.1002/1098-108X.
42. Garner DM. Eating disorder Inventory-3 professional manual. Odessa, FL: Psychological Assessment Resources; 2004.
43. Brown TA, Cash TF, Mikulka PJ. Attitudinal body-image assessment: factor analysis of the body-self relations questionnaire. J Pers Assess. 1990;55:135–44. https://doi.org/10.1080/00223891.1990.9674053.
44. Bailey KA, Gammage KL, van Ingen C. How do you define body image? Exploring conceptual gaps in understandings of body image at an exercise facility. Body Image. 2017;23:69–79. https://doi.org/10.1016/j.bodyim.2017.08.003.
45. Edwards A. The impact of body image on Fitbit use: a comparison across genders. Health Inf Libr J. 2017;34:247–51. https://doi.org/10.1111/hir.12188.
46. Hortensius J, Kars MC, Wierenga WS, Kleefstra N, Bilo HJ, van der Bijl JJ. Perspectives of patients with type 1 or insulin-treated type 2 diabetes on self-monitoring of blood glucose: a qualitative study. BMC Public Health. 2012;12:167.
47. Pols J, Willems D, Aanestad M. Making sense with numbers. Unravelling ethico-psychological subjects in practices of self-quantification. Sociol Health Illn. 2019;41:98–115.
48. Hilviu D, Rapp, A. Narrating the Quantified Self. In: Adjunct Proceedings of the 2015 ACM International Joint Conference on Pervasive and Ubiquitous Computing and Proceedings of the 2015 ACM International Symposium on Wearable Computers (UbiComp/ISWC'15 Adjunct). 2015. New York: ACM, 1051–1056. https://doi.org/10.1145/2800835.2800959.
49. Mehling WE, Price C, Daubenmier JJ, Acree M, Bartmess E, Stewart A. The Multidimensional Assessment of Interoceptive Awareness (MAIA). 2012;7. https://doi.org/10.1371/journal.pone.0048230.
50. Ferentzi E, Horváth Á, Köteles F. Do body-related sensations make feel us better? Subjective Well-being is associated only with the subjective aspect of interoception. Psychophysiology. 2019;56 https://doi.org/10.1111/psyp.13319.
51. Garfinkel SN, Seth AK, Barrett AB, Suzuki K, Critchley HD. Knowing your own heart: distinguishing interoceptive accuracy from interoceptive awareness. Biol Psychol. 2015;104:65–74. https://doi.org/10.1016/j.biopsycho.2014.11.004.
52. Farb NAS, Daubenmier J, Price CJ, Gard T, Kerr C, Dunn BD, et al. Interoception, contemplative practice, and health. Front Psychol. 2015;6:886. https://doi.org/10.3389/fpsyg.2015.00763.
53. Danner U, Avian A, Macheiner T, Salchinger B, Dalkner N, Fellendorf FT, Birner A, Bengesser SA, Platzer M, Kapfhammer HP, Probst M, Reininghaus EZ. "ABC"–the awareness-body-chart: a new tool assessing body awareness. PLoS One. 2017;12:1–13. https://doi.org/10.1371/journal.pone.0186597.
54. Damasio AR. Mental self: the person within. Nature. 2003;423 https://doi.org/10.1038/423227a.
55. Pylvänäinen P, Lappalainen R. Change in body image among depressed adult outpatients after a dance movement therapy group treatment. Arts in Psychotherapy. 2018;59:34–45. https://doi.org/10.1016/j.aip.2017.10.006.
56. Valenzuela-Moguillansky C, Reyes-Reyes A, Gaete MI. Exteroceptive and interoceptive body-self awareness in fibromyalgia patients. Front Hum Neurosci. 2017;11:117.
57. Sharon T, Zandbergen D. From data fetishism to quantifying selves: self-tracking practices and the other values of data. New Media Soc. 2017;19:1695–709. https://doi.org/10.1177/1461444816636090.

58. Garcia-Cordero I, Esteves S, Mikulan EP, Hesse E, Baglivo FH, Silva W, et al. Attention, in and out: scalp-level and intracranial EEG correlates of interoception and exteroception. Front Neurosci. 2017;11:411. https://doi.org/10.3389/fnins.2017.00411.
59. Astin JA, Shapiro SL, Eisenberg DM, Forys KL. Mind-body medicine: state of the science, implications for practice. J Am Board Fam Pract. 2003;16:131–47.
60. Ainley V, Tsakiris M. Body conscious? Interoceptive awareness, measured by heartbeat perception, is negatively correlated with self-objectification. PLoS One. 2013;8:1–10. https://doi.org/10.1371/journal.pone.0055568.
61. Silk M, Millington B, Rich E, Bush A. (re-)thinking digital leisure. Leis Stud. 2016;35:712–23.
62. Berardi F. The soul at work: from alienation to autonomy, trans. Francesca Cadel and Giuseppina Mecchia. New York: Semiotext (e); 2009.
63. Zhao S. The digital self: through the looking glass of telecopresent others. Symb Interact. 2005;28:387–405.
64. Cooley CH. Human nature and the social order. Routledge [1902] (2017).
65. RSPH. Social media and young people's mental health and wellbeing. 2017. http://www.rsph.org.uk/our-work/policy/social-media-and-young-people-s-mental-health-and-wellbeing.html. Accessed 4 January 2020.
66. Eikey EV, Reddy MC. "It's definitely been a journey" a qualitative study on how women with eating disorders use weight loss apps. In: Proceedings of the 2017 CHI conference on human factors in computing systems. 2017. pp. 642–654.
67. Rapp A, Tirabeni L. Self-tracking while doing sport: comfort, motivation, attention and lifestyle of athletes using personal informatics tools. Int J Human-Comp Stud. 2020;140(102434):1–14. https://doi.org/10.1016/j.ijhcs.2020.102434.
68. Esmonde K. Training, tracking, and traversing: digital materiality and the production of bodies and/in space in runners' fitness tracking practices. Leis Stud. 2019;38:804–17.
69. Reeder B, David A. Health at hand: a systematic review of smart watch uses for health and wellness. J Biomed Inform. 2016;63:269–76.
70. Mencarini E, Rapp A, Tirabeni L, Zanacanaro M. Designing wearable Systems for Sport: a review of trends and opportunities in human-computer interaction. IEEE Transactions on Human-Machine Systems. 2019;49(4):314–25. https://doi.org/10.1109/THMS.2019.2919702.
71. Shin G, Jarrahi M, Fei Y, Karami A, Gafinowitz N, Byun A, Lu X. Wearable activity trackers, accuracy, adoption, acceptance. And health impact: a systematic literature review. J Biomed Inform. 2019;93:103153.
72. Rapp A, Cena F, Kay J, Kummerfeld B, Hopfgartner F, Plumbaum T, Larsen JE, Epstein DA, Gouveia R. New frontiers of quantified self 2: going beyond numbers. In: Proceedings of the 2016 ACM International Joint Conference on Pervasive and Ubiquitous Computing: Adjunct (UbiComp '16). New York: ACM; 2016. p. 506–9. https://doi.org/10.1145/2968219.2968331.
73. Rapp A, Cena F. Self-monitoring and technology: challenges and open issues in personal informatics. In: Proceedings of the HCI International Conference. In Universal Access in Human-Computer Interaction. Design for All and Accessibility Practice Lecture Notes in Computer Science, vol. 8516. Springer; 2014. p. 613–22. https://doi.org/10.1007/978-3-319-07509-9_58.
74. Rapp A. Gamification for self-tracking: from world of Warcraft to the Design of Personal Informatics Systems. In: Proceedings of the 2018 CHI Conference on Human Factors in Computing Systems (CHI '18). New York, NY, USA, Paper 80, 15 pages: ACM; 2018. https://doi.org/10.1145/3173574.3173654.
75. Wac K. Quality of life technologies. In: Gellman M, editor. Encyclopedia of behavioral medicine. New York, NY, USA: Springer; 2020. https://doi.org/10.1007/978-1-4614-6439-6_102013-1.

Chapter 10
Your Smartphone Knows you Better than you May Think: Emotional Assessment 'on the Go' Via TapSense

Surjya Ghosh, Johanna Löchner, Bivas Mitra, and Pradipta De

Introduction

Emotions have an enormous impact on our momentary performance, health, and way of relating to others, hence on the quality of a persons' life. In particular, the experience of unpleasant (or pleasant) emotions is directly related to an individual's well-being. Emotions are influenced by subjective experiences and memories and the context the individual is in, and it seems almost impossible to measure this phenomenon objectively, reliably, and validly. Indeed, capturing human emotional states has been a challenging task for researchers for decades, leading to numerous theories about emotions, moods, and feelings. Specifically, psychometrics focuses on the theory and techniques of psychological measurements, including the QoL measurements. The emerging field of affective computing promises to overcome some methodological difficulties that lead to limitations in traditional methods of psychometrics. Affective computing is the study of technologies that can quantitatively measure human emotion from different clues. It is based on the hypothesis that an individual's digital footprint is highly correlated with their perceptions, feelings, and resulting behaviors and that extracting and analyzing this data collected

S. Ghosh (✉)
BITS Pilani, Goa, India
e-mail: surjyag@goa.bits-pilani.ac.in

J. Löchner
German Youth Institute, Munich, Germany
e-mail: loechner@dji.de

B. Mitra
Indian Institute of Technology Kharagpur, Kharagpur, West Bengal, India
e-mail: bivas@cse.iitkgp.ac.in

P. De
Microsoft Corporation, Atlanta, USA
e-mail: prade@microsoft.co

© The Author(s) 2022
K. Wac, S. Wulfovich (eds.), *Quantifying Quality of Life*, Health Informatics,
https://doi.org/10.1007/978-3-030-94212-0_10

over time can prove that "your smartphone knows you better than you may think." In addition to the fact that people use their digital devices extensively, being the first and last thing used during a normal day [1]—this statement becomes even more valid.

In this chapter, we discuss the current development within the affective computing area while focusing on assessing emotions via personal technologies. Firstly, we define emotions as a complex interplay of different components (sensory, cognitive, physiological, expressive, motivational) over time. Several emotion theories have been developed in the past years, along with emotions being classified into three dimensions: valence, arousal, and dominance [2]. Among these emotion theories, the Component Process Model [3] stands out, revealing the emotional process that leads to an individual's perception and processing of negative and positive life experiences. As an extension of this model, the Emotional Competence Model [4] hypothesizes that mental well-being and adverse psychopathology (e.g., anxiety, depression) greatly depends on a well-functioning emotional process. This depends on the individual's experienced emotional response, perception of the situation, adequate appraisal, and emotional regulation. Hence, emotional competence plays a key role in maintaining a person's quality of life. Furthermore, the knowledge and perception of emotions are basic abilities that may elicit more adaptive emotion regulation strategies (like an acceptance of an uncontrollable stressor).

To capture individuals' emotions (e.g., for clinical or research reasons), mostly self-reports for negative and positive emotions are conducted using psychometrically validated questionnaires of concepts as, e.g., stress, depression, and well-being with a recall period of days to months. These assessment instruments face the problem that (i) self-reports are rather subjective and often biased by, e.g., time and motivation to fill out often many questions and (ii) they are influenced to a great deal by the current psychological state interfering with the recall of someone's mood days to weeks ago. To overcome these issues, new assessment methods arose in the past years using personal digital technologies. For example, the Experience Sampling Method (ESM), also known as Ecological Momentary Assessment (EMA) [5–7], is increasingly used in psychology to trigger self-reports for emotions and behaviors momentarily, i.e., as closely as possible to the subject's daily life experiences, periodically (randomly or at fixed intervals) or in an event-driven fashion [6, 8].

Traditionally, different modalities such as facial expression [9–13], speech prosody [14–20], physiological signals like ECG, EEG, HR, GSR blood, brain, posture, [21–27] are explored for emotion assessment. Additionally, other sources can be used to extract emotions using smartphones and internet usage, which is discussed in this chapter. To determine the emotion states, often these affect-aware systems deploy a machine learning model. Therefore, conventional machine learning models like Support Vector Machine (SVM), Random Forest, Bayesian approaches were used in affective computing [28, 29]. In these approaches, first, a set of features (which can distinguish one emotion from another) were extracted manually from, e.g., the physiological signals. These features were correlated with the emotion ground truth labels (self-reports) to construct the emotion inference model. With the latest advances in the Deep Neural Network models, the conventional approaches were replaced by state-of-the-art AI models such as Convolutional Neural Network (CNN), Long Short Term Memory (LSTM), Recurrent Neural Network (RNN) [28, 29]. The advances in this field have helped to overcome the manual feature engineering effort and helped obtain very high classification performance in the affect classification. The advances in affective computing have led to many affect-aware applications such as emotion-aware music player, affective tutor, mood monitor, which influence the quality of life [30–34]. The key working principle of such emotion-aware applications is to collect physiological and behavioral data from different modalities and to train a machine learning model for emotion inference.

Given the current accelerating scale of developments in personal technologies, new assessment techniques for emotions emerged. We present the design and development of a smartphone keyboard interaction-based emotion assessment application. Specifically, among different types of interactions performed in smartphones, keyboard interactions are highly interesting. They represent the input/output interaction between the user and the phone for, e.g., information, communication, or entertainment [35]. Additionally to the interaction itself, the interactive content may be of high interest. The research shows that not only individuals often express momentary emotions on social networks' platforms [36, 37], but also a person's language was found to reveal his/her momentarily psychological state [38].

In our research, we focus on the smartphone interaction itself. Hence, we have designed and implemented an Android application TapSense, which can unobtrusively log users' typing patterns (without actual content) and trigger self-reports for four types of emotion (happy, sad, stressed, relaxed) leveraging the ESM method. Different typing features like typing speed and error rate are extracted from the typing data and correlated with the emotion self-reports to develop a personalized emotion assessment model. However, as the conventional ESM-driven self-report collection for model construction is labor-intensive and fatigue-inducing, we also investigate how the self-report collection approach can be further optimized for suitable probing moments and reduced probing rate. So, we also have developed an adaptive 2-phase ESM schedule (integrated into TapSense), which balances the probing rate and self-report collection time and probes the individual at the opportune moments. The first phase balances between probing rate and self-report collection time and trains an 'inopportune moment assessment' model. The second phase operationalizes such a model so that no triggering is done at an inopportune moment. We investigate the implications of this ESM design on the emotion classification performance.

We evaluate the proposed approach in a 3-week 'in-the-wild study involving 22 participants. Our first important result demonstrates that using smartphone keyboard interaction; we can determine the emotion states (happy, sad, stressed, relaxed) with an average AUCROC of 78%. The next major result shows the performance of the proposed ESM approach. It demonstrates that the proposed 2-phase ESM schedule (a) assesses inopportune moments with an average accuracy (AUCROC) of 89% (b) reduces the probing frequency by 64% (c) while enabling the collection of the self-reports in a more timely manner, with a reduction of 9% on average in elapsed time between self-report sampling and event occurrence. The proposed design also helps to improve self-report response quality by (a) improving valid response rate to 96% and (b) yielding a maximum improvement of 24% in emotion classification accuracy (AUCROC) over the traditional ESM schedules.

The chapter is organized as follows. Firstly, in Sect. 2, we discuss the definitions and the nature of emotions and their importance for the daily life experience, which influences an individual's quality of life. In Sect. 3, we present traditional psychometric assessment instruments for emotions and different assessment methods leveraging affective computing developments and beyond. The background, study design, and empirical evaluation of TapSense—a smartphone-based approach for assessing emotions are presented in Sect. 4. We discuss and conclude the chapter findings in Sect. 5.

Background and Related Work

This section presents the definition and models covering the concept of emotions (2.1), their importance for the quality of life (2.2). It attempts to capture emotions via traditional (2.3) and more novel, data-driven approaches (2.3).

Definition of Relevant Domain Concepts

Emotions, moods, and psychological states play a key role in our personal, professional, and social life. Also, interpersonal relationships, professional success, and mental well-being depend greatly on how we cope with stressful events and navigate adverse emotional experiences. Often the terms: emotions, effects, feelings, and moods are confused in language usage. *Feelings* most commonly involve invariably a direct response of the autonomic nervous system (ANS) involving organ functions (e.g., change in respiration pattern, adrenaline rush). At the same time, an umbrella term refers to all basic senses of feelings, ranging from unpleasant to pleasant (*valence*) and from excited to calm (*arousal*). *Moods* differ from feelings, emotions, and affects in that they are experienced as extended in time (c.f., mood stability by Peters, 1984) but are also subject to certain situational fluctuations [39]. Very similar is the psychological concept of *state*, referring to a person's mental state at a certain point in time, introduced by Cattell and Scheier [40] as a counterpart to the concept of timely persisting (personality, motivational, cognitive) *traits*. In contrast, *emotions* are a much more complex mental construct consisting of several components as the physiological response and lasting from minutes to hours.

Research on emotions is usually based on the central evolutionary importance of emotions for human survival. It defines emotions as "*a genetic and acquired motivational predisposition to respond experientially, physically and behaviorally to certain internal a02nd external variables*" [41]. In the context of survival, emotions imply complex communication patterns and information [42–44], as the feedback of an individual's inner state on different levels enables a biological adaptation to the physical and psychosocial environment. Therefore, emotions are further viewed as complex, genetically anchored behavioral chains that contribute to an individual's homeostasis through various feedback loops [45]. Since the research of Ekman [46], it has become known that elementary emotions such as fear, joy, or sadness show themselves independently of the respective culture. These basic emotions are closely coupled to simultaneously occurring neuronal processes. However, how people communicate and express visible parameters, such as facial expression, are influenced by the values, roles, and socialization practices that vary across cultures [47, 48], age, and gender [49].

Emotions can be divided categorically into primary, secondary, and combined forms: primary emotions are fundamental, while secondary emotions are emotions about emotions (such as guilt over gloating). Ekman distinguishes six basic emotions: happiness, sadness, anger, fear, surprise, and disgust [46]. In contrast, Izard [50] speaks of ten fundamental emotions (1) interest/excitement, (2) pleasure/joy, (3) surprise/fright, (4) sorrow/pain, (5) anger/rage, (6) disgust/repugnance, (7) disdain/contempt, (8) fear/contempt, (9) shame/shyness/humiliation, and (10) guilt/repentance. Another way to categorize emotions relates to their highly variable multidimensional nature: emotions can be categorized into the dimensions as positive or negative (polarity/valence), strong or weak (intensity/arousal), easy or hard to arouse (reactivity), and based on the situation they occur (idiosyncratic vs. universal situation) [51]. Following partly those dimensions, Russel's Circumplex model is the most commonly used emotion model to capture emotion on a continuous scale [52]. It represents every emotion as a tuple of valence and arousal. There exist also a valence, arousal, and dominance model (more commonly known as the VAD model), which captures every emotion as a set of triplets (valence, arousal, dominance) in a continuous scale [2] (see Fig. 10.1).

Following the earlier mentioned, there exists an empirically validated theoretical process model of emotion—the Component Process Model (CPM) [54] and the Emotional Competence Process Model (ECP), further implying the adaptive and maladaptive emotional functioning [4]. In the CPM, emotions are identified within the overall process in which low-level cognitive appraisals, particularly relevance processing, trigger physical reactions, behaviors, and subjective feelings (Fig. 10.2). As a foundation, CPM provides a differentiated theory of the various dimensions of emotions. Emotions are interpreted here as synchronizing several components that interact over time during a defined process regarding the emergence, appraisal, awareness, regulation, and knowledge of emotions. For example, an emotional situation emerges due to a specific trigger (e.g., a job interview), evoking in the appraisal ("I hope I perform well—the interviewer does not look friendly at all") which results in a certain emotional reaction (e.g., adrenaline release, sweating, nervousness, tension). At the moment, a person is aware of this condition and the fact that they can regulate their condition (e.g., by taking a deep breath). They can categorize their emotional reaction (and those of others in return).

The emotional response components cover sensory, cognitive, physiological, motivational, and expressive components [55]. Initially, the **sensory component**

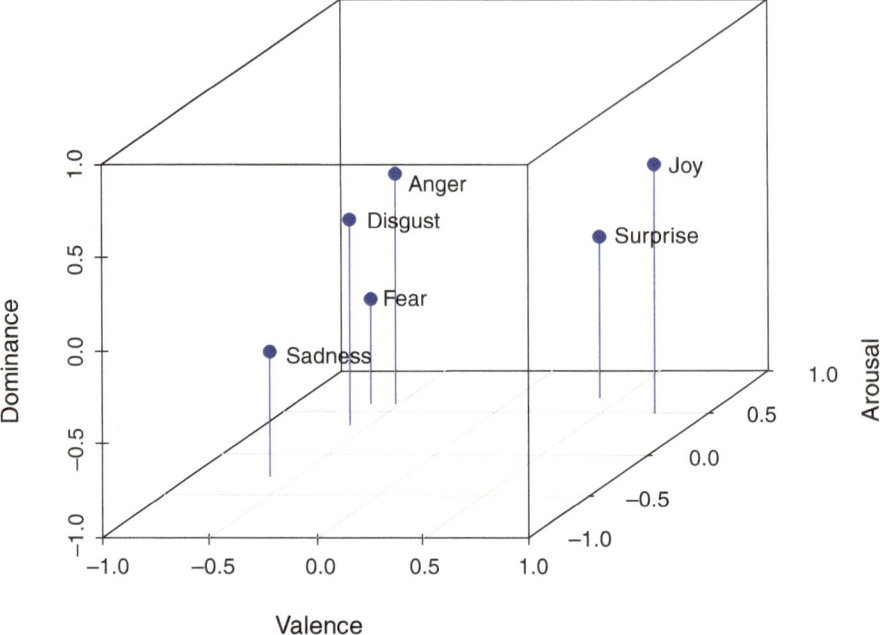

Fig. 10.1 The VAD (Valence-Arousal-Dominance) model spanned across the six basic emotions [53]

enables a subject to recognize an emotional event through the senses (e.g., see, feel). Through the **cognitive component**, the individual can identify possible relationships between itself and the event based on its subjective experiences. The individual then makes a subjective evaluation of the perception of the event (appraisal). This goes along with the two-factor theory of emotions. As early as the 1960s, Schachter and Singer pointed to the cognitive evaluation of a physical response as key to the subsequent emotional sensation [56]. A subject can react to the same event with a different evaluation—depending on his personal world view, value system, and current physiological state. It is resulting in different physical responses to feelings. Depending on the subjective evaluation outcome, the individual reacts by releasing certain neurotransmitters and hormones, thus changing its physiological state (**physical component**). This altered state corresponds to the experience of an emotion. According to Lazarus' appraisal theory [57], the emotional experience first arises through cognitive evaluation and interpreting an emotional stimulus as manageable or not. The **motivational component** follows the event's evaluation and is modulated by the current physiological (or emotional) state. The motivation to a certain action of a person is oriented to an actual-target comparison and the prediction of the effect of conceivable actions. For example, the emotion of anger can result in both the motivation for an attack action (e.g., in the case of a supposedly inferior opponent) and the motivation for a flight action (e.g., in the case of a

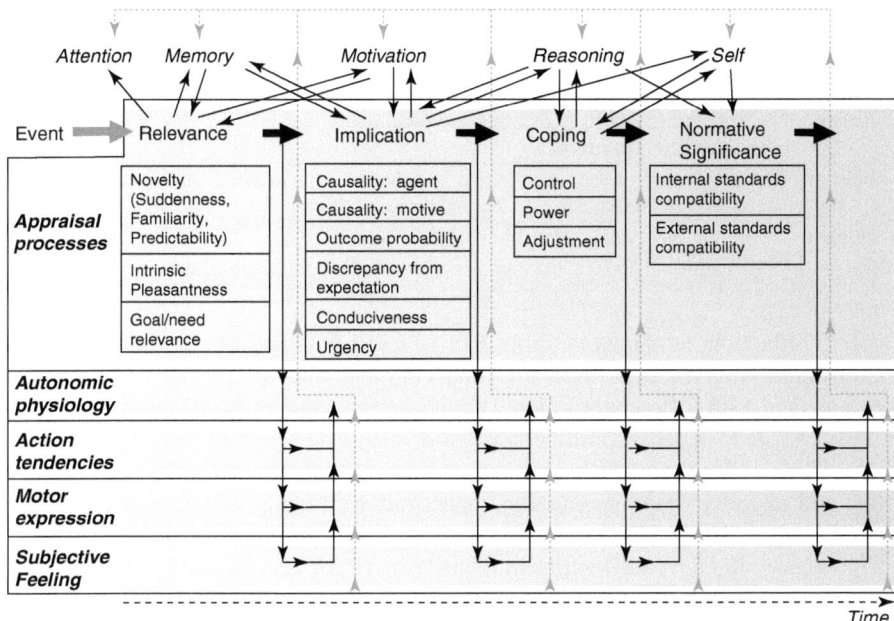

Fig. 10.2 Comprehensive illustration of the Component Process Model of Emotion (adapted from Scherer 2001) [58]

supposedly superior opponent). The **expressive component** refers to the way emotion is expressed. This primarily concerns nonverbal behavior, such as facial expressions, speech, and gestures.

Importance of Feelings Moods, Psychological States

From an evolutionary perspective, emotions play a very important role in motivation, behavior, and attention: they make us act, direct our attention to certain stimuli that might have pleasant or unpleasant consequences for us, and give us signals so that we adjust our behavior to obtain or avoid those consequences. In a modern context, it is not only about our survival and related incentives but also about secondary incentives, such as money, status, entertainment, or others. Moreover, emotions regulate the intensity and duration of different behaviors and cause the learning of those behaviors that were successful under certain conditions (e.g., joy has a pleasant effect on us and motivates us to repeat the behavior) and mark in memory (e.g., via disgust, anger) those that led to failure. In the same way, emotions function in regulating our social interactions, being reinforced by a pleasant (or non-pleasant) effect. This leads to the formation of bonds or rivalries, which provide us with orientation in the social structure. Some distinct emotions may have even a more

Table 10.1 Example Functions of Emotions

Emotion	Function
Fear	Protective function: Avoidance of danger, injury, creating distance to source of danger
Anger	Destruction function: Overcoming something that stands in the way of need satisfaction (e.g. nutrition)
Pleasure	Intake function: Receiving positive stimulus from the external world that supports or nourishes the individual
Disgust	Rejection function: Excrete harmful substance
Surprise	Orientation function: Increased attention to take in information about a new stimulus

specific function, as shown in Table 10.1 [59, 60]. It is part of our daily lives to be confronted with the experience of such emotions and to respond to them. How people deal with different emotional events varies widely. The distinction between "positive" and "negative" emotions is questionable due to emotions' functionality.

As depicted in the ECP, there is a great body of literature showing that both positive and negative emotions greatly impact our well-being and mental health (see meta-analysis [61]). It has been demonstrated, for example, that people suffering from depression have difficulties in identifying (Rude & McCarthy, 2003), bearing and accepting their emotions [62, 63]. In numerous disorders, the presence of undesirable affective states (such as anxiety or depressed mood) in an inappropriate intensity or duration is among a diagnostic criterion of the disorder (e.g., in anxiety disorders or depression). Also, a whole range of cognitive and behavioral symptoms of mental disorders can be understood as dysfunctional attempts to avoid or terminate such undesirable states. Examples include alcohol or substance abuse, self-injurious behavior, or eating attacks.

Given the nature of variety in the emotional response and experience regarding intensity, duration, personality, and situational aspects, it is poorly defined when a certain emotional phenomenon can be considered inappropriate, abnormal, or even psychopathological [64]. This poses a challenge to psychometricians, researchers, and clinical diagnosticians to make the most valid (clinical) judgment or classification and eventually initiate appropriate and effective treatment.

Assessment of Emotions

Traditional assessment of emotions is based on self-reports via questionnaire and interview (3.1).[1] For many years now, other approaches aiming to capture a person's emotions more objectively and independently than subjective self-reports were developed. In the following, assessment methods deriving from physiology-based

---

[1] Of note, not all psychometric approaches for assessing the whole emotional process (or parts of it) can be displayed here. Further research attempts to improve questionnaires' reliability, like eye-tracking experiments (while filling in a questionnaire), appraisal biases, and knowledge of emotions.

emotion assessment such as blood and brain (3.2), expression-based emotion assessment such as facial expression, speech, and posture (3.3) are discussed. Moreover, digital footprints of emotions can be approached via social network platforms and smartphone data collection (3.4), including momentary ecological assessment. In the final part of this section, we compare the advantages and disadvantages of the presented assessment methods (3.5).

Self-Report-Based Methods for Emotion Assessment

In the traditional psychometric approaches, emotions are usually measured with self-report questionnaires, leveraging time- and cost-efficient instruments and enabling access to the cognitive component of someone's emotions. It is noteworthy that those instruments aim to assess single emotions and approach "concepts" as stress, depression, or well-being, indicating well-functioning (or dysfunctional) emotional processes over time. Questionnaires differ on how they are evaluated regarding their standardized psychometric properties as test objectivity, reliability, and validity [65]. Usually, factor analyses are conducted to evaluate the underlying constructs' factor structure in the designed questionnaire. Standardization is usually carried out using large and representative samples, which allow the classification of individual test results compared to the norm sample. Ideally, t-values or percentiles are defined for this purpose based on extensive psychometric studies. For the distinction between pathological and healthy reactions and clinical diagnosis, clinical samples are also collected.

To capture the emotion self-reports in the general population, often different scales are used, guided by the above-mentioned emotion models. In Table 10.2, an overview of some well-established questionnaires, including the item number and recall period, is given. For example, the Self-Assessment Manikin (SAM) is a nonverbal pictorial assessment technique that directly measures the pleasure, arousal, and dominance associated with a person's affective reaction [77]. Similarly, there is an Affect Balance Scale (ABS), based on a model that posits the existence of two independent conceptual dimensions—positive effect (PAS) and negative affect (NAS)—each related to overall psychological well-being by an independent set of variables [68]. The more widely accepted PANAS scale is a 10-item mood scale that comprises the Positive and Negative Affect Schedule (PANAS) [67]. Additionally, a flourishing scale determines psychological flourishing and feelings [92] in relevant areas such as purpose in life, relationships, self-esteem, feelings of competence, and optimism.

To make a clinical diagnosis of an affective disorder, clinical questionnaires are based on the clinical diagnostic criteria: symptom description, duration, intensity, distress, and psychosocial consequences based on the classification criteria of DSM-V [93] and ICD-10 [94]. However, for a valid diagnosis, *a structured clinical interview* is the gold standard (e.g., SKID [95]. Nevertheless, clinical questionnaires are frequently used to validate the diagnosis and evaluation of the treatment process

Table 10.2 Overview of example, well-established questionnaires assessing emotions

Construct	Scale Name	Number of Items	Recall Period	Reference
Affect	The affect balance scale (ABS)	10	Past few weeks	Bradburn (1969) [66]
Affect	The positive and negative affect scale (PANAS)	20	n/a	Watson et al. (1988) [67]
Affect	Affect balance scale	10	Past few weeks	Moriwaki (1974) [68]
Anxiety	Manifest anxiety scale	50/28	n/a	Taylor (1953) [69]
Anxiety	Self-rating anxiety scale	20	Overall	Zung (1965) [70]
Anxiety	Beck anxiety inventory	21	n/a	Beck et al., (1988) [71]
Anxiety, depression	The depression anxiety stress scales	21	Past week	Antony, Bieling, Cox, Enns, & Swinson (1998) [72]
Anxiety	State-trait anxiety inventory	40	Current	Spielberger (1989) [73]
Anxiety	Generalized anxiety Disorder-7 (GAD-7)	7	Past 2 weeks	Spitzer et al., (2006) [74]
Anxiety	The Hamilton anxiety rating scale	14	n/a	Hamilton, (1959) [75]
Coherence	Sense of coherence scale	29	Overall	Antonosky, (1993) [76]
Emotions	Self-assessment manikin	3	Current	Bradley & Lang (1994) [77]
Depression	Patient health questionnaire (PHQ-9)	9	Past 2 weeks	Kroenke, Spitzer, & Williams, 2001 [78]
Depression	The hospital anxiety and depression scale	14	Past week	Zigmond & Snaith, (1983) [79]
Depression	Beck depression inventory	21	Past 2 weeks	Beck, Steer, & Brown (1996) [80]
Depression	Zung self-rating depression scale	20	Current	Zung, (1965) [70]
Depression	Center for Epidemiologic Studies Depression Scale (CES-D)	20	Past week	Lewinsohn et al., (1997) [81]
Depression	Carroll rating scale (CRS) for depression	52	n/a	Carroll, Feinberg, Smouse, Rawson, & Greden, (1981) [82]
Emotional competence	Geneva emotional competence test (GECo)	110	Current	Schlegel & Mortillaro (2019) [83]
Emotion recognition	Geneva emotion recognition test (GERT/-S)	83/42	Current	Schlegel, Grandjean, & Scherer (2014) [84], [85]
Psychopathology	Symptom-checklist SCL-20-R	90	Past week	Hamilton, (1959) [75]
Rumination	Ruminative response scale	22	n/a	Treynor, Gonzalez, & Nolen-Hoeksema, (2003) [86]

Table 10.2 (continued)

Construct	Scale Name	Number of Items	Recall Period	Reference
Stress	Perceived stress scale perceived stress test (PSS)	14	Past month	Cohen, Kamarck, Mermelstein, (1983) [87]
Well-being	General health questionnaire	60	Past few weeks	Goldberg & Hillier (1979) [88]
Well-being	Warwick-Edinburgh mental Well-being scale—WEMWBS	14	Past 2 weeks	Stewart-Brown, et al. (2011) [89]
Well-being	EuroQol instrument EQ-5D-3L	15	Current	Herdman et al.,(2011) [90]
Worrying	Penn State worry questionnaire—(PSWQ)	16	n/a	Hopko et al., (2003) [91]

and follow-up. For example, the Beck's Depression Inventory (BDI-II, [80]) is commonly used in research to diagnose depression. The BDI-II contains 30 items referring to depressive symptoms someone may have experienced the past two weeks with different intensity. Assessing depressive symptoms in nine items, the PHQ-9 corresponds to the depression module of the Patient Health Questionnaire (PHQ, [78]). Unlike many other depression questionnaires, the PHQ-9 captures one of the nine DSM-IV criteria for diagnosing "major depression" with each question. Again, due to depression classification criteria, the recall period is two weeks. The Depression Anxiety Stress Scale (DASS, [72]) covers 42 Items on those three related emotional states in the last week by not focusing on one specific affective disorder. In contrast to assessing mental illness, there are fewer instruments for assessing mental well-being. For example, the Warwick-Edinburgh Mental Well-Being Scale (WEMWBS) aims to assess positive feelings an individual may have experienced to a certain extent in the past two weeks.

Overall, questionnaires to assess emotions (for clinical use) have the advantage of not requiring experienced experts, lead to scalable, comparable results, and are time and cost-efficient than clinical interviews. Besides, the object is an individual's emotional experience, and therefore the subjectivity of self-reports makes sense to capture someone's psychological strain. However, self-report questionnaires face big shortcomings in the assessment of emotions over time. Most instruments (covering the classification criteria of DSM-V and ICD-10) refer to a recall period of up to two weeks (see Table 10.2 for examples). Therefore, they are greatly biased by memory effects [96] and the motivation of an individual [97]. Studies are showing the questionnaire results are more negative when filled on Mondays while Saturday mornings are "the happiest moment during the week" and consequently provoke more positive test results [98, 99]. Other bias factors are fatigue and other non-assessed personal conditions (e.g., bad news occurring just that specific assessment day), as well as misunderstanding of the items and social desirability (aiming "to look good," even in anonymous surveys) [100, 101]. This specific subjectivity is wanted to a somewhat extent since it also informs the examiner about a person's perception and interpretation. Nevertheless, the objectivity of assessment is always limited greatly by these interpretation biases.

Physiology-Based Emotion Assessment

As mentioned above, emotions are represented not only by the cognitive but also physical and expressive components. While the cognitive aspect of emotion cannot be observed directly, the physical and expressive aspects are often manifested in different bodily signals. An emotional state influences underlying human biology and psychophysiology, as well as the resulting behaviors. For example, stress can be manifested in hormonal changes; anxiety can be manifested in terms of a high pulse rate, while happiness can be expressed via laughter. As we point out in the following subsections, some of these manifestations are captured and modeled to determine the emotion in affective computing.

Emotional Assessment from Blood

The emotional state of the human may be assessed via blood-based analytics, as the emotional state influences the individual's hormonal status. Especially in the field of biological psychiatry research, plasmatic biomarkers have been leading endeavors. The five most named plasmatic biomarkers (BDNF, TNF- alpha, IL-6, C-reactive protein, and cortisol) are classically used to predict psychiatric disorders like schizophrenia, major depressive disorder, or bipolar disorder [21]. In a meta-analysis, patterns of variation of those features were identified between those most important psychiatric disorders. The results indicated robust variations across studies but also showed similarities among disorders. The authors conclude that the implemented biomarkers may be interpreted as transdiagnostic systemic consequences of psychiatric illness rather than diagnostic markers. This is in line with another review of Funalla et al. showing evidence for diagnostic biomarkers associated with obsessive-compulsive disorder (OCD) but did not show diagnostic specificity [102]. A commonly used indicator for stress is the cortisol concentration found in human blood or saliva [103]. Overall, in this chapter, we do not focus on plasmatic biomarkers; we mention these for completeness.

Emotional Assessment from Brain

Other biological features to determine emotions with a rather high accuracy can be extracted from brain electroencephalography (EEG) [22]. With the help of EEG, the assessment of the brain's summed electrical activity is enabled by recording the voltage fluctuations on the surface of the head. Emotion extraction is consequently based on arousal and categorized into valence and excitation. EEG evaluation is traditionally performed by pattern recognition by the trained evaluator or by an automatic evaluation. For emotions, the patterns of alpha and beta waves are key classifiers. The alpha wave is associated with mild relaxation or relaxed alertness, with eyes closed (frequency range between 8 and 13 Hz). A beta wave has different causes and meanings and may occur during constant tensing of a muscle or during active concentration (frequency range between 13 and 30 Hz). According to Choppin (2000), high valence is associated with high beta power in the right parietal lobe and high alpha power in the brain's frontal lobe [104]. High beta power in the parietal lobes is associated with higher arousal in emotions, while the alpha activity is lower but also located in the parietal lobes. More specifically, negative emotions are represented by activity in the right frontal lobe, whereas positive emotions result in high power in the brain's left frontal part. EEG was found to achieve 88.86% accuracy for four emotions: sad, scared, happy, and calm [105]. After assessing the EEG waves and extracting the particular emotional features, classifiers are trained for emotion identification. Popular is the Canonical Correlation Analysis (CCA) [106], Artificial Neural Network (ANN) [107], Fisher linear discriminant projection [24], and Adaptive Neuro-Fuzzy Interference System (ANFIS) [108]. Using K-Nearest Neighbor (KNN) [109], and Support Vector Machine (SVM) [105, 110] Mehmood and Lee (2016) used five frequency bands (beside alpha and beta, delta, theta, gamma waves) and identified the four emotions *sad*, *scared*, *happy* and *calm* with an accuracy rate of 55% (KNN) and 58% (SVM).

In a more clinical setting, Khodayari-Rostamabd [111] used EEG data to predict the pharmaceutical treatment response of schizophrenic patients. A set of features was classified using the kernel partial least squares regression method to perform response prediction on the positive and negative syndrome scale (PANSS) with 85% accuracy. In another sample aiming to predict psychopharmacological treatment response (SSRI), the same research group extracted candidate features from the subject's pre-treatment EEG using a mixture of factor analysis (MFA) model in a sample with patients suffering from depression [112]. The proposed method's specificity is 80.9%, while sensitivity is 94.9%, for an overall prediction accuracy of 87.9%. Besides EEG, also functional Magnetic Resonance Imaging (fMRI) is used to assess emotions, especially by exploring the amygdala activity [113, 114]. Zhang et al. [114] analyzed connectivity change patterns in an fMRI data-driven approach in 334 healthy participants before and after inducing stress. Besides, the participants' cortisol level was taken to classify pre- and post-stress states. The machine learning model revealed that the discrimination relied on activation in the dorsal anterior cingulate cortex, amygdala, posterior cingulate cortex, and precuneus and with a 75% accuracy rate. The advantage of using EEG for assessing emotions is that the data extraction is independent of facial or verbal expression that could be impaired due to, e.g., paraplegia, facial paralysis, but the necessity of a lab and the complex, costly installation and maintenance of equipment is a big disadvantage not only for the practical field but also research projects [22].

Emotional Assessment from Physiological Signal Collection

A lot has been written about assessment of the emotional state of the individual from the physiological state—via, e.g., EEG (mentioned above), Electrocardiogram (ECG), Electromyography (EMG), Electrooculography (EOG), Galvanic Skin Response (GSR), Heart Rate (HR), Body Temperature (T), Blood Oxygen Saturation (OXY), Respiration Rate (RR), or Blood Pressure analytics (BP) [115–118]. From the technical perspective, the existing physiological signal-driven emotion assessment methods can also be divided into three categories (a) traditional machine learning methods, (b) deep neural network-based method, and (c) sequence-based models.

Conventional Machine Learning Approach In the case of traditional machine learning-based approaches, first, a set of features are extracted from the captured data, and then different algorithms are used for model construction. Apart from the time domain characteristics, to leverage the spectral-domain characteristics such as power spectral density (PSD), spectral entropy (SE) is computed using Fast Fourier Transform (FFT) or Short-term Fourier Transform (STFT).

SVM is probably the most widely used in physiological signal base emotion recognition among various machine learning algorithms. Das et al. extracted Welch's PSD of ECG and Galvanic Skin Response (GSR) signals for emotion recognition [119]. Liu et al. extracted a set of features from EEG and eye signals and used a linear SVM to determine three emotion states [120]. However, as regular SVM does not work in the imbalanced dataset, Liu et al. constructed an imbalanced support vector machine to solve the imbalanced dataset problem, which increased the punishment weight to the minority class and decreased the punishment weight majority class [121]. A few authors also used KNN ($K = 4$) to classify four emotions with the four features extracted from ECG, EMG, GSR, and RR [121]. In [115], the authors collected 14 features of 34 participants as they watch three sets of 10-min film clips eliciting fear, sadness, and neutrality, respectively. Analyses used sequential backward selection and sequential forward selection to choose different feature sets for 5 classifiers (QDA, MLP, RBNF, KNN, and LDA). Wen et al. used RF to classify five emotional states with features extracted from OXY, GSR, and HR [122].

Deep Learning-Based Approach Among different deep learning-based approaches, CNN is one of the most widely used. Martinez et al. trained an efficient deep convolution neural network (CNN) to classify four cognitive states (relaxation, anxiety, excitement, and fun) using skin conductance and blood volume pulse signals [123]. Giao et al. used the Convolutional Neural Network (CNN) for feature abstraction from EEG signal [124]. In [125], several statistical features were extracted and sent to the CNN and DNN. Song et al. used dynamical graph convolutional neural networks (DGCNN), which could dynamically learn the intrinsic relationship between different EEG channels represented by an adjacency matrix to facilitate feature extraction [126]. DBN is also widely used for emotion recognition. It learns a deep input feature through pre-training. Zheng et al. introduced a recent advanced deep belief network (DBN) with differential entropy features to classify two emotional categories (positive and negative) from EEG data, where a Hidden Markov Model (HMM) was integrated to accurately capture a more reliable emotional stage switching [127]. Huang et al. extracted a set of features and applied DBN in mapping the extracted feature to the higher-level characteristics space [128]. In the work of [129], instead of the manual feature extraction, the raw EEG, EMG, EOG, and GSR signals directly inputted to the DBN, where the high-level features according to the data distribution could be extracted.

Sequence-Based Models To capture the temporal aspects of the physiological signals, often sequence-based models are used. For example, Li et al. applied CNN first to extract features from EEG and then applied LSTM to train the classifier, where the classifier performance was relevant to the output of LSTM in each time step [130]. In the work of [131], an end-to-end structure was proposed, in which the raw EEG signals in 5 s-long segments were sent to the LSTM networks, in which autonomously learned features. Liu et al. proposed a model with two attention mechanisms based on multi-layer LSTM for the video and EEG signals, which combined temporal and band attentions [132].

Overall, as the captured signals are noisy, several pre-processing techniques are often used to eliminate the noise introduced from different sources such as crosstalk, measurement error, and instrument interference. The commonly used preprocessing techniques include filtering [133], Discrete Wavelet Transform (DWT) [134], Independent Component Analysis (ICA) [135], Empirical mode decomposition (EMD) [136].

The overall psychophysiological approach for emotional assessment is cumbersome and not straightforward to be deployed in real-time for real-time accurate emotion recognition. Besides, the required complex laboratory setup (e.g., EEG) is time and cost-intensive.

Expression-Based Emotion Assessment

Following the cognitive and physical components of emotions, we focus on expressing emotions in this section. We describe facial and verbal emotion recognition, as well as how posture may reflect human emotional states.

Facial Emotion Recognition (FER)

We broadly divide facial emotion recognition-related works into the following two groups (a) conventional FER approach and (b) deep learning-based FER approach.

Conventional FER Approach For automatic FER systems, various types of conventional approaches have been studied. First, all these approaches assess the face region and then extract a set of geometric features, appearance features, or a hybrid of geometric and appearance features on the target face. For the geometric features, the relationship between facial components is used to construct a feature vector for training [137, 138]. For example, Ghimire and Lee [138] used two types of geometric features based on 52 facial landmark points' position and angle. First, the angle and Euclidean distance between each pair of landmarks within a frame are calculated. Second, the distance and angles are subtracted from the corresponding distance and angles in the video sequence's first frame. For the classifier, two approaches are used, using multi-class AdaBoost with dynamic time warping, and SVM on the boosted feature vectors.

The appearance features are usually extracted from the global face region [139] or different face regions containing different types of information [140–143]. An example of using global features includes the exploration by Happy et al. [139]. The authors utilized a local binary pattern (LBP) histogram of different block sizes from a global face region as the feature vectors. They classified various facial expressions using a principal component analysis (PCA). This method's classification performance is poor as it cannot reflect local variations of the facial components to the feature vector. A few explorations also used features from different face regions as they may have different levels of importance, unlike a global-feature-based approach. For example, the eyes and mouth contain more information than the forehead and cheek. Ghimire et al. [144] extracted region-specific appearance features by dividing the entire face region into domain-specific local regions. An incremental search approach is used to identify important local regions, reducing feature vector size, and improving classification performance.

For hybrid features, some approaches [144, 145] have combined geometric and appearance features to complement the two approaches' weaknesses and provide even better results in certain cases.

Deep Learning-Based FER Approach The most adopted one in deep neural network-based FER is CNN. The main advantage is to completely remove or highly reduce the dependence on physics-based models and/or other pre-processing techniques by enabling "end-to-end" learning directly from input images [146]. Breuer and Kimmel [147] investigated the suitability of CNNs on different FER datasets and showed the capability of networks trained on emotion assessment and FER-related tasks. Jung et al. [148] used two different types of CNN: the first extracts temporal appearance features from the image sequences. The second CNN extracts the temporal geometry features from temporal facial landmark points. These two models are combined using a new integration method to boost the performance of facial expression recognition.

However, as CNN-based methods are not suitable for capturing the temporal sequence, a hybrid approach combining both CNN (for spatial features) and LSTM (for temporal sequence) was developed. LSTM is a special type of RNN capable of learning long-term dependencies. Kahou et al. [149] proposed a hybrid RNNCNN framework for propagating information over a sequence using a continuously valued hidden-layer representation. In this work, the authors presented a complete system for the 2015 Emotion Recognition in the Wild (EmotiW) Challenge [150]. They proved that a hybrid CNN-RNN architecture for a facial expression analysis could outperform a previously applied CNN approach using temporal averaging for aggregation. Kim et al. [151] utilized representative expression-states (e.g., the onset, apex, and offset of expressions), specified in facial sequences regardless of the expression intensity. Hasani and Mahoor [152] proposed the 3D Inception-ResNet architecture followed by an LSTM unit that together extracts the spatial relations and temporal relations within the facial images between different frames in a video sequence. Graves et al. [153] used a recurrent network to consider the temporal dependencies present in the image sequences during classification. This study compared the performance of two types of LSTM (bidirectional LSTM and unidirectional LSTM) and proved that a bidirectional network provides significantly better performance than a unidirectional LSTM.

In summary, hybrid CNN-LSTM (RNN) based FER approaches combine an LSTM with a deep hierarchical visual feature extractor such as a CNN model. Therefore, such a hybrid model can learn to recognize and synthesize temporal dynamics for tasks involving sequential images. Each visual feature determined through a CNN is passed to the corresponding LSTM, and it produces a fixed or variable-length vector representation. The outputs are then passed into a recurrent sequence-learning module. Finally, the predicted distribution is computed by applying softmax [154, 155]. A limitation of this approach is the challenge of capturing the facial expression in real-time in daily life as a natural reaction to an emotional experience. Additionally, privacy issues can be a problem if a person does not want to be visually recorded during such intimate moments.

Speech Based Emotion Recognition (SER)

The existing literature on speech emotion recognition (SER) is also broadly divided into the following two categories (a) Conventional SER approach and (b) deep learning-based SER approach.

Conventional SER Approach   In traditional SER systems, there are mainly three steps—(a) signal pre-processing, (b) feature extraction, and (c) classification. At first, acoustic pre-processing such as denoising, segmentation is carried out to determine relevant units of the speech signal [156–158]. Once the pre-processing is done, several short-term characteristics of the signal such as energy, formants, and pitch are extracted, and short-term classification of the speech segment is done [159]. On the contrary, for long-term classification, mean, standard deviation is used [160]. Among prosodic features, the intensity, pitch, rate of spoken words, and variance play an important role in identifying various types of emotions from the input speech signal [161]. The relationship between different vocal parameters and their relation to emotion is often explored in SER. Parameters such as intensity, pitch, and rate of spoken words, and quality of voice are frequently considered [162]. The intensity and pitch are often correlated to activation so that the value of intensity increases along with high pitch and vice versa [163, 164]. Factors that affect the mapping from acoustic variables to emotion include whether the speaker is acting, there are high speaker variations, and the individual's mood or personality [165, 166].

In the existing SER literature, there are two types of classifiers—linear and non-linear. Linear classifiers usually perform classification based on object features with a linear arrangement of various objects [166]. In contrast, non-linear classifiers are utilized for object characterization in developing the non-linear weighted combination of such objects [167–170]. The GMMs are utilized for the representation of the acoustic features of sound units. The HMMs, on the other hand, are utilized for dealing with temporal variations in speech signals [171].

Deep Learning-Based SER Approach In SER approaches, different types of deep neural networks are used. Senigallia et al. used a 2D CNN with Phoneme data as input data to determine 7 emotion states [172]. Zhao et al. combined Deep 1D and 2D CNN for high-level learning features from input audio and log-mel spectrograms for emotion classification [173]. Convolutional Neural Network (CNN) also uses the layer-wise structure and can categorize the seven universal emotions from the defined speech spectrograms [174]. In [175], an SER technique based on spectrograms and deep CNN is presented. The model consists of three fully connected convolutional layers for extracting emotional features from the speech signal's spectrogram images. Pablo et al. obtained emotional expressions that are spontaneous and can easily be classified into positive or negative [176]. Mao et al. trained the CNN to learn affect salient features and achieved robust emotion recognition with the variational speaker, language, and environment [177].

The hybrid networks consisting of CNN and RNN are also used in SER [178–180]. This enables the model to obtain both frequency and temporal dependency in a given speech signal. Sometimes, a reconstruction-error-based RNN for continuous speech emotion recognition is also used [181]. SER algorithms based on CNNs and RNNs have been investigated in [180]. The deep hierarchical CNNs architecture for feature extraction has also been combined with LSTM network layers. It was found that CNN's have a time-based distributed network that provides results with greater accuracy. Zhao et al. used a hybrid RCNN model to determine basic emotion [182]. Wootaek et al. used a deep hierarchical feature extraction architecture of CNNs combined with LSTM network layers for better emotion recognition [180]. Like FER, capturing the speech in real-time as a reaction to the emotional experience is a challenging issue. In addition to interfering with privacy, someone may express emotions without using expressed language because they are alone or do not want to speak, especially in emotional moments. Although individuals tend to adapt quickly to being observed, the awareness of being recorded might interfere with someone's speech and expression of emotions.

Emotional Assessment from Posture

In contrast to research on automatic emotion recognition focusing on facial expressions or physiological signals, little research has been done on exploiting body postures. However, they can be useful for emotion recognition and even more accurate than facial features [23]. Bodily postures refer to the physical expression component of emotions and an important channel of communication. Since it is challenging to categorize the expressed posture to discrete emotions due to the variety of validated emotion poses, researchers in this field focus first on defining features, aiming for understanding the cohesion and dimensional ratings with high validity [183]. Therefore, studies recorded actors displaying concrete, discrete emotions, and the recordings were then rated by the study participants categorizing these emotions. For example, Lopez et al. asked study participants to categorize emotion postures depicting five emotions (joy, sadness, fear, anger, and disgust) and to rate valence and arousal for each emotion pose. Besides a successful categorization of all emotion categories, participants accurately identified multiple distinct poses within each emotion category. The dimensional rating of arousal and valence showed interesting overlaps and distinctions, increasing further granularity of distinct emotions. Similarly, the Emotion Recognition Test (GERT) [85] was developed to test the emotional recognition ability using video clips with sound simultaneously presenting facial, vocal, and emotional expression based on the posture by using various data and video material and relatively large samples to validate. Individuals using the GERT test material are asked to watch video clips and rate the displayed emotions to assess their emotion recognition ability.

Postures are also captured using movement sensor data from smartwatches or mobile phones [184]. Quiroz et al. observed the movement sensor logged by the smartwatches of 50 participants differentiated between the emotions *happy*, *sad*, and *neutral* as a response to an emotional stimulus in an experimental setting. Furthermore, the response was validated by additional data from the Positive Affect and Negative Affect Schedule Questionnaire (PANAS). Emotional states could be assessed well by self-report and data obtained from the smartwatch with high accuracy across all users for classification of happy versus sad emotional states. Although the categories here depict only two emotions and other categorization difficulties need to be evaluated, movement sensors' usage still appears promising for emotion recognition purposes. Another smartphone-based approach also uses information about postures for recognizing emotions but uses self-reported body postures [185]. A mobile application was developed that classifies (based on the nearest neighbor algorithm) inserted poses into Ekman's six basic emotion categories and a neutral state. Emotion recognition accuracy was evaluated using poses reported by a sample of users.

Although digital devices may be included to capture postures in real-time, this data collection is challenging to conduct in a person's daily life and close to their natural expression of emotions. Due to the necessity of leveraging, e.g., the use of cameras, this method is obtrusive and implies the same privacy issues we discussed in the context of FER and SER.

Internet-Use Based Emotion Assessment

This subsection presents the emotion assessments based on the internet usage of an individual. Firstly, analysis on social networks (3.4.1) is discussed, followed by smartphone-based emotion assessment (3.4.2) and smartphone-based experience sample methods (3.4.3).

Social Network Analysis

Digital records of an individual's behavior are extracted, including linguistic style, sentiment, online social networks, and other activity traces, which can be used to infer the individual's psychological state. In particular, social networking platforms are becoming increasingly popular. They have recently been used more extensively to study emotions, as they are easily accessible to users, and researchers can collect the necessary information with the users' consent. Based on this approach, Chen et al. aimed to identify users with depression or at risk of depression by assessing the individual's expressed emotions from Twitter posts over time [36]. In another study, voluntarily shared Facebook Likes for $N = 58,000$ users were used to predict several highly sensitive personality attributes [37]: sexual orientation, ethnicity, religious and political views, personality traits, intelligence, happiness, use addictive substances, parental separation, age, and gender. All attributes were predicted with high accuracy, especially the ethnic origin and gender. Other emerging approaches focus on Spotify music or Instagram picture extraction as a feature to predict personality or mood outcomes [186].

Furthermore, studies focus on language content in social media networks or messaging systems to discover depressive symptoms, the so-called Natural Language Processing (NLP) [187]. The depressive language was characterized by more negative and extreme words such as "always, everybody, never" [188]. The challenge with this data is that communication on Twitter, Facebook may be heavily distorted by aspects of social desirability and specific motivations that drive someone to express themselves on the Internet. Furthermore, data collection from someone's account may face privacy issues.

Smartphone-Based Emotion Assessment

Smartphones are personal devices that individuals carry around with them almost all the time [189]. They include a plethora of onboard sensors (e.g., accelerometer, gyroscope, GPS) and can sense different user activities passively (e.g., mobility, app usage history) [190]. In this subsection, we review the smartphone-based methods for emotion assessment in its user's natural daily environments. We consider the usage-based assessment methods, as well as touch-based ones.

Usage-Based Emotion Recognition Methods The smartphone provides numerous data sources for collecting real-world data about emotions. For example, defined FER and SER assessments can also be extracted from the smartphone's camera and microphone. Other smartphone-based sensing sources are connectivity (WIFI on/off), smartphone status (screen, battery, power-saving mode), calls (type, duration), text messages (type, length), notifications (apps, category), calendar (initial query, logging of new entries), technical data (anonymized user ID, IP address, mobile phone type) [191]. Harari et al. [192] sensed conversations, phone calls, text messages, and messaging and social media applications for individual trait assessment. Namely, they collected sensing data in five semantic categories (communication & social behavior, music listening behavior, app usage behavior, mobility, and general day- & night-time activity) and used a machine learning approach (random forest, elastic net) to predict personality traits.[2] MoodScope proposed to infer mood exploiting multiple information channels, such as SMS, email, phone call patterns, application usage, web browsing, and location [143]. In EmotionSense, Rachuri et al. used multiple Emotional Prosody Speech and Transcripts library features to train the emotion classifier [195]. In the same vein, researchers also demonstrated that aggregated features obtained from smartphone usage data could indicate the Big-Five personality traits [196]. We also find that there are multiple works, which use different information sources to infer the presence of a particular emotional state. For example, Pielot et al. tried to infer boredom from smartphone usage patterns like call details, sensor details, and others [197]. In their work on assessing stress, Lu et al. built a stress classification model using several acoustic features [198]. Similarly, Bogomolov et al. showed that daily happiness [199] and daily stress [200] could be inferred from mobile phone usage, personality traits, and weather data.

Touch-Based Emotion Recognition Methods Widespread availability of touch-based devices and a steady increase [35] in the usage of instant messaging apps open a new possibility of inferring emotion from touch interactions. Therefore, research groups started to focus on typing patterns (Shapsough et al., 2016) using a built-in sensor (a smart keyboard) and using machine learning techniques to assess emotions based on different aspects of typing. For example, Lee et al. designed a Twitter client app and collected data from various onboard sensors, including typing (e.g., speed), to predict one user's emotion in the pilot study [201]. Similarly, Gao et al. used multiple finger-stroke-related features to identify different emotional states during touch-based gameplay [202]. Ciman et al. assessed stress conditions by analyzing multiple features of smartphone interaction, including swipe, scroll, and text input interactions [203]. Kim et al. [204] proposed an emotion recognition framework analyzing touch behavior during app usage, using 12 attributes from 3

---

[2] Personality traits are generally described as relatively stable patterns of thought, feelings, and behaviors and are therefore well related to emotional theories [193, 194].

onboard smartphone sensors. Although focused on narrow application scenarios, all of these works point to the value of touch patterns in emotion assessment.

Smartphone-Based Experience Sampling Method Design

One of the key requirements to develop a smartphone-based emotion assessment system is to collect emotion ground truth labels, which are typically collected as emotion self-reports by deploying an Experience Sampling Method (ESM), also known as an Ecological Momentary Assessment (EMA). The Experience Sampling Method (ESM) is a widely used tool in psychology and behavioral research for in-situ sampling of human behavior, thoughts, and feelings [5, 205]. The ubiquitous use of smartphones and wearable devices helps in the more flexible design of ESM, aptly termed as mobile ESM (mESM) [206–208]. It allows the collection of rich contextual information (e.g., sensor information, application usage data) along with behavioral data at an unprecedented scale and granularity. Frameworks like Device Analyzer, UbiqLog, AWARE, ACE, MobileMiner, or MQoL-Lab [190] have been designed to infer user's context based on sensor data, application usage details of the smartphones [142, 209–213]. While these frameworks help in the automatic logging of sensor data, self-reports related to various aspects of human life (like emotion) still require direct input from the user.

Balancing Probing Rate and Self-Report Timeliness In the ESM studies, the participant burden mainly arises from repeatedly answering the same survey questions. Time-based, event-based schedules are the most commonly used ESM schedules [214]. Time-based approaches aim to reduce probing rate (at the cost of fine granularity), while event-driven ones try to collect self-report timely (at the cost of a high probing rate). Recently, hybrid ESM schedules are designed combining time-based and event-based ones to trade-off between probing rate and self-report timeliness [215]. With the proliferation of smartphones, and other wearable devices, more intelligent and less intrusive survey schedules, including these limiting the maximum number of triggers, increasing the gap between two consecutive probes, have been designed. Several open-source software platforms, like ESP [216], my experience [217], psychology [218], Personal Analytics Companion [219], are available on different mobile computing platforms to cater to ESM experiments.

Maintaining Response Quality Via Interruptibility-Aware Designs Recent advancements in interruptibility-aware notification management recommend several strategies to probe at opportune moments leveraging contextual information (e.g., placing between two activities like sitting and walking, after completing one task such as messaging or reading a text on mobile) [220–222]. In [223], the authors showed that features like the last survey response, phone's ringer mode, and user's proximity to the screen could predict whether the recipient will see a notification within a few minutes. Leveraging these findings, intelligent notification strategies were developed, which resulted in a higher compliance rate and improved response quality [224, 225]. However, one of the major challenges of using such details in mobile-based ESM design is resource overhead and privacy. Designing ESM sched-

Table 10.3 Summarization of the existing ESM design approaches

ESM Schedule	Probing rate	Timeliness	Opportune probing
Time-based (TB) (e.g. [31], [144])	✓	✗	✗
Event-based (EB) (e.g. [194], [226])	✗	✓	✗
Hybrid [214]	✓	✓	✗
Interruptibility-aware (e.g. [8], [220], [223], [227])	✗	✗	✗
Ideal ESM schedule	✓	✓	✓

ules based on the underlying study can overcome such limitations [226]. Ideally, an ESM schedule shall optimize the probing rate (like time-based schedules), reduce latency (like event-based schedules), and probing moments (like interruptibility-aware schedules) (Table 10.3).

Pros and Cons in Different Emotion Assessment Approaches

In the previous sections, we have presented different approaches to assess emotions. Table 10.4 summarizes the pros and cons of the different self-report and sensing methods of data collection. We pointed out that self-report questionnaires have the advantage of being rather a time and cost-efficient (for assessors) and enable to reveal cognitions that are otherwise hard to capture. Furthermore, the subjectivity of an individual's view on their emotions expressed via self-reporting might be wanted in some contexts (e.g., for clinical diagnostics). However, self-reports are challenged by numerous confounding variables as fatigue, interpretation and memory biases, non-assessed personal conditions, misunderstanding of the items, and social desirability [100, 101]. Some physiological assessment methods might be more objective (like EEG or blood) but require a laboratory and complex setup and controlled environment. Due to this limitation, real-time assessment of emotions close to an individual's everyday life experience is not possible.

Additionally, some research is based on induced emotional states. Emotional reactions can be induced in experimental settings. However, the transfer and generalizability of such results into an individual's real-life is doubtable. Besides, the period of data collection is often limited, and collecting a high volume of data from a large number of participants is difficult. Finally, the participants need to participate actively and contribute to the data collection effort (via self-reports).

In most cases, the data collection cannot be done passively and, consequently, lacks unobtrusiveness. Moreover, most of the discussed methods focus only on one data collection source (e.g., speech, EEG alpha waves, or social network analysis). They are, therefore, very limited regarding the complex emotional process described in the CPM [4].

In the context of emotional assessment, novel, personal sensing methods embedded in daily life via wearables and smartphones promise to overcome some of those issues by providing real-time observations [143, 197, 207, 209–211]. These devices can capture data passively from different modalities without user intervention, log app usage behavior, and leverage different computational models on the device for

Table 10.4 Pro's and Con's of Emotion Assessment Methods

Method	Pro	Con
Questionnaire-based Emotion Assessment		
Self-report questionnaires	– Easy to conduct – Access to cognitions – Time-efficient for examiners – Established and accepted in practical fields – Privacy is given – Standardization for the classification of individual results available	– Several biases (memory, interpretation, social desirability) – Time consuming for participants (not examiners)
Physiology-based emotion assessment		
Blood	– Objective assessment of autonomous nervous system (ANS) reaction – Standardization for the classification of individual results available	– Intrusive – Not applicable outside a laboratory – Focus on a single factor
Brain	– Objective assessment of brain activity as emotional response – Standardization for the classification of individual results available	– Not applicable outside a laboratory – High costs – High setup up requirements – Focus on a single factor
Posture	– Naturalistic expression – Setup needed (cameras, posture logging devices) – Important channel of communication	– Categories vary greatly, no standardization – Interpretation of postures is limited – Focus on one class of emotion expression only
Expression-based emotion assessment		
Facial emotion recognition	– Objective assessment of expression of emotions – No language needed – Passive data collection possible	– Recording in real-time in daily life may be difficult – Privacy issues – Camera needs to capture facial expression – Awareness of being recorded might interfere natural reaction
Speech emotion recognition	– No laboratory setup needed, objective – No body language and physical parameters needed – Passive data collection possible	– Recording in real-time in daily life may be difficult – Privacy issues – Awareness of being recorded might interfere natural reaction – No standardization
Smartphone—And internet-based emotion assessment		

Table 10.4 (continued)

Method	Pro	Con
Questionnaire-based Emotion Assessment		
Social network	– Time and cost efficient – No special laboratory or technical equipment needed – Real-time behaviour accessible – Passive data collection possible	– Only specific part of daily expressed emotions visible as embedded in a social network use context – Focusing on just a single class of behaviour – Biased by social desirability and other motivations to communicate on social platforms – Objectivity and/or standard values not given
Ecological momentary assessment	– No specific setup needed to conduct – Unobtrusive – Real time assessment in daily life – Access to cognitions	– Self-report with potential biases (social desirability, interpretation bias) – Data protection issues – No standardization
Smartphone-based sensing	– No specific setup needed to conduct – Unobtrusive – Real time assessment in daily life – Passive data collection possible – Great variety of data sources possible – Execute different models for emotion inference	– Data protection issues – Perceived as "scary" by participants – No standardization

emotion inference. As a result, these devices are very promising to determine emotion-based behavior based on different usage patterns. However, the approach is still novel, and some participants are concerned about their privacy [227]. These concerns must be taken into account seriously but contrasted by the fact that personal information is shared openly on the Internet nowadays. This ambiguity is labeled as *Privacy Paradox* and known since 2006 [228].

Tapsense: Smartphone Typing Based Emotion Assessment

This chapter specifically focuses on assessing the individual's emotional state from the smartphone usage patterns via the authors' TapSense study. Therefore, in this section, we first describe the overall research approach of the keyboard interaction study. We focus on a typing-based emotion assessment scenario, which helps to identify the key requirements to design the emotion assessment model and the self-report collection approach using an Experience Sampling Method (ESM) (Sect. 5.1.). In Sect. 5.2, the TapSense field study and data analysis are presented, and in Sect. 5.3. evaluated. In Sect. 5.4, the study is discussed.

Fig. 10.3 Overall
approach of the TapSense
requirements gathering,
study design, data analysis,
performance evaluation,
and discussion

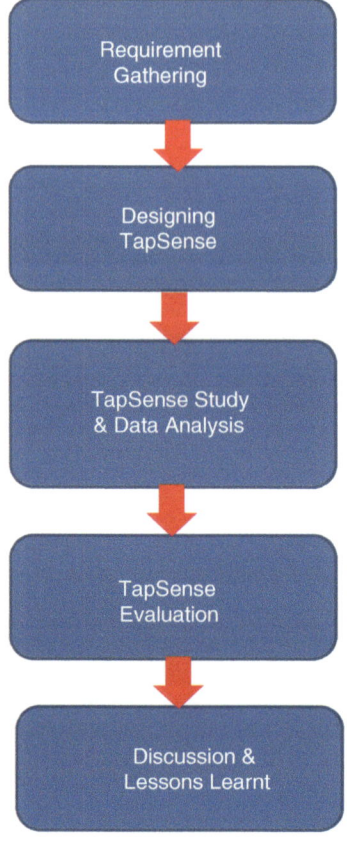

Background

The overall approach for the TapSense study is shown in Fig. 10.3. First, we gathered a set of requirements to design the keyboard interaction-based emotion assessment tool, followed by the actual design and implementation of the TapSense application. We discuss the study in detail and analyze the collected data. Finally, we evaluate the performance of the TapSense application and discuss the lessons learned from this study.

Requirements

TapSense study relies on the users' smartphone usage patterns. We explain the scenario of typing-based emotion assessment in Fig. 10.4. As the user performs typing activity, we extract his/her typing sessions and the amount of time he/she stays in a single mobile application without. For example, when a user uses

WhatsApp without switching to other applications from t1 till t2, then we define elapsed time between t1 and t2 as a Typing Session. Once the user completes the session, he/she is probed via an ESM, i.e., emotion self-report, which is considered the emotion ground truth. Later, several features are extracted from the typing sessions and correlated with the emotion self-report to develop an emotion assessment model. This scenario suggests consideration of the following requirements,

- *Trace keyboard interaction for emotion assessment:* The key requirements while determining emotions from the typing sessions is to make sure that (a) the typing details are captured correctly so that the relevant features can be extracted (b) the emotion ground truths are collected (ESM) and (b) an accurate emotion assessment model is constructed. We discuss these aspects further in this section.
- *ESM design for self-report collection:* Probing a user after every session may induce fatigue due to many probes. So, the probing moments should be chosen in such a manner that it captures the user's response accurately (i.e., before it fades away from the user's memory) and at the same time, the probing rate is not too high. We discuss in detail the ESM design further in this section.

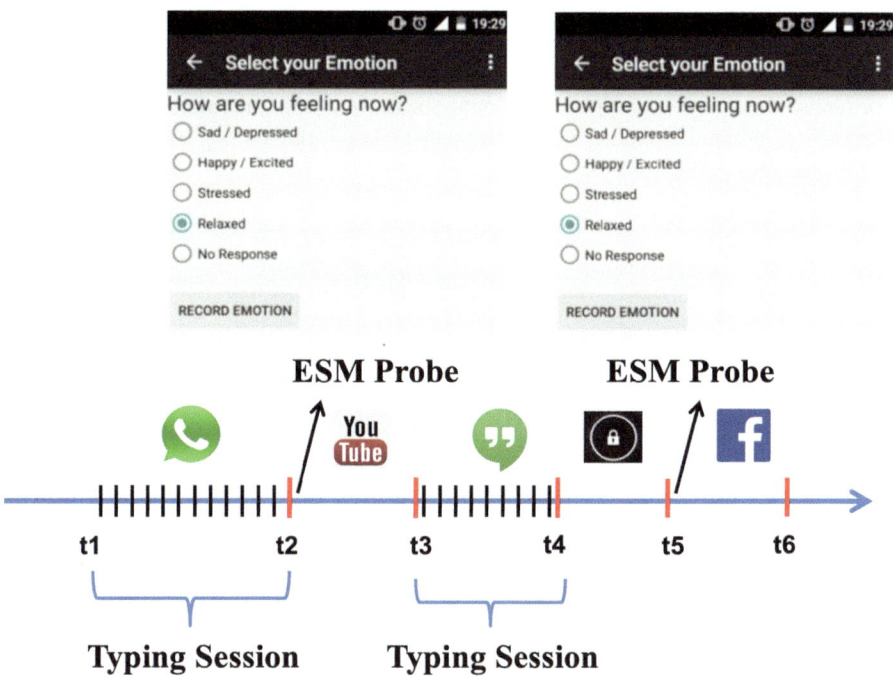

Fig. 10.4 TaspSense Approach: Typing Session-based ESM-triggering scenario

Design and Implementation of TapSense

TapSense consists of the following key components as in Fig. 10.5. TapLogger records the user's typing activity. It implements a virtual keyboard for tracing keyboard interactions. ProbeEngine runs on the phone to generate the user's ESM notifications and collects the ESM responses. The typing details and the associated emotion self-reports are made available at the server via the Uploader module that synchronizes with the server occasionally, or, if the user is offline, once the user connects to the Internet. The emotion assessment model is constructed on the server-side to determine the different emotional states from the typing details and the emotion self-reports. In parallel, a set of typing features is also extracted to construct the inopportune moment assessment model, which feeds back the ProbeEngine to optimize the probe generation. Next, we discuss the two key components of TapSense (a) emotion assessment from keyboard interaction, (b) ESM design for the emotion self-report collection.

TapLogger: Keyboard Interaction Collection

The TapLogger module of TapSense implements an Input Method Editor (IME) [229] provided by Android OS, and we refer to it as the TapSense keyboard (Fig. 10.6). It is the same as any QWERT keyboard; it provides similar functionalities as any Google keyboard. We have selected a standard keyboard because we aimed to provide similar functionalities. The user's keyboard interaction experience does not deviate much from what he/she is used to. It differs from others, as it has the additional capability of logging user's typing interactions, which, for security reasons, is not available in Google keyboard. To ensure user privacy, we do not store or record the characters typed. The logged information is the timestamp of each tap event, i.e., when a character is entered and the key input's categorical type, such as an alphanumeric key or delete key.

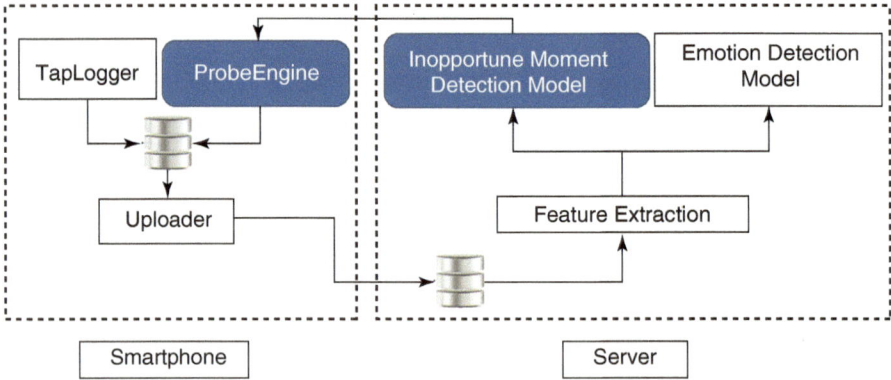

Fig. 10.5 TapSense High-level System Architecture

Fig. 10.6 TapSense
Keyboard

Fig. 10.7 Emotion
Self-report UI

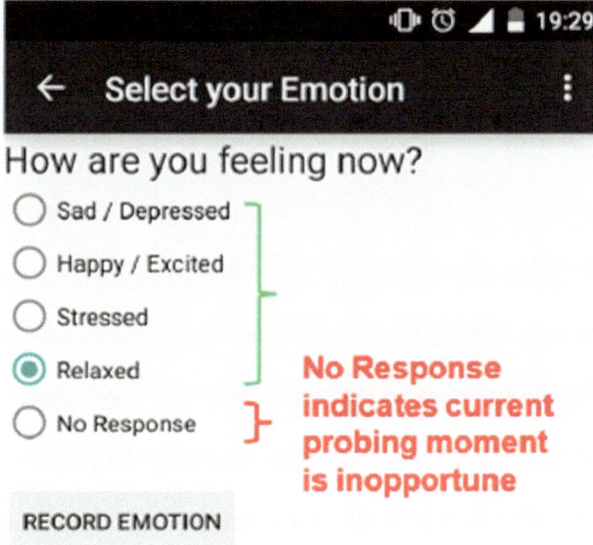

ProbeEngine: Emotion Self-Report Collection (ESM)

The ProbeEngine module of TapSense issues the ESM self-report probes by delivering a self-report questionnaire (Fig. 10.7). This survey questionnaire provides the option (happy, sad, stressed, relaxed) to record ground truth about the user's emotion while typing. This captures four largely represented emotions from four

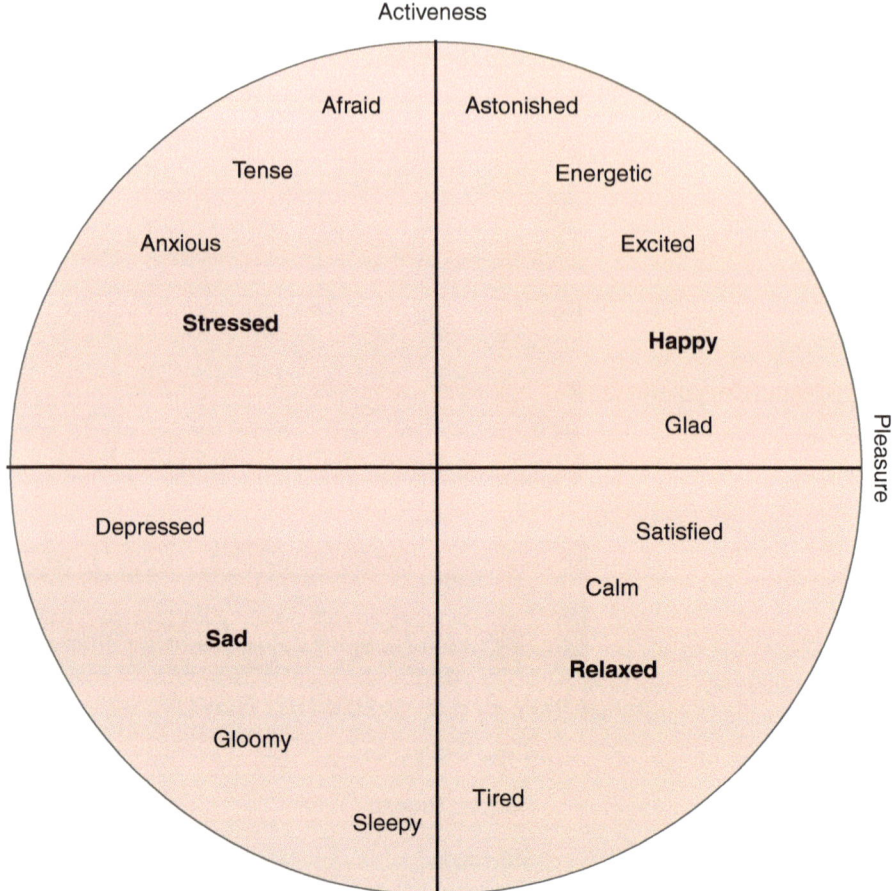

Fig. 10.8 Circumplex Model

different quadrants of the Circumplex model [52], as shown in Fig. 10.8. We select these discrete emotions as their valence-arousal representation is unambiguous on the Circumplex plane. Any discrete emotion and its unambiguous representation on the valence-arousal plane are equivalents [230]. We also include the "No Response" option to select this option to indicate the current probing moment is inopportune.

Emotion Assessment Model Construction

The emotion assessment model in TapSense is responsible for determining the four emotion states based on the keyboard interaction pattern. This is implemented on the server-side once the typing interaction details and the emotion self-reports details are available.

Table 10.5 Features used to Construct TapSense Emotion Assessment Model

Feature Name	Feature Description
Session typing speed (MSI)	Average of all ITDs present in the typing session
Session length	Number of characters typed in the typing session
Session duration	Time duration of the typing session
Backspace percentage	Fraction of backspace and delete keys typed in the typing session
Special character percentage	Fraction of special characters (non-alphanumeric) typed in the typing session
Last ESM trigger response	Emotion label as provided by the user

ITD elapsed time between two consecutive keypress events [ms]

Emotion Assessment Features From raw data collected within every Typing Session, we extract a set of typing features as defined in Table 10.5. For every session, we compute the ITDs, i.e., the elapsed time between two consecutive keypress events for all the presses. We derive the mean of all ITDs in the session and use it as typing speed. We define it as the Mean Session ITD (MSI). We compute the backspace and delete keys present in a session and use it as a feature. This is used as the representation of typing mistakes made in a session.

Similarly, we use the fraction of special characters in a session, session duration, and typed text length in a session as features. Any non-alphanumeric character is considered a special character. We use the last emotion self-report as a label for the model [215, 231]. However, at the later stage, when the TapSense model is operational, we use the predicted emotion for the last session as the feature value for the current session.

Emotion Assessment Model Trees-based machine learning approaches have been accurate in the context of emotion assessment in the past [201, 232]. We design a Random Forest (RF) based personalized multi-state emotion assessment model using the features described in Table 10.5 to assess the emotions. As typing patterns vary across individuals, derived, features will vary. Hence we construct a personalized model. We implement these models in Weka [233], building 100 Random Forest decision trees with a maximum depth of the tree set as 'unlimited' (i.e., the tree is constructed without pruning). We then derive the RF models' performance by deriving the mean and variability of the accuracy for the 100 RF-based models.

Experience Sampling Method Design

The ESM used in TapSense is optimized in two phases. Phase 1 balances probing rate and timeliness of self-report collection, and Phase 2 tries to probe at the opportune moments when the user's attention is available. We achieve this by designing a two-phase ESM [234]. We summarize it in Fig. 10.9. In Phase 1, we combine policy-based schedules to balance probing rate and timeliness and learn the inopportune moment assessment model. In Phase 2, we make the inopportune moment assessment model operational. We discuss both phases in detail now.

Phase 1: Balancing ESM Probing Frequency and Timeliness The collection of ESM emotion self-reports at the end of every typing session would help collect the labels close to the event, but it would lead to the generation of too many probes and user burden. To trade off these two conflicting requirements, we first assess the quality of the session itself, i.e., we make sure that there is a sufficient amount of typing done in a typing session for it to be considered. We issue the ESM probe only (a) if the user has performed a sufficient amount of typing, i.e., a minimum L = 80 characters in a typing session, and (b) a minimum time interval, i.e., W = 30 minutes has elapsed since the last ESM probe. To ensure the labels are collected close to the typing session, we use the polling interval parameter (T = 15 seconds) to check if the user has performed a sufficient amount of typing within a session. We describe the selection of threshold values based on initial field trials in Appendix 1. We name this ESM schedule the *Low Interference High Fidelity (LIHF)* ESM schedule (Fig. 10.9 (Phase 1)).

Phase 2: Inopportune Moment Assessment Model As we collect self-reports, we obtain both "No Responses" and valid emotion responses. We leverage these labels to build the inopportune moment assessment model (Table 10.6).

We use typing session duration and the typing length in a session as features since lengthy and longer typing sessions may indicate high user engagement and not be the ideal moment for triggering a probe. Besides, there may be some types of

Fig. 10.9 Balancing ESM Probing Frequency and Timeliness: ESM Triggering Steps

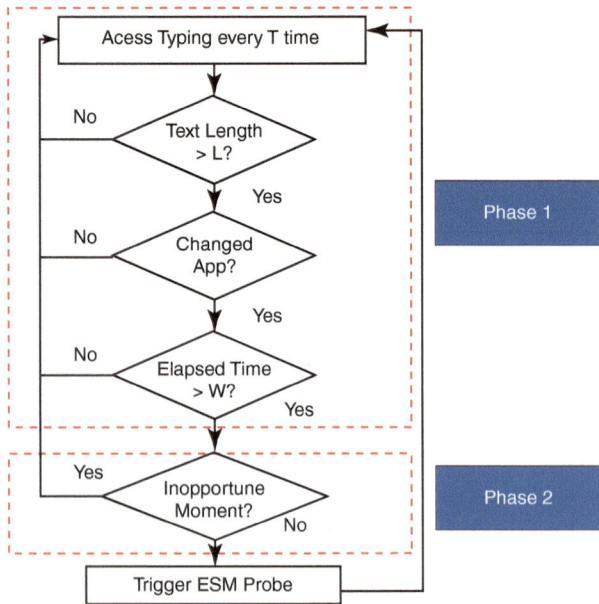

Table 10.6 Features Used To Detect Inopportune Moments

Feature Name	Feature Description
Session duration	Duration of the typing session
Session length	Length of the text in the typing session
App category	Category of the application
Last ESM trigger response	Last ESM trigger response

applications like media, games when the users may not be interrupted for probing. So, we include the application type also as a feature. We categorize the applications into one of the 7 categories: Browsing, Email, Media, Instant Messaging (IM), Online Social Network (OSN), SMS, and 'Misc,' following the application's description in the Google Play Store. Moreover, we use the label of the last ESM probe response as a feature. We use it to determine whether the user continues to remain occupied in the current session and if he/she marked the previous session with "No Response." However, once the model is operational and deployed, we use the predicted value of the inopportune moment for the last session as the current session's feature value. Table 6 summarizes the features used to implement the model. We construct a Random Forest-based prediction model to assess the inopportune moments for all the users. The model is augmented with the LIHF schedule to assess and eliminate inopportune probes (Fig. 10.9 [Phase 2]).

TapSense: Field Study and Data Analysis

In this section, we discuss the TapSense field study and the dataset collected from the study.

Study Participants

We recruited 28 university students (22 males, 6 females, aged 24–35 years) to evaluate TapSense. We installed the application on their smartphones and instructed them to use it for 3 weeks. Three participants left the study in between, and the other three participants have recorded less than 40 labels. We have discarded these 6 users and collected data from the remaining 22 participants (18 males, 4 females). The ethics committee approved the study under the approval order IIT/SRIC/SAO/2017.

Instruction and Study Procedure

During the field study, we executed only Phase 1, where we implement the LIHF schedule for self-report collection. We instructed participants to select the TapSense keyboard as the default keyboard. We informed the participants that when they switch from an application after completing typing activity, they may receive a survey questionnaire as a pop-up to record their emotions. We also advised the participants not to dismiss the pop-up if they are occupied; instead, they were asked to record "No Response" if they do not want to record emotion at that moment.

Collected Dataset

We have collected 4609 typing sessions during this study period, which constitute close to 200 hours of typing labeled with an emotional state of all the participants (N = 22). Out of these sessions, we record 642 "No Response" sessions, which is nearly 14% of all recorded sessions. Notably, the actual number of ESM triggers is less than the number of typing sessions because, as per the LIHF policy, if two sessions are close (as defined by W in Fig. 10.8), only one ESM will be triggered to cater to both the sessions. We summarize the final dataset in Table 10.7.

EMA Self-Report Analysis

The users have reported two types of responses (a) One of the four valid emotions or (b) "No Response." While the valid emotion labels are used to construct the emotion assessment model, the "No Response" labels are important to design the inopportune moment assessment model for the ESM.

Emotion Labels Analysis We show the distribution of different emotion states for every user in Fig. 10.10. We have observed that 'relaxed' is the most dominant emotional state for most of the users. Overall, we have acquired 14%, 9%, 30%, 47% sessions tagged with happy, sad, stressed, and relaxed emotion states.

Table 10.7 Collected Data Summary

Number of participants	N = 22 (18 m, 4 f)
Total typing events	942,827
Total typing sessions	4609
Total typing duration (in Hr.)	199.1
Mean typing sessions (per user)	209 (std. dev 167.2)
Minimum number of typing sessions for a user	46
Maximum number of typing sessions for a user	549
Total ESM triggers	2554
Mean ESM trigger (per user)	116.1 (std. dev 71.9)

No Response Analysis We show the user-wise distribution of "No Response" sessions in Fig. 10.11a. Although for most users, the fraction of "No Response" labels is relatively low, for a few users, it is more than 40%. We observe the application-wise distribution of "No Response" sessions in Fig. 10.11b; the majority of the "No Response" labels are associated with Instant Messaging (IM) applications like WhatsApp. We also compare the distribution of total "No Response" and total valid emotion labels at weekday, weekend, working hour (9 am-9 pm), and non-working hour in Fig. 10.11c. We infer the working hour based on the timestamp of the ESM response. We compute the percentage of total "No Response," and the percentage of total other sessions is recorded at these times. However, in our dataset, we do not observe any major differences among these distributions. We also explore the time-wise distribution of No Response sessions in Fig. 10.11d, which indicates that a small number of No Response sessions were recorded during the late-night from 3 am onwards. This can be attributed to overall less engagement during late night.

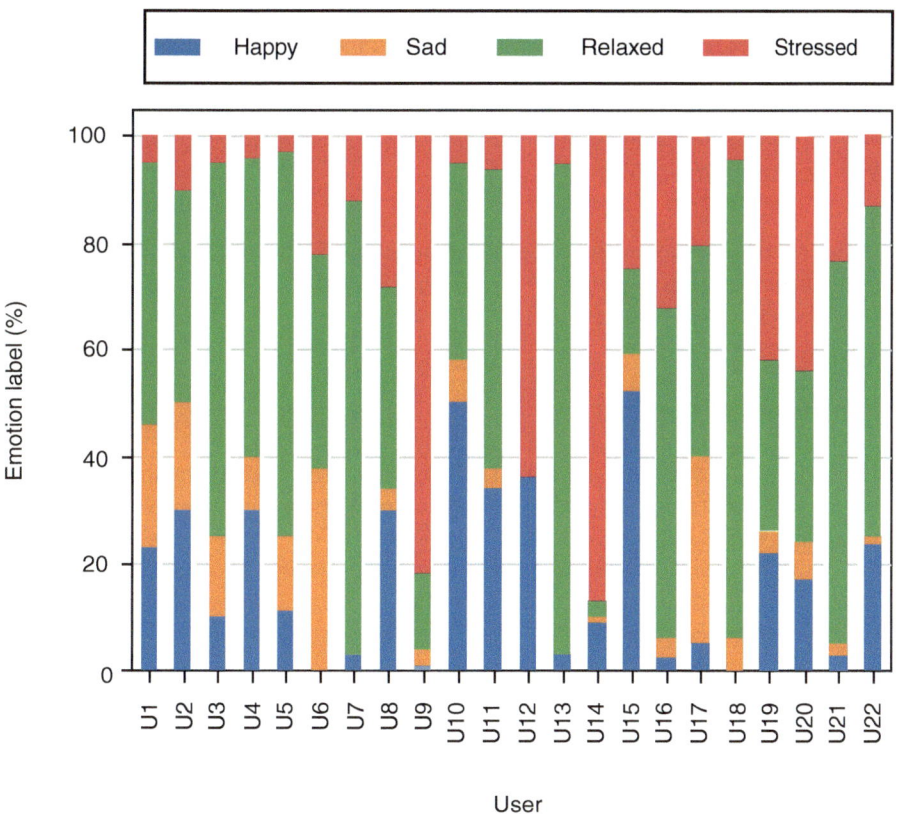

Fig. 10.10 Distribution of emotion labels for every user. All but 5 users (6,7,12, 13, 18) recorded all four emotions

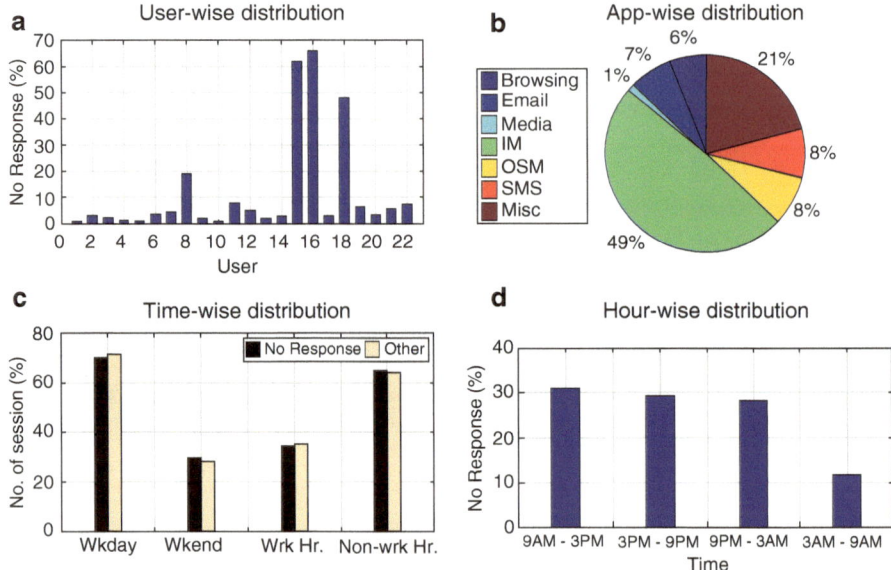

Fig. 10.11 Distribution of No Response sessions (**a**) User-wise (**b**) App-wise (**c**) Time-wise (**d**) Hour-wise

TapSense Evaluation

In this section, first, we discuss the experiment setup. Then we evaluate the emotion classification performance and the ESM performance. Finally, we discuss the limitations of the study.

Experimental Setup

During the field study, we used the LIHF ESM schedule for collecting self-reports. However, to perform a comparative study across different policies, we require data from time-based and event-based ESM schedules under identical experimental conditions from every participant. In the actual deployment, identical conditions are impossible to repeat over different time frames. Hence, we generate traces for the other policy-based schedules from the data collected using LIHF ESM. We outline the generation steps for these traces in Appendix 2. We show the distribution of emotion labels obtained from different schedules after trace generation in Fig. 10.12.

Baseline ESM Schedules

Different ESM schedules, listed in Table 10.8, used for comparison are described.

- *Policy-based ESM:* We focus on the Phase 1 approach (i.e., without optimizing the triggering) and use three policy-based ESM schedules—Time Based (TB),

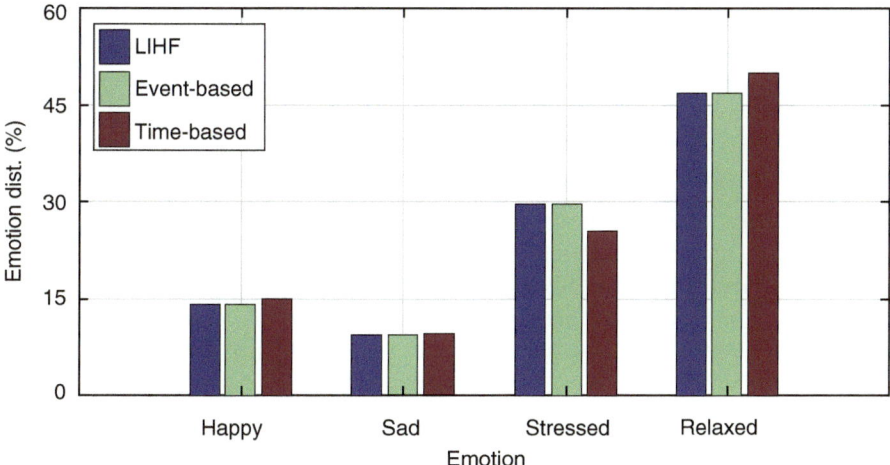

Fig. 10.12 Frequency distribution of different emotions across various ESM schedules

Table 10.8 Different ESM Schedules based on the Policy used in Phase 1 and Usage of the Model in Phase 2

ESM Schedule	Phase 1	Phase 2
TB	Time-based	No model is used.
EB	Event-based	No model is used.
LIHF	LIHF	No model is used.
TB-M	Time-based	Inopportune moment detection model
EB-M	Event-based	Inopportune moment detection model
LIHF-M	LIHF	Inopportune moment detection model

Event-Based (EB), and LIHF. In the case of TB, probes are issued at a fixed interval (3 hours). In EB's case, after every typing session, a probe is issued while LIHF implements the LIHF policy. These approaches do not use an inopportune moment assessment model. Comparing these schedules helps to understand their effectiveness in reducing the probing rate and collecting self-reports timely.

- *Model-based ESM:* We use the following model-based ESM schedules—TB-M, EB-M, and LIHF-M. These ESM schedules implement TB, EB, and LIHF schedules in Phase 1, respectively, followed by the inopportune moment assessment model operational in Phase 2. In all these schedules, the model is constructed using the same set of features (Table 10.6) extracted from relevant trace (i.e., for TB-M, the model is constructed from the trace of TB and similarly). Comparison of these model-driven schedules helps to understand the efficacy of the model in assessing the inopportune moments and whether applying the model with any off-the-shelf ESM is good enough to improve survey response quality.

Overall Performance Metrics

We use the classification accuracy to measure the emotion classification performance, and as for the ESM performance, we assess it along with the probing rate index and its reduction wrt. The classical self-report approach, timely self-report collection, inopportune moment identification, and valid response rate collection.

Emotion Assessment: Classification Accuracy (Weighted AUCROC) The performance of supervised learning algorithms highly depends on the quality of labels [235]. The label quality can adversely impact classification accuracy [236, 237]. In our research, we use Typing Session emotion classification accuracy. We measure it in terms of the weighted average of AUCROC (auc_{wt}) using AUCROC from four different emotional states. Let f_i, auc_i indicate the fraction of samples and AUCROC for emotion state i respectively, then $auc_{wt} = \sum_{\forall i \in \{happy, sad, stressed, relaxed\}} f_i * auc_i$.

ESM Performance Metrics

Probe Frequency Index (PFI) We compare the probing frequencies of different ESM schedules using PFI, defined as follows. Let there be different ESM schedules ($e \in E$) and N_i^e denotes the number of probes issued for the user i for an ESM schedule e, then PFI for user i for ESM schedule e is

expressed as, $\text{PFI}_i^e = \dfrac{N_i^e}{\forall e, \max\left(N_i^e\right)}$.

The Recency of Label (RoL) The timeliness of self-report response collection is measured using RoL defined as follows. Let there be different ESM schedules ($e \in E$), and d_i^e denotes average elapsed time between typing and probing for user i for an ESM schedule e, then RoL for user i for ESM schedule e is expressed as, $\text{RoL}_i^e = \dfrac{d_i^e}{\forall e, \max\left(d_i^e\right)}$.

Inopportune Moment Identification We measure Precision, Recall and F-score for inopportune moment assessment. We also compute the weighted AUCROC (auc_{wt}) for the inopportune and opportune moments. Let f_i, auc_i indicate the fraction of samples and AUCROC for class i respectively, then $auc_{wt} = \sum_{\forall i \in \{inopportune, opportune\}} f_i * auc_i$.

Valid Response Rate (VRR) We also compare the percentage of valid emotion labels for different ESM schedules. Let there be different ESM schedules ($e \in E$), and nr_e denotes the fraction of No Response sessions recorded for ESM e, then Valid Response Rate for ESM e is expressed as $VRR_e = (1 - nr_e) * 100$.

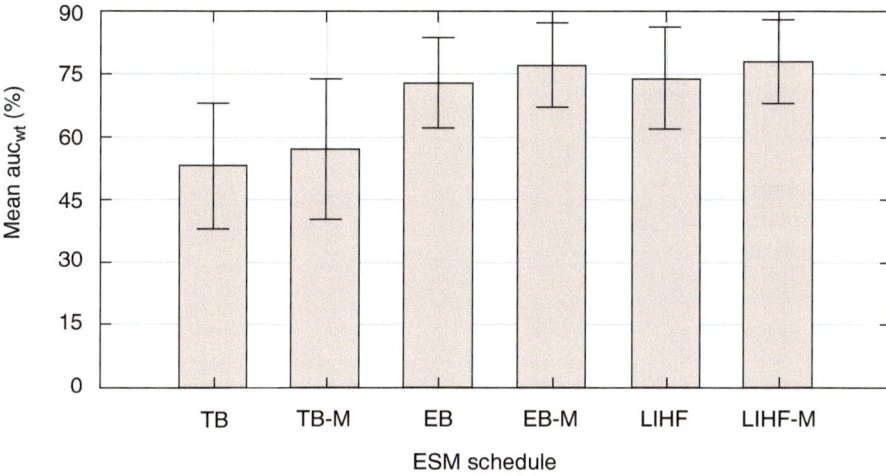

Fig. 10.13 Emotion classification accuracy (AUCROC) using different ESM schedules

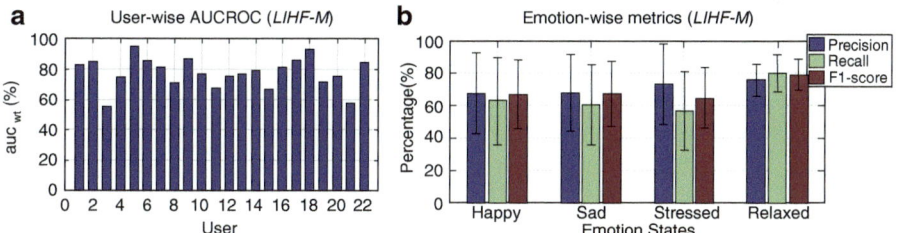

Fig. 10.14 Emotion classification (**a**) AUC_{wt} for LIHF-M per user (**b**) Emotion-wise accuracy LIHF-M

Emotion Assessment: Classification Performance

The emotion classification accuracy for different ESMs is shown in Fig. 10.13. We observe that the LIHF-M outperforms other schedules with a mean AUCROC of 78%. It returns a maximum improvement of 24% with respect to TB and an improvement of 5% with respect to EB. We also observe that after applying the inopportune moment assessment model, the mean AUCROC (auc_{wt}) improves (by 4%) for each corresponding schedule (TB, EB, LIHF).

We also show the user-wise emotion assessment AUCROC (auc_{wt}) corresponding to the LIHF-M schedule in Fig. 10.14a. The quality of the prediction for each emotion category is presented in Fig. 10.14b. The emotion states are identified with an average f-score between 54% and 74%. We observe that the relaxed state is identified with the highest f-score, followed by sad, stressed, and happy states, respectively. As data volume increases, as in the case of the relaxed state, the performance metrics improve.

Table 10.9 Ranking of Features Used to Construct TapSense Emotion Assessment Model

Feature name	Rank	Average IG.
Last ESM trigger response	1	0.468
Session typing speed	2	0.376
Backspace percentage	3	0.270
Session length	4	0.231
Special character percentage	5	0.203
Session duration	6	0.181

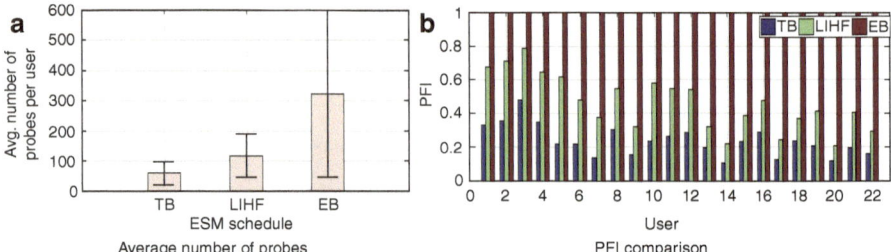

Fig. 10.15 Probing rate across ESM schedules: Average number of probes and the Probe Frequency Index. (**a**) Average number of probes (**b**) PFI comparison

Influence of Emotion Assessment Features

We find the importance of the input features used for emotion assessment using the 'InfoGainAttributeEval' method from Weka. We compute the average Information Gain (IG) of every feature and rank them in Table 10.9. We observe that the last ESM response is the most discriminating feature, followed by features like typing speed and backspace percentage. All the features are found to have an input into the model for the emotion assessment.

ESM Performance

In this section, we evaluate the ESM's performance in terms of the three parameters (ESM probing rate, self-report timeliness, and opportune probing moments).

Probing Rate Reduction

We compare the average number of probes issued by each ESM schedule in Fig. 10.15a. We observe that time-based ESM (TB) issues the minimum number of probes, event-based ESM (EB) issues the maximum number of probes, while LIHF ESM lies in between. It is observed that the average number of probes is reduced by 64% for LIHF ESM policy.

We also perform the user-wise comparison using the Probe Frequency Index (PFI) metric in Fig. 10.15b. For all users, PFI for LIHF ESM is lower than that of event-based ESM. Across all users, there is an average improvement of 54% in PFI. Time-based ESM is the best in PFI but does not capture self-reports timely, as

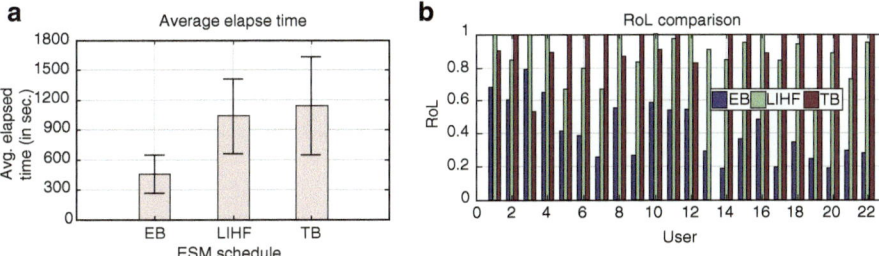

Fig. 10.16 Timeliness of the ESM collection across schedules: Average elapsed time and the Recency of Label. (**a**) Average elapse time (**b**) RoL comparison

shown later. LIHF ESM schedule enforces a minimum elapsed time between two successive probes; it generates fewer probes and reduces probing rate compared to event-based ESM.

Timely Self-Report Collection

We measure how close to the event (i.e., typing session completion) the ESM schedule collects the self-report. We compare the average elapsed time between typing completion and self-report collection for different ESM schedules in Fig. 10.16a. The average elapsed time is the least for event-based ESM, highest for time-based ESM, while for the LIHF, it lies in between. The average elapsed time for label collection is reduced by 9% for LIHF.

We also compare the recency of labels using RoL in Fig. 10.16b. We observe that for every user, RoL is minimum for EB, and for most of the users, RoL is maximum in the case of TB, while for LIHF, the RoL lies in between. In the case of EB, we issue the probe as soon as the typing event is completed; it can collect self-reports very close to the event, resulting in the lowest RoL. On the contrary, in TB, we perform probing at an interval of 3 hours. As a result, there is often a large gap between typing completion and self-report collection, resulting in high RoL. However, in the case of LIHF, we keep accumulating events and separate two consecutive probes by at least half an hour; we compromise to some extent in the label recency, yet less than in the case of TB.

Inopportune Moment Assessment

We compare the inopportune moment classification performance of three model-based approaches in Fig. 10.17a. We observe that the LIHF-M attains an accuracy (AUCROC) of 89%, closely followed by EB-M (88%), while TB-M (75%) performs poorly. We also note the precision, recall, and F-score values of identifying inopportune moments in Fig. 10.17b using the LIHF-M schedule. We also report the recall rate of inopportune moments for every user in Fig. 10.17c. We observe that for 14% of the users, the recall rate is greater than 75%, and for 60% of the users, the recall rate is greater than 50%. It is observed that users with many "No Response" (Fig. 10.8a) get more benefit using the inopportune moment assessment model. In summary, the proposed model combined with LIHF ESM performs best, while other ESM schedules also assess the inopportune moments accurately with this model.

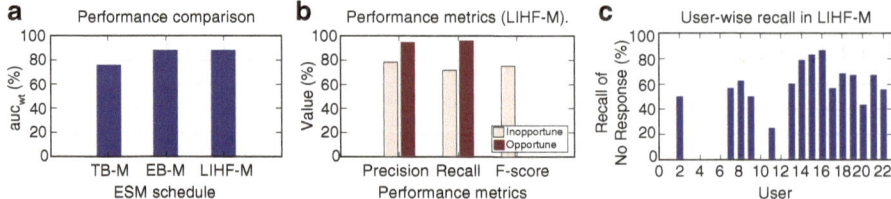

Fig. 10.17 Inopportune moment assessment performance (**a**) different ESM schedules (**b**) LIHF-M schedule (**c**) User-wise LIHF-M schedule

Table 10.10 Ranking of Features Used to Construct Inopportune Moment Model

Feature Name	Rank	Average IG
Last ESM trigger response	1	0.669
App category	2	0.053
Session length	3	0.019
Session duration	4	0.012

Influence of Inopportune Moment Assessment Features

We find the importance of every feature by ranking them based on the information gain (IG) achieved by adding it for predicting the inopportune moment. We use the InfoGainAttributeEval method from Weka [233] to obtain the information gain of each feature. Our results (in Table 10.10) show that the last ESM probe response is the most important feature, followed by the application category.

Valid Response Collection

We compare the valid response rate (VRR) for LIHF, LIHF-M schedules in Fig. 10.18. We do not consider other schedules as those labels were generated synthetically. The VRR for LIHF is 86%, and the same for LIHF-M is 96%. This further proves the effectiveness of the inopportune moment assessment model. As the model is in place for LIHF-M, it assesses and skips probing at the inopportune moments, thereby improving the number of valid emotion responses..

TapSense Study Discussion and Lessons Learnt

In our research, we leverage the user's smartphone for accurate and timely emotion assessment. We designed, developed, and evaluated TapSense, which passively logs typing behavior and develops a personalized machine learning model for multi-state emotion detection. We log the keyboard interactions (typing patterns and not the actual content) of the user and infer four types of emotions (*happy, sad, stressed, relaxed*). We also proposed an intelligent ESM-based self-report collection method

Fig. 10.18 Comparing the number of valid responses for LIHF, LIHF-M schedules

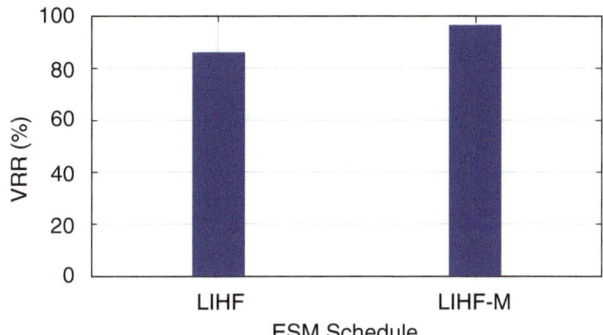

and integrated the same with TapSense, optimizing the manual self-report collection. We evaluate the emotion classification performance and the ESM performance of TapSense in a 3-week in-the-wild study involving 22 participants. The empirical analysis reveals that the TapSense can infer emotions with an average accuracy of 78%. It also demonstrates the efficacy of the proposed ESM in terms of probing rate reduction (on avg. 24%), self-report timeliness (on avg. 9%), and probing at opportune moments (on avg. 89%); all of which improve the emotion classification performance.

However, a few factors need to be considered before deploying sensing technologies as TapSense as an emotion assessment tool. First, it is crucial to consider keyboard interaction experience should not be impacted while using the TapSense keyboard as most of the participants are conversant with the Google keyboard. However, we do not observe a significant effect in the app usage due to this, as we record 86% valid emotion labels and, on average, 209 typing sessions per user. Second, the model-driven probing strategy at opportune moments may not perform well for some users if the number of "No Response" labels are very few (less than 4% of all sessions). If the number of no response labels is very less, then the model may not perform well and may not detect all the inopportune moments.

Another factor to consider is which ESM strategy is to be adopted during self-report collection. We recommend using the LIHF strategy to reduce survey fatigue compared to fixed event-based (EB) schedules, but it may suffer from latency in the self-report collection. Time-driven schedules may not be used if long-time-interval separates two probes. This may miss capturing the fine-grain event details, which are likely to carry emotional signatures. Finally, during self-reporting, if the participants have skipped the pop-up instead of selecting "No Response," we could not capture those moments in our study. However, this can be easily incorporated by logging the skipping events.

The study we have carried out involving TapSense has several limitations that may limit the results' generalization. First of all, the study has been of small size, as only 22 users have been engaged in the study for only 3 weeks. On the other hand, given the number of self-reports and user typing sessions, we may assume we have captured a representative sample of the general population, with 25 minutes a day of interaction being logged and labeled for an emotional expression. As we have shown, the representation of emotional states was diverse. The additional limitation may stem from the Android OS only study; the iOS participants may have different traits and emotional states. The emotional modeling may have led to different results. Nevertheless, we present the results as indicative and will follow future research in this context, given a larger population sample and longer study duration.

Conclusive Remarks

In this chapter, we focused on emotions as indicators of quality of life due to the association of positive/negative life experiences and positive/negative emotional states and other aspects of quality of life, such as health, safety, economic and mental well-being. We highlighted functions and the importance of emotions in a persons' daily life and the challenge of assessing emotions in traditional vs. novel methods of assessing emotions. The advantages and disadvantages of the diverse methods are especially rooted in objectivity, required assessment setup, self-report bias, privacy, real-time measurement, obtrusiveness, and inclusion of a wide range of emotion components.

In more depth, in this chapter, we have leveraged smartphone interactions to assess a user's mental state. As smartphones have become a true companion of our daily life, they can passively sense the usage behavior and mental state. With numerous typing-based communication applications on a smartphone, typing characteristics provide a rich source to model user emotion. In the specific case, the users' keyboard interactions predicted four different emotion categories with an average accuracy of 78%, confirming and outperforming other approaches of smartphone-based emotion-sensing technologies. Hence, emotional assessment is feasible to conduct via different technologies.

TapSense study shows promising results for minimally obtrusive, smartphone-based emotional state assessment, which may be leveraged for further studies and, if largely improved, in clinical practice for a 'companion assessment' of individuals' mental health, accompanying the current gold standard assessment methods and approaches. It is in line with the recent research results showing a potential for co-calibration of the self-reported, gold-standard approaches with the technology-reported ones [238, 239]. The aspects of minimal obtrusiveness may be of interest, especially for the leaders in the self-assessment space—the co-called QuantifiedSelfers [190, 240] who leverage diverse self-assessment technologies for better self-knowledge and optimization of daily life activities for better well-being, health, and other outcomes in the long term. Overall, the TapSense may be seen as an example of the emerging Quality of Life Technologies [241, 242] to assess the individual's behavioral patterns for better life quality. Once proven accurate and timely, and highly reliable in the context of daily life assessment, TapSense, and similar technologies may pave the way for technologies for a better understanding of the physical, mental, and emotional well-being of populations at large.

Appendices

Appendix 1: Parameter Threshold Value

We use three parameters L, W, T as defined in details in sect. 5.1 (Phase 1: Balancing ESM Probing Frequency and Timeliness) to balance between probing frequency and timeliness in label collection. L is defined as the minimum amount of typing performed in a typing session. W is defined as the minimum time elapsed since last ESM trigger, and T is defined as the polling interval (i.e. how frequently the typing session will be checked for sufficient amount of typing). Based on our initial dataset, we observe the CDF of session length (L) in Fig. 10.19a, which reveals that frequency distribution of session length is highly skewed. So, we select 66th percentile value as the threshold so that two-third values are less than this value. We observe similar CDF and frequency distribution (Fig. 10.19b) for inter-session gap (W) and use the 66th percentile value as the threshold.

However, polling interval (T) is to be chosen in such a way that for most of the sessions, the event of interest is captured within this interval. In this case, the event is change of application after typing in a session. For this purpose, we measure the elapsed time between two successive key pressing events (ITD) in a session. We note the CDF of all ITD values from all sessions in Fig. 10.20. We observe that 99% of the inter-tap duration (ITDs) are less than 15 seconds i.e. for most of the sessions the application change happens after 15 seconds. So, we decide to use 15 seconds as the threshold for T.

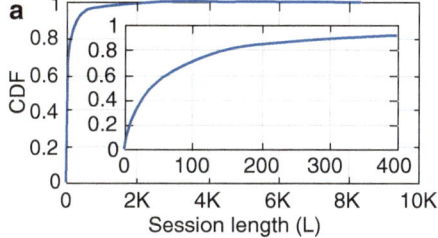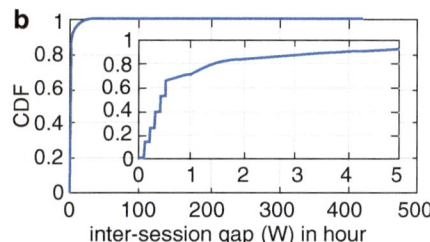

Fig. 10.19 CDFs for session length (L) and inter-session gap (W) reveal that the distribution is skewed. 66th percentile value is chosen as threshold so that two-third values are less than threshold. (**a**) CDF for session length (L). (**b**) CDF for inter-session gap (W)

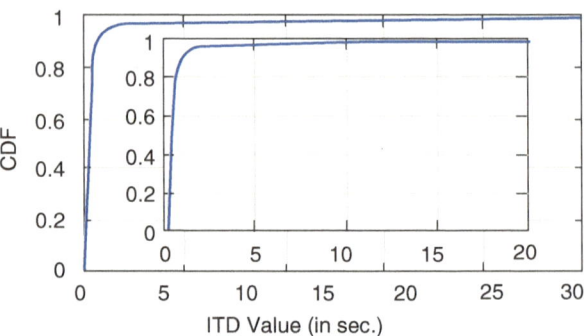

Fig. 10.20 CDF of polling interval (T). It reveals the distribution is skewed and 99% ITD values are close to 15 seconds

Appendix 2: The ESM Trace Generation

In this section, we discuss in detail the steps followed to generate trace for Time-based (TB) ESM and Event-based (EB) ESM schedule from the data collected using LIHF ESM schedule. In Fig. 10.21, a schematic is given to depict the same. Ei denotes the application switching event after sufficient typing. In case of LIHF ESM, there are 6 such events, however only 5 probes were issued (Fig. 10.21a). No probe is issued after E3 because it occurs within time-window (W = 30 minutes) since last probe (Probe 2). In order to generate the corresponding Time-based trace, probes are considered at 3 hour interval. As a result, there will be only one probe Probe 1 and all events E1 to E6 will be labeled with the single emotion response collected via it (Fig. 10.21b). But in case of conversion to Event-based ESM, all events are treated separately, as a result there will be in total 6 probes and the emotion labels will be assigned accordingly to the respective events (Fig. 10.21c). Next, we define the formal procedure for trace generation.

Generation of Time-based Trace

We take the trace collected from LIHF schedule »C.lihf p_5 as p _ 5 matrix where p denotes the total number of key press events. We generate the respective Time based trace »C.time p_5 following the Algorithm 1.We consider the sampling interval of Time-based ESM as 3 hours. We parse through (line 5–12) the LIHF trace »C. Lihf p_5 and all key press events. As in case of LIHF, two responses may be recorded less than 3 hour interval, we may need to down-sample, which is performed in following way. If two emotion responses for key press events are collected within 3 hours, both are considered as a part of single session and the later is labeled with the previous emotion. Otherwise, they belong to different session and the new emotion response is considered (line 7–9).

Generation of Event-based Trace

We design Algorithm 2 to generate the corresponding event-based trace »C.event p_5 from the collected LIHF trace »C. lihf p_5 .We consider changing application after typing as an event. We parse through (line 5–12) the trace obtained from LIHF schedule and all key press events. If two consecutive key press events are associated with different application, they belong to separate session (line 6–7). Otherwise,

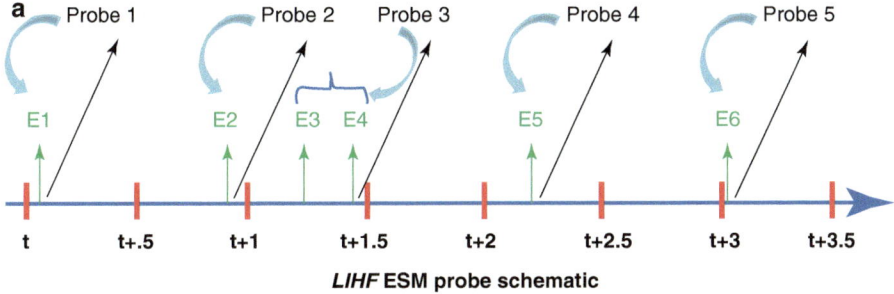

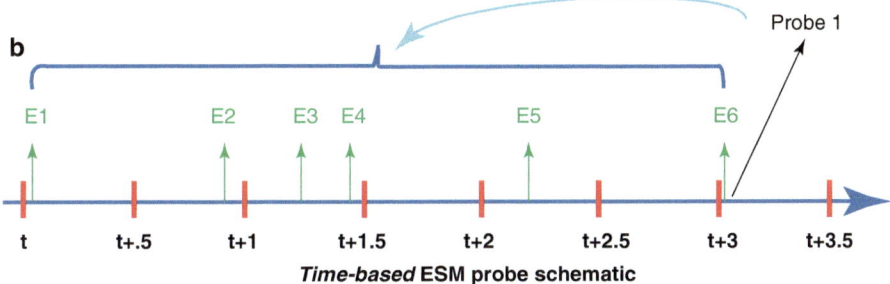

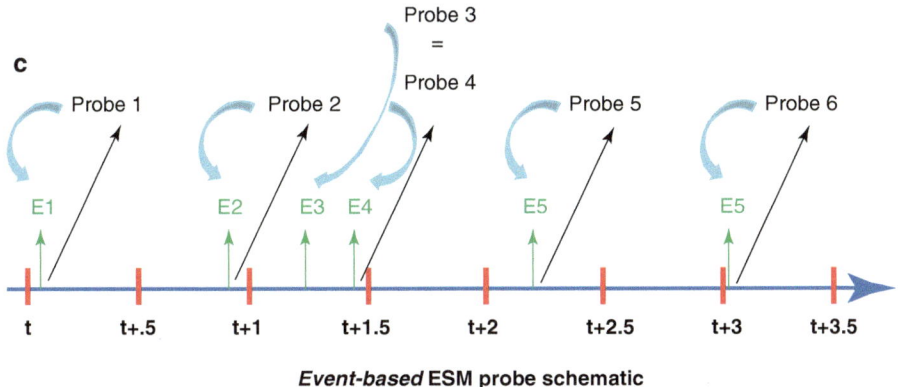

Fig. 10.21 Schematic to display how Event-based and Time-based traces are generated from LIHF ESM

they are considered as part of the same session. In both these cases, no emotion response is dropped (unlike time-based), they are associated with different sessions. In case of LIHF, multiple sessions are grouped and tagged with single emotion, but in case of event-based schedule, this grouping is not done and every session is labeled with the same response. This is how the over-sampling is done in case of event-based schedule.

Algorithm 1: Time-based trace generation

Input: $[\mathbb{C}]_{p \times 5}^{lihf}$, key pressing details of a user as obtained from LIHF schedule; $[\mathbb{C}]_{:,1}^{lihf}$ contain session

number, $[\mathbb{C}]_{:,2}^{lihf}$ contain the application names, $[\mathbb{C}]_{:,3}^{lihf}$ contain the timestamp of the key press,

$[\mathbb{C}]_{:,4}^{lihf}$ contain emotion recording timestamp for the session, $[\mathbb{C}]_{:,5}^{lihf}$ contain the associated

emotion,; p denotes the total number of key press performed by the user.

Output: $[\mathbb{C}]_{p \times 5}^{time}$; Corresponding time-based trace

/* Copy the first row from LIHF schedule (session number, application name, emotion and other)   */

1 $session^{time} \leftarrow [\mathbb{C}]^{lihf}(1,1)$
2 $app^{time} \leftarrow [\mathbb{C}]^{lihf}(1,2)$
3 $emotion^{time} \leftarrow [\mathbb{C}]^{lihf}(1,5)$
4 $[\mathbb{C}]^{time}(1,:) \leftarrow [\mathbb{C}]^{lihf}(1,:)$
5 **for** $i \leftarrow 2$ **to** p **do**
   /* Find elapsed time between two consecutive emotion recording timestamp   */
6 $\quad \delta \leftarrow$ time_difference($[\mathbb{C}]^{lihf}(i,5), [\mathbb{C}]^{lihf}(i-1,5)$)
7 $\quad$ **if** $\delta \geq 3$ *hours* **then**
8 $\quad\quad session^{time} \leftarrow session^{time} + 1$
9 $\quad\quad emotion^{time} \leftarrow [\mathbb{C}]^{lihf}(i,5)$
10 $\quad [\mathbb{C}]^{time}(i,1) \leftarrow session^{time}$
11 $\quad [\mathbb{C}]^{time}(i,2:4) \leftarrow [\mathbb{C}]^{lihf}(i,2:4)$
12 $\quad [\mathbb{C}]^{time}(i,5) \leftarrow emotion^{time}$
13 **return** $[\mathbb{C}]^{time}$

Algorithm 2: Event-based trace generation

Input: $[\mathbb{C}]_{p \times 5}^{lihf}$, key pressing details of a user as obtained from LIHF schedule; $[\mathbb{C}]_{:,1}^{lihf}$ contain session

number, $[\mathbb{C}]_{:,2}^{lihf}$ contain the application names, $[\mathbb{C}]_{:,3}^{lihf}$ contain the timestamp of the key press,

$[\mathbb{C}]_{:,4}^{lihf}$ contain emotion recording timestamp for the session, $[\mathbb{C}]_{:,5}^{lihf}$ contain the associated

emotion,; p denotes the total number of key press performed by the user.

Output: $[\mathbb{C}]_{p \times 5}^{event}$; Corresponding event-based trace

/* Copy the first row from LIHF schedule (session number, application name, emotion and other)   */

1 $session^{event} \leftarrow [\mathbb{C}]^{lihf}(1,1)$
2 $old\_app^{event} \leftarrow [\mathbb{C}]^{lihf}(1,2)$
3 $[\mathbb{C}]^{event}(1,:) \leftarrow [\mathbb{C}]^{lihf}(1,:)$
4 **for** $i \leftarrow 2$ **to** p **do**
   /* Find current application name   */
5 $\quad curr\_app^{event} \leftarrow [\mathbb{C}]^{lihf}(i,2)$
6 $\quad$ **if** $curr\_app^{event} \neq old\_app^{event}$ **then**
7 $\quad\quad session^{event} \leftarrow session^{event} + 1$
8 $\quad [\mathbb{C}]^{event}(i,1) \leftarrow session^{event}$
9 $\quad [\mathbb{C}]^{event}(i,2:5) \leftarrow [\mathbb{C}]^{lihf}(i,2:5)$
10 **return** $[\mathbb{C}]^{event}$

References

1. Harari GM, Müller SR, Aung MS, Rentfrow PJ. Smartphone sensing methods for studying behavior in everyday life. Curr Opin Behav Sci. 2017;18:83–90.
2. Mehrabian A. Pleasure-arousal-dominance: a general framework for describing and measuring individual differences in temperament. Curr Psychol. 1996;14(4):261–92.
3. Scherer KR. The dynamic architecture of emotion: evidence for the component process model. Cogn Emot. 2009;23(7):1307–51.
4. Mehu M, Scherer KR. Normal and abnormal emotions–the quandary of diagnosing affective disorder. Emot Rev. 2015;7(Special issue)
5. Larson R, Csikszentmihalyi M. The experience sampling method. In: Flow and the foundations of positive psychology. Springer; 2014. p. 21–34.
6. Conner TS, Tennen H, Fleeson W, Barrett LF. Experience sampling methods: a modern idiographic approach to personality research. Soc Personal Psychol Compass. 2009;3(3):292–313.
7. Shiffman S, Stone AA, Hufford MR. Ecological momentary assessment. Annu Rev Clin Psychol. 2008;4:1–32.
8. Pejovic V, Musolesi M. InterruptMe: designing intelligent prompting mechanisms for pervasive applications. In: Proceedings of ACM UbiComp; 2014. p. 897–908.
9. Bartlett MS, Littlewort GC, Sejnowski TJ, Movellan JR. A prototype for automatic recognition of spontaneous facial actions. In: Advances in neural information processing systems; 2003. p. 1295–302.
10. Cohn JF, Reed LI, Ambadar Z, Xiao J, Moriyama T. Automatic analysis and recognition of brow actions and head motion in spontaneous facial behavior. In: 2004 IEEE International Conference on Systems, Man and Cybernetics (IEEE Cat. No. 04CH37583), vol. 1; 2004. p. 610–6.
11. Kapoor A, Burleson W, Picard RW. Automatic prediction of frustration. Int J Hum Comput Stud. 2007;65(8):724–36.
12. Littlewort GC, Bartlett MS, Lee K. Faces of pain: automated measurement of spontaneous-allfacial expressions of genuine and posed pain. In: Proceedings of the 9th International Conference on Multimodal interfaces; 2007. p. 15–21.
13. Bartlett MS, Littlewort G, Frank M, Lainscsek C, Fasel I, Movellan J. Recognizing facial expression: machine learning and application to spontaneous behavior. In: 2005 IEEE Computer Society Conference on Computer Vision and Pattern Recognition (CVPR'05), vol. 2; 2005. p. 568–73.
14. Banse R, Scherer KR. Acoustic profiles in vocal emotion expression. J Pers Soc Psychol. 1996;70(3):614.
15. Lee CM, Narayanan SS. Toward detecting emotions in spoken dialogs. IEEE Trans speech audio Process. 2005;13(2):293–303.
16. Devillers L, Vidrascu L. Real-life emotions detection with lexical and paralinguistic cues on human-human call center dialogs. In: Ninth International Conference on Spoken Language Processing; 2006.
17. Schuller B, Stadermann J, Rigoll G. Affect-robust speech recognition by dynamic emotional adaptation. In: Proc. Speech Prosody 2006, Dresden; 2006.
18. Litman D, Forbes-Riley K. Predicting student emotions in computer-human tutoring dialogues. In: Proceedings of the 42nd Annual Meeting of the Association for Computational Linguistics (ACL-04); 2004. p. 351–8.
19. Schuller B, Villar RJ, Rigoll G, Lang M. Meta-classifiers in acoustic and linguistic feature fusion-based affect recognition. In: Proceedings.(ICASSP'05). IEEE International Conference on Acoustics, Speech, and Signal Processing, 2005, vol. 1; 2005. p. I–325.
20. Fernandez R, Picard RW. Modeling drivers' speech under stress. Speech Commun. 2003;40(1–2):145–59.

21. Pinto J, Moulin T, Amaral O. On the transdiagnostic nature of peripheral biomarkers in major psychiatric disorders: a systematic review. In: Transdiagnostic Nat. Peripher. Biomarkers major Psychiatr. Disord. A Syst. Rev., p. 086124; 2016.
22. Egger M, Ley M, Hanke S. Emotion recognition from physiological signal analysis: a review. Electron Notes Theor Comput Sci. 2019;343:35–55.
23. Aviezer H, Trope Y, Todorov A. Body cues, not facial expressions, discriminate between intense positive and negative emotions. Science (80-). 2012;338(6111):1225–9.
24. Healey J, Picard R. Digital processing of affective signals. In: ICASSP, IEEE Int. Conf. Acoust. Speech Signal Process.–Proc, vol. 6; 1998. p. 3749–52.
25. AlZoubi O, Calvo RA, Stevens RH. Classification of EEG for affect recognition: an adaptive approach. In: Australasian Joint Conference on Artificial Intelligence; 2009. p. 52–61.
26. Bashashati A, Fatourechi M, Ward RK, Birch GE. A survey of signal processing algorithms in brain–computer interfaces based on electrical brain signals. J Neural Eng. 2007;4(2):R32.
27. Lotte F, Congedo M, Lécuyer A, Lamarche F, Arnaldi B. A review of classification algorithms for EEG-based brain–computer interfaces. J Neural Eng. 2007;4(2):R1.
28. Shu L, et al. A review of emotion recognition using physiological signals. Sensors. 2018;18(7):2074.
29. Ko BC. A brief review of facial emotion recognition based on visual information. Sensors. 2018;18(2):401.
30. Lane ND, et al. Bewell: a smartphone application to monitor, model and promote wellbeing. In: 5th International ICST Conference on pervasive computing technologies for healthcare; 2011. p. 23–6.
31. Wang R, et al. StudentLife: assessing mental health, academic performance and behavioral trends of college students using smartphones. In: Proceedings of the ACM UbiComp; 2014. p. 3–14.
32. D'mello S, Graesser A. AutoTutor and affective autotutor: learning by talking with cognitively and emotionally intelligent computers that talk Back. ACM Trans Interact Intell Syst. 2013;2(4)
33. Zheng Y, Mobasher B, Burke RD. The role of emotions in context-aware recommendation. Decis RecSys. 2013;2013:21–8.
34. Tkalcic M, Odic A, Kosir A, Tasic J. Affective labeling in a content-based recommender system for images. IEEE Trans. Multimed. 2013;15(2):391–400.
35. Lee U, et al. Hooked on smartphones: an exploratory study on smartphone overuse among college students. In: Proceedings of the 32nd annual ACM conference on human factors in computing systems; 2014. p. 2327–36.
36. Chen X, Sykora MD, Jackson TW, Elayan S. What about mood swings: identifying depression on twitter with temporal measures of emotions. In: Web Conf. 2018–Companion World Wide Web Conf. WWW 2018; 2018. p. 1653–60.
37. Kosinski M, Stillwell D, Graepel T. Private traits and attributes are predictable from digital records of human behavior. Proc Natl Acad Sci U S A. 2013;110(15):5802–5.
38. Guntuku SC, Yaden DB, Kern ML, Ungar LH, Eichstaedt JC. Detecting depression and mental illness on social media: an integrative review. Curr Opin Behav Sci. 2017;18:43–9.
39. Peters UW. Wörterbuch der Psychiatrie und medizinischen Psychologie, vol. 4., überar; 1990.
40. Cattell RB, Scheier IH. The meaning and measurement of neuroticism and anxiety. Oxford, England: Ronald; 1961.
41. Carlson JG, Hatfield E. Psychology of emotion. In: Jovanovich HB, editor. Psychology of emotion; 1992.
42. Frijda NH, Mesquita B. The analysis of emotions: dimensions of variation. What Dev Emot Dev. 1998:273–95.
43. Reisenzein R. Emotionen. In: Lehrbuch Allgemeine Psychologie. Bern: Huber; 2005. p. 435–500.
44. Greenberg S, Safran JD. Emotional-change processes in psychotherapy. Emot Psychopathol Psychother. 1990:59–85.
45. Plutchik R. The nature of emotions: human emotions have deep evolutionary roots, a fact that may explain their complexity and provide tools for clinical practice. Am Sci. 2001;89(4):344–50.

46. P. Ekman, "Gefühle lesen : wie Sie Emotionen erkennen und richtig interpretieren.," pp. XIX, 389 S.:Ill.; 19 cm, 2010.
47. Matsumoto D, et al. Culture, emotion regulation, and adjustment. J Pers Soc Psychol. 2008;94(6):925–37.
48. de Leersnyder J, Mesquita B, Kim HS. Where do my emotions belong? A study of immigrants' emotional acculturation. Personal Soc Psychol Bull. 2011;37(4):451–63.
49. Murphy NA, Isaacowitz DM. Age effects and gaze patterns in recognising emotional expressions: an in-depth look at gaze measures and covariates. Cogn. Emot. 2010;24(3):436–52.
50. Izard CE. Die Emotionen des Menschen: Eine Einführung in die Grundlagen der Emotionspsychologie; 1981. p. 530.
51. Stanley R, Burrows G. Varieties and functions of human emotion. In: Payne RL, Cooper CC, editors. Emotions at work. Theory, research and applications for management. John Wiley & Sons; 2003. p. 3–20.
52. Russell JA. A circumplex model of affect. J Pers Soc Psychol. 1980;39(6):1161–78.
53. Bălan O, Moise G, Petrescu L, Moldoveanu A, Leordeanu M, Moldoveanu F. Emotion classification based on biophysical signals and machine learning techniques. Symmetry (Basel). 2020;12(1):1–22.
54. Scherer KR. Component models of emotion can inform the quest for emotional competence. Sci Emot Intell Knowns Unknowns. 2007:101–26.
55. Mees U. Zum Forschungsstand der Emotionspsychologie–eine Skizze. Emot und Sozialtheorie Disziplinäre Ansätze. 2006:104–24.
56. Schachter S, Singer J. Cognitive, social, and physiological determinants of emotional state. Psychol Rev. 1962;69(5):379.
57. Lazarus RS. From psychological stress to the emotions.Pdf. Annu Rev Psychol. 1993;44:1–21.
58. Lewis MD. Bridging emotion theory and neurobiology through dynamic systems modeling. Behav Brain Sci. 2005;28(2):169–94.
59. Petta P, Trappl R. Emotions and agents; 2001. p. 301–16.
60. Averill JR. Anger and aggression: an essay on emotion, vol. 8, no. 2. New York: Springer; 1982.
61. Sánchez-Álvarez N, Extremera N, Fernández-Berrocal P. The relation between emotional intelligence and subjective Well-being: a meta-analytic investigation. J Posit Psychol. 2016;11(3):276–85.
62. Campbell-Sills L, Barlow DH, Brown TA, Hofmann SG. Effects of suppression and acceptance on emotional responses of individuals with anxiety and mood disorders. Behav Res Ther. 2006;44(9):1251–63.
63. Aldao A, Nolen-Hoeksema S, Schweizer S. Emotion-regulation strategies across psychopathology: a meta-analytic review. Clin Psychol Rev. 2010;30(2):217–37.
64. Elizabeth KG, Watson D. Measuring and assessing emotion at work. In: Payne Roy L, Cooper CL, editors. Emotions at work. Theory, research and applications for management. John Wiley & Sons; 2003. p. 21–44.
65. Jonkisz E, Moosbrugger H, Brandt H. Planung und Entwicklung von psychologischen Tests und Fragebogen; 2008. p. 27–72.
66. Bradburn NM. The structure of psychological Well-being. Chicago: Aldine; 1969.
67. Watson D, Clark LA, Tellegen A. Development and validation of brief measures of positive and negative affect: the PANAS scales. J Pers Soc Psychol. 1988;54(6):1063.
68. Moriwaki SY. The affect balance scale: a validity study with aged samples. J Gerontol. 1974;29(1):73–8.
69. Taylor JA. A personality scale of manifest anxiety. J Abnorm Soc Psychol. 1953;48(2):285–90.
70. W. W. K. Zung, "A self-rating depression scale," Arch Gen Psychiatry, vol. 12, no. 1, pp. 63–70, Jan. 1965.
71. "An inventory for measuring clinical anxiety: psychometric properties."
72. Antony MM, Cox BJ, Enns MW, Bieling PJ, Swinson RP. Psychometric properties of the 42-item and 21-item versions of the depression anxiety stress scales in clinical groups and a community sample. Psychol Assess. 1998;10(2):176–81.
73. Spielberger CD. State-trait anxiety inventory: bibliography. 2nd ed. Palo Alto, CA: Consulting Psycholgoists Press; 1989.

74. Spitzer RL, Kroenke K, Williams JBW, Löwe B. A brief measure for assessing generalized anxiety disorder: the GAD-7. Arch Intern Med. 2006;166(10):1092–7.

75. Hamilton M. The assessment of anxiety states by rating. Br J Med Psychol. 1959;32(1):50–5.

76. Antonosky A. The structure and properties of the sense of coherence scale. Soc Sci Med. 1993;36:725–33.

77. Bradley MM, Lang PJ. Measuring emotion: the self-assessment manikin and the semantic differential. J Behav Ther Exp Psychiatry. 1994;25(1):49–59.

78. Kroenke K, Spitzer RL, Williams JBW. The PHQ-9: validity of a brief depression severity measure. J Gen Intern Med. 2001;16(9):606–13.

79. Zigmon AS, Snaith RP. The hospital anxiety and depression scale. Acta Psychiatr Scand. 1983;67:367–70.

80. A. T. Beck, R. A. Steer, and G. Brown, Beck depression inventory–II, Database. 1996.

81. "Center for Epidemiologic Studies Depression Scale (CES-D) as a screening instrument for depression among community-residing older adults."

82. Carroll BJ, Feinberg M, Smouse PE, Rawson SG, Greden JF. The Carroll rating scale for depression. I. Development, reliability and validation. Br J Psychiatry. 1981;138:194–200.

83. Schlegel K, Mortillaro M. The Geneva emotional competence test (GECo): an ability measure of workplace emotional intelligence. J Appl Psychol. 2019;104(4):559–80.

84. Schlegel K, Scherer KR. Introducing a short version of the Geneva emotion recognition test (GERT-S): psychometric properties and construct validation. Behav Res Methods. 2016;48(4):1383–92.

85. Schlegel K, Grandjean D, Scherer KR. Introducing the Geneva emotion recognition test: an example of Rasch-based test development. Psychol Assess. 2014;26(2):666–72.

86. Treynor W, Gonzalez R, Nolen-Hoeksema S. Rumination reconsidered: a psychometric analysis. Cognit Ther Res. 2003;27(3):247–59.

87. Cohen S, Mermelstein R, Kamarck T. A global measure of perceived stress. J Health Soc Behav. 2016;24(4):385–96.

88. Goldberg DP, Hillier VF. A scaled version of the general health questionnaire. Psychol Med. 1979;9:139–45.

89. Stewart-Brown S, et al. The Warwick-Edinburgh mental Well-being scale (WEMWBS): a valid and reliable tool for measuring mental Well-being in diverse populations and projects. J Epidemiol Community Heal. 2011;65(Suppl 2):A38–9.

90. Herdman M, et al. Development and preliminary testing of the new five-level version of EQ-5D (EQ-5D-5L). Qual Life Res. 2011;20(10):1727–36.

91. Hopko DR, et al. Assessing worry in older adults: confirmatory factor analysis of the Penn State worry questionnaire and psychometric properties of an abbreviated model. Psychol Assess. 2003;15(2):173–83.

92. Diener E, et al. New Well-being measures: short scales to assess flourishing and positive and negative feelings. Soc Indic Res. 2010;97(2):143–56.

93. American Psychiatric Association, Diagnostic and Statistical Manual of. 4th ed. Washington DC; 2013.

94. World Health Organization, ICD-10, the ICD-10 classification of mental and behavioural disorders: diagnostic criteria for research. Geneva, 1993.

95. Wittchen H-U, Zaudig M, Fydrich TH. SKID–Strukturiertes Klinisches Interview für DSM-IV. Achse I und II Handanweisungen [Structured Clinical Interview for DSM-IV]. Göttingen: Hogrefe; 1997.

96. Schmier J, Halpern MT. Patient recall and recall bias of health state and health status. Expert Rev Pharmacoeconomics Outcomes Res. 2004;4(2):159–63.

97. Walter SD. Recall bias in epidemiologic studies. J Clin Epidemiol. 1990;43(12):1431–2.

98. Stone AA, Schneider S, Harter JK. Day-of-week mood patterns in the United States: on the existence of 'blue Monday', 'thank god it's Friday' and weekend effects. J Posit Psychol. 2012;7(4):306–14.

99. Ryan RM, Bernstein JH, Brown KW. Weekends, work, and Well-being: psychological need satisfactions and day of the week effects on mood, vitality, and physical symptoms. J Soc Clin Psychol. 2010;29(1):95–122.

100. Rosenman R, Tennekoon V, Hill LG. Measuring bias in self-reported data. Int J Behav Healthc Res. 2011;2(4):320.
101. Berinsky AJ. Can we talk? Self-presentation and the survey response. Polit Psychol. 2004;25(4):643–59.
102. Fullana MA, et al. Diagnostic biomarkers for obsessive-compulsive disorder: a reasonable quest or ignis fatuus? Neurosci Biobehav Rev. 2020;118:504–13.
103. Galatzer-Levy IR, Ma S, Statnikov A, Yehuda R, Shalev AY. Utilization of machine learning for prediction of post-traumatic stress: a re-examination of cortisol in the prediction and pathways to non-remitting PTSD. Transl Psychiatry. 2017;7(3)
104. Choppin A. EEG-based human Interface for disabled individuals : emotion expression with neural networks submitted for the master degree. Emotion. 2000;
105. Mehmood RM, Lee HJ. A novel feature extraction method based on late positive potential for emotion recognition in human brain signal patterns. Comput Electr Eng. 2016;53:444–57.
106. Li L, Chen JH. Emotion recognition using physiological signals from multiple subjects. In: Proc.–2006 Int. Conf. Intell. Inf. Hiding multimed. Signal process. IIH-MSP 2006; 2006. p. 355–8.
107. Uyl DMJ, Kuilenburg VH. The FaceReader: online facial expression recognition TL–30. FaceReader Online facial Expr. Recognit. 2005;30 VN-r(September):589–90.
108. Jang JR. ANFIS: adaptive-network-based fuzzy inference system. IEEE Trans Syst Man Cybern. 1993;23(3):665–85.
109. Rani P, Liu C, Sarkar N, Vanman E. An empirical study of machine learning techniques for affect recognition in human-robot interaction. Pattern Anal Appl. 2006;9(1):58–69.
110. Naji M, Firoozabadi M, Azadfallah P. Classification of music-induced emotions based on information fusion of forehead biosignals and electrocardiogram. Cognit Comput. 2014;6(2):241–52.
111. Khodayari-Rostamabad A, Hasey GM, MacCrimmon DJ, Reilly JP, de Bruin H. A pilot study to determine whether machine learning methodologies using pre-treatment electroencephalography can predict the symptomatic response to clozapine therapy. Clin Neurophysiol. 2010;121(12):1998–2006.
112. Khodayari-Rostamabad A, Reilly JP, Hasey GM, de Bruin H, MacCrimmon DJ. A machine learning approach using EEG data to predict response to SSRI treatment for major depressive disorder. Clin Neurophysiol. 2013;124(10):1975–85.
113. Robinson S, Hoheisel B, Windischberger C, Habel U, Lanzenberger R, Moser E. FMRI of the emotions: towards an improved understanding of amygdala function. Curr Med Imaging Rev. 2005;1(2):115–29.
114. Zhang W, et al. Discriminating stress from rest based on resting-state connectivity of the human brain: a supervised machine learning study. Hum Brain Mapp. 2020;41(11):3089–99.
115. Kolodyazhniy V, Kreibig SD, Gross JJ, Roth WT, Wilhelm FH. An affective computing approach to physiological emotion specificity: toward subject-independent and stimulus-independent classification of film-induced emotions. Psychophysiology. 2011;48(7):908–22.
116. Wac K, Tsiourti C. Ambulatory assessment of affect: survey of sensor systems for monitoring of autonomic nervous systems activation in emotion. IEEE Trans Affect Comput. 2014;5(3):251–72.
117. Goshvarpour A, Abbasi A, Goshvarpour A. An accurate emotion recognition system using ECG and GSR signals and matching pursuit method. Biom J. 2017;40(6):355–68.
118. Hart B, Struiksma ME, van Boxtel A, van Berkum JJA. Emotion in stories: facial EMG evidence for both mental simulation and moral evaluation. Front Psychol. 9(APR):2018.
119. Das P, Khasnobish A, Tibarewala DN. Emotion recognition employing ECG and GSR signals as markers of ANS. In: 2016 Conference on Advances in Signal Processing (CASP); 2016. p. 37–42.
120. Liu W, Zheng W-L, Lu B-L. Emotion recognition using multimodal deep learning. In: International conference on neural information processing; 2016. p. 521–9.

121. Wang Y, Mo J. Emotion feature selection from physiological signals using tabu search. In: 2013 25th Chinese Control and Decision Conference (CCDC); 2013. p. 3148–50.
122. Wen W, Liu G, Cheng N, Wei J, Shangguan P, Huang W. Emotion recognition based on multi-variant correlation of physiological signals. IEEE Trans Affect Comput. 2014;5(2):126–40.
123. Martinez HP, Bengio Y, Yannakakis GN. Learning deep physiological models of affect. IEEE Comput Intell Mag. 2013;8(2):20–33.
124. Qiao R, Qing C, Zhang T, Xing X, Xu X. A novel deep-learning based framework for multi-subject emotion recognition. In: 2017 4th International Conference on Information, Cybernetics and Computational Social Systems (ICCSS); 2017. p. 181–5.
125. Salari S, Ansarian A, Atrianfar H. Robust emotion classification using neural network models. In: 2018 6th Iranian Joint Congress on Fuzzy and Intelligent Systems (CFIS); 2018. p. 190–4.
126. Song T, Zheng W, Song P, Cui Z. EEG emotion recognition using dynamical graph convolutional neural networks. IEEE Trans Affect Comput. 2018;
127. Zheng W-L, Zhu J-Y, Peng Y, Lu B-L. EEG-based emotion classification using deep belief networks. In: 2014 IEEE International Conference on Multimedia and Expo (ICME); 2014. p. 1–6.
128. Huang J, Xu X, Zhang T. Emotion classification using deep neural networks and emotional patches. In: 2017 IEEE International Conference on Bioinformatics and Biomedicine (BIBM); 2017. p. 958–62.
129. Kawde P, Verma GK. Deep belief network based affect recognition from physiological signals. In: 2017 4th IEEE Uttar Pradesh Section International Conference on Electrical, Computer and Electronics (UPCON); 2017. p. 587–92.
130. Li X, Song D, Zhang P, Yu G, Hou Y, Hu B. Emotion recognition from multi-channel EEG data through convolutional recurrent neural network. In: 2016 IEEE international conference on bioinformatics and biomedicine (BIBM), vol. 2016. p. 352–9.
131. Alhagry S, Fahmy AA, El-Khoribi RA. Emotion recognition based on EEG using LSTM recurrent neural network. Emotion. 2017;8(10):355–8.
132. Liu J, Su Y, Liu Y. Multi-modal emotion recognition with temporal-band attention based on lstm-rnn. In: Pacific Rim Conference on Multimedia; 2017. p. 194–204.
133. Jerritta S, Murugappan M, Wan K, Yaacob S. Emotion recognition from facial EMG signals using higher order statistics and principal component analysis. J Chinese Inst Eng. 2014;37(3):385–94.
134. Cheng Y, Liu G-Y, Zhang H. The research of EMG signal in emotion recognition based on TS and SBS algorithm. In: The 3rd International Conference on Information Sciences and Interaction Sciences; 2010. p. 363–6.
135. Valenza G, Lanata A, Scilingo EP. The role of nonlinear dynamics in affective valence and arousal recognition. IEEE Trans Affect Comput. 2011;3(2):237–49.
136. Patel R, Janawadkar MP, Sengottuvel S, Gireesan K, Radhakrishnan TS. Suppression of eye-blink associated artifact using single channel EEG data by combining cross-correlation with empirical mode decomposition. IEEE Sensors J. 2016;16(18):6947–54.
137. Suk M, Prabhakaran B. Real-time mobile facial expression recognition system-a case study. In: Proceedings of the IEEE Conference on Computer Vision and Pattern Recognition Workshops; 2014. p. 132–7.
138. Ghimire D, Lee J. Geometric feature-based facial expression recognition in image sequences using multi-class adaboost and support vector machines. Sensors. 2013;13(6):7714–34.
139. Happy SL, George A, Routray A. A real time facial expression classification system using local binary patterns. In: 2012 4th International conference on intelligent human computer interaction (IHCI); 2012. p. 1–5.
140. Siddiqi MH, Ali R, Khan AM, Park Y-T, Lee S. Human facial expression recognition using stepwise linear discriminant analysis and hidden conditional random fields. IEEE Trans Image Process. 2015;24(4):1386–98.
141. Khan RA, Meyer A, Konik H, Bouakaz S. Framework for reliable, real-time facial expression recognition for low resolution images. Pattern Recogn Lett. 2013;34(10):1159–68.

142. Srinivasan V, Moghaddam S, Mukherji A, Rachuri KK, Xu C, Tapia EM. Mobileminer: mining your frequent patterns on your phone. In: Proceedings of the 2014 ACM International Joint Conference on Pervasive and Ubiquitous Computing; 2014. p. 389–400.

143. LiKamWa R, Liu Y, Lane ND, Zhong L. MoodScope: building a mood sensor from smartphone usage patterns. In: Proceeding of the ACM Mobisys; 2013. p. 389–402.

144. Ghimire D, Jeong S, Lee J, Park SH. Facial expression recognition based on local region specific features and support vector machines. Multimed Tools Appl. 2017;76(6):7803–21.

145. Fabian Benitez-Quiroz C, Srinivasan R, Martinez AM. Emotionet: an accurate, real-time algorithm for the automatic annotation of a million facial expressions in the wild. In: Proceedings of the IEEE conference on computer vision and pattern recognition; 2016. p. 5562–70.

146. Walecki R, Pavlovic V, Schuller B, Pantic M. Deep structured learning for facial action unit intensity estimation. In: Proceedings of the IEEE Conference on Computer Vision and Pattern Recognition; 2017. p. 3405–14.

147. R. Breuer and R. Kimmel, "A deep learning perspective on the origin of facial expressions." *arXiv Prepr. arXiv1705.01842*, 2017.

148. Jung H, Lee S, Yim J, Park S, Kim J. Joint fine-tuning in deep neural networks for facial expression recognition. In: Proceedings of the IEEE international conference on computer vision; 2015. p. 2983–91.

149. Ebrahimi Kahou S, Michalski V, Konda K, Memisevic R, Pal C. Recurrent neural networks for emotion recognition in video. In: Proceedings of the 2015 ACM on International Conference on Multimodal Interaction; 2015. p. 467–74.

150. Ng H-W, Nguyen VD, Vonikakis V, Winkler S. Deep learning for emotion recognition on small datasets using transfer learning. In: Proceedings of the 2015 ACM on international conference on multimodal interaction; 2015. p. 443–9.

151. Kim DH, Baddar WJ, Jang J, Ro YM. Multi-objective based spatio-temporal feature representation learning robust to expression intensity variations for facial expression recognition. IEEE Trans Affect Comput. 2017;10(2):223–36.

152. Hasani B, Mahoor MH. Facial expression recognition using enhanced deep 3D convolutional neural networks. In: Proceedings of the IEEE Conference on Computer Vision and Pattern Recognition Workshops; 2017. p. 30–40.

153. Graves A, Mayer C, Wimmer M, Schmidhuber J, Radig B. Facial expression recognition with recurrent neural networks. In: Proceedings of the International Workshop on Cognition for Technical Systems; 2008.

154. Donahue J, et al. Long-term recurrent convolutional networks for visual recognition and description. In: Proceedings of the IEEE conference on computer vision and pattern recognition; 2015. p. 2625–34.

155. Chu W-S, la Torre F, Cohn JF. Learning spatial and temporal cues for multi-label facial action unit detection. In: 2017 12th IEEE International Conference on Automatic Face & Gesture Recognition (FG 2017); 2017. p. 25–32.

156. Deng J, Frühholz S, Zhang Z, Schuller B. Recognizing emotions from whispered speech based on acoustic feature transfer learning. IEEE Access. 2017;5:5235–46.

157. Demircan S, Kahramanli H. Feature extraction from speech data for emotion recognition. J Adv Comput Networks. 2014;2(1):28–30.

158. Anagnostopoulos C-N, Iliou T, Giannoukos I. Features and classifiers for emotion recognition from speech: a survey from 2000 to 2011. Artif Intell Rev. 2015;43(2):155–77.

159. Dellaert F, Polzin T, Waibel A. Recognizing emotion in speech. In: Proceeding of Fourth International Conference on Spoken Language Processing. ICSLP'96, vol. 3; 1996. p. 1970–3.

160. Zhou Y, Sun Y, Zhang J, Yan Y. Speech emotion recognition using both spectral and prosodic features. In: 2009 International Conference on Information Engineering and Computer Science; 2009. p. 1–4.

161. Haq S, Jackson PJB, Edge J. Audio-visual feature selection and reduction for emotion classification. In: Proceeding International Conference on Auditory-Visual Speech Processing (AVSP'08), Tangalooma, Australia; 2008.

162. Alpert M, Pouget ER, Silva RR. Reflections of depression in acoustic measures of the patient's speech. J Affect Disord. 2001;66(1):59–69.
163. Ververidis D, Kotropoulos C. Emotional speech recognition: resources, features, and methods. Speech Commun. 2006;48(9):1162–81.
164. Mozziconacci S. Prosody and emotions. In: Speech prosody 2002, international conference; 2002.
165. J. B. Hirschberg *et al.*, "Distinguishing deceptive from non-deceptive speech," 2005.
166. Neiberg D, Elenius K, Laskowski K. Emotion recognition in spontaneous speech using GMMs. In: Ninth international conference on spoken language processing; 2006.
167. Dileep AD, Sekhar CC. HMM based intermediate matching kernel for classification of sequential patterns of speech using support vector machines. IEEE Trans Audio Speech Lang Processing. 2013;21(12):2570–82.
168. Vyas G, Dutta MK, Riha K, Prinosil J. An automatic emotion recognizer using MFCCs and hidden Markov models. In: 2015 7th International Congress on Ultra Modern Telecommunications and Control Systems and Workshops (ICUMT); 2015. p. 320–4.
169. Pan Y, Shen P, Shen L. Speech emotion recognition using support vector machine. Int J Smart Home. 2012;6(2):101–8.
170. Schuller B, Rigoll G, Lang M. Speech emotion recognition combining acoustic features and linguistic information in a hybrid support vector machine-belief network architecture. In: 2004 IEEE International Conference on Acoustics, Speech, and Signal Processing, vol. 1; 2004. p. I–577.
171. Zhao J, Mao X, Chen L. Speech emotion recognition using deep 1D & 2D CNN LSTM networks. Biomed Signal Process Control. 2019;47:312–23.
172. Yenigalla P, Kumar A, Tripathi S, Singh C, Kar S, Vepa J. Speech emotion recognition using Spectrogram & Phoneme Embedding. Interspeech. 2018:3688–92.
173. Zhao J, Mao X, Chen L. Learning deep features to recognise speech emotion using merged deep CNN. IET Signal Process. 2018;12(6):713–21.
174. Zhang Y, Liu Y, Weninger F, Schuller B. Multi-task deep neural network with shared hidden layers: breaking down the wall between emotion representations. In: 2017 IEEE International Conference on acoustics, speech and signal processing (ICASSP); 2017. p. 4990–4.
175. Badshah AM, Ahmad J, Rahim N, Baik SW. Speech emotion recognition from spectrograms with deep convolutional neural network. In: 2017 International Conference on platform technology and service (PlatCon); 2017. p. 1–5.
176. Barros P, Weber C, Wermter S. Emotional expression recognition with a cross-channel convolutional neural network for human-robot interaction. In: 2015 IEEE-RAS 15th International Conference on Humanoid Robots (Humanoids); 2015. p. 582–7.
177. Mao Q, Dong M, Huang Z, Zhan Y. Learning salient features for speech emotion recognition using convolutional neural networks. IEEE Trans Multimed. 2014;16(8):2203–13.
178. Lakomkin E, Zamani MA, Weber C, Magg S, Wermter S. On the robustness of speech emotion recognition for human-robot interaction with deep neural networks. In: 2018 IEEE/RSJ International Conference on Intelligent Robots and Systems (IROS); 2018. p. 854–60.
179. Tzirakis P, Trigeorgis G, Nicolaou MA, Schuller BW, Zafeiriou S. End-to-end multimodal emotion recognition using deep neural networks. IEEE J Sel Top Signal Process. 2017;11(8):1301–9.
180. Lim W, Jang D, Lee T. Speech emotion recognition using convolutional and recurrent neural networks. In: 2016 Asia-Pacific signal and information processing association annual summit and conference (APSIPA), vol. 2016. p. 1–4.
181. S. Sahu, R. Gupta, G. Sivaraman, W. AbdAlmageed, and C. Espy-Wilson, "Adversarial auto-encoders for speech based emotion recognition.," *arXiv Prepr. arXiv1806.02146*, 2018.
182. Zhao Y, Jin X, Hu X. Recurrent convolutional neural network for speech processing. In: 2017 IEEE International Conference on Acoustics, Speech and Signal Processing (ICASSP); 2017. p. 5300–4.
183. Lopez LD, Reschke PJ, Knothe JM, Walle EA. Postural communication of emotion: perception of distinct poses of five discrete emotions. Front Psychol. 2017;8(MAY)

184. Quiroz JC, Geangu E, Yong MH. Emotion recognition using smart watch sensor data: mixed-design study. J Med Internet Res. 2018;20(8)
185. García-Magariño I, Cerezo E, Plaza I, Chittaro L. A mobile application to report and detect 3D body emotional poses. Expert Syst Appl. 2019;122:207–16.
186. Stachl C, et al. Predicting personality from patterns of behavior collected with smartphones. Proc Natl Acad Sci U S A. 2020;117(30):17680–7.
187. Calvo RA, Milne DN, Hussain MS, Christensen H. Natural language processing in mental health applications using non-clinical texts. Nat Lang Eng. 2017;23(5):649–85.
188. Eichstaedt JC, et al. Facebook language predicts depression in medical records. Proc Natl Acad Sci U S A. 2018;115(44):11203–8.
189. Dey AK, Wac K, Ferreira D, Tassini K, Hong JH, Ramos J. Getting closer: an empirical investigation of the proximity of user to their smart phones. In: UbiComp'11–Proc. 2011 ACM Conf. Ubiquitous Comput; 2011. p. 163–72.
190. Berrocal A, Manea V, de Masi A, Wac K. MQOL lab: step-by-step creation of a flexible platform to conduct studies using interactive, mobile, wearable and ubiquitous devices. Procedia Comput Sci. 2020;175:221–9.
191. Schoedel R, Oldemeier M. Basic protocol: smartphone sensing panel. Leibniz Inst für Psychol Inf und Dokumentation. 2020;
192. Harari GM, et al. Sensing sociability: individual differences in young adults' conversation, calling, and texting, and app use behaviors in daily life. J Pers Soc Psychol. 2019;
193. Ozer DJ, Benet-Martínez V. Personality and the prediction of consequential outcomes. Annu Rev Psychol. 2006;57:401–21.
194. Roberts BW, Kuncel NR, Shiner R, Caspi A, Goldberg LR. The power of personality: the comparative validity of personality traits, socioeconomic status, and cognitive ability for predicting important life outcomes. Perspect Psychol Sci. 2007;2(4):313–45.
195. Rachuri KK, Musolesi M, Mascolo C, Rentfrow PJ, Longworth C, Aucinas A. EmotionSense: a Mobile phones based adaptive platform for experimental social psychology research. In: Proceedings of ACM UbiComp; 2010.
196. Chittaranjan G, Blom J, Gatica-Perez D. Who's who with big-five: analyzing and classifying personality traits with smartphones. In: Wearable Computers (ISWC), 2011 15th Annual International Symposium on; 2011. p. 29–36.
197. Pielot M, Dingler T, Pedro JS, Oliver N. When attention is not scarce-detecting boredom from mobile phone usage. In: Proceedings of the ACM UbiComp; 2015. p. 825–36.
198. Lu H, et al. Stresssense: detecting stress in unconstrained acoustic environments using smartphones. In: Proceedings of ACM UbiComp; 2012.
199. Bogomolov A, Lepri B, Pianesi F. Happiness recognition from Mobile phone data. In: Proceedings of the IEEE International Conference on Social Computing (SocialCom); 2013.
200. Bogomolov A, Lepri B, Ferron M, Pianesi F, Pentland A. Daily stress recognition from Mobile phone data, weather conditions and individual traits. In: Proceedings of the 22nd ACM International Conference on Multimedia; 2014.
201. Lee H, Choi YS, Lee S, Park IP. Towards unobtrusive emotion recognition for affective social communication. In: IEEE Consumer Communications and Networking Conference (CCNC); 2012.
202. Gao Y, Bianchi-Berthouze N, Meng H. What does touch tell us about emotions in touchscreen-based gameplay? ACM Trans Comput Hum Interact. 2012;19(4):Dec.
203. Wac K, Ciman M, Gaggi O. iSenseStress: assessing stress through human-smartphone interaction analysis. In: 9th International Conference on Pervasive Computing Technologies for Healthcare-PervasiveHealth; 2015. p. 8.
204. Kim H-J, Choi YS. Exploring emotional preference for smartphone applications. In: IEEE Consumer Communications and Networking Conference (CCNC); 2012.
205. Hektner JM, Schmidt JA, Csikszentmihalyi M. Experience sampling method: measuring the quality of everyday life. Sage. 2007;
206. Pejovic V, Lathia N, Mascolo C, Musolesi M. Mobile-based experience sampling for behaviour research. In: Emotions and personality in personalized services. Springer; 2016. p. 141–61.

207. Van Berkel N, Ferreira D, Kostakos V. The experience sampling method on Mobile devices. ACM Comput Surv. 2017;50(6):93.
208. Hernandez J, McDuff D, Infante C, Maes P, Quigley K, Picard R. Wearable ESM: differences in the experience sampling method across wearable devices. In: Proceedings of ACM MobileHCI; 2016. p. 195–205.
209. Wagner DT, Rice A, Beresford AR. Device analyzer: understanding smartphone usage. In: International Conference on Mobile and Ubiquitous Systems: Computing, Networking. and Services; 2013. p. 195–208.
210. Rawassizadeh R, Tomitsch M, Wac K, Tjoa AM. UbiqLog: a generic mobile phone-based life-log framework. Pers ubiquitous Comput. 2013;17(4):621–37.
211. Rawassizadeh R, Momeni E, Dobbins C, Gharibshah J, Pazzani M. Scalable daily human behavioral pattern mining from multivariate temporal data. IEEE Trans Knowl Data Eng. 2016;28(11):3098–112.
212. Ferreira D, Kostakos V, Dey AK. AWARE: mobile context instrumentation framework. Front ICT. 2015;2:6.
213. Nath S. ACE: exploiting correlation for energy-efficient and continuous context sensing. In: Proceedings of the 10th international conference on Mobile systems, applications, and services; 2012. p. 29–42.
214. Consolvo S, Walker M. Using the experience sampling method to evaluate ubicomp applications. IEEE Pervasive Comput. 2003;2(2):24–31.
215. Ghosh S, Ganguly N, Mitra B, De P. Towards designing an intelligent experience sampling method for emotion detection. In: Proceedings of the IEEE CCNC; 2017.
216. Barrett LF, Barrett DJ. An introduction to computerized experience sampling in psychology. Soc Sci Comput Rev. 2001;19(2):175–85.
217. Froehlich J, Chen MY, Consolvo S, Harrison B, Landay JA. MyExperience: a system for in situ tracing and capturing of user feedback on mobile phones. In: Proceedings of the 5th Mobisys; 2007.
218. Gaggioli A, et al. A mobile data collection platform for mental health research. Pers Ubiquitous Comput. 2013;17(2):241–51.
219. "Personal Analytics Companion.".
220. Sahami Shirazi A, Henze N, Dingler T, Pielot M, Weber D, Schmidt A. Large-scale assessment of mobile notifications. In: Proceedings of the ACM SIGCHI; 2014. p. 3055–64.
221. Fischer JE, Greenhalgh C, Benford S. Investigating episodes of mobile phone activity as indicators of opportune moments to deliver notifications. In: Proceedings of ACM MobileHCI; 2011. p. 181–90.
222. Ho J, Intille SS. Using context-aware computing to reduce the perceived burden of interruptions from mobile devices. In: Proceedings of ACM SIGCHI; 2005. p. 909–18.
223. Pielot M, de Oliveira R, Kwak H, Oliver N. Didn't you see my message?: predicting attentiveness to mobile instant messages. In: Proceedings of the ACM SIGCHI; 2014. p. 3319–28.
224. Kushlev K, Cardoso B, Pielot M. Too tense for candy crush: affect influences user engagement with proactively suggested content. In: Proceedings of the 19th International Conference on Human-Computer Interaction with Mobile Devices and Services (MobileHCI '17). New York, NY, USA: ACM; 2017.
225. Weber D, Voit A, Kratzer P, Henze N. In-situ investigation of notifications in multi-device environments. In: Proceedings of the 2016 ACM International Joint Conference on Pervasive and Ubiquitous Computing; 2016. p. 1259–64.
226. Turner LD, Allen SM, Whitaker RM. Push or delay? Decomposing smartphone notification response behaviour. In: Human behavior understanding: 6th international workshop, HBU, vol. 2015; 2015.

227. Gerber N, Gerber P, Volkamer M. Explaining the privacy paradox: a systematic review of literature investigating privacy attitude and behavior. Comput Secur. 2018;77:226–61.
228. T. Dienlin, "Das privacy paradox aus psychologischer Perspektive.," 2019.
229. "No Title." .
230. Mauss IB, Robinson MD. Measures of emotion: a review. Cogn. Emot. 2009;23(2):209–37.
231. Verduyn P, Lavrijsen S. Which emotions last longest and why: the role of event importance and rumination. Motiv Emot. 2015;39(1):119–27.
232. Ciman M, Wac K. Individuals' stress assessment using human-smartphone interaction analysis. IEEE Trans Affect Comput. 2018;9(1):51–65.
233. Hall M, Frank E, Holmes G, Pfahringer B, Reutemann P, Witten IH. The WEKA data mining software: an update. SIGKDD Explor Newsl. 2009;11(1):10–8.
234. Ghosh S, Ganguly N, Mitra B, De P. Designing an experience sampling method for smartphone based emotion detection. IEEE Trans Affect Comput. 2019:1–1.
235. Tarasov A, Delany SJ, Cullen C. Using crowdsourcing for labelling emotional speech assets. Proc W3C Work Emot Markup Lang. 2010;
236. Zhu X, Wu X. Class noise vs. attribute noise: A quantitative study. Artif Intell Rev. 2004;22(3):177–210.
237. Frénay B, Verleysen M. Classification in the presence of label noise: a survey. IEEE Trans neural networks Learn Syst. 2014;25(5):845–69.
238. Manea V, Wac K. Co-calibrating physical and psychological outcomes and consumer wearable activity outcomes in older adults: an evaluation of the coqol method. J Pers Med. 2020;10(4):1–86.
239. Vidal Bustamante CM, Rodman AM, Dennison MJ, Flournoy JC, Mair P, McLaughlin KA. Within-person fluctuations in stressful life events, sleep, and anxiety and depression symptoms during adolescence: a multiwave prospective study. J Child Psychol Psychiatry Allied Discip. 2020;61(10):1116–25.
240. Wac K. From quantified self to quality of life; 2018. p. 83–108.
241. Wac K. Quality of life technologies. Encycl Behav Med. 2020:1–2.
242. Wac K, Fiordelli M, Gustarini M, Rivas H. Quality of life technologies: experiences from the field and key challenges. IEEE Internet Comput. 2015;19(4):28–35.

Chapter 11
The Elusive Quantification of Self-Esteem: Current Challenges and Future Directions

Stefano De Dominicis and Erica Molinario

Introduction

The concept of self-esteem has been central in the social-psychological literature since the late eighteenth century and it can be arguably considered one of the most important constructs in psychology. A quick database search of PsychINFO reveals a striking 52,126 results in March 2020, and 53,248 results in October 2020, showing how central this topic was and still is (with an estimate increase of 2000 hits per year) for scientists and practitioners alike. William James introduced this topic more than one hundred years ago, and more recently Rhodewalt and Tragakis [1, p. 66] stated that self-esteem is one of the "top three covariates in personality and social psychology research". Perhaps, the relevance of self-esteem can be easily understood if we consider that this construct is linked to all levels of human existence, from mental illness to mental wellbeing: indeed, on the one hand, low self-esteem is related to various mental disorders, such as depression and anxiety (e.g., [2–5]); on the other hand, high self-esteem is related to various proxies of mental wellbeing, such as success, happiness, agency, and motivation (e.g., [6–8]). It is therefore fundamental to understand what self-esteem is, how it is assessed, and why it is so important for people's quality of life.

S. De Dominicis (✉)
Coaching Psychology Unit, Department of Nutrition, Exercise and Sports,
University of Copenhagen, Copenhagen, Denmark
e-mail: sdd@nexs.ku.dk

E. Molinario
Department of Psychology and The Water School, Florida Gulf Coast University,
Fort Myers, FL, USA

© The Author(s) 2022
K. Wac, S. Wulfovich (eds.), *Quantifying Quality of Life*, Health Informatics,
https://doi.org/10.1007/978-3-030-94212-0_11

269

Definition of Self-Esteem

In a way, we all know what self-esteem "really is": indeed, we can have a fairly good understanding of what is meant by self-esteem through introspection and observation of the behavior of others [9]. After all, self-esteem is a human phenomenon; yet, it is hard to put that understanding into precise words. In fact, as soon as we begin to examine self-esteem more closely, the understanding of this construct becomes quite problematic. The issue of defining self-esteem is crucial because definitions help shape what to focus on, which methods to choose and use, and what standards should be adopted to accept or reject evidence or conclusions [10]. Nevertheless, to find a concise and overarching definition of self-esteem is challenging because it encompasses different aspects and levels of analysis related to the context and its time stability or fluctuation. Self-esteem can be understood in terms of values (such as self-enhancement and openness to change values; [11]), feelings or affective dimensions (such as pride and shame; [12]), motivational (such as the desire to protect, maintain, and enhance feelings of self-worth; [13]), cognitive (such as evaluative components of self; [14]) and behavioral factors (such as being more independent or assertive, or more willing to exercise to gain fitness instead of reducing dissatisfaction with one's body image; [6, 15, 16]).

Historically, self-esteem has been conceptualized in terms of (a) self-competence, or the ratio of a person's successes over her failures in areas of life that are relevant to personal identity (dating back to William James work 1890/1983; James, Burkhardt, Bowers, Skrupskelis, and James, 1981; [17]) and (b) self-worth, or the affect concerning the degree to which one feels good about oneself [18]. Therefore, self-esteem is considered an evaluative psychological process that reflects both the extent to which people accept and like themselves, and believe they are competent (e.g., [19, 20]). Such evaluation can occur in relation of a specific point on time (state self-esteem) or as an overall and more stable evaluation of the self (global self-esteem). Scholars have suggested that this evaluation may involve the assessment of several dimensions of the self. For instance, Tafarodi and Swann [20, 21] suggested that self-esteem involves perceptions of self-worth (i.e., self-liking) and personal efficacy and self-regard of one's capabilities (i.e., self-competence); or as suggested by O'Brien and Epstein [22], it encompasses several dimensions related to worthiness, competence, and global self-esteem.

Favorable views of the self, vs. evaluations of the self that are either uncertain or negative would then be the manifestations of high vs. low self-esteem, respectively [23]. It is relevant to note that self-esteem reflects subjective perceptions rather than objective reality, and therefore could be either accurate or not [24]. We suggest that, to fully understand self-esteem, it should be conceptualized in its social-psychological context. Therefore, we define self-esteem as *the extent to which one person accepts and likes herself (in a specific point on time or overall) according to socially and personally defined standards, as well as believes of being competent in specific areas of life which are relevant to her personal and social identity.*

To further corroborate this perspective, we should highlight that our definition is in line with sociometer theory [25], which basically consider self-esteem as the output of a psychological meter, or instrument, that monitors the quality of people's relationships with others [26]. This psychological instrument is used by the self to monitor and respond to threats to the basic need to belong—i.e., that innate human need to form and maintain at least a minimum quantity of interpersonal relationships [25]. What follows, is that the degree to which a person perceives others to regard their relationship with herself as valuable, important, or close—i.e., her own relational evaluation—might change in specific situations or through the lifespan. Accordingly, when a person's relational evaluation is changing, the sociometer puts her attention to the related social acceptance threat or reward and motivates her to deal with it. Thus, the sociometer output is the affectively-charged self-appraisal that we typically perceive as self-esteem [27]:

> At its core, self-esteem is one's subjective appraisal of how one is faring with regard to being a valuable, viable, and sought-after member of the groups and relationships to which one belongs and aspires to belong. [26, p. 2]

This understanding of self-esteem appears coherent and comprehensive, as it can indeed explain why self-esteem is a relatively stable, but by no means immutable, psychological trait, as well as why it appears that self-esteem trait might have a specific trajectory across the individual's lifespan [28].

In light of the above-mentioned arguments, the assessment of self-esteem becomes a critical issue because it lies at the heart of empirical research. Accordingly, the focus of this chapter is to synthetize previous work detailing the assessment of self-esteem and to link this work with future possibilities, especially those arising from the beginning of the digital age.

Assessing Self-Esteem

With more than 200 different scales that ostensibly assess self-esteem [29], the critical issue of assessing self-esteem is far from simple. One of the reasons of this enormous effort lies in a critical element of self-esteem, namely the fact that self-esteem is, by definition, a subjective construct which is not tied to objective standards [30]. Indeed, because its subjective nature, self-esteem has been assessed mostly by self-report scales [31], and to a smaller extent by implicit measures [32]. It is based on subjective, affective-laden evaluations of one's own self, which however can occur with respect to specific domains (such as work, athletics or physical appearance) or to more broad and general level (such as overarching evaluation of the self as a whole).

A time specific self-evaluations represent the state self-esteem, which refers to the "feeling" aspect of self-esteem, that is an individual's affectively loaded self-evaluation in a given, specific situation [26]. Whereas, evaluation of the self as a whole refers to global, general or trait self-esteem [31]: it is one's long-term, characteristic and somewhat stable affectively charged self-evaluation. Because of trait self-esteem is a person's "summary" self-evaluation, it may or may not reflect her state self-esteem in a particular situation.

Finally, self-esteem may capture self-evaluations in specific domains: fluctuations of state self-esteem around a person's self-esteem typical level can be understood in terms of contingencies of self-worth [33]. Central to this model is the controversy that the way events and circumstances impact self-esteem lies on the perceived relevance of those events and circumstances to one's contingencies of self-worth: in other words, contingency self-worth represents a self-evaluation particularly vulnerable in which a failure or rejection is devastating to our sense of self-esteem [13, 34].

Historically, scholars involved in self-esteem research have focused mainly on individual differences in dispositional self-esteem [26]. However, from a practitioner's perspective, it is worth studying both stable self-esteem *and* its fluctuations, not only for the obvious advantage of better defining the construct itself, but also due to its possible applications. From the clinical perspective, measuring fluctuation of self-esteem allows to assess the efficacy of clinical interventions quantifying changes in self-esteem. Additionally, it can be used as a valid manipulation check in experimental design in which self-esteem is enhanced or diminished in a specific point in time. Finally, it can be used to understand confounded relations with other constructs. In the subsections below we discuss the assessment of both trait and state self-esteem (2.1), as well as situational self-esteem (2.2). In Table 11.1 we report the main characteristics of the reviewed scales measuring self-esteem.

Trait, Global, and State Self-Esteem Measures

As James [37] argues self-esteem is open to momentary changes, thus it raises and falls as a function of one's aspirations and success experiences. Within this framework—developed from the wide body of research that supports self-esteem as an enduring yet flexible concept [18, 38–40]—*state self-esteem* refers to how we evaluate or feel about ourselves in a given situation or at a given point in time. State self-esteem is therefore considered as a series of transitory states, momentary fluctuations, short-lived changes in one's own global self-esteem [36]. However, although momentary self-evaluations may be context dependent, there is a self-feelings people have the tendency to maintain and a level of self-esteem that derives by averaging feelings about themselves at one time across a number of different social situations: *trait* or *global self-esteem*.

Trait and Global Self-Esteem Measures

To assess trait self-esteem, scholars have developed several measures which differ in number of items and latent dimensionality. Here we discuss the *Single-Item Self-Esteem Scale* (SISE; [35]), the *Rosenberg Self-Esteem Scale* (RSE; [18]), and the

Self-Liking/Self-Competence Scale-Revised (SLSC-R; [20, 21]) which are among the most common tools used to assess trait self-esteem.

Single-Item Self-Esteem Scale (SISE). The simplest way to assess trait self-esteem is to ask individuals whether they have high self-esteem. In fact, this is the core of the Single Single-Item Self-Esteem Scale, which aims to assess the global self-worth or the overall attitude that one holds about oneself (SISE; [35]). The SISE was developed from the Rosenberg Self-Esteem Scale (described in the next section of this chapter) to assess self-esteem in contexts where time or other constraints severely limit the possibility of using or administering more complex or comprehensive measures of self-esteem. The SISE, as suggested by its name, consists of a single item and assesses a person's explicit knowledge about her global self-evaluation. This very brief, standardized measure of global self-esteem is valid

Table 11.1 Measures of self-esteem reviewed in this chapter

Name of the measure	References	Number of items and Response scale	Definition and Components of self-esteem	Type of self-esteem
Single-Item Self-Esteem Scale (SISE)	[35]	1 item 5-point Likert scale	Global self-worth: Overall attitude that one holds about oneself	Trait/ Global
Rosenberg Self-Esteem Scale (RSE)	[18]	10 items 4-point Guttman scale or 5-point Likert scale	Global self-worth: Overall attitude that one holds about oneself	Trait/ Global
Self-Liking/ Self-Competence Scale-Revised (SLSC-R)	[21]	16 items 5-point Likert scale	Self-esteem involves a personal sense of worth (self-liking) and a personal sense of efficacy (self-competence): (a) Self-liking: feeling positive towards one's own self; (b) Self-competence: feeling capable and in control, and to believe that one will be successful in the future	Trait/ Global
State Self-Esteem Scale (SSES)	[36]	20 items 5-point Likert scale	Self-esteem is the overall evaluation of the self and comprises: (a) Performance self-esteem: the extent to which a person feels her performance is worthy; (b) Social self-esteem: the extent to which a person feels self-conscious, foolish, or embarrassed about her public image; (c) Appearance self-esteem: the extent to which a person feels about her physical appearance.	State

(continued)

Table 11.1 (continued)

Name of the measure	References	Number of items and Response scale	Definition and Components of self-esteem	Type of self-esteem
Multidimensional Self-Esteem Inventory (MSEI)	[22]	116 items 5-point Likert scale	Global self-esteem (satisfaction with the self and confidence); identity integration (the self's internal integrity); and Defensive self-esteem (defensive reaction of changes in one's own self-esteem). Components of self-esteem: (a) Lovability (ability to express and receive affection) (b) Likeability (feeling accepted and liked by others) (c) Moral self-approval (satisfaction with one's moral values and acting accordingly) (d) Body appearance/physical attractiveness (being satisfied with one's body image) (e) Competence (ability to master new tasks) (f) Personal power (being assertive and able to influence others) (g) Self-control (being disciplined, persistent, able to set, and reach one's goals) (h) Body functioning/vitality (motor coordination and the feeling of being fit)	Trait/ Global and Specific
Contingency of Self-Worth scale (CSWs)	[33]	35 items 7-point Likert scale	Seven domains of contingent self-worth: 1. Academic competence 2. Physical appearance 3. Virtue 4. Having God's love 5. Having love and support from family 6. Outdoing others in competition 7. Obtaining others' approval	Contingent

and reliable [31] and asks participants to rate whether to have high self-esteem ("*I have high self-esteem*") is true for them (on a 5-point Likert scale, ranging from 'not very true of me' to 'very true of me.'). Although very simple to use, single-items measures are often less reliable than other multi-items measures, especially when constructs are heterogeneous [41]. Specifically within the realm of self-esteem and compared to multi-item measures, single-item measures are more susceptible to a person's biased knowledge of her own explicit feelings of specific or global self-worth, and are also more vulnerable to acquiescence and social desirability [31, 35]. Yet, SISE demonstrated to have a high convergent validity with other measures of self-esteem [31, 36] and therefore can be used without reservations in situations and research contexts that would require a single-item measure of self-esteem.

Rosenberg Self-Esteem Scale (RSE). Perhaps the most commonly used scale to assess trait self-esteem is the Rosenberg Self-Esteem Scale (RSE; [18]). The global self-esteem measured by the RSE is defined as the overall attitude one holds about oneself. This mono-dimensional definition of self-esteem implies the assumption that one might believe to be 'good enough' (high self-esteem) or not—meaning, to occur in self-rejection and to lack self-respect. This 10-item, easy-to-administer, self-esteem scale (example question: *"I feel that I have a number of good qualities"*), is originally designed as a 4-point Guttman scale but is often measured on a 5-point Likert-type scale, ranging from (1) Strongly disagree to (5) Strongly agree. It quickly became the "gold standard" for self-esteem research [20]. Although RSE is the most widely used assessment of self-esteem in research with a high reliability and validity [31], there has been discussions about its mono-factorial or multi-factorial structure [42]. However, it seems that a prominent global self-esteem factor consistently explains a considerable amount of variance in the RSE items [43–45], supporting the hypothesis of the mono-dimensionality of the RSE items [46].

Self-Liking/Self-Competence Scale-Revised (SLSC-R). To solve the single- vs. multiple-factor composition of the measures of self-esteem, a specific scale measuring two distinct components of global self-esteem was developed. Specifically, the Self-Liking/Self-Competence Scale (SLSC; and its Revised version SLSC-R) measures the two dimensions of self-esteem corresponding to a personal sense of worth (i.e., self-liking) and to a sense of personal efficacy (i.e., self-competence; [20, 21]). More specifically, the authors define self-liking as feeling positive towards one's own self, while self-competence refers to feeling capable and in control, and to believe that one will be successful in the future. The SLSC is a 16-item self-report scale, measured on a 5-point Likert-type response scale, which encompasses two 8-item subscales measuring each of the two components of self-esteem. The two dimensions (self-liking and self-competence) are related, but substantially distinct [31]. This scale has shown high reliability and convergent, discriminant and construct validity.

State Self-Esteem Measures

As mentioned, a momentary self-evaluation is represented by the state self-esteem, also called self-esteem feeling, which indicates a person's affectively laden self-assessment in a given situation [26]. The development of a measure of state self-esteem stemmed by the need for an instrument designed exclusively for assessing the momentary self-evaluations and self-esteem fluctuation [36, 47]. Here we discuss the *State Self-Esteem Scale* (SSES, [36]) and the *Multidimensional Self-Esteem Inventory* (MSEI; [22]).

State Self-Esteem Scale (SSES). The State Self-Esteem Scale (SSES) is a 20-item Likert-type scale designed for measuring temporary changes in individual self-esteem and is composed by three subscales including performance, social, and appearance self-esteem [36]: the performance component measures the extent to which subjects feel their performance is worthy (example question: "*I feel confident about my abilities*"); the social factor assesses the extent to which people feel self-conscious, foolish, or embarrassed about their public image (example question: "*I feel self-conscious*" [reversed item]); finally, the appearance element is instead related to physical appearance (example question: "*I feel good about myself*").

The Multidimensional Self-Esteem Inventory (MSEI; [22]) is an extensive self-report inventory which aims to define a respondent's profile across eight categories: four categories related to worthiness (lovability, likability, moral self-approval, and body appearance) and four related to competence (competence, personal power, self-control, and body functioning)—which are understood to impact self-esteem. Additionally, the MSEI includes three dimensions related to global measures of self-esteem, sense of identity, and defensiveness—which are understood as overall characteristics of or related to self-esteem. This inventory, initially developed as a clinical test for measuring and treating self-esteem and self-esteem related issues [24], has been widely used and validated across a great variety of domains both in research and clinical work (e.g., abuse, substance abuse and harassment; positive psychology, adjustment and emotional intelligence; academic, work and sport performance; goal attainment; childhood and adolescence wellbeing; emotional awareness, expression, reactivity and regulation; health psychology and physical wellbeing; mood, eating and personality disorders; stress and coping, trauma, anxiety; treatment, prevention, psychotherapy, self-help; for a complete list see [48]).

The MSEI consists of 116-item rated on a 5-point Likert scale: ('strongly agree' to 'strongly disagree'; or, 'hardly ever' to 'very often') which profiles the respondent's self-esteem on the following 11 scales: lovability (ability to express and receive affection), likeability (feeling accepted and liked by others) moral self-approval (satisfaction with one's moral values and acting accordingly), body appearance/physical attractiveness (being satisfied with one's body image), competence (ability to master new tasks), personal power (being assertive and able to influence others), self-control (being disciplined, persistent, able to set, and reach one's goals), body functioning/vitality (motor coordination and the feeling of being fit), global self-esteem (satisfaction with the self and confidence), identity integration (the self's internal integrity), and defensive self-esteem (defensive reaction of increasing one's self-esteem). This instrument was developed by gathering and categorizing thousands of incidents that participants reported as impacting on their self-esteem [24]. Accordingly, the emerged eight areas of life impacting self-esteem can be considered the components of self-esteem. In addition, the global measures of self-esteem and of secure sense of identity provide information about the overall feelings of worthiness and sense of security in one's identity; and the assessment of defensive self-enhancement differentiates between secure vs insecure self-esteem:

high scores express a biased self-presentation which denies weaknesses and claims strengths and which may not correspond to genuine self-evaluations [24].

The MSEI can be compared to other measures of self-esteem (especially with the SLSC-R) because the eight components can be construed as corresponding to the realms of worthiness or competence. Specifically, the categories of lovability, likability, moral self-approval, and body appearance imply that the source of self-esteem is more dependent on acceptance and worth, and in fact they are based in part on a judgment of acceptance and value (worthiness). Instead, the categories of competence, personal power, self-control, and body functioning imply that the source of self-esteem is more dependent on one's own ability to proactively impact the world, and are in fact based in part of one's own agency and efficacy (competence)—namely, the pillars of perceived self-efficacy [49]. However, despite the validity, applicability and the comprehensiveness of the MSEI [10, 24, 48, 50], it is not always possible to use such instrument out of clinical settings where time constrains or other impediments might hinder the likelihood of its usage [22].

Nevertheless, it is worth noting that MSEI is able to provide important insights that other instruments cannot grasp. First, global measures of self-esteem may not provide sufficient information specificity on which facet or dimension of self-esteem is particularly problematic (or functional) for a given person. Measures that are too specific or contingent are also problematic since a particular domain (e.g., athletic performance) may be relevant for someone but not for others [26]. Therefore, the mid-level components of self-esteem identified by the MSEI represent a level of analysis useful for effective applied interventions: on the one hand, these components are general enough to be related to a general trait measure of self-esteem (to which they have strong correlations; [24]) and thus changes in these component can have an impact on overall self-esteem; on the other hand, they are specific enough to provide a deeper, idiosyncratic understanding about how to build on one's own strengths and overcome weaknesses in the likelihood of an intervention for increasing self-esteem [24].

Second, the dimension of defensiveness assessed by the MSEI is a critical element in assessing self-esteem: it represents a person's social desirability bias, namely a bias in self-appraisal, in which she claims rare virtues and denies common human weaknesses (e.g., gladly accepting criticism vs. never trying to avoid unpleasant responsibilities). Evaluating defensiveness scores on the MSEI allows insight into self-presentation and projection to others of an overly positive image of themselves, or in other words of a false self-esteem.

Contingency of Self-Esteem

Expanding upon James' [37] idea that individuals differ in the domains on which they base their self-esteem, scholars have proposed several measures that capture domain-specific self-worth. The domains represent the context in which we are not only most likely to pursue self-esteem, but also are most vulnerable—in which a failure or rejection is devastating to our sense of self-worth [34]. Based on the work carried out by Crocker, Luhtanen, Cooper, and Bouvrette [51], which we discuss next, researchers have developed measures of contingent self-worth in several domains, such as friendships [52], romantic relationships [53, 54], and body weight [55].

Contingency of Self-Worth scale (CSWs). According to Crocker and Wolfe [33], individuals differ in the areas on which they base their self-worth and they proposed a model that examined self-worth in specific domains. CSWs is a 35-item measure divided into seven domains of contingent self-worth: academic competence, physical appearance, virtue, having God's love, having love and support from family, outdoing others in competition, and obtaining others' approval [51]. A key prediction of the CSWs model is that the impact of life events on self-esteem and affect is proportionate to the relevance of such events to one's contingency of self-worth. In other words, our self-esteem is more affected (positively or negatively) by events that occur in areas of life that are relevant for us—and for our identity. For example, students who strongly based their self-worth on academic competence experienced lower state self-esteem when they performed poorly on academic tasks, received lower-than-expected grades, or were rejected from graduate schools, compared to those whose self-worth was less based on this domain or did not experience self-threat [56, 57].

The Motivational Force of Self-Esteem

It is clear, at this point, that self-esteem is a complex psychological variable and thus there is not a unique way to assess it. In order to effectively operationalize self-esteem, it is important to have clear what aspect of self-esteem is important to a given research question and be aware of the objective limits of the measures available in the literature.

Perhaps, one of the most worrisome issues related to the abovementioned measures is their self-report nature, which is also a wider issue in psychological research. Self-report measures entangle several advantages but limitations too, such as memory limitations, recall biases, and social desirability sensitiveness, as in the case of the SISE which was found susceptible to acquiescence and social desirability [31, 35]. To overcome these measurement problems, some researchers have assessed implicit self-esteem (Implicit Association Test-IAT; [58]) by assessing automatic associations of self (through reaction times) with positive or negative valence [59]. However, the studies revealed unclear results, as construct divergence between implicit and explicit measures of self-esteem emerged. This result might seem contradictory, yet it highlights that the self-report measures of self-esteem developed to this point, although reliable, valid and all somehow different from each other, have all a common limitation. The operationalization of self-esteem so far developed takes into consideration its cognitive (evaluation of oneself, e.g., RSE), affective (emotional responses, e.g., SSES), behavioral (competence, e.g., SLSC-R), and value components (domain important to the individual, e.g. CSWs), yet they neglect an important aspect of self-esteem: namely, its motivational component. Understanding self-esteem as a fundamental psychological need implies that people are motivated to gain and maintain high levels of self-esteem: in other words, self-esteem is a goal in and of itself [60]. According to this conceptualization, self-esteem is a self-motive: it provides both a standard and a direction for behavior.

However, although the idea that this motive underlies human behavior has been a central theme in psychological theorizing (e.g., [33, 37, 61–65]), self-esteem has been traditionally assessed as a status (situational or stable) that characterizes the individual and does not capture the *desire* for a high self-esteem. In defining self-esteem, Crocker and Wolfe [33] recognize the motivational component of such psychological construct; yet, their operationalization of self-esteem does not capture whether the individual is striving to increase the level of self-esteem. Indeed, this operationalization of self-esteem as a motivational concept is, perhaps, the missing link to understand the behavioral strategies with which the individual might decide to engage to pursue or regain optimal levels of self-esteem.

As mentioned, many scholars have indeed assumed that people possess a motive or need to maintain self-esteem (e.g., [27]). The abovementioned sociometer theory grasps very well this idea, by conceptualizing the motivational force of self-esteem (namely, the self-esteem motive) as the human impulse of minimizing the likelihood of relational devaluation, or in other words, of rejection [25, 26]. This, in turn, leads to the pursue and preservation of high self-esteem. In fact, people typically act in ways that they believe will be of help in increasing their social acceptance by increasing their relational value: when this process is achieved, the individual's self-esteem will be enhanced. On the contrary, when a given situation, a behavior, or even an anticipated, potential, or irrational consequence of a behavior, might hinder the social or relational value of a person, her self-esteem is reduced or, at least, perceived to be threatened. Indeed, events that are known (or potentially known) to be "public", and therefore are more socially laden, have great effects on self-esteem; rather, "private" events, that are lived (or perceived to be lived) only by the individual, are usually less influential on self-esteem: if self-esteem was determined exclusively by private self-judgments, socially laden events should have no greater impact on self-esteem than private ones [27].

The Dark Side of Self-Esteem

Perhaps, one of the most significant and influential consequence of this conceptualization lies in the understanding of the crucial role of self-esteem in undesirable behaviors. Indeed, although self-esteem has been historically associated with desirable outcomes (e.g., better performance; [6]), it also recognized that high self-esteem can be associated with certain dysfunctional psychological processes and undesirable behaviors, such as egotism, narcissism, and violence: the dark side of self-esteem [66].

Within this conceptualization, the defensive aspect of self-enhancement grasped by the MSEI gains significant relevance. As mentioned, this inventory differentiates between secure vs insecure self-esteem, with high scores expressing a biased self-presentation which tends to always denies weaknesses and claims strengths. This biased self-presentation may correspond to a non-genuine self-evaluation [24]. Therefore, this non-authentic self-assessment (being conscious or not) might cover other negative patterns of self-evaluations, such as those related to negative, instable, contingent or narcissistic self-appraisals. In this conceptualization, the definition of self-esteem as a motive potentially could explain the heterogeneity of individuals with high self-esteem, encompassing people who honestly acknowledge their good (and bad) qualities along with narcissistic, defensive, and arrogant individuals [6]. Simply put, the definition of self-esteem as a motive would presuppose that people would strive for high self-esteem, no matter if it is non-genuine or authentic.

Indeed, high scores in overall self-esteem coupled with high variability in the worthiness dimensions of the MSEI are often associated with high scores of narcissism and aggression [67]. What follows, is that *high* and *optimal* self-esteem actually are two distinct constructs: the former can be fragile or secure, while the latter is characterized by genuine, true, stable, and congruent (with implicit self-esteem) high self-esteem [67, 68]. Likewise, contingent self-esteem studies (e.g., [13, 69]) have also argued for the existence of the dark side of positive self-appraisal, in which individuals become psychologically vulnerable due to their dependence on external validation for their self-esteem.

General principles from this work may be useful for clinical and non-clinical consideration alike. Yet, no specific implications have been developed or tested for assessing or intervening to address contingent self-esteem in a way that could consider positive authentic and dark self-esteem simultaneously, as well as its motivational component.

Future Directions in Assessing Self-Esteem

As we described beforehand, researchers have developed several kinds of measures targeting different features of self-esteem. To sum up, the literature provides measures to assess the state self-esteem, trait self-esteem, and domain-specific self-esteem. However, the quantitative psychological studies carried out to develop such measures relied heavily on surveys and laboratory experiments, which have well-known and long-endured limitations. In fact, on the one hand, surveys require people to make retrospective and often generalized judgments, which tend to be affected by memory limitations and recall biases [70, 71]. On the other hand, laboratory experiments do not take into account the context of a person's daily life which can influence her states and responses [72, 73]. Taken together, these considerations raise questions about the ecological validity of theoretical and methodological conclusions if not coupled with field data [74].

Furthermore, it seems that among several possible theorizations and measures of self-esteem, there is still a lack of consensus upon the most adequate ones. Obviously, it would be too simplistic and trivial to presuppose *a priori* which theoretical model and measure should be considered the finest to grasp the most precise definition and operationalization of self-esteem. We suggest that the best possible way to overcome such limitations—at least in part—and therefore to select the most efficient and accurate measure of self-esteem, is threefold: (a) to consider carefully the aspects of the psychological process under scrutiny; (b) to understand which facets of self-esteem would be relevant to the main research question; and (c) to take into account which scientific method (e.g., correlational, experimental, etc.) will be most suitable and effective in a given setting.

Nevertheless, as Baumeister et al. [75] argued, more attention should be given to how individuals actually behave in various situations rather than rely on what they claim, recall having done, or believe they would do in hypothetical situations. In addition, more attention should be given to the interaction between individual and social idiosyncrasies of self-esteem [26], as a growing amount of evidence shows that different constructs related to the *self* should indeed incorporate individual, social and cultural components (e.g., [76]).

Within this perspective, the widespread use of mobile technologies opens up new opportunities of collecting (social-) psychological data in specific contexts and situations. For example, mobile technology is already widely used to optimize health behavior change interventions (e.g., [77]). More specifically, the use of new digital technologies can help to overcome the limitations of the existing measures of self-esteem, giving the opportunity of looking at changes in self-esteem that occur in a field setting and therefore within the daily life moments. Such new data could not only be relevant to the understanding of self-esteem per sè, but could be coupled with other types of data collected through various types of sensors (e.g., global positioning systems-GPS, microphones, cameras, activity and sleep monitors, heart rate monitors, etc.) or software (e.g., social networks activity, mobile apps, etc.). Therefore, thanks to the means provided by new digital and wearables technologies, for the first time self-reported measures of self-esteem can be combined (above and beyond other psychological measures) to physiological and behavioral measures.

Along this line of research, Quantified Self-enabled data collection procedures could be advantageous. The quantification of the self (Quantified Self—QS) is a form of self-tracking an one's own daily life activities and behaviors, which eventually allow for the analysis of behavioral trends and patterns across a variety of life domains (e.g., physical activity, nutrition, weight; [78]). Generally, QS helps the individual to reflect upon such patterns and trends, and potentially leads to the application of behavior change strategies built upon these data. Theoretically, the process of self-monitoring and self-tracking one's own data could be useful to kick-start a process of behavior change across different life domains by enabling the person to change her relation to her own body and health, and to better control health-related decisions [79]. However, although some evidence seems to corroborate these hypotheses (e.g., [80, 81]), caution should be used in implementing interventions based on QS. First, recent research shows that an excessive quantification of one's own actions and behaviors could lead to psychological distress [82, 83]. Second, and perhaps most importantly, an enormous amount of research shows that in order to promote long-lasting behavior change one will need to detach from the monitoring of external objects and data, and rather should "monitor" his/her own internal physiological and psychological states and processes: in fact, by focusing on internal rather than external rewards and therefore by increasing intrinsic motivation in the new behavior (e.g., [84]), the person will in turn develop to greater self-awareness, presence in the moment, values and meanings, and perhaps identity change in the long term [85–87].

New Technologies and Social-Psychological Processes and Constructs

However, as mentioned before, the opportunities arising from the digital era should not be overlooked. In the last couple of decades, plenty of research has shown the link between various technological tools (both hardware and software) and a plethora of psychological processes and characteristics. Since the spread of mobile phones and smartphone, one track of research aimed to study the relationship between the (mis)use of technology and psychological processes (e.g., [88–90]). For instance, Andreassen et al. [88] examined the effect of personality traits such as narcissism, socio-demographic characteristics, and self-esteem on addictive social media. Among the others, and accordingly to previous research [91–93], they found that positive self-concept (i.e., high self-esteem) was negatively related to social media addiction, indicating that the Internet may provide a different social arena from the in the face-to-face life to enhance self-esteem. This track of research is extremely useful and much needed, since it helps clarifying (and will continue to do so) the psychological processes which can cause, be related, or be caused by the functional or dysfunctional use of different technology-based tools. For example, personality traits can predict Facebook use [94], while technology-based self-help therapies—which relies on minimal contact with an actual therapist—have proven effective, low-cost interventions at least for anxiety and mood disorders [95].

Furthermore, together with the development of social media, wearable devices, machine learning, big data, data mining technologies, internet of things, etc., another track of research has been consistently growing and aims to leverage these new technologies and related digital tools in various healthcare-related domains [96, 97]. More specifically within the social-psychological domain, this track of research is extremely interesting since it could unveil a totally new research filed which will transform the way we measure social-psychological constructs and processes: it is indeed possible to leverage new technologies to measure such constructs and processes—which have been historically assessed by self-reported measures exclusively—through behavioral data collected in a unobtrusively and longitudinally manner (via wearables, smartphone use, smart home and smart office, smart meters, etc.; e.g., [98]). Some research has already succeeded in this goal. For example, aggregated smartphone usage features can predict the Big-Five personality traits

[99]; Facebook likes can predict, among other sensitive personal attributes, personality traits, intelligence, happiness and addictions [100]; and, the combination of mobile phone usage and sensor data can predict with 75% accuracy low and high perceived stress [101]. It is therefore clear that the relationship between automatically extracted physiological and behavioral characteristics derived from rich data, and self-reported measures of social-psychological constructs and processes, could be disentangled eventually, and could potentially lead to new, behavior-based, ecologically valid measures of such facets. For example, Sun et al. [102], correlated gait patterns with self-reported self-esteem measured via the RSS: first, all participants completed the RSS; then they walked for 2 min on a rectangular carpet, while their gait data were recorded using a Kinect sensor. Based on machine learning, the authors were able to build predicting models to recognize self-esteem, with significant results ($r = 0.45$ for males; $r = 0.59$ for females; both $p < 0.001$.)

Toward New Methods of Quantifying and Conceptualizing Self-Esteem

Based on the abovementioned considerations, and specifically in relation to self-esteem, we propose a new approach to measuring such construct which could shed light, on the one side, on its theoretical understanding and definition and, on the other side, on its applied implications and applications to promote greater psychological wellbeing and to enhance quality of life. Drawing upon recent developments in applied social psychology (e.g., [98]) the suggested approach would combine both self-reported and digital technology-based tools. Specifically in the context of Quality of Life Technologies (QoLT)—technologies for assessment or improvement of a person's quality of life [103]—we believe that the contingent measurement of self-esteem via self-report and digital technology-based tools, could potentially shed light on its facets and processes above and beyond what has been understood so far through the standalone implementation of self-report measures. In turn, this could potentially expand the goals of QoLT above and beyond its already defined aims [103]: QoLT could aim at clarifying and expanding the theoretical understating of latent variables included in the WHO definition of QoL, such as self-esteem.

In our view, the combination of self-reported and digital technology-based tools would allow researchers and practitioners to take into account the two missing factors that seem important to assess self-esteem, namely, the *momentary experience* and the *social context*. With reference to the former, we suggest that future research should investigate via digital tools the *momentary experience of self-esteem*, namely its daily fluctuations. Indeed, past research has shown that self-esteem: (a) is most closely connected to a specific class of emotions—rather than to the whole spectrum of emotions—that relate to how people feel about themselves in a specific moment [12]; (b) that self-esteem indeed is open to momentary changes, raising and falling as a function of one's aspirations and failure/success experiences [36]; and that (c) one of the simplest way to measure state self-esteem is directly to ask individuals how they feel about themselves at that given moment via a single-item, valid and reliable measure [26]. In this vein, new technologies may be used to monitor *the momentary experience of self-esteem*, through collecting visual data (e.g., posture, clothes worn, facial expression and or posture in 'selfies' etc.), audio data (e.g., voice volume/pitch, etc.), text data for content analysis (e.g., text, social media posts, etc.), emoticon usage for sentimental analysis (e.g., number/type/of emoticon).

With reference to the latter, the social context, we believe that longitudinal information derived from new technologies usage (such as behavior shown on social media—likes, comments, selfies posted, etc.; data gathered from GPS enabled devices; physiological data such as heart rate variability, etc.), if combined with validated measures of self-esteem, can be indeed used to indirectly measure both the affective (e.g., self-liking) and behavioral (e.g., competence) components of self-esteem and to couple such components to the *specific social context* in which self-esteem components are assessed. In fact, previous research shows that: (a) self-esteem can be considered a monitor of social acceptance and therefore it motivates people to behave in ways that would make them valuable in specific social context which is perceived to be relevant by the individual [26, 27]; (b) the self-esteem motive, functioning as a tool to avoid social devaluation and rejection, should be measured in-context by contingent measures of such construct [51]; and (c) in-context measures of self-esteem can shed light on how self-esteem is implicated in affect, cognition, self-regulation of behavior and social processes, and can possibly unveil solutions to debates about the nature and functioning of self-esteem [34, 69]—and, in turn, suggest how self-esteem is causally related to mental health, wellbeing, performance and quality of life [6].

According to the abovementioned evidence, we believe that it would be possible to leverage new technologies specifically in two (or, arguably, one) methods of research: the Experience-Sampling Method [104] also known as the Ecological Momentary Assessment [74, 105]. Certain technology-based applications of such methods are already in use and show important mechanisms and processes that are relevant to the assessment and treatment of mental health—e.g., PsyMate™ [106, 107]—, such as the use of machine learning in predicting therapeutic outcomes in depression [108]. Along the same line of research, initial evidence reveals the efficacy of passive sensors for predicting self-esteem: by using machine learning techniques, it has been possible to relate performance, social and appearance self-esteem to several mobile-based digital behavior categories (such as calls, texts, conversations, and physical activity; [109]); furthermore, preliminary data indicate that is could be possible to correlate gait data to self-esteem with a fairly good criterion validity [102], suggesting that posture and perhaps other body-language characteristics could be taken as a good supplementary method to measure self-esteem.

Conclusion

In conclusion, the present chapter shows that far more research is needed to really uncover the potential of a variety of digital and technological tools in the broader field of applied psychology—such as mobile and wearable technologies used in different contexts of applications for the promotion of better quality of life [110–112]. Within this realm, new measures and conceptualizations of self-esteem should likely depend on both intrapersonal and interpersonal/social factors. We suggest that academics and practitioners alike should try to take advantage of the opportunities provided by the digital age in trying to further understand and measure the concept of self-esteem, by specifically capturing the level of self-esteem in context and anchoring such level to specific behaviors. Such new methods might make both the intrapersonal and social influence of self-esteem salient and decipherable.

References

1. Rhodewalt F, Tragakis MW. Self-esteem and self-regulation: toward optimal studies of self-esteem. Psychol Inquiry. 2003;14(1):66–70.
2. Battle J. Relationship between self-esteem and depression. Psychol Rep. 1978;42(3):745–6. https://doi.org/10.2466/pr0.1978.42.3.745.
3. Greenberg J, Solomon S, Pyszczynski T, Rosenblatt A, Burling J, Lyon D, Simon L, Pinel E. Why do people need self-esteem? Converging evidence that self-esteem serves an anxiety-buffering function. J Pers Soc Psychol. 1992;63(6):913–22. https://doi.org/10.1037/0022-3514.63.6.913.
4. Schmitz N, Kugler J, Rollnik J. On the relation between neuroticism, self-esteem, and depression: results from the National Comorbidity Survey. Compr Psychiatry. 2003;44(3):169–76. https://doi.org/10.1016/S0010-440X(03)00008-7.
5. Sowislo JF, Orth U. Does low self-esteem predict depression and anxiety? A meta-analysis of longitudinal studies. Psychol Bull. 2013;139(1):213–40. https://doi.org/10.1037/a0028931.
6. Baumeister RF, Campbell JD, Krueger JI, Vohs KD. Does high self-esteem cause better performance, interpersonal success, happiness, or healthier lifestyles? Psychol Sci Public Interest. 2003;4(1):1–44. https://doi.org/10.1111/1529-1006.01431.
7. Cheng H, Furnham A. Personality, self-esteem, and demographic predictions of happiness and depression. Personal Individ Differ. 2003;34(6):921–42. https://doi.org/10.1016/S0191-8869(02)00078-8.
8. Lyubomirsky S, Tkach C, DiMatteo MR. What are the differences between happiness and self-esteem. Soc Indic Res. 2006;78(3):363–404. https://doi.org/10.1007/s11205-005-0213-y.
9. Mecca AM, Smelser NJ, Vasconcellos J, editors. The social importance of self-esteem. Berkeley: University of California Press; 1989.
10. Mruk CJ. Self-esteem research, theory, and practice: toward a positive psychology of self-esteem [Computer software]. 3rd ed. New York: Springer; 2006.
11. Lönnqvist J-E, Verkasalo M, Helkama K, Andreyeva GM, Bezmenova I, Rattazzi AMM, Niit T, Stetsenko A. Self-esteem and values. Eur J Soc Psychol. 2009;39(1):40–51. https://doi.org/10.1002/ejsp.465.
12. Brown JD, Marshall MA. Self-esteem and emotion: some thoughts about feelings. Personal Soc Psychol Bull. 2001;27(5):575–84. https://doi.org/10.1177/0146167201275006.

13. Crocker J, Park LE. The costly pursuit of self-esteem. Psychol Bull. 2004;130(3):392–414. https://doi.org/10.1037/0033-2909.130.3.392.
14. Greenwald AG, Banaji MR, Rudman LA, Farnham SD, Nosek BA, Mellott DS. A unified theory of implicit attitudes, stereotypes, self-esteem, and self-concept. Psychol Rev. 2002;109(1):3–25. https://doi.org/10.1037/0033-295X.109.1.3.
15. Furnham A, Badmin N, Sneade I. Body image dissatisfaction: gender differences in eating attitudes, self-esteem, and reasons for exercise. J Psychol. 2002;136(6):581–96. https://doi.org/10.1080/00223980209604820.
16. Mruk CJ. Self-esteem and positive psychology: research, theory, and practice. 4th ed. New York: Springer; 2013.
17. Zeigler-Hill V, editor. Self-esteem. Psychology Press; 2013.
18. Rosenberg M. Society and the adolescent self-image. Princeton: Princeton University Press; 1965. https://doi.org/10.1515/9781400876136.
19. Brown JD. The self. New York: Routledge, Taylor & Francis Group; 1998.
20. Tafarodi RW, Swann WB Jr. Self-linking and self-competence as dimensions of global self-esteem: initial validation of a measure. J Pers Assess. 1995;65(2):322–42. https://doi.org/10.1207/s15327752jpa6502_8.
21. Tafarodi RW, Swann WB. Two-dimensional self-esteem: theory and measurement. Personal Individ Differ. 2001;31(5):653–73. https://doi.org/10.1016/S0191-8869(00)00169-0.
22. O'Brien EJ, Epstein S. The multidimensional self-esteem inventory. Psychological Assessment Resources; 1988.
23. Campbell JD, Trapnell PD, Heine SJ, Katz IM, Lavallee LF, Lehman DR. Self-concept clarity: measurement, personality correlates, and cultural boundaries. J Pers Soc Psychol. 1996;70(1):141–56. https://doi.org/10.1037/0022-3514.70.1.141.
24. Mruk CJ, O'Brien EJ. Changing self-esteem through competence and worthiness training: a positive therapy. In: Zeigler-Hill V, editor. Self-esteem. Psychology Press; 2013. p. 173–89. https://doi.org/10.4324/9780203587874-13.
25. Baumeister RF, Leary MR. The need to belong: desire for interpersonal attachments as a fundamental human motivation. Psychol Bull. 1995;117(3):497–529. https://doi.org/10.1037/0033-2909.117.3.497.
26. Leary MR, Baumeister RF. The nature and function of self-esteem: sociometer theory. In: Advances in experimental social psychology, vol. 32. Elsevier; 2000. p. 1–62. https://doi.org/10.1016/S0065-2601(00)80003-9.
27. Leary MR. Making sense of self-esteem. Curr Dir Psychol Sci. 1999;8(1):32–5. https://doi.org/10.1111/1467-8721.00008.
28. Orth U, Robins RW. The development of self-esteem. Curr Dir Psychol Sci. 2014;23(5):381–7. https://doi.org/10.1177/0963721414547414.
29. Scheff TJ, Fearon DS. Cognition and emotion? The dead end in self-esteem research. J Theory Soc Behav. 2004;34(1):73–90. https://doi.org/10.1111/j.1468-5914.2004.00235.x.
30. Blascovich J, Tomaka J. Measures of self-esteem. In: Robinson JP, Shaver PR, Wrightsman LS, editors. Measures of personality and social psychological attitudes, vol. 1. New York: Academic; 1990.
31. Donnellan MB, Trzesniewski KH, Robins RW. Measures of self-esteem. In: Measures of personality and social psychological constructs. Elsevier; 2015. p. 131–57. https://doi.org/10.1016/B978-0-12-386915-9.00006-1
32. Buhrmester MD, Blanton H, Swann WB. Implicit self-esteem: nature, measurement, and a new way forward. J Pers Soc Psychol. 2011;100(2):365–85. https://doi.org/10.1037/a0021341.
33. Crocker J, Wolfe CT. Contingencies of self-worth. Psychol Rev. 2001;108(3):593–623. https://doi.org/10.1037/0033-295X.108.3.593.
34. Crocker J. Contingencies of self-worth: implications for self-regulation and psychological vulnerability. Self Identity. 2002;1(2):143–9. https://doi.org/10.1080/152988602317319320.
35. Robins RW, Hendin HM, Trzesniewski KH. Measuring global self-esteem: construct validation of a single-item measure and the Rosenberg self-esteem scale. Personal Soc Psychol Bull. 2001;27(2):151–61. https://doi.org/10.1177/0146167201272002.

36. Heatherton TF, Polivy J. Development and validation of a scale for measuring state self-esteem. J Pers Soc Psychol. 1991;60(6):895–910. https://doi.org/10.1037/0022-3514.60.6.895.
37. James W. The principles of psychology. New York: Henry Holt; 1890.
38. Janis IL, Field PB. Sex differences and factors related to persuasibility. In: Hovland CI, Janis IL, editors. Personality and persuasibility. New Haven: Yale University Press; 1959. p. 55–68.
39. Savin-Williams RC, Demo DH. Situational and transituational determinants of adolescent self-feelings. J Pers Soc Psychol. 1983;44(4):824–33. https://doi.org/10.1037/0022-3514.44.4.824.
40. Wells AJ. Variations in mothers' self-esteem in daily life. J Pers Soc Psychol. 1988;55(4):661–8. https://doi.org/10.1037/0022-3514.55.4.661.
41. Postmes T, Haslam SA, Jans L. A single-item measure of social identification: reliability, validity, and utility. Br J Soc Psychol. 2013;52(4):597–617. https://doi.org/10.1111/bjso.12006.
42. Heatherton TF, Wyland CL. Assessing self-esteem. In: Lopez SJ, Snyder CR, editors. Positive psychological assessment: a handbook of models and measures. American Psychological Association; 2003. p. 219–33. https://doi.org/10.1037/10612-014.
43. Donnellan MB, Kenny DA, Trzesniewski KH, Lucas RE, Conger RD. Using trait–state models to evaluate the longitudinal consistency of global self-esteem from adolescence to adulthood. J Res Pers. 2012;46(6):634–45. https://doi.org/10.1016/j.jrp.2012.07.005.
44. Schmitt DP, Allik J. Simultaneous administration of the Rosenberg self-esteem scale in 53 nations: exploring the universal and culture-specific features of global self-esteem. J Pers Soc Psychol. 2005;89(4):623–42. https://doi.org/10.1037/0022-3514.89.4.623.
45. Tafarodi RW, Milne AB. Decomposing global self-esteem. J Pers. 2002;70(4):443–84. https://doi.org/10.1111/1467-6494.05017.
46. Slocum-Gori SL, Zumbo BD, Michalos AC, Diener E. A note on the dimensionality of quality of life scales: an illustration with the satisfaction with life scale (SWLS). Soc Indic Res. 2009;92(3):489–96. https://doi.org/10.1007/s11205-008-9303-y.
47. Bozorgpour F, Salimi A. State self-esteem, loneliness and life satisfaction. Procedia Soc Behav Sci. 2012;69:2004–8. https://doi.org/10.1016/j.sbspro.2012.12.157.
48. O'Brien EJ. Bibliography of references to the multidimensional self-esteem inventory (MSEI). Psychological Assessment Resources; 2010.
49. Bandura A. Self-efficacy mechanism in human agency. Am Psychol. 1982;37(2):122–47.
50. Mruk CJ. Self-esteem and positive psychology: research, theory, and practice [Computer software]. 4th ed. New York: Springer; 2013.
51. Crocker J, Luhtanen RK, Cooper ML, Bouvrette A. Contingencies of self-worth in college students: theory and measurement. J Pers Soc Psychol. 2003;85(5):894–908. https://doi.org/10.1037/0022-3514.85.5.894.
52. Cambron MJ, Acitelli LK, Steinberg L. When friends make you blue: the role of friendship contingent self-esteem in predicting self-esteem and depressive symptoms. Personal Soc Psychol Bull. 2010;36(3):384–97. https://doi.org/10.1177/0146167209351593.
53. Knee CR, Canevello A, Bush AL, Cook A. Relationship-contingent self-esteem and the ups and downs of romantic relationships. J Pers Soc Psychol. 2008;95(3):608–27. https://doi.org/10.1037/0022-3514.95.3.608.
54. Park LE, Sanchez DT, Brynildsen K. Maladaptive responses to relationship dissolution: the role of relationship contingent self-worth: responses to romantic breakup. J Appl Soc Psychol. 2011;41(7):1749–73. https://doi.org/10.1111/j.1559-1816.2011.00769.x.
55. Clabaugh A, Karpinski A, Griffin K. Body weight contingency of self-worth. Self Identity. 2008;7(4):337–59. https://doi.org/10.1080/15298860701665032.
56. Crocker J, Karpinski A, Quinn DM, Chase SK. When grades determine self-worth: consequences of contingent self-worth for male and female engineering and psychology majors. J Pers Soc Psychol. 2003;85(3):507–16. https://doi.org/10.1037/0022-3514.85.3.507.
57. Crocker J, Sommers SR, Luhtanen RK. Hopes dashed and dreams fulfilled: contingencies of self-worth and graduate school admissions. Personal Soc Psychol Bull. 2002;28(9):1275–86. https://doi.org/10.1177/01461672022812012.

58. Greenwald AG, McGhee DE, Schwartz JLK. Measuring individual differences in implicit cognition: the implicit association test. J Pers Soc Psychol. 1998;74(6):1464–80. https://doi.org/10.1037/0022-3514.74.6.1464.
59. Greenwald AG, Farnham SD. Using the implicit association test to measure self-esteem and self-concept. J Pers Soc Psychol. 2000;79(6):1022–38. https://doi.org/10.1037/0022-3514.79.6.1022.
60. Cast AD, Burke PJ. A theory of self-esteem. Soc Forces. 2002;80(3):1041–68. https://doi.org/10.1353/sof.2002.0003.
61. Fein S, Spencer SJ. Prejudice as self-image maintenance: affirming the self through derogating others. J Pers Soc Psychol. 1997;73(1):31–44. https://doi.org/10.1037/0022-3514.73.1.31.
62. Horney K. The neurotic personality of our time. New York: W W Norton & Co.; 1937.
63. Kernis MH, Waschull SB. The interactive roles of stability and level of self-esteem: research and theory. In: Zanna MP, editor. Advances in experimental social psychology, vol. 27. Academic; 1995. p. 93–141. https://doi.org/10.1016/S0065-2601(08)60404-9.
64. Sullivan HS. The interpersonal theory of psychiatry. W W Norton & Co.; 1953.
65. Tesser A. Toward a self-evaluation maintenance model of social behavior. In *Advances in experimental social psychology*, vol. 21. Elsevier; 1988. p. 181–227. https://doi.org/10.1016/S0065-2601(08)60227-0.
66. Baumeister RF, Smart L, Boden JM. Relation of threatened egotism to violence and aggression: the dark side of high self-esteem. Psychol Rev. 1996;103(1):5–33. https://doi.org/10.1037/0033-295X.103.1.5.
67. Kernis MH. Author's response: optimal self-esteem and authenticity: separating fantasy from reality. Psychol Inq. 2003;14(1):83–9. https://doi.org/10.1207/S15327965PLI1401_03.
68. Kernis MH. Target article: toward a conceptualization of optimal self-esteem. Psychol Inq. 2003;14(1):1–26. https://doi.org/10.1207/S15327965PLI1401_01.
69. Crocker J, Brook AT, Niiya Y, Villacorta M. The pursuit of self-esteem: contingencies of self-worth and self-regulation. J Pers. 2006;74(6):1749–72. https://doi.org/10.1111/j.1467-6494.2006.00427.x.
70. Kahneman D, Riis J. Living, and thinking about it: two perspectives on life. In: Huppert FA, Baylis N, Keverne B, editors. The science of well-being. Oxford University Press; 2005. p. 284–305. https://doi.org/10.1093/acprof:oso/9780198567523.003.0011.
71. Schwarz N. Retrospective and concurrent self-reports: the rationale for real-time data capture. In: Stone AA, Shiffman S, Atienza AA, Nebeling L, editors. The science of real-time data capture: self-reports in health research. New York: Oxford University Press; 2007.
72. Hammond JS, Keeney RL, Raiffa H. The hidden traps in decision making. Harvard Business Review, September–October 1998. 1998. https://hbr.org/1998/09/the-hidden-traps-in-decision-making-2
73. Wilhelm P, Perrez M, Pawlik K. Conducting research in daily life: a historical review. In: Mehl MR, Conner TS, editors. Handbook of research methods for studying daily life. Guilford Press; 2012. p. 62–86.
74. Shiffman S, Stone AA, Hufford MR. Ecological momentary assessment. Annu Rev Clin Psychol. 2008;4(1):1–32. https://doi.org/10.1146/annurev.clinpsy.3.022806.091415.
75. Baumeister RF, Vohs KD, Funder DC. Psychology as the science of self-reports and finger movements: whatever happened to actual behavior? Perspect Psychol Sci. 2007;2(4):396–403. https://doi.org/10.1111/j.1745-6916.2007.00051.x.
76. De Dominicis, S., Schultz, P. W., & Bonaiuto, M. (2017). Promoting collective pro-environmental action through self-enhancing motivators. Front Psychol.
77. Walsh JC, Groarke JM. Integrating behavioral science with Mobile (mHealth) technology to optimize health behavior change interventions. Eur Psychol. 2019;24(1):38–48. https://doi.org/10.1027/1016-9040/a000351.
78. Wac K. From quantified self to quality of life. In: Rivas H, Wac K, editors. Digital health: scaling healthcare to the world. Springer International Publishing; 2018. p. 83–108. https://doi.org/10.1007/978-3-319-61446-5_7.
79. Petit A, Cambon L. Exploratory study of the implications of research on the use of smart connected devices for prevention: a scoping review. BMC Public Health. 2016;16(1) https://doi.org/10.1186/s12889-016-3225-4.

80. Goyal S, Morita P, Lewis GF, Yu C, Seto E, Cafazzo JA. The systematic design of a behavioural mobile health application for the self-management of type 2 diabetes. Can J Diabetes. 2016;40(1):95–104. https://doi.org/10.1016/j.jcjd.2015.06.007.
81. Shull PB, Jirattigalachote W, Hunt MA, Cutkosky MR, Delp SL. Quantified self and human movement: a review on the clinical impact of wearable sensing and feedback for gait analysis and intervention. Gait Posture. 2014;40(1):11–9. https://doi.org/10.1016/j.gaitpost.2014.03.189.
82. Andersen TO, Langstrup H, Lomborg S. Experiences with wearable activity data during self-care by chronic heart patients: qualitative study. J Med Internet Res. 2020;22(7):e15873. https://doi.org/10.2196/15873.
83. De Dominicis S. Active breaks as a tool to promote physiological and psychological wellbeing: a randomized block experiment with office workers (in press).
84. Ryan RM, Deci EL. Self-determination theory and the facilitation of intrinsic motivation, social development, and well-being. Am Psychol. 2000;55(1):68–78. https://doi.org/10.1037/0003-066X.55.1.68.
85. Stelter R. A guide to third generation coaching. Dordrecht: Springer; 2014. https://doi.org/10.1007/978-94-007-7186-4.
86. Stelter R. Working with values in coaching. In: Bachkirova T, Spence G, Drake D, editors. The SAGE handbook of coaching. London: Sage; 2017. p. 333–47.
87. Stelter R, Andersen V. Coaching for health and lifestyle change: theory and guidelines for interacting and reflecting with women about their challenges and aspirations. Int Coach Psychol Rev. 2018;13(1):61–71.
88. Andreassen CS, Pallesen S, Griffiths MD. The relationship between addictive use of social media, narcissism, and self-esteem: findings from a large national survey. Addict Behav. 2017;64:287–93. https://doi.org/10.1016/j.addbeh.2016.03.006.
89. Butt S, Phillips JG. Personality and self reported mobile phone use. Comput Hum Behav. 2008;24(2):346–60. https://doi.org/10.1016/j.chb.2007.01.019.
90. Devaraj S, Easley RF, Crant JM. How does personality matter? Relating the five-factor model to technology acceptance and use. Inf Syst Res. 2008;19(1):93–105. https://doi.org/10.1287/isre.1070.0153.
91. Armstrong L, Phillips JG, Saling LL. Potential determinants of heavier internet usage. Int J Hum-Comput Stud. 2000;53(4):537–50. https://doi.org/10.1006/ijhc.2000.0400.
92. Kim H-K, Davis KE. Toward a comprehensive theory of problematic Internet use: evaluating the role of self-esteem, anxiety, flow, and the self-rated importance of Internet activities. Comput Hum Behav. 2009;25(2):490–500. https://doi.org/10.1016/j.chb.2008.11.001.
93. Niemz K, Griffiths M, Banyard P. Prevalence of pathological internet use among university students and correlations with self-esteem, the general health questionnaire (GHQ), and disinhibition. Cycberpsychol Behav. 2005;8(6):562–70. https://doi.org/10.1089/cpb.2005.8.562.
94. Moore K, McElroy JC. The influence of personality on Facebook usage, wall postings, and regret. Comput Hum Behav. 2012;28(1):267–74. https://doi.org/10.1016/j.chb.2011.09.009.
95. Newman MG, Szkodny LE, Llera SJ, Przeworski A. A review of technology-assisted self-help and minimal contact therapies for anxiety and depression: is human contact necessary for therapeutic efficacy? Clin Psychol Rev. 2011;31(1):89–103. https://doi.org/10.1016/j.cpr.2010.09.008.
96. Bhatt C, Dey N, Ashour AS, editors. Internet of Things and big data technologies for next generation healthcare, vol. 23. Cham: Springer International Publishing; 2017. https://doi.org/10.1007/978-3-319-49736-5.
97. Jara AJ, Bocchi Y, Genoud D. Social Internet of Things: the potential of the Internet of Things for defining human behaviours. 2014 International Conference on Intelligent Networking and Collaborative Systems; 2014. p. 581–5. https://doi.org/10.1109/INCoS.2014.113
98. De Dominicis S, Sokoloski R, Jaeger CM, Schultz PW. Making the smart meter social promotes long-term energy conservation. Palgrave Commun. 2019;5(1):51. https://doi.org/10.1057/s41599-019-0254-5.

99. Chittaranjan G, Blom J, Gatica-Perez D. Mining large-scale smartphone data for personality studies. Pers Ubiquit Comput. 2013;17(3):433–50. https://doi.org/10.1007/s00779-011-0490-1.

100. Kosinski M, Stillwell D, Graepel T. Private traits and attributes are predictable from digital records of human behavior. Proc Natl Acad Sci. 2013;110(15):5802–5. https://doi.org/10.1073/pnas.1218772110.

101. Sano A, Picard RW. Stress recognition using wearable sensors and mobile phones. In: 2013 Humaine Association Conference on Affective Computing and Intelligent Interaction; 2013. p. 671–6. https://doi.org/10.1109/ACII.2013.117

102. Sun B, Zhang Z, Liu X, Hu B, Zhu T. Self-esteem recognition based on gait pattern using Kinect. Gait Posture. 2017;58:428–32. https://doi.org/10.1016/j.gaitpost.2017.09.001.

103. Wac K. Quality of life technologies. In: Gellman M, editor. Encyclopedia of behavioral medicine. Springer; 2019. p. 1–2. https://doi.org/10.1007/978-1-4614-6439-6_102013-1.

104. Csikszentmihalyi M, Larson R. Validity and reliability of the experience-sampling method. In: Csikszentmihalyi M, editor. Flow and the foundations of positive psychology. Cham: Springer; 2014. p. 35–54. https://doi.org/10.1007/978-94-017-9088-8_3.

105. Stone AA, Shiffman S. Ecological momentary assessment (Ema) in behavioral medicine. Ann Behav Med. 1994;16(3):199–202. https://doi.org/10.1093/abm/16.3.199.

106. Kramer I, Simons CJP, Hartmann JA, Menne-Lothmann C, Viechtbauer W, Peeters F, Schruers K, van Bemmel AL, Myin-Germeys I, Delespaul P, van Os J, Wichers M. A therapeutic application of the experience sampling method in the treatment of depression: a randomized controlled trial. World Psychiatry. 2014;13(1):68–77. https://doi.org/10.1002/wps.20090.

107. Myin-Germeys I, Birchwood M, Kwapil T. From environment to therapy in psychosis: a real-world momentary assessment approach. Schizophr Bull. 2011;37(2):244–7. https://doi.org/10.1093/schbul/sbq164.

108. Lee Y, Ragguett R-M, Mansur RB, Boutilier JJ, Rosenblat JD, Trevizol A, Brietzke E, Lin K, Pan Z, Subramaniapillai M, Chan TCY, Fus D, Park C, Musial N, Zuckerman H, Chen VC-H, Ho R, Rong C, McIntyre RS. Applications of machine learning algorithms to predict therapeutic outcomes in depression: a meta-analysis and systematic review. J Affect Disord. 2018;241:519–32. https://doi.org/10.1016/j.jad.2018.08.073.

109. Morshed MB, Saha K, De Choudhury M, Abowd GD, Plötz T. Measuring self-esteem with passive sensing. 14th EAI International Conference on Pervasive Computing Technologies for Healthcare (PervasiveHealth 20), Atlanta; 2020.

110. Burke LE, Shiffman S, Music E, Styn MA, Kriska A, Smailagic A, Siewiorek D, Ewing LJ, Chasens E, French B, Mancino J, Mendez D, Strollo P, Rathbun SL. Ecological momentary assessment in behavioral research: addressing technological and human participant challenges. J Med Internet Res. 2017;19(3):e77. https://doi.org/10.2196/jmir.7138.

111. Cohn AM, Hunter-Reel D, Hagman BT, Mitchell J. Promoting behavior change from alcohol use through mobile technology: the future of ecological momentary assessment: promoting behavior change from alcohol use through mobile technology. Alcohol Clin Exp Res. 2011;35(12):2209–15. https://doi.org/10.1111/j.1530-0277.2011.01571.x.

112. Runyan JD, Steinke EG. Virtues, ecological momentary assessment/intervention and smartphone technology. Front Psychol. 2015;6 https://doi.org/10.3389/fpsyg.2015.00481.

Chapter 12
Yoga and Meditation for Self-Empowered Behavior and Quality of Life

Gerlinde Kristahn

> *Do not believe in anything simply because you have heard it.*
> *Do not believe in anything simply because it is spoken and*
> *rumored by many. Do not believe in anything simply because it*
> *is found written in your religious books. Do not believe in*
> *anything merely on the authority of your teachers and elders.*
> *Do not believe in traditions because they have been handed*
> *down for many generations. But after observation and analysis,*
> *when you find that anything agrees with reason and is*
> *conducive to the good and benefit of one and all, then accept it*
> *and live up to it.*[1]
>
> *Buddha*

Introduction

Quality of life (QoL) is defined for the purposes of this chapter along the World Health Organization (WHO) definitions as an individual's physical and psychological health, their social relationships and the environment in which they live [1]. According to the criteria used in the Scales of General Well-Being (SGWB) questionnaire, the main contributing factors to an individual's QoL are well-being criteria such as happiness, vitality, calmness, optimism, involvement, self-awareness, self-acceptance, self-worth, competence, development, purpose, significance, self-congruence and connection [2]. The importance of an individual's spirituality and personal beliefs for their well-being, QoL and health status have been largely

---

[1] *(This citation has been taken from the website Knowing Buddha: https://www.knowingbuddha. org/unique-knowledge)*

The original version of this chapter was revised. The correction to this chapter can be found at https://doi.org/10.1007/978-3-030-94212-0_26

G. Kristahn (✉)
University of Geneva, Geneva, Switzerland
e-mail: Gerlinde.kristahn@unige.ch

confirmed through numerous scientific studies [3, 4]. Personal beliefs are what an individual believes to be the truth, or the beliefs that define their individual world-view [5]. Additionally, individuals have a tendency to believe that others "feel, think and act" as they do [5]. Personal beliefs [6] are crucially linked to individual well-being and the individual's "engagement to explore—and deeply and meaningfully connect one's inner self—to the known world and beyond." [7] This definition focuses on a personal search for meaning and exploration of QoL [7]. The definition provided by WHO the focuses on the "person's personal beliefs and how these affect quality of life".

Besides personal beliefs and their connection to a good QoL and happiness, this chapter also explores possibilities regarding self-empowered behavior change. Self-empowerment describes the way individuals "promote beliefs and attitudes favor-able to deferring immediate reward for more substantial future benefit."[2] [8] We consider behavior mainly as habit, or as an individual's unconscious acts; therefore, producing a change in behavior involves repeated adjustment, as "changing behav-iors is not about changing one act; it is about altering the routines in which the acts are embedded." [9]

Another central component of personal-level QoL is the notion of happiness. Research has shown happiness to be a key aim for individuals [10, 11]. It has been widely shown that happiness plays a crucial role in the search for the meaning of life in religion and spiritual understanding, as well as in one's personal beliefs (see also below). Numerous scientific studies have explored what makes individuals happy and the main factors for experiencing bliss, self-realization and QoL. However, there are important differences between how to be happy and what makes individu-als happy [12]. In this chapter, happiness is defined as a constant positive feeling or subjective attitude towards one's life [10–12] and is viewed on a continuum from happy to unhappy. This definition differs from others that define happiness as spe-cial moments where individuals feel bliss [10].

Happiness, as an essential component of QoL and well-being, does not only depend on the environment in which one lives. Individuals with enormous wealth and comfortable living conditions can be depressed, whereas others living in adverse conditions can manage their life circumstances to achieve a high degree of inner happiness [13–18]. Objective variables (e.g., income or lack of traumatic experi-ences) are far less significant determiners of happiness than intuition and the way one views their life experience [13–17, 19]. Indeed, social pressure and stress can occur when humans primarily consider happiness as a life achievement and

---

[2] To learn more:

"There are four factors which Tones (1993) considers central to the concept of empowered action for the individual: (1) The environmental circumstances which may either facilitate the exercise of control or, conversely, present a barrier to free action. (2) The extent to which individu-als actually possess competencies and skills which enable them to control some aspects of their lives, and perhaps overcome environmental barriers. (3) The extent to which individuals believe themselves to be in control. (4) Various emotional states or traits which typically accompany dif-ferent beliefs about control—such as feelings of helplessness and depression, or feelings of self-worth." p. 1274.

unhappiness as a failure [20–22]. What is less clear, however, is how striving for happiness influences human belief systems [23–27].

In this chapter, we explore the role of thoughts, personal beliefs and emotions in the development of behavioral habits. Considering stress as an example of a factor influencing QoL that is shaped by one's habits, we propose yoga and meditation as practices that can help individuals achieve self-empowerment by altering their thoughts, beliefs, emotions and, ultimately, their behaviors, thus resulting in a reduction of stress and an improvement in QoL. Furthermore, we discuss the potential use of technological devices and other tools for quantifying thoughts, beliefs, emotions and behaviors as a means of further empowering behavioral change.

The chapter is structured as follows. First, we provide a short overview of some relevant concepts such as the influence of stress and emotion on the human body and mind (Section "Stress and Emotions' Influence on Psychophysiological Factors"). Second, adopting an interdisciplinary perspective, we explore how human behavior, and therefore QoL, is influenced by the mind (thoughts), by one's belief in the thought and by the emotions produced by one's thoughts (Section "Thoughts, Emotions, Beliefs and Behaviors"). Next, the chapter offers a discussion that can contribute to readers' understanding of how stress-relieving techniques such as yoga and meditation can improve QoL by changing an individual's thinking, emotions, beliefs and related behaviors (Section "Behavior: Yoga and Meditation as Interventions for Stress Management"). Specifically, yoga and meditation are explored as techniques to (1) observe thoughts (the mind), (2) separate beliefs from thoughts (personal beliefs) and (3) understand the connections that lead from a belief to a feeling (emotion) and (4) from a conscious or unconscious feeling to a behavior (behavior). Based on several case studies, we then suggest new technology-enabled ways of quantifying thought, belief, emotion and behavior that can be used by practitioners of yoga and meditation to assess the impact of their practice on these factors that influence QoL (Section "Quantifying Thoughts, Beliefs, Feelings and Behaviors"). The possibilities of self-empowered behavior change through yoga and meditation, as well as psychophysiological assessment tools, are discussed in section "Yoga and Meditation as Techniques for Achieving Self-Empowered Behavior". Finally, a conclusion to the chapter is provided in section "Concluding Remarks".

Stress and Emotions' Influence on Psychophysiological Factors

Stress can be defined as "the non-specific response of the body to any demand made upon it." [28] Individuals consider all subjective experiences to be either pleasant or unpleasant to some extent. Therefore, an individual's subjective experience can have a very positive or very negative impact on their overall well-being, health and QoL [29]. The mechanisms of stressful thoughts leading to disease [28] are

Table 12.1 Most common categories of emotions [36]

Negative Emotions	Positive Emotions	Emotions without clear Valence Connotation
(a) anger (approach-oriented anger, withdrawal-oriented anger, anger in defense of other, anger in self-defense, indignation) (b) anxiety (dental anxiety, performance anxiety, agitation) (c) disgust (disease-related disgust, food-related disgust) (d) embarrassment (social anxiety, shame, social rejection) (e) fear (threat) (f) sadness (achievement failure, dejection, depression)	(a) affection (love, tenderness, sympathy) (b) amusement (humor, mirth, happiness in response to slapstick comedy) (c) contentment (pleasure, serenity, calmness, peacefulness, relaxation) (d) happiness (except happiness in response to slapstick comedy) (e) joy (elation) (f) anticipatory pleasure (appetite, sexual arousal) (g) pride (h) relief (safety)	(a) surprise (wonder) (b) suspense

becoming an increasingly important field of scientific research. Studies have shown that an individual's attitude towards stress plays a crucial role in their health and well-being. However, the absence of stress in one's life can have an equally negative influence on one's health as can an excess of stress, since only an adequate balance between the two generally leads to greater well-being [30].

Stress itself is not an emotion, although it is often created by certain feelings [31–33]. Emotions "typically unfold dynamically" [32] and "have a beginning and an end," [32] whereas feelings are present for a longer period of time [32]. Humans usually distinguish positive and negative feelings as those that create emotions and impact the "fluidity or strain of life process(es)." [34] Often, feelings are considered to lead to negative emotions when they throw life out of balance and cause struggle. Feelings are regarded as producing positive emotions when life is harmonious, efficient and flowing [34]. Familiar patterns often allow life to continue smoothly, whereas changing patterns create an unfamiliar situation that "generates feelings and emotions." [35] The emotions that have been evaluated most often in the psychological studies are presented in Table 12.1 [36], with each one categorized according to its valence (positive, negative or unassigned).[3] [37]

Although there are numerous studies that have demonstrated the difficulties involved in relating emotions to psychophysiological reactions in the human body

---

[3] Emotional valence (Definition).

"The concept of emotional valence, or simply valence, has been used among emotion researchers to refer to the positive and negative character of emotion or some of its aspects, including elicitors (events, objects), subjective experiences (feeling, affect) and expressive behaviors (facial, bodily, verbal). The valence of feelings has been argued to be a pivotal criterion for demarcating emotion and a core dimension of the subjective experience of moods and emotions."

[36], there are examples from the literature that provide persuasive evidence regarding particular correspondences between physiology and emotional states. In particular, the studies in the latter group indicate that, on the physical level, the body reacts to any situation in its environment that creates psychological flow or strain. For example, if one part of the body is changed or influenced by a situation, as in the case where one consciously slows down one's breath, the other body parts (such as one's heart rate) adapt to the new situation. This reaction is indicative of "synchronized interactions among multiple systems," [38] other examples of which, relating to heart rate, are shown in Fig. 12.1 and explained further below.

Overall, common repetitive behaviors, such as inactivity, poor nutritional habits and alcohol and tobacco consumption, may affect the way an individual manages daily life stress [40] and can determine up to 40% of the individual's future health state and life quality [41]. Meanwhile, there are various psychophysiological factors that can be signs of stress (e.g., galvanic skin response, respiration, blood pressure and heart rate) [36]. In this section, we have chosen to discuss heart rate as a main indicator of stress, as it has been demonstrated that one's heart rate responds directly to momentary emotional states [42]. While an abnormally low 24-hour heart rate can indicate increased risk of heart disease and premature mortality, it is also a strong marker of stress and emotions [43], as is shown in Fig. 12.1, and thus demonstrates the "synchronized interactions between different systems in the body" [44] alluded to above. It can be observed that heart rate patterns in particular—that is, the patterns of the curves rather than specific heart rate values—often directly reflect and synchronize with an individuals' emotional states [44].

Furthermore, the analysis of heart rate and heart rate pattern provides an indicator of neurocardiac fitness and autonomic nervous system function [42]. Some studies have even found that heart rate pattern "covaried with emotions in real time." [44] As heart rate patterns often directly reflect emotional states, various techniques to reduce stress, such as meditation techniques, produce specific heart rate patterns;

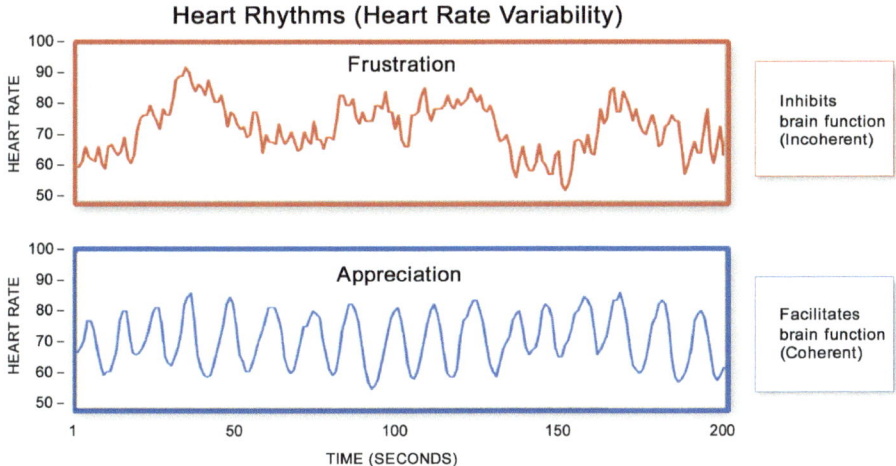

Fig. 12.1 Example of Emotions reflected in Heart Rhythm Patterns [39]

in particular, such techniques have been found to produce a coherent, fine, smooth, "wave-like-pattern in the heart rhythms." [45] Scientists have also found that a person practicing a technique to maintain positive emotional feelings displays heart rhythms that are synchronized with their brainwaves influenced by their emotional states (see Figs. 12.2 and 12.3). In comparison with individuals who experience a high degree of stress, those who apply such techniques also exhibit more regular and stable heart rate patterns.

Recent research has revealed that the body is also able to develop coherence or synchronization with the physiological processes of others. For instance, a mother thinking about her in utero baby can lead to the synchronization of her brainwaves with her baby's heartbeat [48]. It has also been shown that couples that have lived together for many years can have coherent heart rate patterns while they sleep [49].

To summarize this section, we observe that there is a synchronization between brain and cardiac activities [50, 51]. There is particular evidence of a relation

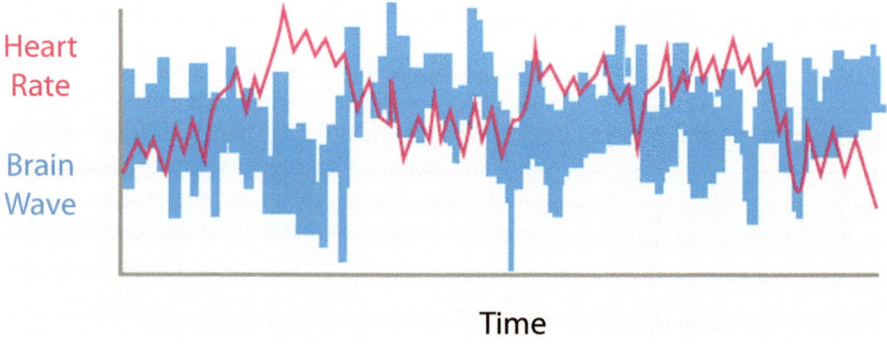

Fig. 12.2 Image of Dissonant Heart and Brainwave Rhythm [46]

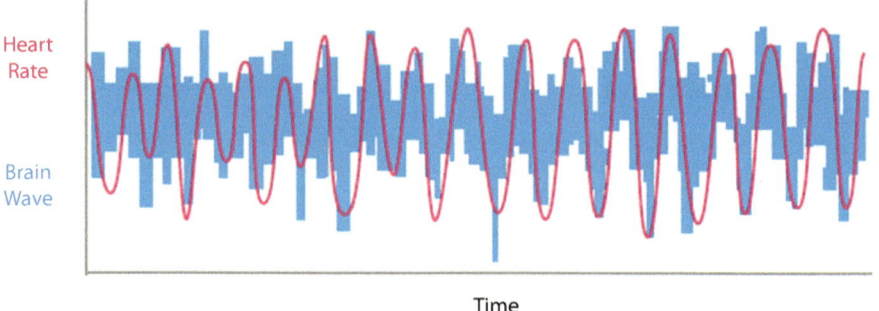

Fig. 12.3 Image of Heart Rhythm and Brainwaves in sync during a Meditative State [47]

between heart rate patterns and positive and negative emotions [42]. Additionally, body stress regulation techniques enable a more regular and stable heart rate pattern.[4]

Thoughts, Beliefs, Emotions, and Behaviors

Behavior can be defined as "the internally coordinated responses (actions or inactions) of whole living organisms (individuals or groups) to internal and/or external stimuli." [54] As indicated in the conceptual model presented in Fig. 12.4, behavior is influenced by (a) thoughts, (b) belief in thoughts and (c) feelings affected by beliefs and (d) decisions motivated by feelings [55]. Behavior can be influenced by "individual-level attributes as well as by the conditions under which people live." [62] Thus, human behavior depends on the context in which it takes place [63]. Additionally, some studies state that as much as 99.9% of the decisions made by humans are shaped by other people [64]. Bargh provides further evidence of environmental influences on human behavior, stating that "our psychological reactions from moment to moment... are 99.44% automatic." [65] Therefore, in order to better understand how an individual makes decisions, one needs to observe them within their natural environment [66] and consider contextual factors, as is indicated in Fig. 12.4. This principle must be applied when one is assessing the influence of human thinking on individual belief systems and emotions, leading to behaviors.

As mentioned above, this chapter considers behavior in relation to stress-management. In section "Stress and Emotions' Influence on Psychophysiological Factors", it was seen that emotions and feelings represent an important factor in stress management. The model presented in Fig. 12.4[5] conceptualizes the steps by which an individual adopts a behavior that can either increase or reduce stress, beginning with a change in one's thoughts, beliefs or feelings.

---

[4] Recent studies found out that individuals can influence and measure these impacts on emotions, for example psychophysiological functions like the heart rate. According to the HeartMath-Institute, "These positive emotion-focused techniques help individuals learn to self-generate and sustain a beneficial functional mode known as psychophysiological coherence, characterized by increased emotional stability and by increased synchronization and harmony in the functioning of physiological systems." [43]. This tool helps restructuring emotions, and takes usually 5–15 min.
 For in-depth descriptions of these techniques, see [52, 53].

[5] Disclaimer: the illustration visualizes what has been described hereafter, but is not yet empirically validated. Individual & Contextual Factors of Behavior (own graph based on empirical confirmed findings from): [55–61].

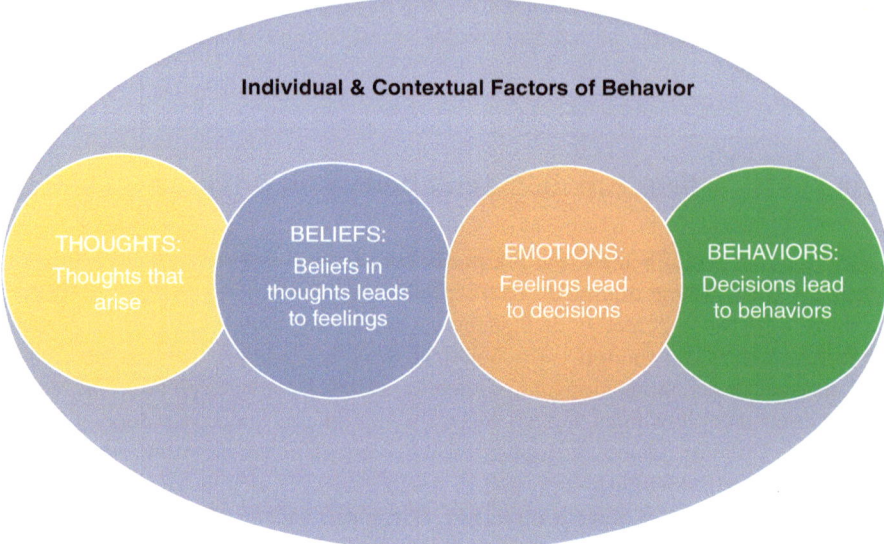

Fig. 12.4 Individual and contextual factors of behavior

The Influence of the Mind on Human Behavior

Corresponding to the first circle (Thought) in Fig. 12.4, thoughts are related to "one's own individual experience." [67] Most psychological research on decision-making over the past 30 years has focused on the logic of rational, controlled decisions (i.e., the logic behind behaviors). While logical mechanisms may describe the decision-making processes of scientific pursuits, their analysis is less appropriate for understanding the uncontrolled aspects of humans' ordinary changes in behavior, which are based on both thoughts and feelings [56]. According to Baumeister et al., "it is plausible that almost every human behavior comes from a mixture of conscious and unconscious processing." [60] This also means that a description of one's conscious thoughts alone (or unconscious thoughts alone) is rarely sufficient to explain a specific behavior.

The mind, defined for the purposes of this chapter as an individual's collected thoughts and cognition, is one of the great influencers of behavior [68]. According to scientists such as Anderson, if someone envisions an action, it leads to an "increased intention to do it" [69] and therefore a greater chance that the action will be achieved. According to Tiller, meanwhile, individuals can influence their future with their thoughts in the form of intentions [57]. The systems of thinking, or personal beliefs, that affect one's intentions begin to be developed in one's youth through imitation of parents, relatives, teachers and figures in media, among others. The information one gains in the process serves to "structure that person's understood world and purposive ways of coping in it." [70] When an individual is confronted with a choice, they unconsciously call upon all the past experiences that have been collected and registered in their brain, recalling them as thoughts. This process has been widely studied in psychology as the "theory of planned behavior." [71, 72]

The effect of one's state of mind, including one's thoughts, on the outcome of a task has been demonstrated in a study by Armor and Taylor. In this study, the authors assessed how implemental and deliberative mindsets—the former characterized by plans and thoughts regarding how to implement an action, and the latter by questions concerning the merits of the action—respectively influenced one's performance of a task. As the authors observed, "it appears that the simple difference in perspective that mindsets create—from the uncertain query 'Will I do X?' to the agentic assertion 'I will do X'—can yield substantial differences in how the considered action will be perceived and, subsequently, acted on." [73] The authors concluded that the expected results of a task, as well as one's predictions regarding one's own performance, "become less favorable following deliberation and more favorable following thoughts of implementation." [73, 74]

Additionally, results from various studies have indicated that individuals adopting the perspective of another, as opposed to simply discussing a situation from their own perspective, can be a successful strategy for producing behavioral change [75–77].

In summary, the mind significantly influences behavior across multiple dimensions. As is seen above, the mind retains past experiences that may be triggered by external impulses. These experiences consciously and unconsciously influence the ways individuals think, shaping their beliefs, emotions and actions. In the next subsection, we will explore how beliefs (and unbelieves) can influence human behavior.

Influence of Personal Beliefs on Human Behavior

The second circle of Fig. 12.4, corresponding to beliefs, represents a step in the process leading to behavior in which an individual evaluates whether they believe a thought or not. A belief is not necessarily true; it is simply what an individual believes and claims to be true or false or right or wrong. It is also important to note that not all an individuals' beliefs are derived from their own experience or thinking. As Barth et al. observes, "we all extend the reach and scope of our knowledge immensely, relying on judgements based on whatever criteria of validity we embrace—above all, what others whom we trust tell us they believe." [59]

What an individual hears and considers to be true will have a profound and long-lasting influence on their behavior. As Geraerts et al. have demonstrated, even "the false suggestion of a childhood event can lead to persistent false beliefs that have lasting behavioral consequences." [58] To illustrate this fact, the authors conducted a study where they informed subjects that they had become ill during childhood after eating egg salad. During the period immediately after receiving this false information, and even four months later, these subjects tended to avoid eating egg salad, whereas the control group continued to eat it as usual [58].

The personal belief system that leads to such behaviors is useful, and in some cases it can serve to protect one from harm or save one's life. This is especially important in dangerous situations, where one needs to quickly judge and react, often as part of a fight or flight response [78]. However, making quick judgements of this sort in a broader context may become problematic when an individual does not question their thoughts and beliefs and assumes that their thinking always captures the truth. These individuals might cling to their beliefs because of the familiar feeling that one experiences when the mind recognizes a situation [59]. In such cases, these feelings unconsciously lead the individual to make decisions based on their previously stored knowledge or beliefs, or "what a person employs to interpret and act on the world." [59] However, if such an individual explores the ways they can alter their personal belief system and its impact on their behavior [79], they can ultimately achieve better QoL, especially in areas—such as physical activity, nutrition and sleep—that contribute to their health later in life [80–83].

An important example of personal beliefs includes stereotypes about the differences between different cultural, ethnic and religious groups. Such beliefs, which derive from one's one cultural, religious or social context, are often challenged when individuals from stereotyped groups behave in ways that are unexpected. The practice of sharing beliefs between members of different cultural, ethnic and religious groups can lead to confidence-creating emotions which enable individuals to easily create relationships, as well as a willingness to cooperate with others [84, 85].

Spirituality, as a cultural concept, also significantly influences an individual's beliefs [86]. Scientific research has attempted to measure spiritual well-being using scales, the most common of which are the Spirituality Index of Well-Being, the Spirituality Scale, and the Daily Spiritual Experience Scale.[6] These scales assess spirituality using elements which ask the individual about their beliefs.[7]

Similar to spirituality, relaxation methods aim to help individuals manage stress through practices that affect one's personal beliefs. Many proposed methods for reaching a stress-free mental state include guidelines that instruct readers in changing their behavior. These guidelines are often meant to encourage hopeful, optimistic beliefs and feelings of inner peace [87]. Some methods recommend practices such

---

[6]The Spirituality Index of Well-being (SIWB) states that there is currently no instrument that can measure spirituality in its complexity, but it can deliver "health-related quality of life measures" in a qualitative form. The Spirituality Scale (SS) is a holistic approach measuring beliefs, intuitions, life choices and rituals. It identifies mainly three factors of spirituality: Self-Discovery, Relationships and Eco-Awareness. The Daily Spiritual Experience Scale (DSES) addresses spiritual experiences like awe, joy and a sense of deep inner peace. With this scale, emotions, cognition and behavior may be identified. According to the DSES, daily spiritual experience is related to improved Quality of Life and positive psychosocial status. As stated in the SIWB study, "Life scheme is similar to the construct of sense of coherence, which was described by Antonovsky as a positive, pervasive way of viewing the world, and one's life in it, lending elements of comprehensibility, manageability, and meaningfulness."

[7]For example: "I accept others even if they do things I think are wrong" (Spirituality Scale (SS))

as sitting down every morning to meditate [88] or drinking a cup of lukewarm water before eating [89]. While adopting these behaviors may be effective for some individuals, not everyone will be able to attain from them the change in stress-reduction or health-improvement that they desire. Various approaches and techniques of teaching have therefore been developed, allowing each individual to determine what works best for them and what is most appropriate in specific situations [90].

In the health context, meanwhile, it has been demonstrated that healing beliefs can play an important role in health outcomes. Healing beliefs are a conscious and mindful determination one makes to improve one's health, or an expectation that one's health and well-being are going to improve and that one will live a meaningful, productive life. Healing beliefs are a way of understanding and giving meaning to illness and suffering through the continued belief that the patient will achieve healing and well-being [91].

Additional studies in this area have found that health expectancy (one's personal belief regarding future health outcomes) is an important part of healing processes, especially when there is a close relationship between the healer and the patient [92]. Studies have demonstrated that a healer's capacity to self-heal tends to have a positive influence on a patient's health outcomes. As Schmidt notes, "practicing loving kindness and compassion toward oneself helps develop these qualities in our relationship with others." [93] The self-healing capacity may be observed by the patient unconsciously, stimulating the patient to use the same skills to heal themselves.

In conclusion, there are many studies that demonstrate the impact of personal beliefs on individual behavior. More important than the act of changing one's mind (one's thoughts) in order to bring about behavioral change is changing one's belief of this thought. This occurs on both a conscious and an unconscious level. The next subsection explores the influence of emotions on behavior, especially of those emotions that arise from beliefs.

The Influence of Emotions on Human Behavior

As is shown in Fig. 12.4, believing a thought can create certain feelings in individuals that ultimately influence behavior. Feelings and emotions are interdependent. As mentioned in section "Stress and Emotions' Influence on Psychophysiological Factors", feelings produce emotions and last for a longer time than emotions, which always have a beginning and an end. Emotions influence behavior because an individual is more likely to do something if it is pleasurable and not stressful or if the outcome feels positive [94]. On the other hand, research shows that "anticipated emotion, especially anticipated regret, has been shown to motivate people and change behavior." [60] Both examples illustrate how varied emotions, from positive emotions such as love to negative emotions such as fear, affect human behavior and supply important impulses for actions.

A wide range of theoretical models of behavior have been proposed that demonstrate how behaviors leading to long-term improvements in health and QoL, such as

Fig. 12.5 The effects of induced mood on the frequency of attentional lapses [96]

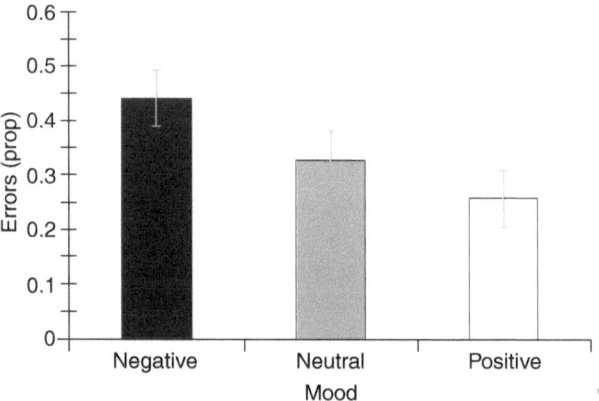

exercise and nutritional practices, can be implemented and adopted [95]. According to these models, individuals have the ability to adopt an alternative perspective or state of mind by altering their thoughts and personal beliefs, ultimately influencing their own emotions and leading to new behaviors. According to a recent study,[8] mood is a factor that has a major influence on how effectively individuals are able to improve their behavior. Moods, which are defined here as states or frames of mind, can last for a brief moment or be present during longer periods [32, 97]. Individuals in a negative mood have been found to exhibit less attention and concentration on tasks than those in a positive frame of mind and to ultimately deliver worse results. Individuals in a negative mood have also exhibited particular difficulty adapting to tasks again after being distracted by their own thoughts. These results demonstrate that those in a negative mood commit more behavioral lapses when engaged in a task, have "task-irrelevant thoughts" with more frequency and exhibit less motivation towards attentional commitments than those in a positive mood [96]. Meanwhile, a positive mood has been found to help individuals adapt to a task after a lapse [97].

A similar study concerning the impact of emotions on behavior has confirmed that emotions have a major influence in this regard (Fig. 12.5) [98]. According to Avey et al., individuals in a positive mood were not only better able to control their own behavior, but were also able to spread positivity to those around them. Their results demonstrate that one individual's positive mood can even influence f.ex. employees in an organizational structure. More specifically, the individual's mood was found to influence their own psychological capital, especially in the areas of hope, efficacy, optimism, resilience and mindfulness, by facilitating relationships and positive emotions, which led in turn to changes in the behavior of the surrounding individuals [99].

---

[8] Fifty-nine undergraduate students (25 males) participated in this experiment for course credit. The mean age of the sample was 21.7 (SD 2.0; range 18–25) years. Individuals were allocated to positive (n 20), negative (n 20), and neutral (n 19) mood induction conditions using a counterbalanced design.

From the study: [96].

When describing the impact of emotions on behavior, it is important to also consider how the way one is received or treated in one's surroundings affects emotional responses and, ultimately, behavior. Positive and negative feedback from one's environment, for instance, have been found to affect an individual's emotions in ways that influence behaviors such as "counterproductive behavior, turnover intentions, citizenship, and affective commitment." [100]

In summary, there is significant evidence regarding the influence of emotions on an individual's behavior. This influence can be seen in the effects of mood on task performance, the impact of positive and negative feedback on behavioral change and the capacity of individuals to spread positivity and alter the behaviors of others, examples that demonstrate the crucial role emotions play in behavior choice and change.

Behavior: Yoga and Meditation as Interventions for Stress Management

Given the above discussions concerning the influence of the mind, belief of thoughts and emotions on behavior, this section presents behavior techniques, particularly those of yoga and meditation, that have been found in research to help reduce stress through changes to an individual's thoughts, beliefs and emotions that lead to behavioral change [101]. Nowadays, yoga and meditation are practiced worldwide to reduce stress and achieve life satisfaction [102–104]. Yoga and meditation techniques are claimed to be especially effective at helping practitioners (1) observe their thoughts (thought), (2) disconnect beliefs from thoughts (belief), (3) understand the connection between personal beliefs and feelings (emotion) and (4) recognize how conscious or unconscious feelings lead to a behavior (behavior).[9]

In this chapter, yoga is defined as "an ancient practice with origins in India, […] incorporat[ing] muscle stretching, muscle relaxing, balancing poses (asanas), breathing (pranayama), guided relaxation, and/or meditation (dhyana) to achieve harmony and spiritual growth." [105, 106] It is a culturally based tool that can become an integrated habit in one's life [107]. A key aspect of yoga and meditation is a concept of embodiment that is focused on self-creation, as "control of the self is a project internal to the person rather than imposed by others, and modern persons actively define themselves through lifestyle choices, rather than being passively defined by their membership in traditional groups and moral orders." [108]

Research on the mindful body, which is an approach to the body that treats mental and physiological processes as aspects of an integrated whole, has revealed significant insight into the effect of emotions, passions and conversational practices on behavioral outcomes. As yoga and meditation seek to alter behavior through

---

[9] Individual & Contextual Factors of Behavior (own combination based on empirical confirmed findings from): [55–61].

deliberate changes to one's emotions, such findings are highly relevant to the analysis of yoga presented in this chapter [109, 110]. Meanwhile, yoga psychology is a field of research and practice focusing on the use of yoga as a psychological tool that integrates behavioral and introspective approaches to facilitate an individual's growth [111]. Several studies have revealed benefits of practicing yoga for a certain period of time. On a psychological level, yoga has been shown to improve emotional functioning [112] and reduce negative emotions in breast cancer patients [113]. In addition, studies of embodied cognition have demonstrated that yoga "may reduce stress by affecting the way individuals appraise stressors." [114] One study has also found that regular practitioners of yoga react to negative emotional stimuli in such a way that it "does not have downstream effects on [their] later mood state." [115]

On a physical level, yoga reduces stress by stimulating exchange and harmonization between the body and the mind. One of the parts of the human anatomy most responsible for communication between body parts is the nervous system, particularly the vagus nerve, which corresponds to the central energy channel called "*Sushumna Nadi*" in yoga [116]. According to experts, the vagus nerve communicates information mainly from the heart, gut and other organs to the brain [117]. This nerve is stimulated in all yoga postures, or *asanas* [118, 119]. Yoga practices are based on a distinction between the two sides of the body: the left, called "*Ida*," which is feminine and lunar, and the right, called "*Pingala*," which is masculine and solar. The aim of yoga is to unite both sides within the Sushumna Nadi or vagus nerve. When the body has been trained to slow its processes using the breath (e.g., via yoga), the heart rate automatically slows down when the individual is at rest. Neurological signals are transmitted from the brain to the heart and other organs via the vagus nerve without any change in the breath being necessary [120].

As is defined above, meditation is part of yoga practice. Meditation is defined as "practices that self-regulate the body and mind, thereby affecting mental events by engaging a specific attentional set." [121] Meditation, like yoga, is considered a health-improving tool, and has been linked to numerous health benefits, such as reductions in pain sensation, short-term increases in blood pressure and improvements to respiratory efficiency, fluid exchange, cardiovascular system capacity and synchronization of the cells [122]. In a study of the efficacy of meditation techniques for treating medical illnesses, Arias et al. reported that "the strongest evidence for efficacy was found for epilepsy, symptoms of the premenstrual syndrome and menopausal symptoms. Benefit was also demonstrated for mood and anxiety disorders, autoimmune illness, and emotional disturbance in neoplastic disease." [123] As Ospina et al. have observed in their survey of meditation research, the effects of meditation include physiological changes such as the reduction of the heart rate, blood pressure and cholesterol. On the neurophysiological side, meditation was also found to increase verbal creativity [124].

The calm mental states reached by practicing mindful meditation have been shown to impact the brain and emotions. As Hayes and Davis note, "There is evidence that mindfulness helps develop effective emotion regulation in the brain." [125–128] The calm mental states associated with meditation can facilitate coherence or synchronization between bodily functions.[10] Coherence tends to produce a sense of peace of mind and emotional balance that increases overall well-being [122] and QoL. The coherence of different psychophysiological functions can be developed through training such that the body, without specific stimulation, later applies these trained techniques unconsciously [129]. A meditation practice that has provided particularly effective for the achievement of calm mental states is Transcendental Meditation. Research on this method has focused especially on its relationship to stress reduction, decreased depression, reduced anxiety and improved sleep quality [130–132]. Other studies, however, have shown that more desirable psychological states can be achieved with the practice of yoga paired with meditation, especially in cases of burn-out syndrome, depression and mental disorders [133, 134].

The Center for Healthy Minds, led by Professor Richard Davidson at the University of Wisconsin–Madison,[11] has stated, after many years of research on the topic, that well-being is a set of skills that can be learned in a way similar to playing the guitar. According to the results of their research, regular meditation can rewire areas of the brain [135] so that, although negative experiences from daily life are recognized, they affect individuals' brain health, thoughts, emotions and actions for a much shorter time than they do for individuals who are not trained to meditate.[12]

As indicated in the previous section, several studies [136] have demonstrated that positive emotions can positively influence human behavior, health and overall well-being [45]. A key factor in this process is the cultivation of coherence between emotional and bodily states. If an individual regularly practices a meditation technique that draws attention to positive emotions (such as Loving-Kindness Meditation [137]) or other techniques that are viewed as stimulating positive emotions (such as progressive muscle relaxation [136]), the body system establishes a habit. Research results indicate [138–140] that past experience "builds within us a set of familiar patterns, which are instantiated in the neural architecture." [141] These patterns manifest as habits that are automatically enacted, even when the person is not actively practicing any meditation technique. External stimuli are then managed by the brain as perceptions, which automatically give rise to habitual feelings and behaviors [141]. According to one study, "the system then strives to maintain [positive emotional habits] automatically, thus rendering coherence a more readily accessible state during day-to-day activities, and even in the midst of stressful or challenging situations." [122] In frequent meditators, stress responses are often automatically reduced the moment stress is experienced [142]. Interestingly, not every meditation technique leads to coherence between different body functions. Techniques that concentrate on positive emotions, for example, have been found to be more successful at facilitating coherent states than those that focus on the

---

[10] Coherence is defined as the "*synchronization or entrainment between multiple waveforms. A constructive waveform produced by two or more waves that are phase- or frequency-locked*".
    Resource: https://www.heartmath.org/research/science-of-the-heart/coherence/

[11] https://centerhealthyminds.org/

[12] https://centerhealthyminds.org/science/overview (Accessed 02.06.2020)

mind and thoughts (such as concentrative meditation[13]) [122]. Therefore, one may assume that integrating the emotions into one's meditation practices, rather than merely concentrating the mind on a given object, can result in more effective physiological and psychological functioning.

The approach to changing behavior through the cultivation of positive emotions is also considered in the understanding of impacts of yoga and meditation. According to the self-development perspective on behavioral change, the only way to change one's behavior is from within oneself [143, 144]. Individuals may need help from outside sources, such as other humans or their external conditions, but managing one's behavioral changes is the responsibility of each individual. This is also confirmed by the results of research on intrinsic behavior change and motivation. Self-determined motivation in particular, an attitude which yoga training aims to cultivate and draw on, has been found to have a direct effect on behavior [145].

Both yoga and meditation are promising tools that can influence well-being and QoL by changing one's relation to one's thoughts, personal beliefs, feelings and emotions and behavior (as conceptualized in Fig. 12.4). Results from a variety of studies have confirmed these benefits; nevertheless, the exact mechanism by which meditation affects QoL is still unclear. It is known that meditation influences cognitive processes; however, there are other factors that also influence well-being [146], such as "cultural differences, socioeconomic status, health, the quality of interpersonal relations, and specific psychological processes." [146]

In summary, the current state of research indicates that yoga and meditation can be helpful techniques for individuals to observe their own thoughts and make sense of the beliefs their thoughts give rise to. Furthermore, yoga and meditation affect the emotions and one's resulting behavior in a positive way by reducing the stress caused by an individual's environment.

Quantifying Thoughts, Beliefs, Feelings and Behaviors[14]

When yoga and meditation are practiced as techniques of self-empowerment, as will be discussed in section "Yoga and Meditation as Techniques for Achieving Self-Empowered Behavior", it can be beneficial for the practitioner to assess changes to their thoughts, beliefs, emotions and behaviors through reliable means. This section therefore discusses how each of these variables can be quantified leveraging recent developments in personal, miniaturized technology such as smartphones and wearables to evaluate and ultimately improve the effectiveness of one's practice. The table (Table 12.2) presents methods of quantifying these concepts

---

[13] Concentrative meditation is the term sometimes used for a type of meditation in which the mind is focused entirely on one thought, object, sound or entity. The intention is to maintain the single-pointed concentration for the duration of the meditation. For an object of concentration, the yogi may use a sound or mantra, their breath or a physical object such as an icon or candle. https://www.yogapedia.com/definition/10427/concentrative-meditation

[14] This section has been mainly written by Prof. Katarzyna Wac.

Table 12.2 Examples of methods for the quantification of thoughts, beliefs, emotions and behaviors via technologies

Data source	Dimension	Mind (thoughts)	Beliefs (personal beliefs)	Emotions	Behaviors
Self-reported	Patient Reported Outcomes (PRO) and/or Ecological Momentary Assessments (EMA)	Overall thoughts: PRO Momentary thoughts: EMA	Overall beliefs: PRO Momentary self-beliefs: EMA	Overall emotional states: PRO Momentary emotional states: EMA	Overall behaviors: PRO Momentary behaviors: EMA
Other/ partner-reported	(ObsRO, PeerMA)	Overall perceived thoughts: ObsRO Momentary perceived thoughts: PeerMA	Overall perceived beliefs: ObsRO Momentary perceived beliefs: PeerMA	Overall perceived emotional states: ObsRO Momentary perceived emotional states: PeerMA	Overall perceived behaviors: ObsRO Momentary perceived behaviors: PeerMA
Technology reported Outcome (TechRO)	Wearable (Psycho-physiology, Movement)	Galvanic Skin Response (GSR) Heart Rate (HR) (note: may be impossible to disentangle from thoughts)	GSR HR (note: may be impossible to disentangle from thoughts)	Body or body-part movements Body postureGSR, HR Body temperature Respiratory rate Voice analysis Face analytics Posture analysis	Body/parts of body movements Body posture GSR HR Respiratory rate Voice analysis
	Smartphone Behavioral Metrics	Analytics of search engine use Social net activity voice (analysis of content, e.g., asking questions)	Analytics of search engine use Social net activity voice (analysis of content, e.g., asking questions)	Context (where, with whom) Interaction with context Interaction with phone	Context (where, with whom) Interaction with context Interaction with phone

using various data sources that are categorized in the manner defined by Mayo et al. in the context of patient care [147]. In this case, the tools and methods are adapted for the use of any individual in or outside of care settings. The first data source that is considered is self-reported data that the individual provides on their own. This includes information on both internal, unobservable states and external, observable states that is collected via patient-reported outcome (PRO) surveys for a given recall period (e.g., the last month) and in shorter ecological momentary assessments (EMAs) [148]. The second data source includes data reported by other individuals, such as a friend, family member or caregiver, on the individual in question's perceived internal and observable states. This information is collected in the form of observer-reported outcome (ObsRO) surveys and/or peer ecological momentary assessments (PeerMAs) [149], the content of which may differ from the PRO surveys and EMAs depending on what is being reported upon (e.g., internal, unobservable states or external, observable states). The third data source includes technology-reported outcomes (TechRO) datasets collected via wearable devices that record aspects of the individual's psychophysiology and their movements and via smartphones that record information concerning the individual's context and their interaction with their context.

At present, the measurement of thoughts, beliefs, emotions and behaviors using the tools mentioned in Table 12.2, especially the technology-reported datasets, is only a hypothetical proposal. Concrete experimentation is still needed to ensure the accuracy, timeliness and reproducibility of the assessments of these variables [36] and the overall quality of the data collected. That said, these methods of quantification indicate a possible set of tools practitioners of meditation, yoga and other behavioral techniques might use to measure the effectiveness of their practices on their behavior and the variables—namely thoughts, beliefs and emotions—that ultimately shape behavior.

Yoga and Meditation as Techniques for Achieving Self-Empowered Behavior

Based on the research presented in this chapter up to this point, this section assesses yoga and meditation as techniques of individual empowerment through which one might change one's health behaviors to live a happier life. In particular, this section focuses on the use of yoga and meditation techniques for stress management as a means of enacting behavioral change. Self-empowered changing of one's health behaviors also implies an attempt to change one's perspective towards health and illness. Such a change might involve adopting the view held by Antonovsky, who stated, "We are coming to understand health not as the absence of disease, but rather as the process by which individuals maintain their sense of coherence (i.e., sense that life is comprehensible, manageable, and meaningful) and ability to function in the face of changes in themselves and their relationships with their environment." [150] A similar definition of health that one might adopt is "the ability to adapt and self-manage in the face of social, physical, and emotional challenges." [151]

Yoga and meditation can improve one's ability to generate thoughts that are capable of leading to positive emotions. Although cognitive methods by themselves can contribute to stress management [152], individuals can achieve better results when they integrate emotions into their methods. Recent studies indicate, for instance, that individuals adapt their behaviors more easily when they integrate emotions into behavioral change techniques [153]. One reason for this is that positive feelings towards a new situation can facilitate physiological and psychological acceptance of a change and of the new situation that results. Overall, it has been shown that when emotions are a focal point of behavioral change practices, individuals are more likely to successfully change their behavior and better able to improve their overall QoL. When practiced regularly, yoga and meditation can help induce unconscious shifts towards stress-reducing behavior and facilitate synchronization (coherence states) of a body's heart rate and brainwaves.

Yoga and meditation can train one to develop an ability to change perspectives by being aware of one's personal beliefs, which allows individuals to make choices regarding how they progress towards change. When individuals are able to change perspectives and show competence in empathy, they are ultimately more likely to change their behavior. For example, it has been demonstrated that an individual who wants to start doing more exercise to improve their QoL, better understands their negative emotions during moments when they are not exercising.

Yoga and meditation can empower individuals to change their perspective towards situations that are considered stressful. Some studies have demonstrated that one's attitude towards stress plays a crucial role in the effort to reduce stress. Those who consider stress as something positive are more likely to manage it successfully than those who view it negatively.

Self-observation with meditation and yoga helps to raise awareness and activate unconscious processes; raised awareness is a result of logging emotional states [154]. Meanwhile, developing an emotion-focused understanding of stress allows one to better understand what is happening when one is dealing with undesired behaviors. Yoga and meditation increase awareness and understanding at critical behavioral moments. Meditation in particular also helps individuals to enter a state in which they are less focused on thoughts so that they can approach their emotions as an observer.

Individuals who wish to empower themselves may utilize technology for the self-assessment of their states and behaviors. As discussed in the previous section, miniaturized devices such as smartphones and wearables can be used to gather relevant quantitative data. The result of using these tools is what is called, in the study of technology, the quantified self [155], a term that equally describes the practitioner of yoga and meditation. There exists a myriad of personal digital devices [156] that can automatically gather information regarding physiological processes and hard-to-observe behaviors [156]. The use of these devices and the data they provide may motivate individuals to better understand the way certain thoughts, beliefs or feelings lead to specific behaviors. Researchers have found that individuals find it easier to internalize such knowledge when using technology and, consequently, are able to realize behavioral changes almost instantly [154]. In general, the idea that technology can enable the assessment of individual behaviors and health and produce improvements in life quality has been expressed before [157]. The next step in the process is to leverage these technologies to potentially focus on the assessment and improvement of the individual's internal states—particularly their thoughts, beliefs and emotions.

Concluding Remarks

In this chapter, we have examined the role of personal beliefs, together with thoughts and emotions, in the development of behavioral habits. Considering stress as a reaction that is influenced by behavior, we discussed yoga and meditation practices as possible interventions to alter personal beliefs and influence behaviors and thereby reduce stress, leading ultimately to an improvement in QoL. In addition, we suggested the use of technological devices and other tools for quantifying thoughts, beliefs, emotions and behaviors as a means of further empowering behavioral change.

Many benefits of yoga and meditation have been uncovered in recent research, including their effect on stress reduction. It has been specifically shown that managing stress and adapting one's behavior to live a more meaningful and fulfilled, healthier and happier life, depends very much on each individual's thoughts, their personal beliefs and related emotions. Yoga and meditation, as techniques in which individuals seek to understand their perceptions and reactions towards stress and the outer world, are promising, learnable skills that can lead towards self-empowered behavioral change. Once the body has learned these skills, it reacts (i.e., thinks, believes, feels and behaves) unconsciously and without effort, even when the individual is not consciously paying attention. Studies have shown that, in such cases, behavioral change can then occur unconsciously.

Scientific research is now beginning to map the bioenergetic communication systems between the "inside" and "outside" of the body (including physiological functions, cognitive processes, emotions and behavior) that influence one's health, happiness and QoL in the long term [158]. More research is expected to be carried out in the future concerning this important, interesting and impactful topic.

Acknowledgements I want to thank Prof. Katarzyna Wac for her input in this chapter regarding technologies and technical tools to assess QoL-related behavior (especially in section "Quantifying Thoughts, Beliefs, Feelings and Behaviors") and, along with Sharon Wulfovich, for providing overall feedback. I also want to thank my yoga teachers, especially Sanjeev Bhanot, Shiva Rea and Anne Macnabb, for their inspiring input and ideas influencing this book chapter. And finally, I would like to thank Prof. Hanna Kienzler, Felix Kirchmeier, Prof. Magali Dubosson and Jane Sabherwal for their helpful feedback.

References

1. Whoqol Group. The World Health Organization quality of life assessment (WHOQOL): position paper from the World Health Organization. Soc Sci Med. 1995;41(10):1403–9.
2. Longo Y, Coyne I, Joseph S. The scales of general well-being (SGWB). Person Individual Diff. 2017;109:148–59.
3. Moreira-Almeida A, Koenig HG. Retaining the meaning of the words religiousness and spirituality: a commentary on the WHOQOL SRPB group's "A cross-cultural study of spirituality, religion, and personal beliefs as components of quality of life"(62: 6, 2005, 1486–1497). Soc Sci Med. 2006;63(4):843–5.
4. Koenig HG, McCullough M, Larson DB. Handbook of religion and health: a century of research reviewed. New York: Oxford University Press; 2001.

5. Krueger J. Personal beliefs and cultural stereotypes about racial characteristics. J Personal Soc Psychol. 1996;71(3):536–48.
6. Peterson M, Webb D. Religion and spirituality in quality of life studies. Appl Res Qual Life. 2006(1):107–16.
7. Kale SH. Spirituality, religion, and globalization. J Macromark. 2004;24(2):92–107.
8. Mackintosh N. Self-empowerment in health promotion: a realistic target? Br J Nursing. 1995;4(21):1273–8.
9. Heimlich JE, Ardoin NM. Understanding behavior to understand behavior change: a literature review. Environ Educ Res. 2008;14(3):215–37.
10. Oishi S, Diener E, Lucas RE. The optimum level of well-being: can people be too happy?. In: The science of well-being. Dordrecht: Springer; 2009. p. 175–200, 347.
11. Atkinson S. Beyond components of wellbeing: the effects of relational and situated assemblage. Topoi. 2013;32(2):137–44.
12. Carlisle S, Henderson G, Hanlon PW. 'Wellbeing': a collateral casualty of modernity? Soc Sci Med. 2009;69(10):1556–60.
13. Lyubomirsky S. Why are some people happier than others? The role of cognitive and motivational processes in well-being. Am Psychol. 2001;56(3):239.
14. Freedman JL. Happy people: what happiness is, who has it, and why. New York: Harcourt Brace Jovanovich; 1978.
15. Diener E, Suh EM, Lucas RE, Smith HL. Subjective well-being: three decades of progress. Psychol Bull. 1999;125(2):276–302.
16. Ryff CD, Dienberg G, Love E, Marilyn J, Singer B. Resilience in adulthood and later life. Defining features and dynamic processes. In: Lomranz J, editor. Handbook of aging and mental health. The Springer series in adult development and aging. Boston, MA: Springer; 1998.
17. Taylor SE, Brown JD. Illusion and well-being: a social psychological perspective on mental health. Psychol Bull. 1988;103(2):193–210.
18. Eysenck MW. Basil Blackwell (1990): The Blackwell dictionary of cognitive psychology, ed.
19. Eysenck MW. Basil Blackwell (1990): The Blackwell dictionary of cognitive psychology, ed., p. 240.
20. Sumner LW. Welfare, happiness, and ethics. New York: Oxford University Press; 1996.
21. Ahmed S. Killing joy: feminism and the history of happiness. Signs: J Women Cult Soc. 2010;35(3):571–94.
22. Ehrenreich B. Smile or die: how positive thinking fooled America and the world. Granta books; 2010.
23. Samman E. Psychological and subjective wellbeing: a proposal for internationally comparable indicators. Oxford: Oxford University Press; 1995; 2007. OPHI Working Paper Series.
24. Ryff CD, Keyes CL. The structure of psychological well-being revisited. J Pers Soc Psychol. 1995;69:719–27.
25. Ryff CD, Lee YH, Essex MJ, Schmutte PS. My children and me: midlife evaluations of grown children and of self. Psychol Aging. 1994;9:195–205.
26. Ryff CD, Dienberg Love G, Urry HL, Muller D, Rosenkranz MA, Friedman EM, Davidson RJ, Singer B. Psychological well-being and ill-being: do they have distinct or mirrored biological correlates? Psychother Psychosom. 2006;75:85–95.
27. Wrosch C, Scheier MF, Miller GE, Schulz R, Carver CS. Adaptive self-regulation of unattainable goals: goal disengagement, goal reengagement, and subjective well-being. Pers Soc Psychol Bull. 2003;29:1494–508.
28. Levi L, editor. Stress and distress in response to psychosocial stimuli: laboratory and real-life studies on sympatho-adrenomedullary and related reactions. Oxford: Pergamon Press; 2016. p. 11.
29. Levi L, editor. Stress and distress in response to psychosocial stimuli: laboratory and real-life studies on sympatho-adrenomedullary and related reactions. Oxford: Pergamon Press; 2016. p. 13.
30. Levi L, editor. Stress and distress in response to psychosocial stimuli: laboratory and real-life studies on sympatho-adrenomedullary and related reactions. Oxford: Pergamon Press; 2016. p. 14.

31. Robinson MD, Johnson JT. Is it emotion or is it stress? Gender stereotypes and the perception of subjective experience. Sex Roles. 1997;36(3–4):235–58.
32. Mulligan K, Scherer KR. Toward a working definition of emotion. Emotion Rev. 2012;4(4):345–57.
33. Wang M, Saudino KJ. Emotion regulation and stress. J Adult Dev. 2011;18(2):95–103.
34. McCraty R, Atkinson M, Tomasino D, Bradley RT. Heart–brain interactions, psychophysiological coherence, and the emergence of system-wide order. HeartMath Research Center, Institute of HeartMath, Publication No. 06-022. Boulder Creek; 2006. p. 3.
35. McCraty R, Atkinson M, Tomasino D, Bradley RT. Heart–brain interactions, psychophysiological coherence, and the emergence of system-wide order. HeartMath Research Center, Institute of HeartMath, Publication No. 06-022. Boulder Creek; 2006. p. 25–6.
36. Kreibig SD. Autonomic nervous system activity in emotion: a review. Biol Psychol. 2010;84(3):394–421.
37. Colombetti G. Appraising valence. J Consciousness Stud. 2005;12(8–9):103–26.
38. McCraty R, Atkinson M, Tomasino D, Bradley RT. Heart–brain interactions, psychophysiological coherence, and the emergence of system-wide order. HeartMath Research Center, Institute of HeartMath, Publication No. 06-022. Boulder Creek; 2006. p. 5.
39. McCraty R, Atkinson M, Tomasino D, Bradley RT. Heart–brain interactions, psychophysiological coherence, and the emergence of system-wide order. HeartMath Research Center, Institute of HeartMath, Publication No. 06-022. Boulder Creek; 2006. p. 10.
40. Mokdad AH, Marks JS, Stroup DF, Gerberding JL. Actual causes of death in the United States, 2000. JAMA. 2004;291(10):1238–45.
41. Naghavi M, Abajobir AA, Abbafati C, Abbas KM, Abd-Allah F, Abera SF, Ahmadi A, et al. Global, regional, and national age-sex specific mortality for 264 causes of death, 1980–2016: a systematic analysis for the global burden of disease study 2016. The Lancet. 2017;390(10100):1151–210.
42. Hollis V, Konrad A, Whittaker S. Change of heart: emotion tracking to promote behavior change. In: Proceedings of the 33rd annual ACM conference on human factors in computing systems; 2015. p. 2643–52.
43. McCraty R, Atkinson M, Tomasino D, Bradley RT. Heart–brain interactions, psychophysiological coherence, and the emergence of system-wide order. HeartMath Research Center, Institute of HeartMath, Publication No. 06-022. Boulder Creek; 2006. p. 2.
44. McCraty R, Atkinson M, Tomasino D, Bradley RT. Heart–brain interactions, psychophysiological coherence, and the emergence of system-wide order. HeartMath Research Center, Institute of HeartMath, Publication No. 06-022. Boulder Creek; 2006. p. 6.
45. McCraty R, Atkinson M, Tomasino D, Bradley RT. Heart–brain interactions, psychophysiological coherence, and the emergence of system-wide order. HeartMath Research Center, Institute of HeartMath, Publication No. 06-022. Boulder Creek; 2006. p. 7.
46. Rea S. Tending the heart fire: living in flow with the pulse of life. Boulder: Sounds True; 2014. p. 61.
47. Rea S. Tending the heart fire: living in flow with the pulse of life. Boulder: Sounds True; 2014. p. 58.
48. McCraty R. Science of the heart, Volume 2: Exploring the role of the heart in human performance: Rollin McCraty Ph. D., Sandy Royall. Boulder Creek: HeartMath Institute; 2015. p. 43.
49. McCraty R. Science of the heart, Volume 2: Exploring the role of the heart in human performance: Rollin McCraty Ph. D., Sandy Royall. Boulder Creek: HeartMath Institute; 2015. p. 42.
50. McCraty R, Atkinson M, Tomasino D, Bradley RT. Heart–brain interactions, psychophysiological coherence, and the emergence of system-wide order. HeartMath Research Center, Institute of HeartMath, Publication No. 06-022. Boulder Creek; 2006. p. 47.
51. McCraty R, Atkinson M, Tomasino D, Bradley RT. Heart–brain interactions, psychophysiological coherence, and the emergence of system-wide order. HeartMath Research Center, Institute of HeartMath, Publication No. 06-022. Boulder Creek; 2006. p. 31.

52. Childre D, Martin H, Beech D. The HeartMath solution: The Institute of HeartMath's Revolutionary Program for Engaging the Power of the Heart's Intelligence. New York: HarperOne; 2000.
53. Childre D, Rozman D. Transforming stress: the HeartMath solution for relieving worry, fatigue, and tension. Oakland: New Harbinger Publications; 2005.
54. Levitis D, William ZL, Glenn F. Behavioral biologists do not agree on what constitutes behavior. Anim Behav. 2009;78:103–10.
55. Romer P, M. Thinking and feeling. Am Econ Rev. 2000;90(2):439–43.
56. Gigerenzer G, Hertwig R, et Pachur T, editors. Heuristics: the foundations of adaptive behavior. Oxford University Press; 2011.
57. Tiller WA. What are subtle energies? In: Journal of Scientific Exploration, 1993; 7(3): Altern Complement Med. 2004;10(2):307–14. 293–304.
58. Geraerts E, Bernstein DM, Merckelbach H, Linders C, Raymaekers L, Loftus EF. Lasting false beliefs and their behavioral consequences. Psychol Sci. 2008;19:749–53.
59. Barth F, Chiu C, Rodseth L, Robb J, Rumsey A, Simpson B, Barth F, et al. An anthropology of knowledge. Curr Anthropol. 2002;43(1):1–18.
60. Baumeister RF, Masicampo EJ, Vohs KD. Do conscious thoughts cause behavior? Annu Rev Psychol. 2011;62:331–61.
61. Gigerenzer G, Brighton H. Can hunches be rational. JL Econ. & Pol'y. 2007;4:155.
62. Cohen DA, Scribner RA, Farley TA. A structural model of health behavior: a pragmatic approach to explain and influence health behaviors at the population level. Prevent Med. 2000;30(2):146–54.
63. Mehl M. Why researchers should think "real-time": a cognitive rationale. In: Handbook of research methods for studying daily life. New York: Guilford Press; 2013.
64. Berger J. Invisible influence: the hidden forces that shape behavior. London: Simon & Schuster; 2016. p. 2.
65. Bargh JA. Reply to the commentaries; 1997. See Wyer 1997, 10:231–46, p. 243.
66. Dhami MK, Hertwig R, Hoffrage U. The role of representative design in an ecological approach to cognition. Psychol Bull. 2004;130(6):959–88.
67. Barth F, Chiu C, Rodseth L, Robb J, Rumsey A, Simpson B, Barth F, et al. An anthropology of knowledge. Curr Anthropol. 2002;43(1):2.
68. Barth F, Chiu C, Rodseth L, Robb J, Rumsey A, Simpson B, Barth F, et al. An anthropology of knowledge. Curr Anthropol. 2002;43(1):10.
69. Anderson CA. Imagination and expectation: the effect of imagining behavioral scripts on personal influences. J Personal Soc Psychol. 1983;45:293–305.
70. Barth F, Chiu C, Rodseth L, Robb J, Rumsey A, Simpson B, Barth F, et al. An anthropology of knowledge. Curr Anthropol. 2002;43(1):1.
71. Armitage CJ, Conner M. Efficacy of the theory of planned behavior: a meta-analytic review. Br J Soc Psychol. 2001;40(4):471–99.
72. Ajzen I, Fishbein M. The influence of attitudes on behaviour. In: Albarracin D, Johnson BT, Zanna MP, editors. The handbook of attitudes. Lawrence Erlbaum Associates; 2005.
73. Armor DA, Taylor SE. The effects of mindset on behavior: self-regulation in deliberative and implemental frames of mind. Person Soc Psychol Bull. 2003;29(1):92.
74. Taylor SE, Lerner JS, Sherman DK, Sage RM, McDowell NK. Are self-enhancing cognitions associated with healthy or unhealthy biological profiles? J Personal Soc Psychol. 2003;85(4):605–15.
75. Galinsky AD, Maddux WW, Gilin D, White JB. Why it pays to get inside the head of your opponent: the differential effects of perspective taking and empathy in negotiations. Psychol Sci. 2008;19:378–84.
76. Galinsky AD, Wang CS, KuG. Perspective-takers behave more stereotypically. J Personal Soc Psychol. 2008b;95:404–19.
77. Ackerman JM, Goldstein NJ, Shapiro JR, Bargh JA. You wear me out: the vicarious depletion of selfcontrol. Psychol Sci. 2009;20:326–32.
78. Bambling M. Mind, body and heart: psychotherapy and the relationship between mental and physical health. Psychother Aust. 2006;12(2)

79. Gustafson A, Ballew MT, Goldberg MH, Cutler MJ, Rosenthal SA, Leiserowitz A. Personal stories can shift climate change beliefs and risk perceptions: the mediating role of emotion. Commun Rep. 2020;33(3):121–35.
80. Ferrini R, Edelstein S, Barrettconnor E. The association between health beliefs and health behavior change in older adults. Prevent Med. 1994;23(1):1–5.
81. Valente TW, Pumpuang P. Identifying opinion leaders to promote behavior change. Health Educ Behav. 2007;34(6):881–96.
82. Usher EL. Personal capability beliefs. Handb Educ Psychol. 2015;3:146–59.
83. Steeves JA, Liu B, Willis G, Lee R, Smith AW. Physicians' personal beliefs about weight-related care and their associations with care delivery: the US National Survey of energy balance related care among primary care physicians. Obes Res Clin Pract. 2015;9(3):243–55.
84. Frank De Jerome D, Frank JB. Persuasion and healing: a comparative study of psychotherapy. JHU Press; 1993.
85. Frank KA. The "ins and outs" of enactment: a relational bridge for psychotherapy integration. J Psychother Integr. 2002;12(3):267–86.
86. Werth JL Jr, Blevins D, Toussaint KL, Durham MR. The influence of cultural diversity on end-of-life care and decisions. Am Behav Sci. 2002;46(2):204–19.
87. O'Connell KA, Skevington SM. Spiritual, religious, and personal beliefs are important and distinctive to assessing quality of life in health: a comparison of theoretical models. WHO Centre for the Study of Quality of Life, Department of Psychology, University of Bath. Br J Health Psychol. 2009;00:1–21.
88. Unsworth S, Palicki SK, Lustig J. The impact of mindful meditation in nature on self-nature interconnectedness. Mindfulness. 2016;7(5):1052–60.
89. Barrett M, Uí Dhuibhir P, Njoroge C, Wickham S, Buchanan P, Aktas A, Walsh D. Diet and nutrition information on nine national cancer organisation websites: a critical review. Eur J Cancer Care. 2020;29(5):e13280.
90. Alvesson M, Willmott H. Identity regulation as organizational control: producing the appropriate individual. J Manage Stud. 2002;39(5):619–44.
91. Ananth S. Developing healing beliefs. 2009;5(6):354–5.
92. Wirth D. The significance of belief and expectancy within the spiritual healing encounter. Soc Sci Med. 1995;41(2):249–60.
93. Schmidt S. Mindfulness and healing intention: concepts, practice, and research evaluation. J Complement Altern Med. 2004;10:S7–S14.
94. Montaño DE, Kasprzyk D. Theory of reasoned action, theory of planned behavior, and the integrated behavioral model. In: Glanz K, Barbara K, Rimer K, editors. Health behavior: theory, research, and practice. 4th ed. Wiley; 2008.
95. Hagger MS, Cameron LD, Hamilton K, Hankonen N, Lintunen T, editors. The handbook of behavior change. Cambridge University Press; 2020.
96. Smallwood J, Fitzgerald A, Miles LK, Phillips LH. Shifting moods, wandering minds: negative moods lead the mind to wander. Emotion. 2009;9(2):271–6.
97. Bennett MR, Hacker PMS. History of cognitive neuroscience. New York: Wiley-Blackwell; 2008. p. 166.
98. Zeelenberg M, Nelissen RM, Breugelmans SM, Pieters R. On emotion specificity in decision making: why feeling is for doing. Judgment Decision Making. 2008;3(1):18.
99. Avey JB, Wernsing TS, Luthans F. Can positive employees help positive organizational change? Impact of psychological capital and emotions on relevant attitudes and behaviors. J Appl Behav Sci. 2008;44(1):48.
100. Belschak FD, Hartog D, Deanne N. Consequences of positive and negative feedback: the impact on emotions and extra-role behaviors. Appl Psychol. 2009;58(2):274.
101. Adhia, Hasmukh, H. R. Nagendra, and B. Mahadevan. "Impact of adoption of yoga way of life on the emotional intelligence of managers." IIMB Management Review. 2010;22(1-2):32–41.
102. Park, Crystal L. et al. "How does yoga reduce stress? A clinical trial testing psychological mechanisms." Stress and Health. 2021;37(1):116–126.

103. Verma, Sudhanshu, Kamakhya Kumar. "Evidence-based comparative study of group and individual consciousness on life satisfaction among adults." Yoga Mimamsa. 2020;52(1):34.
104. Boeger, Gabriella H. YOGA THROUGH A SYSTEMIC LENS: THE IMPACT OF YOGA PRACTICE ON SELF-COMPASSION, COUPLE SATISFACTION, AND FAMILY FUNCTIONING. Diss. Purdue University Graduate School, 2020.
105. Kupershmidt S, Barnable T. Definition of a yoga breathing (pranayama) protocol that improves lung function. Holistic Nursing Pract. 2019;33(4):197.
106. De Michelis E. A history of modern yoga: Patanjali and western esotericism. London: A & C Black; 2005.
107. Ignatow G. Culture and embodied cognition: moral discourses in internet support groups for overeaters. Social Forces. 2009;88(2) (The University of North Carolina Press, 643–70. p. 415).
108. Ignatow G. Culture and embodied cognition: moral discourses in internet support groups for overeaters. Social Forces. 2009;88(2) (The University of North Carolina Press, 643–70)
109. Csordas TJ. Embodiment and experience. Cambridge University Press; 1994.
110. Scheper-Hughes N, Lock MM. The mindful body: a prolegomenon to future work in medical anthropology. Med Anthropol Q. 1987;1(1):6–41.
111. Swami Rama R, Ballentine SA. Yoga and psychotherapy: the evolution of consciousness. Honesdale, PA: Himalayan International Institute; 1976.
112. Menezes CB, Dalpiaz NR, Kiesow LG, Sperb W, Hertzberg J, Oliveira AA. Yoga and emotion regulation: a review of primary psychological outcomes and their physiological correlates. Psychol Neurosci. 2015;8(1):82.
113. Zuo XL, Li Q, Gao F, Yang L, Meng FJ. Effects of yoga on negative emotions in patients with breast cancer: a meta-analysis of randomized controlled trials. Int J Nurs Sci. 2016;3(3):299–306.
114. Francis AL, Beemer RC. How does yoga reduce stress? Embodied cognition and emotion highlight the influence of the musculoskeletal system. Complement Therap Med. 2019;43:170–5.
115. Froeliger B, Garland EL, Modlin LA, McClernon FJ. Neurocognitive correlates of the effects of yoga meditation practice on emotion and cognition: a pilot study. Front Integrative Neurosci. 2012;6:48.
116. Khedikar SG, Erande MP, Shukla Deepnarayan V. Critical comparison of yogic nadi with nervous system. Joinsysmed. 2016;4(2):108–13.
117. Institute of HeartMath: https://www.heartmath.org/research/science-of-the-heart/heart-brain-communication/. Accessed 29 May 2020.
118. Salmon P, Lush E, Jablonski M, Sephton SE. Yoga and mindfulness: clinical aspects of an ancient mind/body practice. Cogn Behav Pract. 2009;16:59–72.
119. Streeter CC, Gerbarg PL, Saper RB, Ciraulo DA, Brown RP. Effects of yoga on the autonomic nervous system, gamma-aminobutyric-acid, and allostasis in epilepsy, depression, and post-traumatic stress disorder. Med Hypotheses. 2012;78:571–9.
120. McCraty R, Atkinson M, Tomasino D, Bradley RT. Heart–brain interactions, psychophysiological coherence, and the emergence of system-wide order. HeartMath Research Center, Institute of HeartMath, Publication No. 06-022. Boulder Creek; 2006. p. 23.
121. Cahn BR, Polich J. Meditation states and traits: EEG, ERP, and neuroimaging studies. Psychol Bull. 2006;132(2):180–211.
122. McCraty R, Atkinson M, Tomasino D, Bradley RT. Heart–brain interactions, psychophysiological coherence, and the emergence of system-wide order. HeartMath Research Center, Institute of HeartMath, Publication No. 06-022. Boulder Creek; 2006. p. 11.
123. Arias AJ, Steinberg K, Banga A, Trestman RL. Systematic review of the efficacy of meditation techniques as treatments for medical illness. J Alternat Complement Med. 2006;12(8):817–32.
124. Ospina MB, Bond K, Karkhaneh M, Tjosvold L, Vandermeer B, Liang Y, Bialy L, Hooton N, Buscemi N, Dryden DM, Klassen TP, editors. Meditation practices for health: state of the research. AHRQ Publications; 2007.

125. Hayes JA, Davis DM. What are the benefits of mindfulness? A practice review of psychotherapy-related research. Psychother Am Psychol Assoc. 2011;48(2):198–208.
126. Farb N, Segal ZV, Mayberg H, Bean J. Attending to the present: mindfulness meditation reveals distinct neural modes of self-reference. Soc Cogn Affect Neurosci. 2008;2(4):313–22.
127. Corcoran KM., Farb N, Anderson A, Segal ZV. Mindfulness and emotion regulation: outcomes and possible mediating mechanisms; 2010.
128. Siegel DJ. Mindfulness training and neural integration: differentiation of distinct streams of awareness and the cultivation of well-being. Soc Cogn Affect Neurosci. 2007;2(4):259–63.
129. Kabat-Zinn J. Mindfulness meditation. Mind-Body-Medicine, earthandspiritcenter.org. 1995.
130. Eppley KR, Abrams AI, Shear J. Differential effects of relaxation techniques on trait anxiety: a meta-analysis. J Clin Psychol. 1989;45:957–74.
131. Barnes VA, Treiber FA, Davis H. Impact of transcendental meditation on cardiovascular function at rest and during acute stress in adolescents with high normal blood pressure. J Psychosomatic Res. 2001;51:597–605.
132. Orme-Johnson DWK, Lonsdorf N. Meditation in the treatment of chronic pain and insomnia. In: National Institutes of Health Technology Assessment Conference on Integration of Behavioral and Relaxation Approaches into the Treatment of Chronic Pain and Insomnia. Bethesda Maryland: National Institutes of Health; 1995.
133. Ray US, Sinha B, Tomer OS, Pathak A, et al. Aerobic capacity & perceived exertion after practice of Hatha yogic exercises. Indian J Med Res New Delhi. 2001;114:215–21.
134. Barrett B, Hayney MS, Muller D, Rakel D, Ward A, Obasi CN, Brown R, Zhang Z, Zgierska A, Gern J, West R, Ewers T, Barlow S, Gassman M, Coe CL. Meditation or exercise for preventing acute respiratory infection: a randomized controlled trial. Ann Fam Med. 2012;10(4):337–46.
135. Siegle GJ, Steinhauer SR, Thase ME, Stenger V, Andrew; Carter, Cameron S. Can't shake that feeling: event-related FMRI assessment of sustained amygdala activity in response to emotional information in depressed individuals. Biol Psychiatry. 2002;51(9):693–707.
136. Fredrickson BL. Cultivating positive emotions to optimize health and well-being. Prevent Treatment. 2000;3(1):1a.
137. Fredrickson BL, Cohn MA, Coffey KA, Pek J, Finkel SM. Open hearts build lives: positive emotions, induced through loving-kindness meditation, build consequential personal resources. J Personal Soc Psychol. 2008;95(5):1045.
138. Pribram KH, Melges FT. Psychophysiological basis of emotion. In: Vinken PJ, Bruyn GW, editors. Handbook of clinical neurology. Amsterdam: North-Holland Publishing Company; 1969. p. 316–41.
139. Bradley RT, Pribram KH. Communication and stability in social collectives. J Soc Evolut Syst. 1998;21(1):29–80.
140. Pribram KH, Bradley RT. The brain, the me and the I. In: Ferrari M, Sternberg R, editors. Self-awareness: its nature and development. New York: The Guilford Press; 1998. p. 273–307.
141. McCraty R, Atkinson M, Tomasino D, Bradley RT. Heart–brain interactions, psychophysiological coherence, and the emergence of system-wide order. HeartMath Research Center, Institute of HeartMath, Publication No. 06-022. Boulder Creek; 2006. p. 25.
142. McCraty R, Atkinson M, Tomasino D, Bradley RT. Heart–brain interactions, psychophysiological coherence, and the emergence of system-wide order. HeartMath Research Center, Institute of HeartMath, Publication No. 06-022. Boulder Creek; 2006. p. 27.
143. NEFF, Kristin D. The role of self-compassion in development: a healthier way to relate to oneself. Hum Dev. 2009;52(4):211–4.
144. Armor DA, Taylor SE. Situated optimism: specific outcome expectancies and self-regulation. In: Zanna Mark P, editor. Advances in experimental social psychology, vol. 30. Academic Press; 1998. p. 309–79.
145. Gardner B, Lally P. Does intrinsic motivation strengthen physical activity habit? Modeling relationships between self-determination, past behavior, and habit strength. J Behav Med. 2013;36:488–97.
146. Dahl J, Lutz A, Davidson RJ. Reconstructing and deconstructing the self: cognitive mechanisms in meditation practice. Trends Cogn Sci. 2015;19(9):515–23.

147. Mayo NE, Figueiredo S, Ahmed S, Bartlett SJ. Montreal Accord on patient-reported outcomes (PROs) use series–Paper 2: terminology proposed to measure what matters in health. J Clin Epidemiol. 2017;89:119–24.
148. Hektner J, Schmidt J, Csikszentmihalyi M. Experience sampling method: measuring the quality of everyday life. Sage; 2006.
149. Berrocal A, Concepcion W, De Dominicis S, Wac K. Complementing human behavior assessment by leveraging personal ubiquitous devices and social links: an evaluation of the perceived momentary assessment method. JMIR mHealth and uHealth. 2020;8(8):e15947.
150. Antonovksy, A. (1994) Unraveling the mystery of health: how people manage stress and stay well. San Francisco: Jossey-Bass, 1987. Cited in: Tresolini, CP and the Pew-Fetzer Task Force. "Health Professions Education and Relationship-Centered Care" San Francisco: Pew Health Professions Commission and the Fetzer Institute, 1994 p. 15.1987.
151. Huber M, Knottnerus JA, Green L, et al. How should we define health? BMJ. 2011;343(2).
152. Raio CM, Orederu TA, Palazzolo L, Shurick AA, Phelps EA. Cognitive emotion regulation fails the stress test. Proc Natl Acad Sci. 2013;110(37):15139–44.
153. Hollis V, Konrad A, Whittaker S. Change of the heart: emotion tracking to promote behavior change. In: Quantified self for humans & pets. Crossings, Seoul; 2015. p. 2643–52.
154. Hollis V, Konrad A, Whittaker S. Change of the heart: emotion tracking to promote behavior change. In: Quantified self for humans & pets. Crossings, Seoul; 2015. p. 2650.
155. Wac K. From quantified self to quality of life. In: Digital health. Cham: Springer; 2018. p. 83–108.
156. Estrada-Galiñanes V, Wac K. Collecting, exploring and sharing personal data: why, how and where. 2020:79–106.
157. Wac K. Quality of life technologies. In: Gellman M, editor. Encyclopedia of behavioral medicine. New York: Springer; 2020.
158. McCraty R, Atkinson M, Tomasino D, Bradley RT. Heart–brain interactions, psychophysiological coherence, and the emergence of system-wide order. HeartMath Research Center, Institute of HeartMath, Publication No. 06-022. Boulder Creek; 2006. p. 56.

Chapter 13
Technology-Enabled Assessment and Improvement of Inclusive Learning and Quality of Life in Higher Education

Maria Toledo-Rodriguez and Thomas Boillat

Introduction

Universities have, for centuries, been the sources of wisdom and knowledge. In order to acquire this knowledge students used to travel to the universities where lecturers, masters in their field, would impart their teachings. Often the transfer of knowledge consisted in masters giving hour's long monologues while the students passively listened and took notes. In order to be successful, this system requires from the student to: (a) attend to all learning sessions, (b) be able to record all the knowledge provided while listening to the lecturer and (c) understand and put in context (within their applications and implications) all information passively received during that unique offering of knowledge. This system puts in clear disadvantage many students who have learning differences, suffer from poor physical or mental health or have additional family/work commitments. Quality of higher education is recognized by researchers and governments to impact the student's quality of life in the long term [1, 2]. The term "inclusive learning" is currently being widely used by academics and governments as a tool to improve the learning in those populations and therefore potentially improve the long-term quality of life of these students and dependants. However, inclusive learning as described by May and Bridger"… *necessitates a shift away from supporting specific student groups through a discrete set of policies or time-bound interventions, towards*

M. Toledo-Rodriguez (✉)
School of Life Sciences, University of Nottingham, Nottingham, UK
e-mail: maria.toledo@nottingham.ac.uk

T. Boillat
Design Lab, Mohammed Bin Rashid University of Medicine and Health Sciences, Dubai, UAE
e-mail: Thomas.boillat@mbru.ac.ae

© The Author(s) 2022
K. Wac, S. Wulfovich (eds.), *Quantifying Quality of Life*, Health Informatics,
https://doi.org/10.1007/978-3-030-94212-0_13

equity considerations being embedded within all functions of the institution and treated as an ongoing process of quality enhancement" [3]. In the following subsections we define the current technological solutions to enable holistic inclusive learning.

This book chapter presents results of a study employing mixed methods and will focus on the role of different digital technology platforms in supporting assessment and improvement of inclusive learning in higher education. However, it does not systematically focus in detail on software specially developed to help students with learning difficulties. Rather, it presents and analyses recent advances in learning technologies that bring opportunities for inclusive learning and enable individuals to tailor the learning experience to their needs and abilities regardless of their degree of disability. Most of these platforms analyzed in here are relatively recent and while they charge per use, the majority also have a free version, which enables any educator to make use of them (with some limitations), as long as students and teachers have internet connection and a mobile device. In this chapter, we present results of our analysis and discuss design implications for technologies, which, if implemented effectively, will influence the assessment and improvement of learning and attainment of educational goals, as well as quality of life in the long term of any student.

Current Assessment Methods for Cognition/Executive Function, Attention, Memory and Learning in Healthy Populations

Before focusing on digital platforms that enable inclusive learning, we should take a quick look at the current assessment methods of cognition, executive function, including attention and memory and learning in healthy populations. The list of currently used tests used is very large and describing it into detail would take us away from the focus of this chapter. Table 13.1 describes a representative number of tests for executive function, attention, intelligence and different types of memory and academic resilience. We focused on tests that measure performance (PerfRO) [32]. Regarding learning, the best methods to test healthy populations are current academic assessments (e.g. written exams, coursework, oral presentations, group presentations, etc.).

Table 13.1 Representative listing of tests commonly used to measure attention, memory, executive functioning and intelligence

	Scale name (sub-scales)	Consist in	S[a]
Executive function	Behavioral Rating Inventory of Executive Function (BRIEF)	Measures executive function impairment (mainly in children)	[4]
	Covert Orienting of Attention Task	Measures the ability of mentally shifting one's focus without moving one's eyes.	[5]
	Go-No-Go Test	Assesses inhibitory control and cognitive processes needed to stop a movement after being initiated (based on a new stimulus received) The participant has to respond to one of the choices but also has to withhold the response to the other alternative.	
	Ross Information Processing Assessment (RIPA) (Organization, Problem Solving, Abstract Reasoning)	Profiles 10 key areas in communicative and cognitive functioning: Immediate Memory, Recent Memory, Temporal Orientation (Recent Memory), Temporal Orientation (Remote Memory), Spatial Orientation, Orientation to Environment, Recall of General Information, Problem Solving and Abstract Reasoning, Organization, and Auditory Processing and Retention.	[6]
	Ruff 2 and 7 Test (Digits)	Measures visual selective attention utilizing different distractor conditions.	[7]
	Stroop Color/and Word Test (SCWT)	Measures the ability to inhibit cognitive interference between two stimulus mismatch between the name of a color and the colour it is printed on (e.g. the word "pink" colored in blue. When asking what is the colour of the word it takes longer and people make more mistakes).	[8]
	Trail Making Test (TMT)	Measures visual attention and task switching (search and processing speed, mental flexibility and executive functioning).	[9]
	Verbal Fluency Test (FAS)	Measures the ability to produce as many words as possible either within a certain semantic category (category fluency) or starting by a specific letter (letter category) within a set time	[10]
	Wechsler Adult Intelligence Scale (WAIS) Similarities	Measures multiple cognitive abilities and intelligence in adults.	[11]
	Wisconsin Card Sorting Task (WCST)	Measures perseverance, abstract thinking and set shifting.	[12]

(continued)

Table 13.1 (continued)

	Scale name (sub-scales)	Consist in	S[a]
Attention	The Attentional Control Scale	A self-report questionnaire that measures individual differences in attentional control.	[13]
	Continuous Performance Task (CPT)	Measures sustained and selective attention. Individuals must maintained attention while performing repetitive and boring tasks.	[14]
	Dot Probe Task (DPT)	Measures selective attention and attentional biases.	[15]
	d2 Test of Attention	Measures selective and sustained attention and concentration of subjects.	
	Digit Symbol Substitution Test (DSST)	Measures associative learning. Consist on matching symbols to numbers, during a given amount of time, according to a key. It is sensitive to changes in cognition in patients with multiple mental health and brain disorders.	[16]
	Multiple Object Tracking (MOT)	Measures selective/distributed and sustained visual attention.	[17]
	Paced Auditory Serial Attention Test (PASAT)	Measures information processing and sustained and divided attention. Consist on a test of serial addition. It is often used to assesses the effects of traumatic brain injury or other neurodegenerative disorders on cognitive functioning.	[18]
	Ruff 2 and 7 Test (Letters)	Measures visual selective attention utilizing different distractor conditions.	[7]
	Stop-Signal Task	Measures response inhibition (ability to suppress unwanted or inappropriate actions)	[19]
	Stroop Color/Word or Interference	Measures the ability to inhibit cognitive interference between two stimulus mismatch between the name of a colour and the colour it is printed on (e.g. the word "pink" coloured in blue. When asking what is the colour of the word it takes longer and people make more mistakes).	[8]
	Trail Making Test (TMT)	Measures visual attention and task switching (search and processing speed, mental flexibility and executive functioning).	[9]
Intelligence	Stanford–Binet Intelligence Scales	Measures cognitive strengths and weaknesses in children and adults: fluid reasoning, knowledge, quantitative reasoning, visual-spatial processing and working memory. Latest edition (SB5) also measures giftedness.	[20]
	Wechsler Adult Intelligence Scale (WAIS)	Measures multiple cognitive abilities and intelligence in adults.	[11]

Memory	ADAS Word List Recall	Alzheimer Disease Assessment Scale-cognitive subscale.	[21]
	Benton Visual Retention Test (BVRT)	Measures visual memory and perception and visoconstructive abilities.	[22]
	Rey Auditory Verbal Learning Test (RAVLT) (Delay, Temporal Order, Faces and Pictures)	Measures ability to encode, combine, store and recover verbal information.	[23]
	Ross Information Processing Assessment (RIPA) (Auditory Processing, Immediate Memory, Recent Memory)	Profiles 10 key areas in communicative and cognitive functioning: Immediate Memory, Recent Memory, Temporal Orientation (Recent Memory), Temporal Orientation (Remote Memory), Spatial Orientation, Orientation to Environment, Recall of General Information, Problem Solving and Abstract Reasoning, Organization, and Auditory Processing and Retention.	[6]
	Sternberg Memory Search Task	Series of tests aimed at determining how we access the information in our short-term memory.	[24]
	Verbal Learning and Memory Test (Delay recall, Direct Recall, Recognition)	Memory test that consist on five repeated auditory presentations of a word list. After each presentation the subject will try to recall as many as possible.	[25]
	Wechsler Memory Scale, IV (WMS-IV) (Facial Recognition, Immediate Figural Memory, Delayed Figural Memory, Immediate Logical Memory, Delayed Logical Memory, Verval Paired Associates, Visual Reproduction)	Series of tests that provided detailed assessments of clinically relevant aspects of memory functioning.	[26]

(continued)

Table 13.1 (continued)

	Scale name (sub-scales)	Consist in	S[a]
Working memory	Automated Working Memory Assessment	Measures active and passive memory (verbal and visuospatial) in children and adolescents.	[27]
	Corsi Block-Tapping Task (CBTT)	Measures visua-spatial working memory, similar to Digit Span Test.	[28]
	Digit Span Backwards	Measures transient number storage capacity.	
	N-Back Task (Auditory, Spatial)	Measures ability to remember whether a stimulus is similar to another presented previously (N trials ago).	[29]
	Self-Ordered Pointing Task (SOPT)	Measures working memory and the contribution of the frontal lobes to it.	[30]
	WAIS Digit Span Task (Letter Number)	Measures multiple cognitive abilities and intelligence in adults.	[16]
Academic Resilience	Academic Resilience Scale (ARS)	Measures the ability to succeed academically in the face of adversity.	[31]

[a] S Source

Visual Tools

Tools	Consist in	S
Executive function		
Approach/avoidance task (ATT)	Measures behavior when faced with a conflicting goal or event (which has both positive and negative characteristics or effects).	[33]
Working memory		
Automated Working Memory Assessment (AWMA)	Fully automated online assessment of working memory.	[34]
Functional capacity/Executive function		
Virtual Reality Functional Capacity Assessment Tool (VRFCAT)	Immersive virtual reality interactive gaming that measures functional abilities using realistic simulations of daily environments.	[35]

In this paper we focus on means of technology-enabled assessment and improvement of inclusive learning and quality of life in higher education.

Common Challenges to Inclusive Learning in Higher Education

Student's ability to learn and fulfil their educational goals is affected by different, physical, mental (as defined by DSM-5 [36]) and situational challenges. It is estimated that on the UK, 3.5% of full time university students had a mental health condition in the 2017–2018 academic year. We will now discuss how each one of them impacts learning.

Physical Health and Illnesses

During their university studies a student might have periods of ill physical health or hospitalization. While in most cases the physical illness has a short duration (e.g., flu), some students might have long periods of illness (e.g., broken arm) or might live with chronic conditions (e.g., Chronic Fatigue Syndrome). In those cases, the student might not be able to attend university for periods of time. As a result, they won't be able to take part in lectures, practical sessions or exams. They can also not engage with in-class activities, such as peer-assessment or debates. Additionally, students will struggle to submit paper-based coursework (although they may be able to submit coursework online). Moreover, if they are in pain (e.g., broken leg) the pain and/or pain medication might interfere with their concentration and sleep patterns. Finally, the injury or illness might prevent students from typing or taking notes.

Mental Health and Illnesses

While it is normal to feel sad or anxious at times, clinical depression and anxiety disorders can be crippling conditions, which can render the student unable to engage with their studies. Recent data shows that the mental health of students in higher education is deteriorating [37]. For example, a recent study from the Institute for Public Policy Research, reports that the number of first year students reporting a mental health problem in 2016 was 3 times higher than 10 years earlier in the UK [38]. As a result, there has been an increase in demand of Counselling and Mental Health services putting huge pressure on the universities' and local welfare resources. In order to deal with this challenge, universities, governments and charities are trying to increase their mental health provision and there is a rising number of preventive campaigns. Additionally, there is an increase interest in initiatives, which aim to embed mental wellbeing within the curriculum [39]. These initiatives can be very varied, encompassing, from university-wide curriculum infusion, such as the Engelhard project at Georgetown University, USA [40] to focus on processes that might cause undue stress or disadvantage to some students (e.g. the use of lecture capture to revise lectures they were not well enough to attend or the use of anonymous online interactive tools to eliminate social anxiety).

The most common mental health disorders experienced by undergraduate students are depressive disorders (e.g. major depressive disorder, bipolar disorder or seasonal affective disorder), and anxiety disorders (e.g. general anxiety disorder or social anxiety disorder) [41]. A small percentage of university-age individuals experience other types of mental health disorders such as schizophrenia [41].

Depression is a common mood disorder that is characterized by constant sadness, lack of hope, motivation and energy, disturbed sleep patterns, and difficulties to make decisions [36]. As a result students feel that doing every day tasks is a constant struggle; they feel worthless and believe that their actions are useless and so their engagement with their studies suffers (they may stop attending lectures or might not be able to keep up with coursework or revision). At times they can become angry because they feel judged by others and themselves and believe they are falling short of expectations all the time. Their view of the world is pessimistic and they believe their situation will never improve. In extreme cases the student might plan and/or attempt to take their own life, although the percentage is lower compared to the general population of similar ages [42].

Bipolar disorder is characterized by periods when the student is depressed (see above for symptoms) alternated with "manic" phases, when the student is overexcited and overactive, but often doing tasks that are not productive nor make sense (e.g., writing for hours an essay about how to tickle ants when the assignment was on the habitat of insects that live in colonies) [36].

Seasonal Affective Disorder is very similar to depression, but usually only happens during a time of the year (e.g. in the European context, commonly during winter, when the days are short and often there is lack of sun) [36, 43]. Often students find particularly difficult to wake up on time in the morning, and have low energy levels during the day. As a result students might stop attending lectures and/ or struggle to keep up with studies.

Generalised anxiety disorder is characterised by constant worry which leads to difficulties in concentration and/or sleep. In extreme cases students might develop panic attacks (for example in an exam) [36].

Students suffering from **social anxiety** worry disproportionately about social activities, such as going to class or speaking in public [36]. They feel judged at all times by others and fear that they will do something in public that is inappropriate. They are scared of criticism and can become paralised when others are watching what they are doing (such as giving a public presentation). In extreme cases students might develop social phobia, where they avoid interacting with others or speaking in public and/or experience panic attacks.

Schizophrenia is a severe mental disorder characterised by relapsing episodes of hallucinations (hearing or seeing things that do not exist outside their mind), disordered thinking, social withdrawal and cognitive impairment [36]. Students with schizophrenia may react in "odd ways", sometimes believing that staff and/or other students are being bullies.

Learning Differences, Neurodiversity

Students' intellectual ability can vary widely resulting in some cases in being diagnosed with a learning difference/disability (neurodiversity). Specific learning disorder is defined by DSM 5 as "*a neurodevelopmental disorder with a biological origin that is the basis for abnormalities at a cognitive level*" [36]. One essential feature of specific learning disorder is persistent difficulties learning keystone academic skills. In the 2015–2016 academic year, the Higher Education Statistics Agency reported 4.95% of all students in the UK enrolled in all higher education had a Specific Learning Difficulty [44]. These disorders are often diagnosed in childhood and are managed with the help of school teachers/advisers and/or specialised software/techniques. However, when students reach university, they might struggle with the transition into a less structured and/or supportive environment. University students with learning differences can use specialized assertive and adaptive technologies, which compensate for the decreased ability (such as Texthelp[1] or ClaroView[2] and ScreenRuler,[3] which can help them reach their full academic potential [45]. There is a spectrum of learning differences as discussed below.

Students in the **Autism spectrum or Aspergers** find it hard to figure out what others are thinking or feeling based on their facial expressions and/or words [36]. Because of this they find socialising difficult. Noise or crowded spaces might be overwhelming for somebody with autism and thus sometimes they need to retire to a quiet space in order to calm down (they may also sit close to the entrance in class

---

[1] https://www.texthelp.com/

[2] https://www.clarosoftware.com/

[3] https://www.clarosoftware.com/

so that they can leave easily or might avoid crowded/noisy classrooms). They can seem awkward or feel uncomfortable in social situations, not making eye contact (resulting in appearing rude or patronising). Often they may take literally what is told to them (so they might struggle to understand jokes or sarcasm). Additionally, changes in their routine can cause them anxiety. Because of this they are very keen on having clear plans for everything that has to happen.

Students with **dyslexia** often mix up letters that have some similarity (like "b" and "d") or words that have very similar spelling (e.g. "from" and "form"). As a result they often make spelling mistakes and are very slow readers/writers. They avoid reading and/or writing whenever they can (e.g. they will prefer a presentation than a written exam or coursework). They also struggle to write coherent notes during lectures (as they struggle to write while trying to understand what the lecturer is explaining and reading from the slides). However students with dyslexia can easily understand information when it is spoken to them and can explain themselves orally much clearer than in writing. Sometimes they struggle to follow directions (sequence of things that have to be done in a certain order) and thus find it difficult to organise and plan, which compound with their writing challenge results in students struggling to meet coursework deadlines [46].

Although **Attention Deficit Hyperactivity Disorder (ADHD)** is not considered a learning disability, it has at times a profound impact on the ability of the student to learn [36]. ADHD is characterised by difficulties in focusing (and lack of attention to details) and hyperactivity (which might make attending to long lecturers, of finishing coursework a challenge). At times, a student with ADHD might have difficulties coping with stress or remembering.

We should note that learning disabilities often do not sit in binary silos where the student either has a learning disability or not (with students often displaying different grades of a learning disability) [36]. Therefore, scholars are starting to strongly argue against the concept of "the average or typical student". They reason that the students traditionally identified as having learning differences belong to the extremes of a continuum [47] and that we could consider that all students abilities are scattered across a spectrum of learning differences [48].

Students Juggling Higher Education with Work or Caring/Family Commitments

The majority of university students are relatively young (between late teens to mid 20s), do not have caring responsibilities (such as caring for young children or elder relatives) and do not need to work in order to fund their studies and daily life [49]. However, due to recent economic downturns and drastic increase in university fees, many students in higher education have to work while they study, in order to provide for themselves and/or finance their studies [50]. Moreover, due to an increased diversity in student population and the development of graduate entry courses, there is a slight rise in the number of university students with family and/or caring commitments, particularly within some vocational disciplines such as nursing [51].

These commitments might at times clash with their studies, preventing them from attending university (as they have to work or take care of somebody). Thus, at times, they may not be able attend lectures, practicals or exams. They might also struggle to submit paper-based assessments and can not benefit from taking part in in-class activities.

Method

In view of answering our research questions 2 and 4, we relied upon a mixed methods research approach based on literature reviews, surveys and interviews as shown in Fig. 13.1.

To get an accurate understanding of the problem definition, the research started with (a) reviewing the most common learning challenges using the Diagnostic and Statistical Manual of Mental Disorders (DSM–5) as primary source and selected literature as secondary source (presented in the previous section). We then built upon a (b) recent literature review [52] to gain a deeper understanding on the role of information technology (IT) in education as well as a description of digital learning platforms and the evolution of learning management systems (LMS). (c) We then surveyed eight faculty members (see table below) to gather current uses of digital learning platforms (Table 13.2). We asked them the three following open-ended questions: (q_1) "What digital platforms or tools do you use for your lectures?" (q_2) "What makes these platforms or tools attractive?" and (q_3) "What other platforms would you recommend and why?" (d) By means of interviews, we then asked the same questions to one course director and three learning designers to gather their experiences of state-of-the-art tools for inclusive learning. The interviews allowed us to derive four digital learning tools categories, categorized independently by the co-authors of this paper. (e) In order to analyze the commonalities and differences across these four categories, we used the Business-Application-Information-Technology (BAIT) model. We then summarized our findings by refining the seven principles of the Universal Instructional Design [53, 54] framework for the context of inclusive learning.

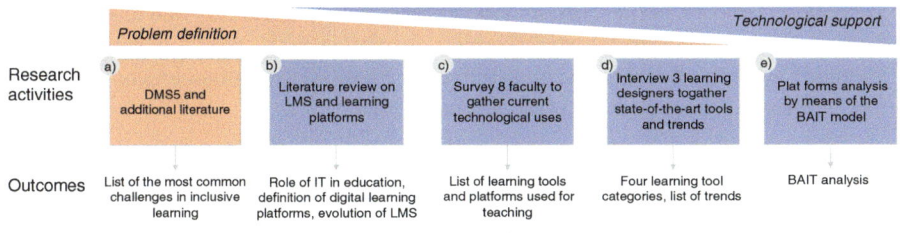

Fig. 13.1 Research methods

Table 13.2 Profiles of survey participants in step (c) and (d)

Academic position	Country	Current class size	Educational area
Associate professor, 2 years of TE[a]	United Arab Emirates	40–65 BSc students	Medicine
Assistant professor, 4 years of TE	United Arab Emirates	25–65 BSc students	Medicine
Professor and chair, 20 years of TE	Switzerland	30–200 BSc and MSc students	Business informatics
Assistant professor, 5 years of TE	Switzerland	30–70 BSc and MSc students	Business informatics
Professor, 12 years of TE	Switzerland	50–300 BSc and MSc students	Economics
Teaching Fellow, 2 years of TE	United Kingdom	100–300 BSc and MSc students	Life Science
Associate Professor, 6 years of TE	United Kingdom	20–200 BSc and MSc students	Life Science
Assistant Professor, 14 years of TE	United Kingdom	15–200 BSc and MSc students	Life Science
Course director, 7 years of TE (+15 years in secondary school teaching)	United Kingdom	45 Foundation year students	Life Science
Senior learning designer, 8 years of TE	United Arab Emirates	300 BSc and MSc students	Health Sciences
Senior learning designer, 6 years of TE	United Arab Emirates	250 BSc and MSc students	Health Sciences
Learning designer, 4 years of TE	United Arab Emirates	250 BSc and MSc students	Health Sciences

[a] *TE* Teaching Experience

Digital Learning and Teaching Platforms (DLTP) Description and Analysis

Along with the almost universal access to personal computers and smartphones, Internet and web applications have disrupted education [52]. Not only have they changed the way knowledge is transferred and learnt, but also the dynamics between lecturers and students as well as among the students. From our interviews with learning designers (Table 13.2), four categories of digital learning and teaching platforms have emerged, as potential enablers for the assessment and improvement of inclusive learning: Learning Management Systems (LMS), Social and collaborative tools, In-class interactions and Out-of-class interactions.

Learning Management Systems

LMS are software applications designed with the specific intent of assisting instructors in meeting their pedagogical goals of delivering learning content to students [55]. Available as web application by means of an Internet browser (Figs. 13.2 and

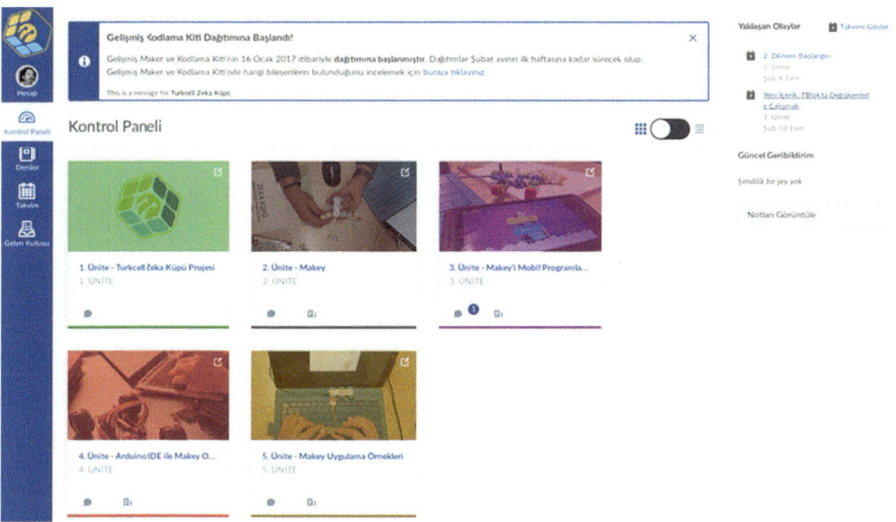

Fig. 13.2 Canvas LMS—Assignments View of a Class (Under the license: https://creativecommons.org/licenses/by-sa/3.0/deed.en)

13.3), LMS have disrupted the way students access (digital) lectures, communicate with their lecturers, classmates and other learning communities, access course materials, take online quizzes and submit their assignments [56]. From the lecturer's point of view, LMS allow to organize classes, publish course materials, create and grade assignments among many additional features. Our interviews revealed two types of software applications: Open-source with Moodle[4] being on top of the list and Graasp[5] in the role of outsider [57]. On the proprietary side, Canvas[6] was the most cited. In our case, LMS were used to structure and describe lectures, upload and share documents, lecture recordings and tutorials as well as post and submit assignments. Uploaded files are presentation slides, PDF and text documents. Our survey revealed that LMS are mostly accessed via desktop or laptop computers.

Social and Collaborative Tools

Social and collaborative tools are defined as (web-based) applications that allow users to create and to share online documents, spreadsheets, presentations, and forms [58]. They are mostly used by students to create and share content such as assignments. Similar to LMS, open-source and proprietary tools are being used. The

---

[4] https://moodle.com

[5] https://graasp.eu

[6] https://www.instructure.com/canvas/

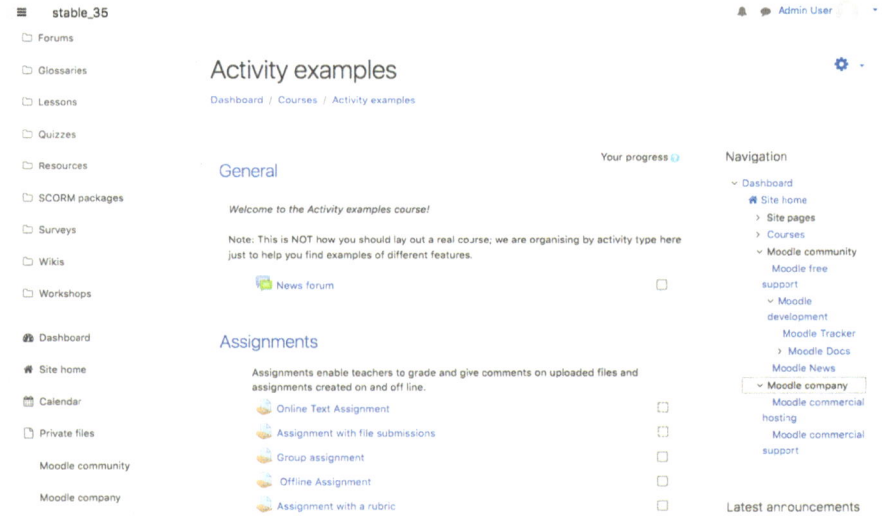

Fig. 13.3 Moodle—Class View

two most used are respectively Google Docs[7] and Microsoft Office 365.[8] They allow students to create different types of documents from presentations to spreadsheets and letters and work simultaneously on the same documents (Fig. 13.4). As part of Microsoft Office 365, Teams offers new opportunities for lecturers to engage with students, and also for creating groups amongst students. It provides instant messaging capabilities, both via text and video, enabling lecturers and students to communicate regardless of their location. In addition, it offers live captions to help students with hearing impairments. Newcomers such as Padlet[9] are increasingly used in class for students to publish their work and collect feedback (Fig. 13.5). These tools are most of the time accessed from a desktop or laptop computer while a mobile interface is also available but difficult to use.

In-Class Engagement

As stated in Holzer et al. 2013 [57], interactivity in person classrooms is considered an important success factor in learning but it remains very challenging to promote. Simple technologies such as the *Clickers* (that combine hardware with software) gained popularity in classrooms for their ability to gather anonymous answers to questions asked in class [59]. It gives the lecturer a chance to immediately assess the

---

[7] https://docs.google.com

[8] https://www.office.com/

[9] https://padlet.com

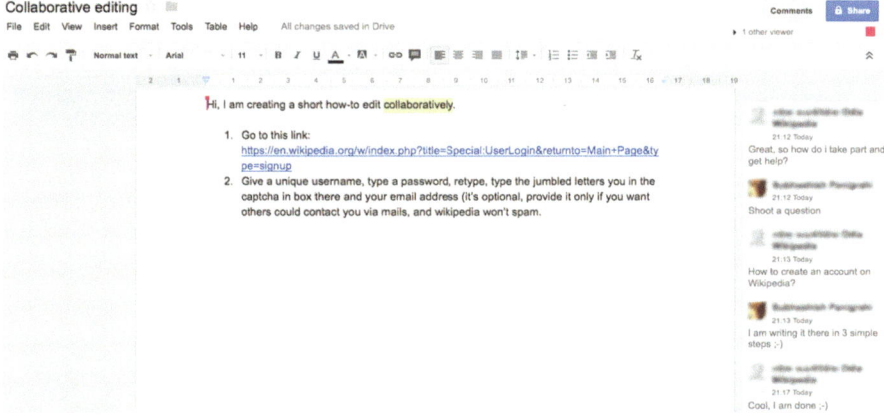

Fig. 13.4 Padlet—Main View (shared amongst groups or students)

Fig. 13.5 Google Doc—Collaborative work (Under the license: https://creativecommons.org/licenses/by-sa/3.0/deed.en)

students' knowledge and quantitatively measure the progress of the class (Fig. 13.6). The in-class engagement tools are seen as the next generation of interactions by allowing lecturers and students to send anonymous text messages in addition to creating polls (Fig. 13.7). They are mostly used by lecturers to gather feedback from students and to encourage them to engage in lectures/workshops/seminars. A few tools were revealed by the survey: Speakup,[10] which was developed by two Swiss

---

[10] http://speakup.info

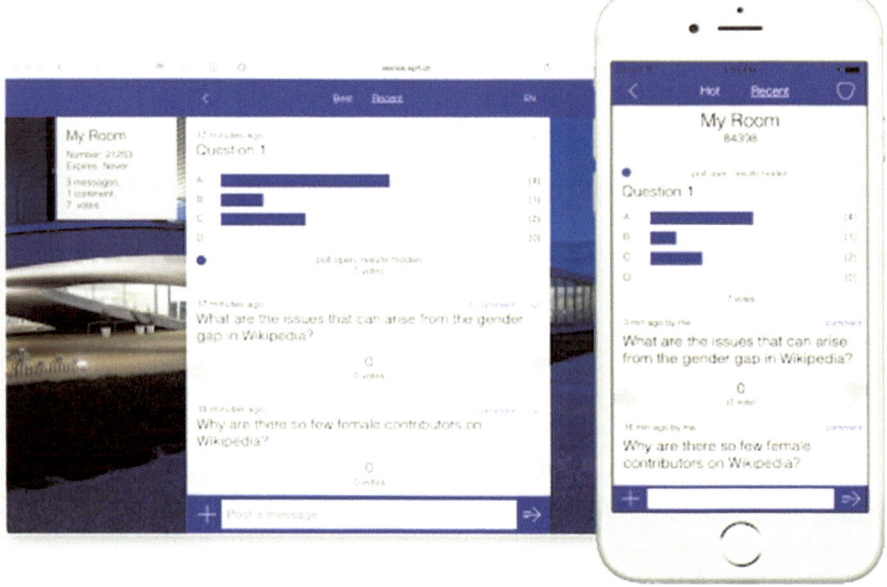

Fig. 13.6 Mentimeter—Live question/answers

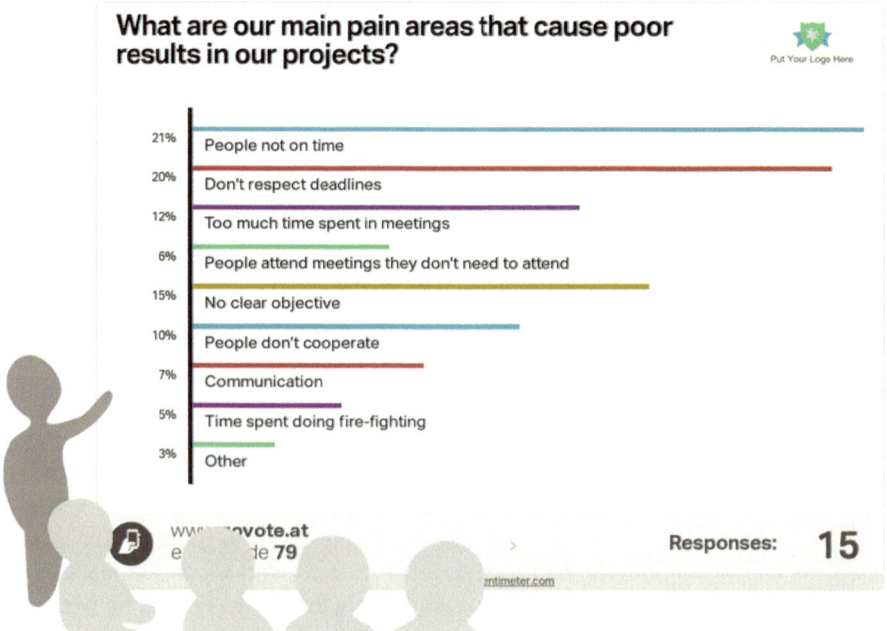

Fig. 13.7 SpeakUp—Live question/answers

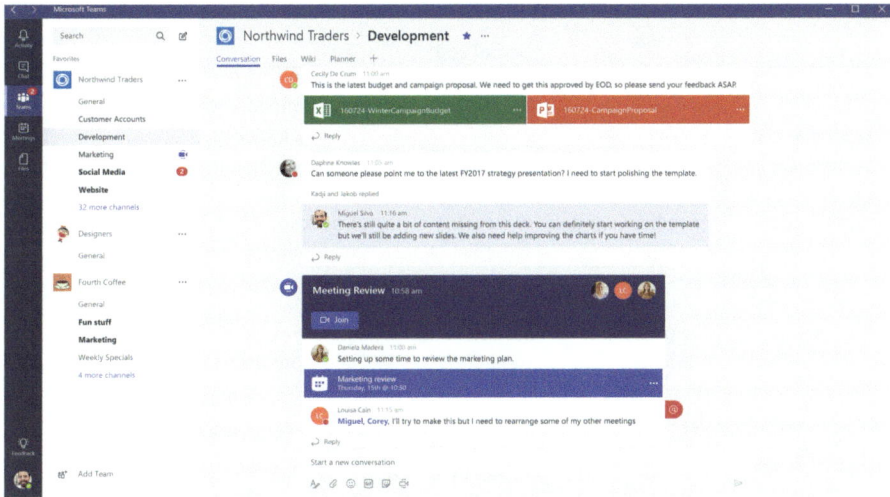

Fig. 13.8 Ms. Teams—Discussion panel

universities and available free of charge; Socrative[11] and Slido[12] that offer similar functionalities than Speakup; Mentimeter[13] that also allows for the creation of pools but requires a subscription for intensive usage; and Kahoot[14] that leverages the concept of gamification. All these tools are available as web and mobile applications and are most commonly used from a smartphone and laptops.

Out-of-Class Interactions

Out-of-class communication channels widely depend upon the university, the faculty/school (discipline) and the culture. Most of the course communication and feedback relies on emails or on LMS (e.g., class forums). Tools such as MS Teams,[15] Speakup or Slido are also used after the class whereby lecturers or students give feedback (Fig. 13.8). Recently Whatsapp[16] and other text message applications have become very popular amongst students to exchange tips and documents. These latter tools are easier to access and to use, thanks to their integration in smartphones.

---

[11] https://www.socrative.com

[12] https://www.sli.do

[13] https://www.mentimeter.com

[14] https://kahoot.com

[15] https://www.microsoft.com/en-ww/microsoft-365/microsoft-teams/group-chat-software

[16] https://www.whatsapp.com

Table 13.3 Platforms listed by the surveyed faculty

	LMS	Social and collaborative tools	In-class engagement	Out-of-class interactions
Teaching processes	Manage classes, documents and assignments	Create and share content	Assess knowledge and understanding	Share or ask course information
Application	Moodle, Canvas, Graasp	Office 365, Google Docs, Padlet	Speakup, mentimeter, Kahoot, Socrative, Slido	LMS, whatsapp, emails
Information	Lectures, readings, tutorials, assignment (description)	Assignments (content)	Course-related questions and answers	Course-related catch up information
Files	PDF, PPT, videos, images	DOC, PPT, PDF	N/A	(text messages) photos, videos
Technologies	Web applications	Web applications	Web applications, mobile applications	Web applications, mobile applications
Mediums	Desktop and laptop computers	Desktop and laptop computers, tablets	Desktop and laptop computers, smartphones	Smartphones (desktop and laptop computers)

Within and Cross DLPT Analysis

Table 13.3 summarizes and analyzes the differences and commonalities of the identified DLTP. As analysis framework, we used the Business-Application-Information-Technology—BAIT model [60]. The latter is used to logically and functionally describe a system and its dependencies. More specifically, the Business layer looks at the business rationale and the processes that follow from it; the Application layer investigates the organization of capabilities and functions; the Information layer looks at the data models; and the Technology layer investigates the implementation of logic, standards, bundling and tooling. For a more distinct analysis across the DLTP, we adapted the BAIT model as follows: The Business layer became the teaching process. We added a File layer to represent the types of file managed by the system and a Medium layer to be able to better differentiate the medium used to execute the application.

Assessment and Improvement of Inclusive Learning with Digital Learning Platforms (DLTP)

The analysis above sheds light on the types of application used by a selected sample of lecturers. Though it helps identify what applications and functionalities are the most popular DLTPs, the following critical question remains: How much do these learning platforms and applications support inclusive learning, and how it can be evaluated based on the data originating from these platforms? To answer it, we proceeded as follows: First, we extracted the challenges linked to inclusive learning as described in the section above. Second, we classified the challenges along three impairments—i.e., physical that refers to a student's physical limitation to follow a course; cognitive that refers to a student's potential troubles to remember, learn new things, concentrate, or take decisions and behavioral that refers to inability for a student to build or maintain satisfactory interpersonal relationships with his or her classmates and lecturers (Table 13.4). Third, we mapped these impairments to functionalities provided by digital learning platforms and tools and reasoned upon their assessment functionalities based on the data originating from the application logs (Table 13.5). Finally, we suggested potential implications and functionalities for some of the unmet challenges based on our focus groups with learning designers (N = 3).

Physical Impairments

Physical impairments are divided into two subcategories: (a) The inability to attend classes or exams, caused by injuries, care commitments or illness, and (b) some limitations that prevent students to hear, type or see. For the former (a), DLTPs should allow students to access teaching materials regardless of their location, as long as they have an Internet access. The policy with regards to video-recording of the lectures depends very much upon the university. In some cases, all classes are recorded, while the opposite is also true. Traditional universities are afraid that students would not attend classes if they had online access to such content. Given that some students might not return to classes for several months, the ability to view lectures and write exams remotely is critical. Software applications such as Examsoft[17] or Rogo[18] that provide an Internet-based safe environment are increasingly used by universities (particularly during the Covid pandemic). On the other hand, students suffering from hearing, typing or visual impairments (b) are not offered much help from DLTPs. Though collaborative tools such as Office 365 embed "read aloud" and "dictate" functionalities for Word and Excel, it is often not the case for slide-decks (e.g., PowerPoint) that require the lecturers to manually add captions. Similar problems can occur with PDFs documents depending on the original format and the conversion process. Moreover, video captions are also very rarely added by lecturers and when added the caption software can make mistakes (particularly when scientific terms and acronyms are used). However, some applications such as Microsoft Teams allow for live captions. Lastly, students who have difficulties using a keyboard do not receive much support from DLTPs. Although

---

[17] https://examsoft.com

[18] https://getrogo.com

Table 13.4 Inclusive learning conditions and system requirements

Inclusive learning conditions		System requirements	R#
Behavioral	Difficulties coping with stress and anxiety	Assessment of stress, anxiety level and exercises such as meditation which might complement medical treatment (e.g. medication or counselling)	1
	Become angry with others	Assessment of anger level and exercises to reduce it. Some might complement medical treatment (e.g. Cognitive Behavioral Therapy (CBT))	2
	Hallucinations	Assessment of the hallucination and treatment to reduce it might complement medical treatment (e.g. medication or CBT)	3
	React in "odds way"	Assessment of the behavior and whether it impairs learning and/or normal life. If it does develop	4
	Anxiety/Worry disproportionally	Assessment of the condition and suggestion of exercises according to the condition, which might complement medical treatment (e.g. anxiolytics or CBT)	5
	Depression	Assessment of the condition and suggestion of exercises according to the condition, which might complement medical treatment (e.g. antidepressants or CBT)	6
	Scared of criticism and feel judged/Perfectionism	Assessment of the condition and suggestion of exercises which will complement treatment (e.g. Counselling or CBT)	7
Physical	Not able to attend lectures	Distance learning capabilities	8
	Not able to attend exams	Distance exam taking and monitoring	9
	Not being able to hear	Generation of content captures (images, schemas, videos) and audio captures (lecturers, students)	10
	Not being able to see	Generation of content captures (text, images, schemas, videos)	11
	Not being able to type	Speech recognition throughout the different applications	12
Cognitive	Attention—Difficulties focusing	Assessment of the attention and exercises to improve focus and attention (some might complement treatment such as medication)	13
	Memory—Difficulties remembering	Assessment of the "memory" and exercises to improve memory	14
	Executive function—Mix up letters	Assessment of the condition (e.g., dyslexia) and exercises to improve and use of specialized software or hardware (e.g. colored filters)	15
	Executive function—Slow readers and writers	Assessment of the impairment (might be one of the symptoms of dyslexia). Alternative input and output modes (e.g., voice) and exercises to improve or use of software to read text and to type text	16
	Executive function—difficulties to follow and organize plans	Assessment of the condition (e.g., dyslexia) and exercises to improve and specialized software (e.g. mindmaps)	17

Table 13.5 Map between system requirements and DLTPs

R#	LMS	Social and collaborative tools	In-class engagement	Out-of-class interaction	Opportunities
1	Asynchronous teaching (e.g. video recording of lectures, so that student can stop/rewind when needed and does do not stress them because they might not understand a detail)	Offline interactions (reduce stress of in-preson interactions)	Software to ask or answer questions anonymously	Use of online tools to interact that do not require in person meeting	Wearables such as Apple Watch[a] or FitBit Sense[b] embed mechanism to detect stress using different biomarkers. There also exists dedicated mobile apps, building on scientific research, to assess and reduce stress [61]. More commercial solutions are also on the market such as Headspace[c]
2	Asynchronous teaching (e.g. video recording of lectures, so that student does not need to come in person to a room where they might become angry with others)				Some scientifically tested mobile applications allow the assessment and management of anger, notably based on Remote Exercises for Learning Anger and Excitation (RELAX) [62].
3	Asynchronous teaching (e.g. video recording of lectures, enable student to study the material when medication has helped reduce the psychosis)				Several mobile apps allow for the assessment and treatment of hallucinations [63].

(continued)

Table 13.5 (continued)

R#	LMS	Social and collaborative tools	In-class engagement	Out-of-class interaction	Opportunities
4,5	Asynchronous teaching (e.g. video recording of lectures, so that student does not need to attend in person, reducing chances to interrupt the study of other students because of their odd behaviour)	Offline interactions (reduce disruption cause by odd behaviour, e.g. tourette syndrome)	Software to ask and answer questions online will enable student to participate regardless of ticks or other odd behaviour	Use of online tools to interact that do not require in person meeting (e.g. small group meetings using MS Teams where cameras are off so student does not have to worry about suppressing ticks or other odd behaviour)	
6			Software to ask or answer questions anonymously		
7					There exist many mobile apps that help assess the condition and suggest digital therapies [64].
7		Distance learning is enabled by means of slide sharing and live sessions.	Students and teachers can interact via video or text messages.		
8,9	Additional modules allow for protected and secured online exams			Bespoke apps/ webpages for online examination (e.g. Rogo)	
10		Voice captures can be automatically generated (from video too), while image captures require manual description.		Text messaging software/tools enable off class interaction with peers	

Table 13.5 (continued)

R#	LMS	Social and collaborative tools	In-class engagement	Out-of-class interaction	Opportunities
11		Platforms such as Office365 embed "read aloud" throughout its applications. Still limited with pictures or other media.			
12					Specialized dictation software[d]
13	Class material (including videos) can be accessed away from the classroom (in a quiet place with less distractions) and at a time when the student is less tired				Mobile apps,[e] scientifically tested can help student focus
14	Class material (including videos) can be accessed multiple times to enable revising multiple times content				Mindmap software to organize the knowledge more visually[f]
					Several apps have shown to be efficient to assess, improve information retention and exercise the memory [65, 66].
15		Software (like Office365) help with spelling and grammar			Apps can support the assessment and provide exercises to improve the condition [67].

(continued)

Table 13.5 (continued)

R#	LMS	Social and collaborative tools	In-class engagement	Out-of-class interaction	Opportunities
16		Platforms such as Office365 embed "read aloud" throughout its applications			Specialized dictation software[g] Content captures as well as speech recognition apps can help with providing alternative modes.
					Specific apps can also be used to work on different aspect of the memory and help develop reading and writing skills [68].
17	Timetables embedded in LMS can help student create their individual timetable	Platforms such as Ms. Teams embed calendar functionality that can be linked to online classes			Time management tools such Rescue Time[h]

[a]https://www.apple.com/ae/watch/
[b]https://www.fitbit.com/global/us/products/smarwatches/sense?sku=512BKBK
[c]https://www.headspace.com/
[d]https://www.nuance.com/en-gb/dragon.html
[e]https://www.frontiersin.org/articles/10.3389/fnbeh.2019.00002/full
[f]http://freemind.sourceforge.net/wiki/index.php/Main_Page
[g]https://www.nuance.com/en-gb/dragon.html
[h]https://www.lifehack.org/articles/technology/top-15-time-management-apps-and-tools.html

there exist additional software applications such as DragonSystems,[19] they would only partially fulfill the tasks of a student. As an alternative, in-class engagement apps can also be used post-lecture to continue discussions that occurred during lectures. To conclude, multimodal access to content is currently missing.

Cognitive Impairments

Cognitive impairments refer to students who have difficulties focusing, remembering and those who are slow readers and mix up letters. It can also be side effects from anxiety, stress, depression or medication taken for physical or mental health

---

[19] https://www.nuance.com/dragon.html

disorders such as pain or schizophrenia. In both cases, online access to teaching materials and video-recordings in particular, brings many advantages. Students can learn at their own pace, moving back and forth until they have acquired the knowledge. However, LMS and other platforms fail at two levels: (a) in providing tailored learning experience, assuming that each student has the same learning capacity, (b) in assessing the condition and providing exercises to support the students. Similar to guidelines for people suffering from physical impairments (e.g., minimum font-size, reduced amount of length, prioritizing images), digital learning platforms have the opportunities to implement machine learning driven quizzes that adapt the questions to the learners as well as adapt the content presentation (e.g., learning section) too. In parallel, students can also use wearables to track their biometrics and thus have a more objective assessment of the impairments.

Behavioral Impairments

Students with disorders such as depression, anxiety, autism or schizophrenia display some behavioral impairments that prevent them from fully engaging with the learning. They sometimes have access to support workers that take notes for them or extra time during assignments. They also sometimes have access to "safe" environments such as dedicated rooms for them to calm down in case of a sensory overload. Aside from this, tools such as Speakup offer safe digital space in which students can anonymously ask and answer questions without the fear of being individually criticized or judged. For students whose behavioral impairments prevent them from attending lectures or can attend only partially, remote access to teaching materials (e.g., video-recordings or slides) significantly increases the students' chance to succeed. In this view LMS plays an important role as central platform for content access but also for handing in assessments and receiving potential feedback and discussions such as in forums.

What Do DLTPs Support? What Remains Unsupported?

In order to analyze the means by which DLTPs address the identified inclusive learning conditions summarized in Table 13.4, we extracted the system requirements from the same table and mapped them to functionalities provided by the currently used, digital learning platforms and tools. We added a column *opportunities* to demonstrate how technologies beyond our scope of analysis can support inclusive learning. The results are presented in Table 13.5. The first column of the table, R#, refers to the system requirements presented in Table 13.4. We also differentiate between technologies that assess or directly address the condition (in dark grey) and those that mitigate (in light grey) the condition. For instance, condition R#—difficulties coping with stress and anxiety: DLTPs offer functionalities to mitigate the condition by means of asynchronous teaching that allows students to review the

Summary of Week by Week schedule

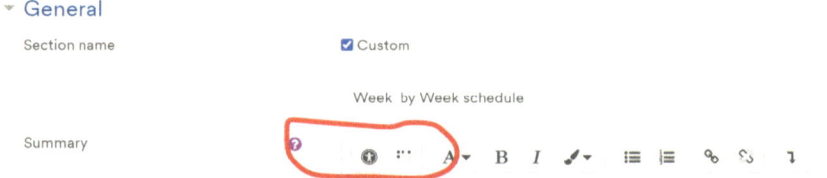

Fig. 13.9 "Accessibility checker" and "screen reader helper" currently embedded within the LSM system Moodle

knowledge as many times as required, and offline and anonymous communication that allow students to be protected. On the other hand, wearables and mobile apps relying on biometrics and self-reporting allow for a quantitative assessment of the level of stress.

When looking at the table above, it is clear that the coverage of inclusive learning's needs by the identified DLTPs is very limited. Most of the platforms rather mitigate (dark grey) than assess and improve the different conditions. Apart from providing distance learning capabilities required for students who cannot access the facilities, the tools fail in supporting behavioral and cognitive impairments. Let us take stress problem as an example: though some of the learning designers we interviewed have setup dedicated pages and modules in LMS to provide links to specialized websites, DLTPs do not support students in assessing and helping with their stress overload. However, while a considerable number of techniques, approaches and a significant number of mobile applications a at the moment, DLTPs do not provide any integration for these applications. For some of the challenges for learning, we identified specialized tools/apps that are already on the market. However sometimes it might be challenging to embed them within the existing learning technologies. Some of the current technologies are starting to address this issue (i.e. Moodle "accessibility checker" and "screen reader helper" as shown below (Fig. 13.9).

Additionally, in some cases the challenges faced by students require medical intervention (e.g. psychosis or depression). In these cases the teaching apps should facilitate learning, but they can not replace medication and/or talking therapies. For the majority of the conditions, assessment capabilities are available with external applications (light grey). Very often these applications and platforms have been scientifically evaluated.

Supporting Inclusive Learning: Implications for Design of Future Platforms

Increasing access to learning has been the focus of a great deal of research. The Universal Instructional Design (UID) is certainly the most comprehensive [69]. It builds on the concept of Universal Design to identify and eliminate unnecessary barriers to teaching and learning while maintaining academic rigor. UID is therefore described as a process that involves considering the potential needs of all learners when designing and delivering instruction. We used the UID as framework in order to refine the seven UID principles for the purpose of inclusive learning. Our inclusive learning principles, presented in Table 13.6, build upon the analysis that we presented above as well as the survey and the interviews we conducted.

Table 13.6 Universal Instructional Design applied to inclusive learning

UID principles	Application in the context of inclusive learning
Accessible and fair (equitable) use: Instructions should be designed to be useful and accessible by people with different abilities, respectful of diversity, and with high expectations for all students.	Universities and lecturers should make use of video-recordings as well as remote access to teaching materials for students who cannot physically attend classes. In addition, they should use (automatic and accurate) captions as much as possible to reach students suffering from hearing impairments.
Flexibility in use, participation and presentation: Learning is most effective when it is multimodal. Instruction is designed to meet the needs of a broad range of learner preferences. Students can interact regularly with the instructor and their peers.	Content should be delivered in different ways including videos (with audio) and in writing. Lecturers should also use digital forums and other in-class engagement tools to foster exchanges with and among students.
Straightforward and consistent: Instruction is designed in a clear and straightforward manner, consistent with user expectations. Tools are intuitive. Unnecessary complexity or distractions that may detract from the learning material or tasks are reduced or eliminated.	Content (e.g., terminologies, description, abbreviations, file names) as well as learning objectives should be presented consistently across all digital platforms used by the lecturers. If applicable, a dictionary of terms shall be accessible. Humor or references to undocumented content should not be used.
Information is explicitly presented and readily perceived: Course expectations are transparent. Instructions are easy to understand. Communication is clear. Any barriers to receiving or understanding are removed. Information may be presented in multiple forms.	One digital platform (e.g., LMS) should be used as a central hub to indicate the learners what to do and where to find the information. Lecturers must choose digital platforms and communication channels accessible by all students (e.g., cross platforms).
Supportive learning environment: Instruction anticipates that students will make mistakes. While instruction recognizes that errors are necessary, and if handled properly, present powerful learning opportunities. It tries to minimize hazards that can lead to irreversible errors and failures. Instruction also recognizes that systems will fail and things can go wrong—thus, a tolerance for error and preparation by way of backup are important so that learning will not be interrupted.	Instructions should be designed so to reduce the chance of mistakes to its minimum. It starts with the use of a widespread vocabulary accessible to everyone and continues with a session design that ensures that students identify potential misunderstandings rapidly. It includes the use of self-assessment tools, group work, in-class interactions and quizzes (which provide a safe environment for mistakes to be used as learning opportunities).

(continued)

Table 13.6 (continued)

UID principles	Application in the context of inclusive learning
Minimize or eliminate unnecessary physical effort or requirements: Instruction is designed to minimize non-essential physical effort (i.e., not related to a learning outcome) in order to allow maximum attention to learning.	Instruction and knowledge must be accessible for students regardless of their impairments. The lecturers should make use of remote access whenever possible. Vocabulary as well as instructions must be as clear as possible. Students should also be able to hand in assignments or exams online, while using a system that integrates word processing whenever appropriate.
Learning space accommodates both students and methods: The learning space is accessible and the environment supports multiple instruction strategies.	Learning spaces (e.g., method, content, discussion, assignment) should be physically and mentally accessible by all students. It should combine spaces for physical and virtual classes, communications and instructions. The learning space should adapt according to the individual student needs and impairments.

Discussion

There is no doubt that information technology is transforming education. When used for assessment and improvement of inclusive learning, it offers a wide range of opportunities for students suffering from behavioral, physical and cognitive impairments. Beyond the application's functionalities and goals, from our data and own observations, we identified five key elements for a successful implementation and uses of digital platforms and tools that assess and improve the individual's inclusive learning. The discussion part ends with some limitations and a note to the significant positive impact that information technology to enable teaching and learning during the Covid19 pandemic.

Assessment vs. mitigation of the conditions Apart from physical impairments that are not necessary expected to be assessed via DLTPs, behavioral and cognitive impairments are not assessed by DLTPs as shown in Table 13.4. They rather provide limited functionalities to mitigate the conditions such as "read aloud" for students suffering from visual impairments. Though logs from DLTPs can be used in order to assess the number of interactions between students and teaching material for instance, this functionality has its limits. It will not inform the lecturer whether the student has read or understood the lecture, but only that he had opened it. However, a huge number of opportunities come from mobile applications as well as wearables that rely on either biometrics or self-reporting to provide objective assessment of the conditions. This is notably the case for stress and anxiety, attention loss and memory retention's problems.

Usability Research has looked into the usability of digital learning platforms and learning management systems in particular. As stated by Harrati et al. (2016), *"Positive user experience is of prime importance for educational learning systems* *playing vital role for the acceptance, satisfaction and efficiency of academic institutions*" [70]. Usability of digital learning platforms has drawn the attention of many researchers after universities and schools complained that the tools do not bring the promised outcomes and that they are not used appropriately. For example, it is not uncommon for universities to install plugins for improving user interface's ease of use. For a successful implementation of digital learning platforms, it is then critical for the solution provider to focus on user-centeredness (e.g., lecturer, student). Though most LMS embed functionalities to interact in classes and create pools to engage with students, all lecturers that we interviewed use additional applications for their high usability and accessibility. It is even more important in the case of inclusive learning due to students' potential impairments.

Integration in the lecture design It is not sufficient to equip lecturers and students with digital learning platforms and tools. The latter must be fully integrated in the course and curriculum design. Video-recordings as well as in-class engagement activities must be part of a wider program aligned with learning objectives.

Integration in the enterprise architecture In-class engagement apps as well as collaborative tools are rarely integrated with learning management systems. Beyond technical challenges, they require taking additional steps (e.g., different logins, user interfaces, terminologies) that become confusing for students with cognitive impairments. Universities therefore have to ensure that the same terminologies and descriptions are used. As mentioned above, one system (e.g., LMS) must play the role of central hub from which students can access every other application.

Platform monitoring Once in place, the adoption and usability of the platforms have to be closely monitored, also serving for an assessment of the inclusive learning path, i.e., the learning path of the individual. For each impairment category, champions amongst students have to be identified and their uses (e.g., logs) analyzed. Together with the lecturers, learning designers and software providers, universities have to collaborate to ensure that the functionalities meet the student needs. Although it might sound trivial, it is much often not the case as highlighted by the learning designers we interviewed.

Assessment and impact on quality of life In their current implementations, digital learning platforms are designed to assist rather than assess students' quality of life. Following the Quality of Life Technologies' definition [71], DLPT are limited in two areas. First, in quantitatively assessing and informing the students with regards to their improvement in any of the three impairments: physical, behavioral and cognitive. To date, only the total (sum up) mark for all assessments in a subject/course/module can be used to assess any type of progression. Though time spent on lectures, number of submission's attempts, number of typos could potentially be used to evaluate a student's progress, this information is not available to students nor educators. Second, in integrating data from sensors, activity trackers and mobile devices, students (and potentially parents) and teachers have a chance to either detect or follow up on some conditions and quality of life in particular using quantitative data [72]. In addition to questionnaires that are given to students at the beginning of the semester to assess e.g., stress and anxiety, the integration of continuous data from sensors in the LMS could help students, parents and schools to quantitatively analyze students' progress and more importantly find correlation between changes in quality of life and impact of academic results.

Digital therapeutics Some behavioral conditions such as stress, anxiety and depression, digital therapeutics are seen as a new type of (mobile) applications that are either used to complement medication or to replace it. Described as *"evidence-based therapeutic interventions driven by high-quality software programs to prevent, manage, or treat a medical disorder or disease*[20]*"*, some of these applications have already been approved by the U.S. Food and Drug Administration. Building on Cognitive Behavioral Therapy, these applications engage with their users on a daily basis while using objective data to assess and customize digital treatments [73]. These applications could be seen as an alternative or a complement to counselling sessions.

Overcoming learning ability? Limitations Technology, and in particular the apps/programs discussed this chapter, is not the only solution to facilitate inclusive learning. Often students benefit from a combination of multiple high-tech and low-tech approaches, tailored to their unique needs. For example, there is software specially developed to help students with learning difficulties (such as Optical Character Recognition software to help students with dyslexia or talking calculators to help students with dyscalculia), which has been widely reviewed on [45, 74]. Another example is granting extension on coursework deadlines to enable students with dyslexia the extra time they may need for proof reading or to provide a student with a broken arm time for their arm to heal, so that they can write the coursework. Additionally, students with mental health disorders (poor mental health) often require medical treatment aimed to heal their condition, such as medication and/or counselling, which has little to do with the platform capabilities. Another word of caution included a comment that *"There are many other apps, but they tend to behave similarly to the ones we analyzed"* (comments from a learning designer), indicating that the platforms need proof of authenticity.

Impact of information technology on teaching and learning during the Covid19 pandemic Although most of the content of this book chapter was written before Covid19, we could not conclude this chapter without referring to the role that information technologies have had in ensuring a seamless transition into online-only teaching during the pandemic. Within weeks universities had to migrate all their

---

[20] https://dtxalliance.org/

learning and teaching activities to online-only provision. From discussions with colleagues across the world we have learn that level of previous familiarity with LMS and applications (and "willingness to change" of members of staff) impacted the speed and success of the migration. However, regardless of previous preparation, all universities had to transfer as much teaching as they could online, with MS Teams suddenly being used by lecturers and students across the globe, often with just hours or minutes to learn how to use it. An extreme case was seen at the School of Veterinary Medicine and Science, Nottingham University (UK), which achieve a full-online start of the academic year for their April 2020 intake, with lecturers having only a few days' notice to migrate online their teaching delivery. At the time of writing of this article, we could see that across the world academics have adapted their courses to distance and Covid-secure teaching, using a mixture of online delivery/assessment for any subject that can be taught remotely and in person Covid-secure teaching for activities that must happen in person, e.g. nursing practical teaching [75].

Conclusion

Universities have a long history of trying to offer education to students who have different needs [76, 77]. However, regardless the technology used or the teaching method, it is interesting to observe that students sometimes complain; for them a "proper" lecture is sitting on a chair for a couple of hours while listening to the lecturer's monologue that tells them what they need to know in order to pass the exam. Though digital learning and teaching platforms do support assessment and improvement of inclusive learning and teaching, which may influence the wellbeing and quality of life of the students, for a sustainable impact, changes in mindsets have to take place for both, lecturers and students. These platforms became an unexpected blessing and saving grace when the world faced the Covid19 pandemic. We are yet to know the real impact of these new practices on the long term learning outcomes and life quality of the individuals influenced by these radical and unexpected changes.

References

1. Jackson D. Factors influencing job attainment in recent Bachelor graduates: evidence from Australia. High Educ. 2014;68(1):135–53.
2. Layer G. Disabled students sector leadership group (DSSLG) inclusive teaching and learning in higher education as a route to excellence. London: Department for Education; 2017.
3. May H, Bridger K. Developing and embedding inclusive policy and practice in higher education. York: Higher Education Academy; 2010.

4. Gioia GA, Isquith PK, Guy SC, Kenworthy L. Behavior rating inventory of executive function professional manual. Florida: Psychological Assessment Resources. Inc; 2000.
5. Wojciulik E, Kanwisher N. The generality of parietal involvement in visual attention. Neuron. 1999;23(4):747–64.
6. Ross-Swain D. RIPA-2: Ross information processing assessment. Pro-Ed; 1996.
7. Ruff RM, Niemann H, Allen CC, Farrow CE, Wylie T. The Ruff 2 and 7 selective attention test: a neuropsychological application. Perceptual Motor Skills. 1992;75(Suppl 3):1311–9.
8. Ridley Stroop J. Studies of interference in serial verbal reactions. J Exp Psychol. 1935;18(6):643.
9. Army Individual Test Battery. Manual of directions and scoring. Washington, DC: War Department, Adjutant General's Office; 1944.
10. Deutsch Lezak M, Howieson DB, Bigler EB, Tranel D. Neuropsychological assessment. New York: Oxford University Press; 2012.
11. Wechsler DA. Wechsler adult intelligence scale. New York, NY: The Psychological Corporation; 1997.
12. Berg EA. A simple objective technique for measuring flexibility in thinking. J Gen Psychol. 1948;39(1):15–22.
13. Derryberry D, Reed MA. Anxiety-related attentional biases and their regulation by attentional control. J Abnormal Psychol. 2002;111(2):225.
14. Enger Rosvold H, Mirsky AF, Sarason I, Bransome ED Jr, Beck LH. A continuous performance test of brain damage. J Consult Psychol. 1956;20(5):343.
15. MacLeod C, Mathews A, Tata P. Attentional bias in emotional disorders. J Abnormal Psychol. 1986;95(1):15.
16. Wechsler D. The measurement of adult intelligence. Baltimore, MD: Williams & Wilkins Co; 1939. https://doi.org/10.1037/10020-000.
17. Pylyshyn ZW, Storm RW. Tracking multiple independent targets: evidence for a parallel tracking mechanism. Spatial Vis. 1988;3(3):179–97.
18. Gronwall DMA. Paced auditory serial-addition task: a measure of recovery from concussion. Perceptual Motor Skills. 1977;44(2):367–73.
19. Logan GD, Van Zandt T, Verbruggen F, Wagenmakers E-J. On the ability to inhibit thought and action: general and special theories of an act of control. Psychol Rev. 2014;121(1):66.
20. Roid GH. Stanford-Binet intelligence scales. 5th ed. Itasca, IL: Riverside Publishing; 2003.
21. Rosen WG, Mohs RC, Davis KL. A new rating scale for Alzheimer's disease. Am J Psychiatry. 1984;141(11):1356–64.
22. Abigail Benton Sivan. Benton visual retention test. San Antonio: Psychological Corporation; 1992.
23. Rey A. L'examen clinique en psychologie [The clinical psychological examination]. Paris: Presses Universitaires de France; 1964.
24. Sternberg S. High speed memory scanning. Science. 1966;133(1966):652–4.
25. Delis DC, Kramer JH, Kaplan E, Ober BA. California verbal learning test. Research Edition Manual. San Antonio: The Psychological Corporation; 1987.
26. Weiss LG, Saklofske DH, Coalson D, Raiford SE. WAIS-IV clinical use and interpretation. Scientist-practitioner perspectives. Academic; 2010.
27. Alloway TP. Automated working memory assessment. London: Pearson Assessment and Information BV Translated and reproduced with ….; 2007.
28. Berch DB, Krikorian R, Huha EM. The Corsi block-tapping task: methodological and theoretical considerations. Brain Cognition. 1998;38(3):317–38.
29. Kirchner WK. Age differences in short-term retention of rapidly changing information. J Exp Psychol. 1958;55(4):352.
30. Petrides M, Milner B. Deficits on subject-ordered tasks after frontal-and temporal-lobe lesions in man. Neuropsychologia. 1982;20(3):249–62.
31. Cassidy S. The academic resilience scale (ARS-30): a new multidimensional construct measure. Front Psychol. 2016;7:1787.

32. Mayo NE, Figueiredo S, Ahmed S, Bartlett SJ. Montréal accord on patient-reported outcomes (PROs) use series–paper 2: terminology proposed to measure what matters in health. J Clin Epidemiol. 2017;89(2017):119–24.
33. Rinck M, Becker ES. Approach and avoidance in fear of spiders. J Behav Therapy Exp Psychiatry. 2007;38(2):105–20. https://doi.org/10.1016/j.jbtep.2006.10.001.
34. Alloway TP, Gathercole SE, Kirkwood H, Elliott J. Evaluating the validity of the automated working memory assessment. Educ Psychol. 2008;28(7):725–34. https://doi.org/10.1080/01443410802243828.
35. Atkins AS, Stroescu I, Spagnola NB, Davis VG, Patterson TD, Narasimhan M, Harvey PD, Keefe RSE. Assessment of age-related differences in functional capacity using the virtual reality functional capacity assessment tool (VRFCAT). J Prev Alzheimers Dis. 2015;2(2):121–7. https://doi.org/10.14283/jpad.2015.61.
36. American Psychiatric Association. Diagnostic and statistical manual of mental disorders (DSM-5®). 5th ed. American Psychiatric Pub; 2013.
37. Brown P. The invisible problem?: improving students' mental health. Oxford: Higher Education Policy Institute; 2016.
38. Thorley C. Not by degrees: improving student mental health in the UK's universities. London: Institute for Public Policy Research; 2017.
39. Houghton A-M, Anderson J. Embedding mental wellbeing in the curriculum: maximising success in higher education. High Educ Acad. 2017;68 (forthcoming).
40. The Engelhard Project. Retrieved December 29, 2020 from http://engelhard.georgetown.edu/
41. McManus S, Bebbington P, Jenkins R, Brugha T. Mental health and wellbeing in England: adult psychiatric morbidity survey 2014. A survey carried out for NHS digital by NatCen social research and the department of health sciences. University of Leicester; 2016.
42. Caul S. Estimating suicide among higher education students, England and Wales: experimental statistics. London: Office for National Statistics; 2018.
43. Melrose S. Seasonal affective disorder: an overview of assessment and treatment approaches. Depress Res Treat. 2015;2015
44. HESA. Higher education student statistics: UK, 2017/18 - Student numbers and characteristics | HESA. Higher Education Statistics Agency, London; 2019. Retrieved June 14, 2020 from https://www.hesa.ac.uk/news/17-01-2019/sb252-higher-education-student-statistics/numbers
45. Perelmutter B, McGregor KK, Gordon KR. Assistive technology interventions for adolescents and adults with learning disabilities: an evidence-based systematic review and meta-analysis. Comput Educ. 2017;114(2017):139–63.
46. Pino M, Mortari L. The inclusion of students with dyslexia in higher education: a systematic review using narrative synthesis. Dyslexia. 2014;20(4):346–69.
47. Treviranus J. Learning differences & digital equity in the classroom. 2018.
48. Rose T. The end of average: how to succeed in a world that values sameness. London: Penguin; 2016.
49. Mantle R. UK, 2017/18 – Student numbers and characteristics. London: Higher Education Student Statistics; 2019.
50. Sanchez-Gelabert A, Figueroa M, Elias M. Working whilst studying in higher education: the impact of the economic crisis on academic and labour market success. Eur J Educ. 2017;52(2):232–45.
51. Cuthbertson BH, Hull A, Strachan M, Scott J. Post-traumatic stress disorder after critical illness requiring general intensive care. Intensive Care Med. 2004;30(3):450–5.
52. Brown LA. Instructor usage of learning management systems utilizing a technology acceptance model. PhD Thesis. Montana State University-Bozeman, College of Education, Health & Human; 2017.

53. Bowe F. Universal design in education: teaching nontraditional students. Greenwood Publishing Group; 2000.
54. Burgstahler S. Universal Design of Instruction (UDI): definition, principles, guidelines, and examples. DO-IT; 2009.
55. Machado M, Tao E. Blackboard vs. Moodle: comparing user experience of learning management systems. In: 2007 37th annual frontiers in education conference-global engineering: knowledge without borders, opportunities without passports, IEEE, S4J-7; 2007.
56. Kakasevski G, Mihajlov M, Arsenovski S, Chungurski S. Evaluating usability in learning management system Moodle. In ITI 2008-30th International Conference on Information Technology Interfaces, IEEE; 2008. p. 613–8.
57. Holzer A, Govaerts S, Ondrus J, Vozniuk A, Rigaud D, Garbinato B, Gillet D. Speakup–a mobile app facilitating audience interaction. In: International conference on web-based learning. Springer; 2013. p. 11–20.
58. Chu SK-W, Kennedy DM. Using online collaborative tools for groups to co-construct knowledge. Online Inf Rev. 2011;2011
59. Yourstone SA, Kraye HS, Albaum G. Classroom questioning with immediate electronic response: do clickers improve learning? Decis Sci J Innov Educ. 2008;6(1):75–88.
60. Prasad G. Dependency-oriented thinking: volume 1 analysis and design. InfoQ; 2013.
61. Dennis TA, O'Toole LJ. Mental health on the go: Effects of a gamified attention-bias modification mobile application in trait-anxious adults. Clin Psychol Sci. 2014;2(5):576–90.
62. Mackintosh M-A, Niehaus J, Taft CT, Marx BP, Grubbs K, Morland LA. Using a mobile application in the treatment of dysregulated anger among veterans. Military Med. 2017;182(11–12):e1941–9.
63. Demeulemeester M, Kochman F, Fligans B, Tabet AJ, Thomas P, Jardri R. Assessing early-onset hallucinations in the touch-screen generation. Br J Psychiatry. 2015;206(3):181–3.
64. Shen N, Levitan M-J, Johnson A, Bender JL, Hamilton-Page M, Jadad AAR, Wiljer D. Finding a depression app: a review and content analysis of the depression app marketplace. JMIR mHealth uHealth. 2015;3(1):e16.
65. Pechenkina E, Laurence D, Oates G, Eldridge D, Hunter D. Using a gamified mobile app to increase student engagement, retention and academic achievement. Int J Educ Technol High Educ. 2017;14(1):1–12.
66. Zhu Y, Jiang H, Hang S, Zhong N, Li R, Li X, Chen T, Tan H, Jiang D, Ding X. A newly designed mobile-based computerized cognitive addiction therapy app for the improvement of cognition impairments and risk decision making in methamphetamine use disorder: randomized controlled trial. JMIR mHealth uHealth. 2018;6(6):e10292.
67. Politi-Georgousi S, Drigas A. Mobile applications. In: An emerging powerful tool for dyslexia screening and intervention: a systematic literature review; 2020.
68. Adefila A, Graham S, Patel A. Fast and slow: using spritz for academic study? Technol Knowl Learn. 2020;2020:1–21.
69. Burgstahler S. Equal access: universal design of instruction. DO-IT, University of Washington; 2008.
70. Harrati N, Bouchrika I, Tari A, Ladjailia A. Exploring user satisfaction for e-learning systems via usage-based metrics and system usability scale analysis. Comput Hum Behav. 2016;61(2016):463–71.
71. Wac K. Quality of life technologies. In: Encyclopedia of behavioral medicine. New York: Springer; 2020.
72. Wac K. From quantified self to quality of life. In: Digital health. New York: Springer; 2018. p. 83–108.
73. Patel NA, Butte AJ. Characteristics and challenges of the clinical pipeline of digital therapeutics. npj Digital Med. 2020;3(1):1–5.
74. Faggella-Luby M, Gelbar N, Dukes L, Madaus J, Lalor A, Lombardi A. Learning strategy instruction for college students with disabilities: a systematic review of the literature. J Postsecondary Educ Disability. 2019;32(1):63–81.

75. Mostyin A, Toledo-Rodriguez M. Online tools for teaching. From inclusive learning to "the way" to teach and assess during the COVID-19 pandemic. Physiol News. 119

76. Elaine Allen I, Seaman J. Changing course: ten years of tracking online education in the United States. ERIC; 2013.

77. Eskey MT, Roehrich H. A faculty observation model for online instructors: observing faculty members in the online classroom. Online J Distance Learn Adm. 2013;16:2.

Chapter 14
Beyond Pen and Paper: Reimagining Assessment of Personal Relationships and Quality of Life Using Digital Technologies

Matej Nakić and Igor Mikloušić

Introduction

Maintaining personal relationships represents an integral part of one's self-care and thus contributes significantly to one's overall health. According to self-determination theory, relatedness is one of the three basic psychological needs humans possess [1]. Maintaining constructive and healthy personal relationships is a way of satisfying this core psychological need. Continuously neglecting one's personal relationships and relatedness with others can potentially cause various psychopathological symptoms and prove detrimental to overall health. This chapter will focus on personal relationships in the context of quality of life (QoL) and new digital technologies, highlighting the potential benefits of implementing these technologies in research on these issues as well as limitations that may slow their widespread adoption.

The World Health Organization (WHO) defines personal relationship quality as "the extent to which people feel the companionship, love, and support they desire from the intimate relationships in their life" [2]. Personal relationships can be further described as a network that is created through the close connections one maintains with others and that implies an emotional investment of some sort [3]. Personal relationships are also highly mutable and are impacted daily by one's immediate social environment [4]. While the perceived quality of one's relationships influences QoL to a significant degree, it is also important to note that the quality of one's social interactions (and not the quantity) is the best predictor of self-reported health [5], like the relationships themselves, the perceived quality of relationships is also susceptible to change. While the perceived quality of one's relationships influences

M. Nakić (✉)
Zagreb School of Economics and Management, Zagreb, Croatia
e-mail: mnakic@zsem.hr

I. Mikloušić
Ivo Pilar Institute of Social Sciences, Zagreb, Croatia
e-mail: igor.miklousic@pilar.hr

K. Wac, S. Wulfovich (eds.), *Quantifying Quality of Life*, Health Informatics,
https://doi.org/10.1007/978-3-030-94212-0_14

QoL to a significant degree and is dependent upon numerous factors. Social exchange theory, for example, argues that the perceived quality of any relationship depends on the (1) the perceived benefit vs. cost ratio (BCR); (2) the individual's comparison level (CL), or what they believe they deserve to get out of a particular relationship in terms of BCR; and lastly, (3) the individual's comparison level for alternative (CLalt), which is the lowest BCR outcome they are willing to accept in a specific relationship given the perceived BCR outcomes of alternative relationships or of merely being alone [6]. Furthermore, equity theory suggests that BCR outcomes determine the quality of a personal relationship to the extent that they affect the relationship's perceived fairness [7].

Research on QoL and personal relationships, along with other complex social science issues, has until now been burdened and held back by technological limitations. Researchers of these topics, for instance, have generally accepted a reliance on self-reported data from small sample sizes. Additionally, due to ethical and financial limitations, collection techniques such as mobility traces and the use of personal call records have been relatively rare. Such limitations have had particularly adverse implications for intricate longitudinal research designs, making it difficult for researchers to engage individual participants over time. The number of iterations or data points researchers have been able to study has also been limited, as methodologies based on pen-and-paper self-reporting were costly and made the logging of daily, weekly, or even monthly changes in behaviors and states procedurally burdensome.

However, with the advent of computers, the Internet, and, most importantly, the nearly universal adoption of smartphones, a more suitable way to record such data emerged. Unlike traditional survey instruments, modern devices such as smartphones, smartwatches, and other wearables allow researchers to monitor behaviors and social interactions effortlessly and without additional engagement from participants. They have also allowed further innovations in the measurement of personal relationships by enabling real-time and synchronous data recording. Moreover, developments in modern data science, including advancements in sensor networks, machine learning, deep learning, and AI, offer novel ways of capturing and analyzing observed phenomena. In accordance with General Data Protection Regulation (GDPR) guidelines, researchers can accumulate large amounts of private data, such as location information, call logs, and messages (from SMS and applications such as Facebook Messenger, WhatsApp, and Viber), provided they first obtain participants' permission, which has permitted novel insights into a range of scientific inquiries. Cloud technologies, in turn, allow for almost unlimited data storage, which removes the limitations on sample size that troubled previous research. Finally, the global availability of smartphones allows sample representativeness in relevant populations to be more easily attained, thereby ensuring studies' external validity. Some pioneering researchers have already applied algorithms to data pertaining to mobile phone usage in order to measure different facets of communication, such as the frequency and duration of virtual communications and their relationship to tie strength [8].

Given these developments, this chapter aims, first, to explore the methodologies that have been used until now in studies focusing on the assessment of QoL and personal relationships, indicating, where relevant, their deficiencies and shortcomings; and, second, to propose a methodology that overcomes these shortcomings by using new digital technologies to capture and analyze relevant data in a reliable and timely manner.

Assessment of Personal Relationships and Their Impact on QoL

In studying personal relationships and their relation to QoL, researchers have relied on a variety of existing validated QoL measures that include assessments of the quantity, quality, and intensity of personal relationships to quantify QoL. As a variable, personal relationships can encompass a diverse range of phenomena, potentially including everything from marital relationship, kinship, and friendship to neighborhood-based relationships and acquaintances with fellow members of a church or club. Some studies, for instance, have considered the concept of relationship to equally imply both sexual and private relations, hence overlapping personal relationships with relationships of sexual intimacy (as covered in the WHOQOL facet "Sexual Activity" [2]). In some studies, personal relationship measures have overlapped with measures from the WHOQOL "Social Support" facet, which focuses more on the social structures an individual belongs to and the social support that is available to them when needed than personal relationships in general [2].

Below we provide a list of psychometrically standardized and validated QoL measures that involve some assessment of personal relationships as a variable. Taken together, these measures (a) apply to a range of demographics, including ages ranging from childhood to adulthood; (b) take various operationalizations of personal relationships into account, ranging from holistic to multi-faceted approaches, and (c) address various socio-environmental contexts, ranging from a particular context to a collective context, such as family.

1. **WHOQOL-100** [9].
   The World Health Organization has made sure to develope and to publish an instrument that measures quality of life in a quantitative manner. It consists of 100 items and therefore is titled WHOQOL-100. This instrument produces various scores, namely related to (a) particular facets of quality of life (e.g., social support and financial resources); (b) larger domains (e.g., physical, psychological and social relationship domain), and (c) overall quality of life and general health. With regards to personal relationships, WHOQOL-100 contains items such as "How satisfied are you with your personal relationships?" Furthermore, WHOQOL-100 is both culturally sensitive and psychometrically sound.

2. **KIDSCREEN-27** [10–13].

The KIDSCREEN project involved 13 European countries in the development of a cross-culturally harmonized QoL measure designed to be administered to children. The resulting KIDSCREEN instrument was based on literature reviews, expert consultation, and focus groups conducted in all the participating countries. The pilot version consisted of 185 items, which, after elimination of some items, was reduced to 52 of the original items. Later, a shorter, 27-item version of the same measure, called "KIDSCREEN-27," was created and validated [14]. Concerning personal relationships and their relation to QoL, KIDSCREEN-27 assesses the quality of personal relationships a child can develop in school. Items such as "Have you spent time with your friends?" "Have you had fun with your friends?" and "Have you and your friends helped each other?" are intended to evaluate the safety net of personal relationships that serve as building blocks in the development of one's self-esteem.

3. **Satisfaction with Life Scale (SWLS)** [15].

A widely used and validated measure [16–18], the SWLS encompasses five broad, global items and allows respondents to weight domains of their lives according to their own beliefs and values. The five-item scale provides a holistic and overarching assessment of an individual's satisfaction with their life. In the area of personal relationships, the SWLS's measure of QoL takes into account items such as "In most ways, my life is close to my ideal," "The conditions of my life are excellent," and "I am satisfied with my life," to which participants respond with a rating from 1 (strongly disagree) to 7 (strongly agree). Although the items do not reference personal relationships directly, individuals with close, fulfilling relationships are much more likely to report being satisfied with their lives or describing the conditions of their life as excellent or ideal. Assessments of the quality of one's relationships can therefore be a useful predictor of an individual's ratings for these three items.

4. **RAND-36** [19–21].

This measure has been validated in multiple studies [22, 23] and represents one of the most common measures of health-related QoL (HRQoL). It generates ordinal level data for each of its 36 items, which are in turn aggregated to subscale scores [22]. When assessing the quality of life with regards to personal relationships, RAND-36 limits the time frame of the retrospection. This can be shown in the following item: "During the past 4 weeks, to what extent has your physical health or emotional problems interfered with your normal social activities with family, friends, neighbors or groups." The participants are then required to choose between responses varying from "all of the time" to "none of the time."

5. **Beach Center Family Quality of Life (FQOL) Scale** [24–26].

The FQOL scale is an inventory consisting of 25 items rated on a five-point Likert-type scale that has been widely used and validated. It measures satisfaction with family life as a significant factor in QoL. In particular, it assesses family interaction, parenting, emotional well-being, physical/material well-being, and disability-related support [24] and contains items such as "My family members talk openly with each other," "My family is able to handle life's ups and downs," and "My family enjoys spending time together."

Apart from these validated measures, some researchers have leveraged non-standardized instruments and interviews to assess QoL, while others have applied more innovative approaches, such as extrapolating QoL measures from tests of social cognition and self-reported measures of social understanding [27]. However, as these approaches are far from widely adopted and often do not consider social relationships, we do not elaborate on these in this chapter.

Assessment of Personal Relationships Via Digital Item Representation

Having discussed the most common approaches to assessing personal relationships in measures of QoL, we now outline ways in which the items included in these measures have been assessed using digital technologies. We define digital item representation as the digitalization of any item of a particular instrument that maintains the format in which the item was traditionally recorded (such as binary or Likert scale formats). For an item assessing how much time an individual spent with their spouse in the past week, for example, this might involve the collection and recording of relevant information via GPS, Cell-ID positioning, WiFi information, or Bluetooth signals acquired directly from both spouses' smartphones and identifying the intervals when the two devices were in close proximity, thus rendering the individual's self-reporting superfluous. The automation that such techniques may increase engagement among study participants and motivate them to provide more data while also reducing dropout rates. It could also be argued that the quantitative and objective data derived from GPS, Cell-ID, WiFi, and Bluetooth technologies is likely to be a more accurate source of information than participants' subjective recollections [28, 29]. Such a position is further supported by research that has demonstrated that the context, wording, and format of questions can significantly impact the responses one receives [30]. Leveraging quantitative and objective datasets in the manner described above allows one to avoid the introduction of error from these sources.

An example of how digital techniques can be used to investigate social relationships is provided in a study conducted by Wiese et al. [8], where the researchers used a computational model to assess and classify personal relationships. Using call and text message logs from smartphones as inputs, Wiese et al. leveraged an algorithm to classify participants' contacts into three relationship categories: family, work, and social relationships. Based on extracted features such as the intensity, regularity, duration, and medium of communication, the authors were able to classify relationship categories with up to 90.5% accuracy [8]. Meanwhile, an ambitious approach to relationship imaging is the social MRI method developed by Aharony et al. [31]. Using credit card records and information on social media and mobile phone use, including calls and text messaging, this method provides an objective means of visualizing social systems. The authors also utilized self-reporting to collect Big Five personality tests and assessments of participants' momentary moods and sleep quality, which were later used to detect correlations between these variables and aspects of participants' social relationships. By incorporating vast amounts of privacy-sensitive data in their model's framework, the authors were able to "help further our understanding of the interconnections and mechanics of human society" [31].

From Paper-Based Surveys to Digital Item Representation

Every instrument mentioned in section "Assessment of Personal Relationships Via Digital Item Representation" operationalizes a specific variable or set of variables corresponding to a specific item in terms of data that can be collected from digital devices. Incorporating digital item representation is a process that involves replacing traditional data recording methods (in particular, the method of self-reporting) with more technologically savvy ones. An initial step towards digital item representation is made with the use of electronic patient-reported outcomes (ePRO; [32]), which adapts traditional, paper-based means of assessing patient states as patient-reported outcomes (PROs; [33]) to a digital format so that patients can complete them electronically. The next step in implementing digital item representation is to replace the self-reported data sources, which are memory-based, subjective, and infrequent, with quantitative context- and sensor-based sources that are objective and frequent. In what follows, we discuss the latter of these steps, focusing in particular on the ways digital item representation might be applied to the QoL measures described in section "Assessment of Personal Relationships and Their Impact on QoL" in order to better quantify the aspects of personal relationships that influence QoL.

To illustrate how digital item representation could be applied in the case of the WHOQOL facets, we take as an example a study by Chang et al. [34] that explored the mediating effect of depression on WHOQOL facets such as positive feeling and social support. In this study, depression was measured via traditional paper-based self-reporting. To assess each individual's depression and the quality of their personal relationships, the researchers used a WHOQOL-100 item "How alone do you feel in your life?" Another WHOQOL-100 item that would also be suitable for digital representation is "How satisfied are you with your personal relationships?"

In this case, the researchers, along with the healthcare practitioners caring for the patient, might have benefited from having additional, more objective data sources regarding the intensity and frequency of the patient's social interactions. It has been demonstrated, for instance, that depressive patients tend to suffer cognitive distortions that lead them to downplay positive events in their life, which further exacerbates feelings of depression [35, 36] and could lead to inaccurate self-reporting. As a step towards digital item representation, collecting GPS, Cell-ID, WiFi, and/or Bluetooth data from the patient would have allowed the researchers to objectively measure the time the patient spent at home or close to those whom they recognize as their significant others. An individual's close friends and contacts could be recognized by training an algorithm to assess the intensity and frequency of communications with specific individuals as well as analyze the individual's social media profiles and use of social media platforms [37]. This approach could provide a reliable way of evaluating the quality of individuals' relationships that is more accurate than asking them directly, leading to more objective assessments of the WHOQOL-BREF facet relating to personal relationships, which asks "How satisfied are you with your personal relationships?". By improving the accuracy of information, it would also allow both practitioners and researchers to better distinguish causal patterns between aspects of personal relationships and QoL. Finally, in the case of depression, as studied by Chang et al., the use of these measurement techniques could significantly impact researchers' or practitioners' assessments regarding the state of an individual's depression, the cause of their depression, and the potential mediating impact of personal relationships on their depression.

The KIDSCREEN-27 measure, meanwhile, includes a facet called "Peers and Social Support" [38]. The traditional paper-based KIDSCREEN-27 tries to operationalize this facet via items such as "Have you and your friends helped each other?" [11]. However, a digital representation of this item would be possible via an algorithm that incorporates sociometric techniques to render a sociogram, which would in turn provide information about specific peer dynamics inside an observed cohort of children [39]. To develop a clear, unbiased picture of this item, this algorithm could monitor the social network activity of a child (including information such as GPS data) and analyze the online content they are posting, sharing, liking, or commenting on (e.g., on their phone, tablet, or computer), as well as other relevant information such as the number of hours the child spends playing outside or in multiplayer games online.

As for the Satisfaction with Life Scale (SWLS; [15]), traditional items that ask participants to retrospectively assess and self-report their satisfaction with their lives could easily be digitalized via a mobile application that would occasionally prompt the individual to engage in a brief QoL assessment on their phones. Such an assessment would leverage methods such as experience sampling methods (ESM; [40, 41]), also referred to as ecological momentary assessment (EMA; [42]), and be designed to capture momentary, self-reported ratings of experiences, moods, thoughts, symptoms, or behaviors that are expected to change over time. Such momentary methods have been shown to be psychometrically superior to the usual, retrospective self-reports with their longer recall periods [43]. The near-ubiquitous availability and affordability of smartphones has contributed significantly to the realization of the ESM/EMA methodology's true potential by facilitating the capture of momentary data. Not only do individuals in most developed countries already possess this potential research hardware [44], but they also carry it at their side throughout the day [45], providing researchers with a potentially unlimited stream of self-reported as well as passive datasets. Despite the benefits of conducting self-reporting with a smartphone, however, there is still room for improvement. In a comprehensive review of the use of mobile phone devices in ESM, Van Berkel et al. [46], have noted that the notifications prompting participants to rate their feelings are often burdensome. These prompts can be adapted by analyzing user-generated data to send the prompts at times when the individual is not occupied with other tasks.

While self-reports provide important information on an individual's situational contexts, they continue to rely on the user and require mobile notifications that are likely to continue being somewhat disruptive. As described above, data concerning one's situational context can be collected passively by leveraging multiple sensors that are embedded within smartphones by default [47]. For example, an automatic, continuously sensing smartphone-based application may incorporate data from a phone's accelerometer, microphone, light sensor, GPS, Cell-ID, WiFi, and Bluetooth activity. Data collected through smartphone sensing—including call logs, messages, and data on sleep, physical activity, and location—could also provide information about individuals' social interactions, activities, and mobility [48]. It is further possible to create a computational model that leverages deep learning technologies for classification tasks on data received from sources such as GPS, Cell-ID, WLAN, and call, SMS and social media logs. In summary, the techniques of digital item representation discussed above, including the use of ESM/EMA-based self-reporting and the collection of passive datasets using smartphones, could be leveraged for SWLS assessments in individuals' daily lives. Using smartphones for momentary self-reporting can improve the reliability of individuals' responses to items such as "In most ways, my life is close to my ideal," "The conditions of my life are excellent," and "I am satisfied with my life." Meanwhile, computational models of individuals' social, physical, and recreational activity developed through passively collected data can help researchers and practitioners better assess the veracity of responses and the factors that are correlated with positive or negative assessments.

Another instrument mentioned above that is used for assessing personal relationships is the RAND-36 [22]. Apart from asking participants to report the same information via a digital platform and thus leveraging the ESM/EMA approach, there are specific methods that can be implemented to complement self-reported RAND-36 data. In section "Assessment of Personal Relationships and Their Impact on QoL" we have already emphasized a specific RAND-36 item pertaining to personal relationships ("During the past 4 weeks, to what extent has your physical health or emotional problems interfered with your normal social activities with family, friends, neighbors or groups"). The scope of this item is fairly broad and we can therefore measure it in various ways. Firstly, it can be assessed through analysis of the specific patterns or content of one's conversations. Conversations can be logged to detect patterns, while text collected from messaging apps and SMS logs can be semantically analyzed. An example of how an algorithm can be deployed to analyze textual output on social media is found in the recent case of Weibo users being monitored amidst the COVID crisis in China. The algorithm identified persons at risk of suicide and alerted responsible volunteers to contact authorities [49]. Secondly, to assess the tangible element of RAND-36's social activities, an algorithm that analyzes textual output could be implemented in combination with GPS, Cell-ID, WiFi, and Bluetooth proximity data to complement self-reported responses.

Thirdly, the emotional element of social activities could be further evaluated based on analysis of the emotional content and intensity of virtual interactions found in messaging app and SMS logs, including the use of emoticons. Self-reporting concerning the facet of positive social interactions could be complemented with GPS, Cell-ID, WiFi, and Bluetooth data that indicates an individual's most frequently visited places (via geolocation) and time spent in proximity to other individuals (via analysis of other devices in one's proximity).

Finally, in the case of the Beach Center FQOL [25], which contains items such as "My family members talk openly with each other," "My family is able to handle life's ups and downs," and "My family enjoys spending time together," digital item representation could be carried out using GPS, Cell-ID, WiFi, and Bluetooth proximity data for the individual and their family members. Logging the locations where family members spend time together and the duration of their interactions could provide insight into familial social relationships and enrich the data gathered from self-reporting. Furthermore the call, SMS, and messaging app logs of family members might be compared to assess individual family members' communication styles. Multiple communication parameters could be recorded, such as frequency, duration, content, expression style (analyzed for emotional expressions), and most preferred recipient. If a family collectively uses a messaging app such as Viber or WhatsApp, a group chat would offer valuable data for assessing individual family members' communication styles while exploring family dynamics.

A summary of suggestions for implementing digital item representation in the QoL measures discussed above is presented in Table 14.1.

Researchers seeking to assess QoL using QoL measures such as those mentioned above can enrich their methodological toolboxes by making use of accurate, timely, and privacy-conscious computational models of personal relationships developed using the techniques mentioned above. Since many researchers continue to use self-report-based, non-standardized, single-item QoL measures that are conceptually broad and fail to establish structural relations between variables or provide in-depth insights (e.g., "Describe your quality of life"; [27, 50–52]), in the following section we discuss the necessity for further research to establish the validity of digital item representations.

Discussion: Limitations of Digital Item Representation

As discussed above, digital assessment via smartphones and wearables can provide researchers with larger sample sizes and render more accurate, synchronous measurements. Problems relating to sample size, representativeness, external validity, and assessment standardization could be singlehandedly solved by using validated digital platforms to mine and store data. An important issue that emerges as researchers move towards collecting data from applications and other smartphone metrics, however, is privacy. There is rising concern among mobile phone users regarding

Table 14.1 Proposed digital item representations of items from selected QoL questionnaires

Instrument	Items	Sources to quantify the factor
WHOQOL-100	"How alone do you feel in your life?"; "How satisfied are you with your personal relationships?"	Smartphone (GPS, Cell-ID, WLAN, data use info for indoor/outdoor activity assessment; call, SMS, social media messenger logs); ESM/EMA (for loneliness assessment)
KIDSCREEN-27	"Have you and your friends helped each other?"	Smartphone (calls, social media activity); gaming consoles (gameplay logs); personal computers (eLearning platforms, gameplay logs)
SWLS	"How would you rate your satisfaction with your own life in the past three months?"; "How often do you feel sad?"	Smartphone (call, SMS, social media messenger logs); ESM/EMA (for sadness assessment)
RAND-36	"During the past 4 weeks, to what extent has your physical health or emotional problems interfered with your normal social activities with family, friends, neighbors or groups."	Smartphone (GPS, Cell-ID, WLAN info; call, SMS, social media messenger logs); ESM/EMA (for quality of personal relationship rating assessment)
Beach Center FQOL	"My family enjoys spending time together"; "My family members talk openly with each other."	Smartphone (GPS, Cell-ID, WLAN info; call, SMS, social media messenger logs for family members)

privacy, data sharing within applications, and security breaches [53–55], the latter of which allow unwanted parties to collect certain information about users and use it for marketing or more sinister purposes. For this reason, the European Union introduced the General Data Protection Regulation (GDPR) in 2018, which focuses on data protection and privacy limitations and outlines necessary data protection practices in the European Union. These regulations thus specify legal and ethical limitations to the data practices researchers in Europe can adopt to complement standard survey practices. As regulations in distinct areas are likely to differ, the ways in which digital research methodologies can be implemented will vary greatly by geographic region.

Furthermore, with the use of data-gathering applications and algorithms, elaborate consent forms and data security protocols need to be implemented to protect participants' privacy. Rather than study the relevant regulations and implement these protocols, many researchers are likely to opt out of using such tools despite their benefits. However, initial steps have been made towards implementing these methods more broadly, while several studies have been able to evaluate users' attitudes towards data sharing, providing future researchers with concise and specific suggestions concerning how to increase the validity and trustworthiness of their digital instruments [56]. When collecting smartphone, wearable, or EMA data from users in order to assess their relationships, one essential requirement is that researchers be clear about the immediate aims of their research, as well as the fact that their ultimate goal is to improve the individual's life quality in the long term. As far as child privacy is concerned, when using instruments such as the KIDSCREEN-27, it is of the utmost importance that researchers acquire parental permission before collecting any data.

The scope of the research presented in this chapter has had certain limitations. We do not offer a systematic literature review of all the instruments used to evaluate personal relationships in relation to QoL, nor is our discussion of the possibilities offered by smartphones and wearables for the digital quantification of the above items comprehensive, as such devices as smartphone gyroscopes and wearable heart rate and galvanic skin response monitors likely present additional research applications. Nevertheless, this chapter has opened a discussion on the digitalization of current methods of assessing personal relationship as a factor in ratings of QoL.

Looking towards the future, recent research results have shown that the use of new, personal, miniaturized technologies is bound to become prevalent. Indeed, there are already individuals who are taking part in the "quantified self" movement, leveraging these technologies for assessment of their own personal relationships [57]. One noteworthy wearable device in this trend is the Filip Smartwatch (https://www.myfilip.com/), which supports family communication and location-based information exchange without the use of a smartphone. Developments in the quantified self community are at the forefront of the trend of "self-knowledge through numbers," and personal relationships are just one of the relevant areas individuals seek to quantify and improve. Overall, technologies used to assess personal relationships are examples of quality of life technologies (QoLT), a term referring to any technologies used for assessing or improving an individual's QoL. These technologies leverage the increasing availability of miniaturized, communication sensor- and actuator-based, context-rich technologies for computation and storage that can be embedded within personal devices such as smartphones and wearables [58]. There is therefore a promising future in the use of QoLT for the assessment of personal relationships based on data collected from daily life environments.

Concluding Remarks

With the further implementation and standardization of digital item representations for the assessment of personal relationships and QoL, researchers and practitioners will have excellent opportunities develop more accurate and timely knowledge about their study participants and patients. As this chapter has discussed, smartphones and wearables can be utilized to perform standardized momentary QoL assessments, which can be administered to individuals through a simple and user-friendly mobile interface. At the same time, the fact that the methods described in this chapter are unobtrusive provides researchers the opportunity to obtain a more nuanced longitudinal, context-based view of individuals' personal relationships and QoL and to identify behavioral correlations. Leveraging both self-assessments and passive datasets, relationship models could be developed that provide researchers with more information while measuring an individual's states and behaviors in real-time and in their present context, allowing for better assessment of questions such as "How alone do you feel in your life?" Finally, similar to the use of standardized tests, objectively acquired data would enable behavioral and computer scientists access to quantitative data that is standardized and comparable, permitting the development of algorithms that provide further insight into the connections between aspects of personal relationships and QoL. Implementing these technologies in new ways can thus lead to new improvements in individuals' everyday lives.

References

1. Deci EL, Ryan RM. Intrinsic motivation and self-determination in human behavior. New York: Plenum; 1985.
2. WHOQOL PROGRAMME. Measuring Quality of Life. The World Health Organization Quality of Life Instruments (The Whoqol-100 and The Whoqol-Bref); 1997. https://www.who.int/mental_health/media/en/68.pdf
3. Allen LF, Babin EA, McEwan B. Emotional investment: an exploration of young adult friends' emotional experience and expression using an investment model framework. J Soc Pers Relat. 2012;29(2):206–27.
4. Duck S, editor. Understanding relationship processes. Dynamics of relationships, vol. 4. Sage; 1994.
5. Fiorillo D, Sabatini F. Quality and quantity: the role of social interactions in self-reported individual health. Soc Sci Med. 2011;73(11):1644–52.
6. Redmond MV. Social exchange theory; 2015. English Technical Reports and White Papers.
7. Messick D, Cook K. Equity theory: psychological and sociological perspectives. Praeger; 1983.
8. Wiese J, Min J-K, Hong JI, Zimmerman J. You never call, you never write: call and SMS logs do not always indicate tie strength. In: Proceedings of the 18th ACM conference of computer supported cooperative work & social computing. ACM; 2015. p. 765–74.
9. Whoqol Group. Development of the World Health Organization WHOQOL-BREF quality of life assessment. Psychol Med. 1998;28(3):551–8.
10. Andersen JR, Natvig GK, Haraldstad K, et al. Psychometric properties of the Norwegian version of the Kidscreen-27 questionnaire. Health Qual Life Outcomes. 2016;14:58.
11. Berman AH, Liu B, Ullman S, Jadbäck I, Engström K. Children's quality of life based on the KIDSCREEN-27: child self-report, parent ratings and child-parent agreement in a Swedish random population sample. PLoS One. 2016;11(3):e0150545.
12. Power R, Akhter R, Muhit M, Wadud S, Heanoy E, Karim T, Badawi N, Khandaker G. Cross-cultural validation of the Bengali version KIDSCREEN-27 quality of life questionnaire. BMC Pediatr. 2019;19 https://doi.org/10.1186/s12887-018-1373-7.
13. Vélez CM, Lugo-Agudelo LH, Hernández-Herrera GN, García-García HI. Colombian Rasch validation of KIDSCREEN-27 quality of life questionnaire. Health Qual Life Outcomes. 2016;14:67.

14. Ravens-Sieberer. Screening for and promotion of children and adolescents health: A European Public Health Perspective (KIDSCREEN); 2008.

15. Diener ED, Emmons RA, Larsen RJ, Griffin S. The satisfaction with life scale. J Pers Assess. 1985;1985(49):71–5.

16. Galanakis M, Lakioti A, Pezirkianidis C, Karakasidou E, Stalikas A. Reliability and validity of the satisfaction with life scale (SWLS) in a Greek sample. Int J Human Soc Stud. 2017;5:120–7.

17. López-Ortega M, Torres-Castro S, Rosas-Carrasco O. Psychometric properties of the satisfaction with life scale (SWLS): secondary analysis of the Mexican health and aging study. Health Qual Life Outcomes. 2016;14:170.

18. Pavot W, Diener E, Colvin CR, Sandvik E. Further validation of the satisfaction with life scale: evidence for the cross-method convergence of well-being measures. J Pers Assess. 1991;57(1):149–61.

19. Hays RD, Morales LS. The RAND-36 measure of health-related quality of life. The Finnish Medical Society Duodecim; 2001.

20. Hays R, Sherbourne C, Mazel R. The Rand 36-item health survey 1.0. Health Econ. 1993;2:217–27.

21. Steward AL, Sherbourne C, Hayes RD, et al. Summary and discussion of MOS measures. In: Stewart AL, Ware JE, editors. Measures functioning and well-being: the medical outcome study approach. Durham: Duke University Press; 1992. p. 345–71.

22. Orwelius L, Nilsson M, Nilsson E, Wenemark M, Walfridsson U, Lundström M, Taft C, Palaszewski B, Kristenson M. The Swedish RAND-36 health survey – reliability and responsiveness assessed inpatient populations using Svensson's method for paired ordinal data. J Patient-Reported Outcomes. 2017;2(1):4.

23. Vanderzee K, Sanderman R, Heyink J, de Haes H. Psychometric qualities of the RAND 36-item health survey 1.0: a multidimensional measure of general health status. Int J Behav Med. 1996;3:104–22.

24. Hoffman L, Marquis J, Poston D, Summers J, Turnbull A. Assessing family outcomes: psychometric evaluation of the beach center family quality of life scale. J Marriage Fam. 2006;68 https://doi.org/10.1111/j.1741-3737.2006.00314.x.

25. Poston D, Turnbull A, Park J, Mannan H, Marquis J, Wang M. Family quality of life: a qualitative inquiry. Ment Retard. 2003;41(5):313–28.

26. Summers JA, Poston DJ, Turnbull AP, et al. Conceptualizing and measuring family quality of life. J Intellect Disabil Res. 2005;49(Pt 10):777–83.

27. Szemere E, Jokeit H. Quality of life is social – towards an improvement of social abilities in patients with epilepsy. Seizure. 2015;26:12–21.

28. Tsiourti C, Wac K. Towards smartphone-based assessment of burnout. In: International conference on mobile computing, applications, and services. Springer; 2013. p. 158–65.

29. Insel TR. Digital phenotyping: technology for a new Science of behavior. JAMA. 2017;318(13):1215–6.

30. Schwarz N. Self-reports: how the questions shape the answers. Am Psychol. 1999;54(2):93.

31. Aharony N, Pan W, Ip C, Khayal I, Pentland A. Social fMRI: investigating and shaping social mechanisms in the real world. Pervasive Mobile Comput. 2011;7(6):643–59.

32. Coons SJ, Eremenco S, Lundy JJ, O'Donohoe P, O'Gorman H, Malizia W. Capturing patient-reported outcome (PRO) data electronically: the past, present, and promise of ePRO measurement in clinical trials. Patient. 2015;8(4):301–9. https://doi.org/10.1007/s40271-014-0090-z.

33. Mayo NE, Figueiredo S, Ahmed S, Bartlett SJ. Montréal accord on patient-reported outcomes (PROs) use series–paper 2: terminology proposed to measure what matters in health. J Clin Epidemiol. 2017;89:119–24.

34. Chang YC, Yao G, Hu SC, Wang JD. Depression affects the scores of all facets of the WHOQOL-BREF and may mediate the effects of physical disability among community-dwelling older adults. PLoS One. 2015;10(5):e0128356.
35. Beck AT. Cognitive therapy and the emotional disorders. International Universities Press. 1976.
36. Beck AT, Rush AJ, Shaw BF, Emery G. Cognitive Therapy of Depression. New York: Guilford Press. 1979.
37. Krakan S, Humski L, Skočir Z. Determination of friendship intensity between online social network users based on their interaction. Vjesn. / Tech. Gaz. [Internet]. 2018;25:655–62.
38. Wehmeier PM, Schacht A, Barkley RA. Social and emotional impairment in children and adolescents with ADHD and the impact on quality of life. J Adolesc Health. 2010;46(3):209–17.
39. Kulawiak PR, Wilbert J. Introduction of a new method for representing the sociometric status within the peer group: the example of sociometrically neglected children. Int J Res Method Educ. 2020;43(2):127–45.
40. Hektner JM, Schmidt JA, Csikszentmihalyi M. Experience sampling method: measuring the quality of everyday life. Sage; 2007.
41. Larson R, Csikszentmihalyi M. The experience sampling method. In: Flow and the foundations of positive psychology. Dordrecht: Springer; 2014. p. 21–34.
42. Shiffman S, Stone AA, Hufford MR. Ecological momentary assessment. Annu Rev Clin Psychol. 2008;4:1–32. https://doi.org/10.1146/annurev.clinpsy.3.022806.091415. PMID: 18509902.
43. Ebner-Priemer UW, Trull TJ. Ecological momentary assessment of mood disorders and mood dysregulation. Psychol Assess. 2009;21(4):463.
44. Pew Research Center. 2021. Mobile fact sheet. [Report]. https://www.pewresearch.org/internet/fact-sheet/mobile/
45. Dey AK, Wac K, Ferreira D, Tassini K, Hong JH, Ramos J. Getting closer: an empirical investigation of the proximity of user to their smart phones. In: Proceedings of the 13th international conference on Ubiquitous computing; 2011. p. 163–72.
46. Van Berkel N, Ferreira D, Kostakos V. The experience sampling method on mobile devices. ACM Comput Surveys (CSUR). 2017;50(6):1–40.
47. Gouveia R, Karapanos E. Footprint Tracker: Supporting Diary studies with lifelogging. In CHI 2013: Proceedings of the SIGCHI Conference on Human Factors in Computing Systems. 2013; pp. 2921–30. https://doi.org/10.1145/2470654.2481405.
48. Harari GM, Lane ND, Wang R, Crosier BS, Campbell AT, Gosling SD. Using smartphones to collect behavioral data in psychological science: opportunities, practical considerations, and challenges. Perspect Psychol Sci. 2016;11(6):838–54.
49. Liu S, Yang L, Zhang C, Xiang YT, Liu Z, Hu S, Zhang B. Online mental health services in China during the COVID-19 outbreak. Lancet Psychiatry. 2020;7(4):e17–8.
50. Twenge JM, King LA. A good life is a personal life: relationship fulfillment and work fulfillment in judgments of life quality. J Res Pers. 2005;39(3):336–53.
51. Canha L, Simões C, Owens L, Matos M. The importance of perceived quality-of-life and personal resources in transition from school to adult life. Procedia Soc Behav Sci. 2012;69:1881–90.
52. Maass R, Kloeckner CA, Lindstrøm B, Lillefjell M. The impact of neighborhood social capital on life satisfaction and self-rated health: a possible pathway for health promotion? Health Place. 2016;2016(42):120–8.
53. Barkhuus L, Dey AK. Location-based services for mobile telephony: a study of users' privacy concerns. Interact. 2003;3:702–12.
54. Klasnja P, Consolvo S, Choudhury T, Beckwith R, Hightower J. Exploring privacy concerns about personal sensing. In: International conference on pervasive computing. Berlin: Springer; 2009. p. 176–83.
55. Okazaki S, Li H, Hirose M. Consumer privacy concerns and preference for degree of regulatory control. J Advert. 2009;38(4):63–77.

56. Gustarini M, Wac K, Dey AK. Anonymous smartphone data collection: factors influencing the users' acceptance in mobile crowdsensing. Pers Ubiquit Comput. 2016;20:65–82.
57. Wac K. From quantified self to quality of life. In: Digital health. Cham: Springer; 2018. p. 83–108.
58. Wac K. Quality of life technologies (definition). In: Gellman M, Turner J, editors. Encyclopedia of behavioral medicine. New York: Springer; 2019.

Part IV
Social Relationships

Chapter 15
The Influence of Technology on the Assessment and Conceptualization of Social Support

John F. Hunter, Nickolas M. Jones, Desiree Delgadillo, and Benjamin Kaveladze

Introduction

Our friends and families define the fabric of our lives, and the supportive network they construct determines much of well-being. Beyond (or perhaps due to) feelings of warmth and belonging, strong social support is related to lower rates of morbidity, mortality, and to better cardiovascular, neuroendocrine, and immune function [1]. The quantity, quality, structure and function of these intricate and dynamic networks is a critical component of quality of life and should be considered carefully when assessing and improving health. Social support is the provision of psychological and material resources from one's social network intended to benefit an individual [2]. There are several ways in which researchers have defined and operationalized social support, but House and colleagues' theoretical framework about the domain of social support is particularly clear and comprehensive [3]. This framework delineates three approaches for understanding the components of social support and aligns well with various assessment techniques.

The simplest and most direct method of assessing social support is to examine the *quantity* of social support. Measuring marital status, number of friends, and community involvement (e.g., church membership) are the most common variables that quantify this concept. Terminology such as social integration, isolation, loneliness, and social embeddedness are used to describe this facet of social support [4]. This type of approach to quantifying social support is relatively objective, reliable and easy to obtain [5]. The Social Network Index (SNI [6]) is the most comprehensive and popular tool for measuring the quantity of social relationships. The SNI is a self-reported questionnaire that quantifies social connections, evaluates the

J. F. Hunter (✉) · D. Delgadillo · B. Kaveladze
Department of Psychological Science, University of California-Irvine, Irvine, CA, USA
e-mail: johnhunter@chapman.edu

N. M. Jones
Department of Psychology, Princeton University, Princeton, NJ, USA

© The Author(s) 2022
K. Wac, S. Wulfovich (eds.), *Quantifying Quality of Life*, Health Informatics,
https://doi.org/10.1007/978-3-030-94212-0_15

frequency of contact, and categorizes individuals on a continuum from socially isolated to socially integrated. The SNI measures marital status, sociability (how many
friends and family members one has and how often they are in contact), participation in a religious group, and participation in other community groups. Some measures, such as the Social Support Questionnaire [7] add another element by assessing
both the number of people respondents feel they can count on as sources of social
support and their satisfaction with the support they receive from each those people.
While this straightforward approach for assessing the existence of social connections is useful and informative, it does not encapsulate the full complexity and variation of social support.

Social support may also be assessed by moving beyond the mere number of
social connections and focusing on the *structure* of relationships [3]. A social network analysis approach broadens the range of relationships considered, includes
both positive and negative influences of relationships, and analyzes the patterns of
relationship structure [8]. This approach provides a richer level of detail because it
assumes that not all relationships are created equal, and different facets of social
connection may have differential impacts on well-being. There are several characteristics of social networks (e.g., size, density, reciprocity, homogeneity) that should
be considered when assessing these social connections [9]. This approach provides
detailed insight into the interconnected webs of social relationships and how multiple levels of influence may impact an individual. In the past, this methodology has
been limited by a lack of data. However, with the advent of the internet and big data
approaches, social network analysis has boomed in recent years.

If one hopes to pinpoint the specific effects of social support on well-being, it
may be advisable to adopt a *functional* approach that emphasizes the differing influence of various types of support [3]. While quasi-objective support measures examining the mere number of connections or structure of those connections are valuable
[10], it is important to look at whether an individual believes that enough support
exists to help them in times of need. The type of support provided, the source of
support, and the manner in which it is delivered are key aspects that determine the
effects of social support on well-being [11]. One type of social support, emotional
support, entails sympathy and love, encouragement, communication or care that
may reduce negative psychological states. Informational support is characterized by
advice, facts or information that may assist in addressing a problem. Instrumental
support is providing tangible assistance such as money, resources or time [12, 13].
Each of these types of support serves its own unique function and it may be informative to delineate the differences when assessing health-relevant impacts.

Many of the most robust and informative social support scales are functional in
nature. The Interpersonal Support Evaluation List (ISEL [13]) is the most widely
used instrument to assess perceptions and functions of support. This is particularly
important because the perception or appraisal of that support is the key element that
drives many of the positive health effects found in the literature [4]. Important social
support measures also include the Multidimensional Scale of Perceived Social
Support [14] (MSPSS), a 12-item instrumental and emotional support scale that
measures the degree to which respondents *feel* or perceive that they are supported
by friends, family or a significant other. Questions used to assess this include *"I*

have a special person who is a real source of comfort to me", "I can talk about my problems with my family" and *"I can count on my friends when things go wrong"* to which respondents rate each statement on a 7-point likert scale ranging from 1 = Very Strongly Disagree to 7 = Very Strongly Agree. Similarly, the Duke-UNC Functional Social Support Questionnaire [15] assesses the amount of support respondents receive and categorizes this support into affective, confidant, and instrumental categories. This is an instrument frequently used by medical professionals since higher levels of social support are often tied to better medical adherence. Questions are rated on a 5-point scale ranging from 1 = "Much less than I would like" to 5 = "As much as I would like" and include items such as, *"I get help when I am sick in bed"* and *"I get useful advice about important things in my life"*.

Researchers assert that functional measures are more informative because they target the specific influences provided by one's social connections [16]. For example, adequate functional support may be derived from one very good relationship, but may not be available to those with multiple superficial relationships. This is particularly important to consider in the digital age because we have so many platforms of communication and many "weak" online relationships. A functional operationalization helps to clarify the psychosocial impacts of the many constantly-evolving ways in which people exchange support online.

How Does Social Support Influence Health?

Provision and perception of social support has been linked to a variety of positive well-being outcomes, particularly in the realm of health [17]. Social support predicts physical health [18] and each type of support (instrumental, informational, emotional) offers a unique blend of benefits [19]. Social support serves as a protective factor against stress and chronic illness, and confers numerous benefits on an individual's psychological and physiological well-being [1]. However, it is important that the type of support properly aligns with the needs of the individual. If the type of support offered matches the type of support desired, then it will most likely lead to positive health outcomes [11].

How does this support actually influence well-being and health? One theory is that social support influences health outcomes primarily through its ability to buffer stress. A strong social support system may reduce the negative effects of stressful experiences by providing a less threatening interpretation of a stressor and allowing the individual to feel that they have the proper resources to cope with the situation [2]. Since high levels of stress are linked to negative physical health outcomes (e.g., allostatic load), reduction of stress through reliance on social support systems may ultimately be beneficial for a variety of health outcomes. Another theory posits that social support may be beneficial in all situations, regardless of whether stress is involved. This main-effect theory states that social integration may directly influence health through things such as social control and normative behavior [16]. Most likely these theories operate in tandem, and both partially explain how social support may promote positive outcomes.

Assessing and Conceptualizing Social Support in the Digital Age

The advent of the internet and smartphone culture has fundamentally transformed the nature of social support. Technological change has altered (1) The ways in which we assess social support, (2) The perception and effects of social support. In the following chapter, we will discuss these two areas at the intersection of technology, social support, and health.

In the first part, we will examine how recent technological innovations have allowed for much more detailed, objective, and accurate assessments of social support. A large portion of one's social interactions are conducted online, and since online activity can be tracked and analyzed, we are able to peer into the window of one's life and gain a better understanding of how their social relationships unfold. In addition, we have developed tools that allow us to capture more fine-grained and real-time data about social interactions that can better inform our understanding of in-vivo social connections.

In the second part, we will discuss how the concept of social support has changed in the age of digital communication. We will focus on how the presence and use of technological devices influences face-to-face interactions, online groups, and family dynamics. To conclude the chapter, we will summarize the current research from a variety of domains about the assessment of social support via digital technology. We will identify gaps in the literature, challenges for researchers and practitioners, and important areas for future directions. Taken together, this chapter will recognize the changes in social assessment afforded by technology and consider several important areas in which technological tools have transformed social support.

How Has Technology Altered Our Assessment of Social Support?

New technologies represent an immense opportunity for clinicians and researchers to study social support's links to well-being and health, as well as the dynamic social interactions that underpin these relationships. Researchers today have access to more data than ever before. The potential for gaining theoretical and actionable insights abound. By leveraging *big data* across several accessible technological platforms, researchers can begin to understand how social support processes unfold in real time and the myriad ways technology can be used to measure meaningful aspects of social support.

But First: What Are Big Data?

In the social sciences, big data typically represent large-scale data comprised of many thousands of people and sometimes hundreds of data points about any single person. In essence, these data are long and wide. A single row could represent one person, or sometimes, a single measurement of one person among thousands of other people. Big data are typically longitudinal, and the granularity of measurement depends on the source from which the data come. Although big data manifest in many forms, there are two types that may be particularly useful for studying social support. The first is to use big *social media* data scraped from platforms like Twitter (est. 2006, 330 million users worldwide) or Facebook (est. 2004, 2.6 billion users worldwide). These data are longitudinal, naturalistic, and involve millions of social interactions, by design. The other is via big *personal* data captured on a device with which most people in the world are now intimately familiar: the smartphone.

Big Social Media Data

Harvesting data from social media platforms to learn about social support in the wild makes good sense. These platforms are social by design. We "friend" and "follow" people we know (or are attracted to or have interest in). Using features of these platforms, we interact with people in our close circles and with people who are far outside the reaches of our immediate friend group, sometimes with complete strangers across the country or the world. We post messages and media to our timelines and walls, sometimes making those posts searchable and visible to anyone who is willing to attend to them. We pour ourselves into a networked digital social sphere, sometimes hoping for connection, sometimes seeking advice, and in any case, because we believe someone will read or listen to the content we post and respond in kind. Because of the accessibility of their data, most researchers study user posts and interactions on Facebook and Twitter. Below, we offer a flavor of possible ways to use data from these platforms to study social support.

Facebook

Facebook is a platform for personal online diary-like posts, containing text, images, and GIFs, organized in timelines, where some of the content is strictly private and by default is shared only with members of an individual's social network of approved friends. In some cases, personal profile posts can be shared publicly such that anyone in the world who navigates to a user's profile can see public content. One critical feature of the Facebook platform is the ability to form groups based on common interests and experiences. For example, cancer survivors who use Facebook can search the platform for existing groups of other survivors and join those groups to connect with strangers who have experienced similar struggles with fighting their particular type of cancer. If the group is designated as public, a researcher can scrape posts and replies on the group's main page to evaluate the provision of social support in this context. Analyzing the content of posts makes it possible to differentiate the provision of emotional, informational, or instrumental support across users, explore the dynamics of support provision over time, and quantitatively analyze the popularity of posts to gauge which types of support are most valued in the group. Moreover, there are tens of thousands of groups on Facebook. This provides a unique opportunity to determine how the provision of social support varies by context. Extending this example a bit further, it would be possible to explore how cancer survivors from different groups across the country provide social support; alternatively, one could compare the provision of social support among groups formed around other health issues, identities, social statuses, political ideologies, etc.

Although there are many opportunities to study the provision of social support on Facebook, there are some important drawbacks to consider. The first is that researchers typically cannot link the receipt of social support from group members to psychological or health outcomes associated with any single user. For example, if a user posted a message about seeking prayers or advice for a loved one in the hospital, there is no way to link the level of emotional or informational support expressed in responses to this post to any outcomes related to the original poster (there are exceptions to this that are not worth mentioning here). The second is that it is difficult to get general user profile information (e.g., Facebook timeline posts) that allow for analyses of long-term outcomes. For example, if support was provided in one moment on a public Facebook page, researchers cannot evaluate how that provision of support predicts a user's positivity in their posts to their personal timeline 6 months later. Without explicit permission from the Facebook user, no access to a personal timeline is granted. Thus, Facebook data are excellent for understanding *how* groups of users interact with one another to provide support to each other through the messages they post. These data present an opportunity to unpack underlying processes in support provision, but unfortunately do not allow us to easily link this provision to important outcomes.

Twitter

Twitter is a social media platform on which users can create micro-diary posts (i.e., tweets) limited to 288 characters but can contain pictures, videos, and links. By default, Twitter profiles are public, meaning that anyone who accesses a profile can see a user's posts. If the user desires, they can convert their profile to private such that only approved followers have access to a timeline content. The ubiquity of Twitter data makes it possible to link support provision with indicators of psychological well-being. This is the case because Twitter data are comprised of user profile information and tweet generated *over time*. Thus, they are more flexible for not only describing social support provision at specific intervals, but potentially delineating short- and long-term outcomes associated with its provision. This can be achieved by using natural language processing (NLP) tools like latent Dirichlet allocation or latent semantic analysis in R (https://www.r-project.org/) or Python (python.org) to explore important topics that emerge over time; alternatively one could use the Linguistic Inquiry and Word count program (LIWC [20]) to explore psychological constructs that appear in a user's tweets (e.g., positive emotion, social words). Several studies reveal that when traumatic events like terrorist attacks [21–23], natural disasters [24, 25], school shootings [26–28], and other large-scale distressing events (i.e. mass communications of impending threats [29]) occur in communities, people express their emotions on Twitter. In such contexts, there is ample opportunity to explore how community members provide social support and ultimately link the receipt of social support to psychological functioning in the weeks and months that follow.

It is also possible to examine the provision of social support in the networked communications between Twitter users. Users engage with each other in conversation by explicitly tagging members of the conversation, or by tweeting to users contained in lists which are user-created and curated for specific purposes. As long as the tweets and lists are public, researchers can scrape these data from the platform. This network of tweets linked between users can give rise to analyses that explore whether receiving social support (via tweets from others) functions to bolster the wellbeing of the users who are targets of support. Using this information in conjunction with user-level profile information (e.g., number of followers, baseline engagement with others in a particular context) can also illuminate how outcomes related to support provision differ by users with a robust social network versus those with a meager network.

Other Social Media Platforms

Facebook and Twitter are only two social networking sites in a sea of hundreds; however, data on most platforms are not accessible to researchers. One exception is Reddit, which has garnered some popularity, as a target for social science research. Reddit is a public message-board platform that is organized into general and user-moderated topical areas (subreddits) in which users post an array of content including text, images and links. As a more traditional topic-related message board platform, there are opportunities for analyzing posts for support provision that mirror much of what has been discussed above.

Big Personal Data

Big data are not exclusive to social media platforms. They are generated moment-to-moment, every day, by many of the internet-enabled devices individuals use regularly. These data are passively logged on a device most of us carry around with us: a smartphone. Hundreds of times each day, individuals pick up their smartphone and engage with it in many ways—they text and call friends and family; interact with strangers and acquaintances on social media applications; map their routes and coordinate all the logistics of their lives. By their very nature, smartphone metrics can serve as proxies for social information about people. The number of phone calls one makes, the number of times a text or messaging application is opened, the characteristics of the physical locations one inhabits when opening a mapping application, all tell a story about how much people interact with others every day and can characterize the type of environments people tend to be in (crowded public spaces versus private homes). Recently, researchers have begun to tap into the information embedded in smartphones to understand daily human sociability [30] but they have not yet linked this information to the experience or perception of social support in peoples' lives.

How, then, can smartphone data be leveraged to understand anything substantive about social support? The answer is muddy at best, but exciting nonetheless. By focusing on contextual factors that can be gleaned from passive sensors on our smartphones (e.g., GPS locations, recorded conversational elements), we can attempt to operationalize certain elements of social support. For example, if a researcher wanted to operationalize the question on the MSPSS that states, "I can talk about my problems with my friends", they could potentially use smartphone sensors to determine proximal locations and linguistic tendencies during social interactions that would provide information about whether the individual of interest does indeed talk to their friends about problems. This type of approach for operationalizing social support is promising for future development, but is limited in its scope by the complexities of social behavior. Smartphone data cannot stand on their own for the simple reason that they are approximations of social behavior and connection. For example, just because someone spends a lot of time using social messaging apps does not inherently signal that they have a lot of social support. In fact, those digital social interactions could potentially be mostly negative. Without explicit access to messages in order to analyze the content or tone of the exchange, relying solely on logged texting behavior could be misleading. The same is true for leveraging smartphones' Bluetooth capabilities to detect nearby people to gauge how often people are around other people. Without more information from the individual about whether an exchange occurred, and the nature of that exchange, not much can be gleaned from looking at these data without supplementing with self-report data.

To remedy this, smartphones do, however, offer a powerful way to reveal social support processes in real time through *ecological momentary assessment* (EMA; [31]), a method that allows researchers to reach people directly through their smartphone. This method involves pinging participants a few times a day (via a notification), at fixed or random intervals, to obtain a snapshot of their emotions, social interactions, and other psychological information of interest to the researcher. Before smartphones, researchers would have to acquire funds to provide participants with handheld devices (e.g., palm pilots) to survey their experiences throughout the day. However, the ubiquity of smartphones today has made it easier than ever to measure the quantity and quality of daily social interactions by prompting individuals to complete brief daily surveys. Imagine, then, coupling EMA data with smartphone (or other wearables) sensor data. When paired, these data can allow researchers to untangle the complexities of daily life in analytic frames that help us understand biopsychosocial processes that unfold in real time. Tapping into the bountiful cornucopia of big personal and social media data represents an exciting opportunity for researchers to quantify social support with unparalleled detail and accuracy.

How Does Technology Influence Our Conceptualization of Social Support?

Technology has fundamentally altered the way we interact with others. We are increasingly using internet-mediated communication platforms as the primary mode of interacting with others. This shift to online social interactions has transformed our conceptualization of social support. We understand social support differently and it impacts our lives differently as we continue to intertwine our lives with technology. In the following sections, we will highlight three areas in which social support is evolving due to technology adoption. We will examine how smartphones can exert positive and negative influences on face-to-face social interaction. Next, we will peer into online communities and explore how social support garnered online is influencing the dynamics of health and social support. Finally, we will focus on how family parent-child relationships are different in the digital age, and what this means for quality of life.

The Effect of Technology on Interpersonal Communications

Personal technological devices, such as smartphones, have become a constant companion in most people's lives and subsequently influence the dynamics of face-to-face social interactions. These devices are cognitively distracting, even when not actively used [32] and often lead to a state of *absent presence* [33] in which an individual is physically present, but their mind is wandering elsewhere. This distraction induced by smartphones is particularly influential during social situations because people tend to associate their devices with external social networks [34] likely because we use smartphones to call, text, message, share, and communicate with our wider social support system. Thus, the mere presence of a smartphone may orient someone to think of people outside the context of their face-to-face conversation and divert their attention away from a conversational partner. Indeed, qualitative evidence has demonstrated that smartphones make social networks more salient and direct attention away from face-to-face conversations [35]. Interestingly, this type of activation of relational schema may take place without a person's awareness [36]. The simple presence of a device, consciously or unconsciously, distracts us and activates representations of social networks that may exert positive and/or negative influences on in-person communication depending on the context of the interaction.

Several psychological experiments have demonstrated that when an individual is distracted from their immediate face-to-face conversation partner(s) due to smartphone presence, the quality of that interaction suffers. Researchers found that the presence of a phone can have a negative influence on closeness, connection, and conversation quality between dyads [37]. In a naturalistic experiment, people in a coffee shop who conversed together without smartphones present reported higher levels of empathetic concern for their partner than those who had phones present [38]. Similarly, groups of friends who ate a meal together without phones present reported less distraction and more enjoyment than those who had phones with them [39]. Finally, using objective assessments of smiling behavior, researchers found that conversation partners who were in the presence of a phone were less likely to smile than those who had no phone present [40]. The detrimental effects of having a smartphone present during potentially positive social interactions have implications for social support and quality of life. By lowering the quality of the interaction, phones may interfere with the formation of new relationships and disrupt the maintenance of existing relationships. In this way, smartphones themselves, even when not used, may have a negative influence on social support in our modern world.

However, the presence of a smartphone may not always lead to negative consequences. In undesirable social situations, such as stressful or isolating interactions, smartphones may actually provide benefits. People can rely on smartphones as an avoidant coping mechanism, as demonstrated by an experiment that showed how the presence of a smartphone can lower an individual's initial reaction to social stress [41]. Another experiment demonstrated that the mere presence of a phone can aid in recovery from a socially stressful situation. Individuals who had their phones with them, but were restricted from using the devices, exhibited sharper declines in physiological stress after they were exposed to a stressful social exclusion paradigm compared to those who had no phone or used their phone [42]. The representational image of our smartphone may increase feelings of perceived support and provide a reminder of the social resources available to cope with a stressor at hand. As discussed earlier, perceived social support is a key predictor of health because of its ability to help us handle stress. So if smartphones symbolize perceived social support, they may be relied upon to help us overcome stressful situations. In this way, phones can be health-protective by serving as something akin to a *digital security blanket* that offers comfort in uncomfortable circumstances.

The distracting pull of smartphones on our attentional awareness is responsible for both the positive and negative effects mentioned above. Specifically, distraction caused by the symbolic connections offered by phones can shift attention away from negative environmental stressors and provide a sense of security. On the other hand, the salience of social networks represented by a phone can pull attention away from a potentially positive interaction and lower the quality of that conversation. In both cases, the key element that allows the mere presence of a phone to exert these positive or negative influences is the symbolic social connections represented by a smartphone. The way we conceptualize social support has altered significantly due to the widespread adoption of smartphones. The digital threads that connect us to our wider social support system are in our pockets at all times, and carrying that symbolic network of friends and families with us can be positive or negative depending on the circumstances of our face-to-face interactions.

Online Social Support

Online communities, such as the aforementioned Facebook groups, provide abundant opportunities for users to give and receive social support. These online interactions—and the support and strain users derive from them—is similar but distinct from that exchanged offline. Much like support exchanged in offline groups like Alcoholics Anonymous and hobby clubs, giving and receiving online social support can be a source of validation contributing to an improved quality of life. Online support can take the form of advice written in a reply to a weight loss forum post, a "like" on an Instagram photograph, or banter amongst high school friends in a groupchat. Likewise, norms around social support exchange differ across online locales; for example, social networking sites (SNSs) like Facebook can be contrasted with anonymous online support groups (OSGs) composed of strangers facing a shared challenge, like the subreddits/depression. Despite the challenge of defining online social support, seeking social support it is one of the foremost reasons that individuals choose to participate in online communities.

The unique dynamics of Internet-mediated social interactions shape the ways that users exchange support. The relatively low stakes of online interactions minimize typical in-person impediments to conversation and relationship formation (e.g., shame, stigma, appearance, and physical inability) [43]. Online interactions enable users to express their 'true selves' more than they would in person [44–46], creating the potential for "hyperpersonal interactions" [47] featuring high openness and liking between parties. Even relationships that exist entirely online can be meaningful sources of social support [48]. However, loosened social norms online also facilitate the misinformation and bullying for which online communities are notorious [49].

Numerous positive impacts of OSGs have been identified, although their impacts on "hard" health outcomes need to be more rigorously investigated [50]. Perceived social support from an OSG is mediated by identification with the community and interpersonal bonds with other members [51]. Similarly, identification with other forum members is a key moderator of the link between positive psychosocial outcomes and participation in online discussion forums [52]. Research has also demonstrated that Facebook-based social support generally improves physical and mental health, and reduces symptomatology related to mental illness [53].

Online sources of social support are particularly crucial for people who struggle to find support offline [54, 55]. One study found that typical inequities in support availability related to race and age are minimized among those that have access to the internet and SNSs [56]. Using massive social networking sites like Twitter, people facing rare or understudied health issues can rapidly connect to exchange first-hand experiences and expert insights. For example, after he was diagnosed with COVID-19 early in the pandemic, the digital health speaker Maneesh Juneja used his twitter to publicly share frequent updates on his recovery and recommend digital resources to others. Another unique benefit of OSGs is that they provide forms of support that are specific to the needs of their users. For example, members of Mood Disorder communities offer one another primarily emotional support, while Compulsive Disorder community members tend to exchange instrumental support in the form of tips for dealing with symptoms [57].

Providing support to others, while often fulfilling, can also be quite taxing. Some SNS users feel overwhelmed by the inundation of support requests they encounter from other members of their online communities (i.e. a Facebook status asking for help moving or a Twitter post venting about a difficult breakup). This feeling of "social overload" is particularly common for users who feel they are obliged to respond to SNS support requests, as well as those who have a greater number of online-only friends, as compared to friends with whom they have offline relationships as well [58]. Similarly, mental-health OSG users often complain of "endless grief loops" from encountering an excess of disheartening stories from other users [59]. These concerns, as well as the prevalence of trolling, bullying, and misinformation under anonymity, present substantial downsides to participation in online spaces for some [60].

Online communities' scale and accessibility enable users to bypass common barriers to giving and receiving social support. These spaces hold particular appeal for people who lack the in-person networks to openly discuss the topics they care about. At their best, online communities offer empowerment through genuine human connection. Yet, because people interact online in such diverse ways, more research is necessary to fully understand how online social support contributes to users' broader social lives.

Parent-Child Relationships

Frank Lloyd Wright's well-known quote, "the hearth is the psychological center of the home" conjures images of quiet reflection, children cuddling with parents, storytelling, and bonding with loved ones. Today's switch to an electronic epicenter of the home may seem abrupt but it has been happening incrementally in many cultures across the globe for decades. Indeed, the popularization of television in the 1950s marked the beginning of the screen as a replacement for the hearth, followed by the rise of the desktop computer in the 1980s, the mobile phone in the 1990s, smartphones in the 2010s, and more recently, social media and online gaming fiercely vies for attention at the family dinner table. Yet, as tempting as it may be to romanticize times passed, it is likely that the drive for social connection has not changed from one generation to the next, rather, it is the vehicle of connection that has radically transformed. Research on the impact of this technological shift on family relationships is quite mixed with some studies suggesting that it is detrimental to the family bond while other studies show that technology, in its various forms, promotes healthy connection. Benefits and detriments alike, most families have invited this virtual guest into the home and technology has established a firm seat at the table.

An electronic third-party is now in near constant attendance, at times enhancing interactions with close loved ones and at other times, detracting from them. Many children are raised by parents tied to mobile devices with popular media and large portions of society expressing concerns regarding the repercussions on children's well-being due to distracted parenting. These concerns are not completely unfounded. Some research has shown that screen time is detrimental to the parent-child relationship. Specifically, distracted parenting due to mobile device use is linked to more child behavior problems, increased risky-behaviors in children, reduced parent-child interaction, and reduced parental sensitivity [61]. Further, children notice parents' use of mobile devices and report feeling excluded and emotionally dissatisfied during these occurrences [62]. However, this evidence should not be over-simplified and is only one piece of a much larger body of literature. Research also shows that a brief distraction may help parents to recharge and re-engage with their children more effectively. For example, one study found that following approximately 15-min of focused mobile device use; parents often initiated exuberant and joyful play with their children [63]. In this sample, parents first ensured that children were engaged in safe play prior to using their mobile phones. It was only then that most parents began smartphone use. After this well-placed distraction, parents re-engaged with more enthusiasm and interest than they displayed pre-distraction. The researchers described an ebb and flow between engagement and disengagement and suggest that the cycle between interactions may provide relational benefits. While mobile phones may foster a type of disengaging recharge for parents, other technologies appear to facilitate parent-child engagement.

The majority of American homes have a television as the focal point of the family room and the TV is one screen designed to host a group experience. Research has found that parental co-use of technology may have advantages [64], for example, co-viewing of television is linked with gains in preschoolers comprehension [65], attenuated fear and aggression among school-aged children [66], direct positive effects on language development in low-income immigrant families [67] and may help with young children's verbal development in well-designed programming [68, 69]. Beyond television, there is also evidence supporting the benefits of parent-child co-use of mobile phones, gaming systems and computers. Research shows that teenagers benefit from parental help with computers and this may promote self-efficacy and technological expertise [70–72]. Further, parents that played video games with teenage daughters had daughters that reported higher parental connection, fewer internalizing problems, and increased prosocial behaviors than those that did not [73].

Advantages and disadvantages alike, technology is a firmly entrenched presence in most homes with the number of smartphone subscriptions surpassing the world population [74]. Even still, parents who use technology with children present bear the brunt of heavy criticism and regular shaming from the popular media. But parental distraction is not a new phenomena and is certainly not unique to the current generation. In times past, parental heads might have been buried in a newspaper or absorbed in home projects, hobbies or social clubs. The fact that the new distraction is digital does not make it inherently harmful and undivided attention from parents to children is not necessarily beneficial for either party. The challenge is found in the ability to discern how to calibrate and modify technological use, so that it facilitates familial well-being. Technology can be used as a vessel that helps to hold relationships in the same space or it can be used to divide. It is neither good nor bad until the user gives it its purpose. One could argue that parents who use mobile devices to play games with their children, photograph memorable moments, and connect with other parents *better* serve their children. Further, the co-use of technology gives parents an opportunity to influence how children navigate their way through virtual networks. Indeed, school-aged children that co-use the internet with parents are more likely to seek out educational websites when compared to children who co-used the internet less [75] and respected organizations are adjusting recommendations based on emerging evidence. In 2013, the American Academy of Pediatrics changed its recommendation from strictly limiting media for young children to encouraging parents to co-use media with their children [76]. Taken together, the collection of information we have discussed suggests that the influence of the digital presence on familial relationships fluxuates based on contextual cues and is more likely a reflection of the psychology of the user rather than the influence of the tool itself.

Discussion

The integration of technology into our daily lives has broadened the scope of interaction with our social networks and expanded the ways in which we can assess the influence of social support on health and quality of life. The abundant and detailed data produced by our online activity and technology-infused lifestyles represents a fertile ground for exploration into the intricacies of social interactions. Since social support is such a critical aspect of quality of life, it is imperative that we continue to develop innovative methods for assessing social interactions. A focus on Quality of Life Technologies (QoLT), the software and hardware that allow us to assess and monitor well-being [77], may allow us to take this next step forward in capturing and disseminating social information about our lives. By leveraging these QoLT, such as the social data recorded on our smartphones, we can hope to better understand social interactions and draw upon the multitude of opportunities for improvement of well-being. The omnipresence in social circles coupled with the hardware and software capabilities of QoLT allow for an unprecedented level of insight into the dynamics of social well-being that can be drawn upon to assess and improve social support. Traditionally, researchers would solely rely on self-report measures to assess social support and its related constructs. And while this data is certainly valuable and important, the objective and unbiased information gathered from technology-enabled methods provides an unprecedented level of detail and insight that uncovers the dynamic and complex nature of how social support unfolds. In tandem with social implications, mobile technologies offer us the ability to track and measure health indices such as sleep, exercise, nutrition, and cardiovascular function. Together, these tools can be employed as electronic observers providing insight into the various ways social behaviors (on and off-line) may impact important health behaviors and outcomes.

By drawing on a multitude of technological resources, researchers are able to leverage big data to examine the complex ways in which social support transpires in the modern age. By focusing on social media activity of groups and individuals via platforms like Facebook and Twitter, researchers can begin to understand how social interactions manifest online. This data can be linked to important individual psychosocial outcomes or health-relevant group concepts that may inform how online social support influences health. Furthermore, the plethora of technological devices that pervade our daily lives can be harnessed to provide big personal data that informs our understanding of social support. Researchers can glean information from smartphone behavior or wearable devices to passively track socially-relevant factors that occur in real-time in the real-world. We can also utilize techniques such as ecological momentary assessment to gather in-vivo measurements of daily social and well-being variables. This information may be used by researchers to update theories and ideas about the biopsychosocial effects of social support, and may also be relied upon to inform health practitioner recommendations or interventions. Additionally, individuals can analyze their own digital social metrics to better quantify their social wellness and recognize areas for potential improvement or change. By engaging in this way in the Quantified Self movement, individuals may be able to augment their quality of life by creating data-driven goals for behavior change. Taken together, the advent of internet-enabled technological devices has opened a never-before-seen window into the intricacies of individuals social lives and health behaviors that allow us to capture a wide array of biopsychosocial information that can be leveraged to better our understanding of these quality of life indicators.

When considering the meaning of this technology derived social data, it is critical that researchers and practitioners also keep in mind the ways in which technology has altered our understanding about the conceptualization, meaning, and influence of social support in the digital age. The intrusion of technology into our social interactions has innumerable positive and negative effects on the quantity, quality and function of social support. Social support is increasingly taking place in the realm of internet-mediated communications, and this transition to virtual communication alters the applicability of and relevance of the traditional ways in which we understand the impact social support on well-being. In this chapter, we focused on three areas in which technology has changed our conceptualization of social support, namely in regard to face-to-face conversations, social support groups, and family interactions.

As smartphones have come to symbolize social networks, due to their use as social communication devices, the presence and/or use of these devices has altered the dynamics of face-to-face conversations. The symbolic social support represented by smartphones distracts us from in-person interactions by unconsciously or consciously drawing our attention away from the present situation. This absent presence often decreases the quality of our face-to-face interactions, especially when that interaction is potentially positive. On the other hand, that same social distraction can be beneficial in aversive situations when the symbolic connections provided by our smartphone provide a temporary *digital security blanket* crutch on which we can rely on to buffer the stress experienced in the undesirable circumstance.

Similarly, our interactions in online communities introduce an extra layer of complexity into how group-level social support influences well-being. Online communities, ranging from massive social media platforms to niche hobby forums, are an evolving and influential social phenomenon. These spaces enable interactions distinct from real-world groups, presenting unique opportunities and challenges. For some, online communities can be a gateway into progressively darker mindsets, yet for others they are a lifeline, offering hope in the form of genuine human interaction and social support. Our online activity deeply influences our lives and society on a global scale, and as such deserves careful attention.

Family dynamics are also undoubtedly shifting due to the widespread adoption of smartphones and the internet. The digital presence in many households can often drive a virtual wedge between family members and lead to developmental or relational problems when devices are overused. Further, whether technology is used to promote the quality of family relationships or not, any overuse of screens could promote sedentary behavior ultimately impacting health and well-being. On the other hand, technology can be used to bring families together, foster communal experiences, and increase familial engagement. As technology cements its seat at the dinner table, it will be critical for parents (and children) to be aware of the positive and/or negative ways in which their digital behavior influences family well-being. Families will need to learn to use technology in moderation and operate this electronic tool with the purpose of fostering positive health behaviors. In sum, a verdict cannot be assigned as to whether technology is "good" or "bad" for families. Like any tool, it can be abused and misused or in can become an instrument boosting the quality of life for many.

Future Directions

There are innumerable possibilities for how technology will influence our assessment and conceptualization of social support in the coming years. Technological change is occurring at a blistering pace and it is nearly impossible to predict exactly how it will impact our society and individual well-being. We remain optimistic that the breadth and depth of our knowledge and understanding of social support will continue to expand.

Technology-enabled methods and techniques for objectively assessing social support will likely flourish in the coming years and provide an unprecedented level of detail about our social lives. Researchers can leverage passive sensors to capture behavioral observations like facial expression, tone of voice, or body posture along with location or usage information to further understand the nature of social interactions. Of course, there are privacy issues that will need to be considered with such approaches. But individuals who are willing to provide this access to their information will help researchers and practitioners capture more fine-grained details about the nature of social support. The amount of data accumulated through online activity, smartphone behavior, wearables, and yet-to-be invented technological devices will paint a complex picture of social life, and the social interactions therein. These data will be leveraged by researchers and practitioners in order to find ways of improving our social relationships and long-term wellbeing through rigorous research methodologies, data mining, and targeted interventions. There are so many different streams of data, some on the individual level and others on the group level, that will accrue information about social happenings across the globe. As wearable technology and smart devices become ever more integrated into our lives, the moment-to-moment details of our existence will inevitably leave a digital trace. If researchers can find a way to funnel this information together to develop comprehensive social profiles of individuals or groups, then that information could be used to provide unprecedented insight into the manifestation of social support.

As seen in the sections above about the influence of technology on social support, the intersection of these devices with our traditionally understood reality can produce mixed results. Depending on contextual factors, individual characteristics, motivational reasons, and conscious or unconscious behaviors, technology can either wreak havoc on the quality of social relationships or supplement our connections and enrich social activity. One factor that seems to be overwhelmingly positive in regard to the use of technology and well-being is when technology is used communally (i.e., watching a video or playing a game together) rather than solitarily. While this kind of shared experience does not necessarily promote conversation in the moment (depending on the viewing choice) it can promote touch, warmth, shared suspense, and excitement when users choose to utilize it this way.

One possible trend of interest to researchers and practitioners that could continue to develop is the displacement of the functions of social support with internet-mediated support. Traditionally, we think of social support providing benefits by offering emotional, informational, or instrumental resources that aid an individual in times of need. For most of human existence, if someone had a stressful problem to deal with they would most likely rely on significant others for advice, comfort or resources. But what if technology itself, rather than another human being, provided this support? Indeed, many people already rely on technology to provide support in times of need. Googling a question about a novel health concern may provide more useful information than asking your neighbor, interacting with a caring avatar in a video game (or a caretaker robot) may be able to give someone a sense of comfort and security that busy friends cannot provide, soliciting strangers to crowdfund for a personal cause may be more effective than asking a family member for a loan. In these instances, direct human contact has been taken out of the equation. As people become more reliant on digital means of communication, individuals may reap the benefits of instrumental, emotional, and tangible support through Internet use, rather than through face-to-face social relationships. By considering this possibility, we can see how technology has opened up the doors to a whole new network of possibilities for support sources that could conceivably replace the traditionally understood social support.

Yet, we do not believe that society will evolve (or devolve) into a place where face-to-face human connection has been subsumed by virtual environments. It is a frightening thought to imagine a future where individuals do nothing more than sit alone and stare endlessly into a glowing screen that seamlessly provides for all their needs and desires. I believe that even as we climb ever higher up the ladder of technological innovations, humankind will always be at the core of it all. People have an innate desire to seek social connection [78], and that desire will ensure that we never stray too far from reliance on our place-based social networks. Even as we rely more heavily on non-human sources of support (i.e., Internet), it is important to remember that *people* are the ones who create the content of the Internet. While attempting to reduce your stress or receive support by asking Google a question might seem like an entirely non-social activity, a human-being was the one who actually wrote the answer that you seek. So instead of thinking about technology replacing social relationships, it may be more appropriate to view technology as a medium that can supplement and broaden the ways in which social connections play out. We can use the Internet to strengthen our current social bonds, to expand our social networks, and most importantly to draw on the worldwide human experience to provide the support that will allow us to flourish.

Conclusions

Technology-enabled methods have allowed for more accurate and detailed quantification of social support. By analyzing the digital traces of individual and group behavior, we are able to better understand the complex dynamics of social interactions and its influence on well-being. Continued advancements in technological approaches will likely further enhance our comprehension of social support and equip individuals, researchers, and practitioners with the necessary knowledge and tools to improve quality of life.

References

1. Uchino BN. Social support and health: a review of physiological processes potentially underlying links to disease outcomes. J Behav Med. 2006;29(4):377–87. https://doi.org/10.1007/s10865-006-9056-5.

2. Cohen S. Social relationships and health. Science. 2004;241(4865):540–5. https://doi. org/10.1126/science.3399889.
3. House JS, Kahn RL, McLeod JD, Williams D. Measures and concepts of social support. Soc Support Health. 1985;83:108. https://doi.org/10.1016/j.jpsychores.2009.10.001.
4. Barrera M. Distinctions between social support concepts, measures, and models. Am J Community Psychol. 1986;14(4):413–45. https://doi.org/10.1007/BF00922627.
5. Donald CA, Ware JE. The quantification of social contacts and resources. Rand Corporation; 1982.
6. Berkman LF, Syme SL. Social networks, host resistance, and mortality: a nine-year follow-up study of Alameda County residents. Am J Epidemiol. 1979;109(2):186–204. https://doi. org/10.16953/deusbed.74839.
7. Sarason IG, Levine HM, Basham RB, Sarason BR. Assessing social support: the social support questionnaire. J Pers Soc Psychol. 1983;44(1):127–39. https://doi. org/10.1037/0022-3514.44.1.127.
8. Wellman B. Applying network analysis to the study of support. Soc Netw Soc Support. 1981;4:171–200.
9. Israel BA. Social networks and health status: linking theory, research, and practice. Patient Couns Health Educ. 1982;4(2):65–79. https://doi.org/10.1016/S0190-2040(82)80002-5.
10. Barger SD, Messerli-Bürgy N, Barth J. Social relationship correlates of major depressive disorder and depressive symptoms in Switzerland: nationally representative cross sectional study. BMC Public Health. 2014;14:273. https://doi.org/10.1186/1471-2458-14-273.
11. Cutrona CE. Stress and social support- in search of optimal matching. J Soc Clin Psychol. 1990;9(1):3–14. https://doi.org/10.1521/jscp.1990.9.1.3.
12. Weiss R. The provisions of social relationships. In: Doing unto others. Englewood Cliffs: Prentice Hall; 1974. p. 17–26.
13. Cohen S, Mermelstein R, Kamarck T, Hoberman HM. In: Sarason IG, Sarason BR, editors. Social support: theory, research and applications. Cham: Springer; 1985. p. 73–94. https://doi. org/10.1007/978-94-009-5115-0_5.
14. Zimet G, Dahlem NW, Zimet SG, Farley GK. The multidimensional scale of perceived social support. J Pers Assess. 1988;52(1):30–41. https://doi.org/10.1207/s15327752jpa5201_2.
15. Broadhead WE, Gehlbach SH, de Gruy FV, Kaplan BH. The Duke-UNC functional social support questionnaire. Measurement of social support in family medicine patients. Med Care. 1988;26(7):709–23. https://doi.org/10.1097/00005650-198807000-00006.
16. Cohen S, Wills T. Stress, social support, and the buffering hypothesis. Psychol Bull. 1985;98(2):310–57. https://doi.org/10.1037/0033-2909.98.2.310.
17. Thoits PA. Mechanisms linking social ties and support to physical and mental health. J Health Soc Behav. 2011;52(2):145–61. https://doi.org/10.1177/0022146510395592.
18. Cohen S. Psychological models of the role of social support in the etiology of physical disease. Health Psychol. 1988;7(3):269–97. https://doi.org/10.1037//0278-6133.7.3.269.
19. Semmer NK, Elfering A, Jacobshagen N, Perrot T, Beehr TA, Boos N. The emotional meaning of instrumental social support. Int J Stress Manag. 2008;15(3):235. https://doi. org/10.1037/1072-5245.15.3.235.
20. Pennebaker JW, Booth RJ, Boyd RL, Frances ME. Linguistic inquiry and word count: LIWC2015. Austin, TX: Pennebaker Conglomerates; 2015. Retrieved from http://www. LIWC.net
21. Gruebner O, Sykora M, Lowe SR, Shankardass K, Trinquart L, Jackson T, Galea S. Mental health surveillance after the terrorist attacks in Paris. Lancet. 2016;387:2195–6. https://doi. org/10.1016/S0140-6736(16)30602-X.
22. Jones NM, Brymer M, Silver RC. Using big data to study the impact of mass violence: opportunities for the traumatic stress field. J Trauma Stress. 2019;32:653–63. https://doi.org/10.1002/ jts.22434.
23. Lin Y-R, Margolin D, Wen X. Tracking and analyzing individual distress following terrorist attacks using social media streams. Risk Anal. 2017;37:1580–605. https://doi.org/10.1111/ risa.12829.

24. Gruebner O, Lowe SR, Sykora M, Shankardass K, Subramanian S, Galea S. A novel surveillance approach for disaster mental health. PLoS One. 2017;12 https://doi.org/10.1371/journal.pone.0181233.

25. Murthy D, Longwell SA. Twitter and disasters: the uses of Twitter during the 2010 Pakistan floods. Inf Commun Soc. 2013;16:837–55. https://doi.org/10.1080/1369118X.2012.696123.

26. Doré B, Ort L, Braverman O, Ochsner KN. Sadness shifts to anxiety over time and distance from the national tragedy in Newtown, Connecticut. Psychol Sci. 2015;26:363–73. https://doi.org/10.1177/0956797614562218.

27. Jones NM, Thompson RR, Dunkel Schetter C, Silver RC. Distress and rumor exposure on social media during a campus lockdown. Proc Natl Acad Sci USA. 2017;144:11663–8. https://doi.org/10.1073/pnas.1708518114.

28. Jones NM, Wojcik SP, Sweeting J, Silver RC. Tweeting negative emotion: an investigation of Twitter data in the aftermath of violence on college campuses. Psychol Methods. 2016;21:526–41. https://doi.org/10.1037/met0000099.

29. Jones NM, Silver RC. This is not a drill: anxiety on twitter following the 2018 Hawaii false missile alert. Am Psychol. 2019. Advance online publication. https://doi.org/10.1037/amp0000495

30. Harari GM, Müller SR, Stachl C, Wang R, Wang W, Bühner M, Gosling SD. Sensing sociability: individual differences in young adults' conversation, calling, texting, and app use behaviors in daily life. J Pers Soc Psychol. 2019; https://doi.org/10.1037/pspp0000245.

31. Shiffman S, Stone AA, Hufford MR. Ecological momentary assessment. Annu Rev Clin Psychol. 2008;4:1–32.

32. Ward A, Duke K, Gneezy A, Bos M. Brain drain: the mere presence of smartphones reduces cognitive capacity. J Assoc Consum Res. 2017;2(2) https://doi.org/10.1017/CBO9781107415324.004.

33. Gergen KJ. The challenge of absent presence. Perpetual Contact: Mobile Communication, Private Talk, Public Performance. 2002;227–241 https://doi.org/10.1017/CBO9780511489471.018.

34. Srivastava L. Mobile phones and the evolution of social behaviour. Behav Inf Technol. 2005;24(2):111–29. https://doi.org/10.1080/01449290512331321910.

35. Turkle S. Alone together: why we expect more from technology and less from each other. New York: Basic Books; 2011.

36. Shah J. Automatic for the people: how representations of significant others implicitly affect goal pursuit. J Pers Soc Psychol. 2003;84(4):661. https://doi.org/10.1037/0022-3514.84.4.661.

37. Przybylski AK, Weinstein N. Can you connect with me now? How the presence of mobile communication technology influences face-to-face conversation quality. J Soc Pers Relat. 2013;30(3):237–46. https://doi.org/10.1177/0265407512453827.

38. Misra S, Cheng L, Genevie J, Yuan M. The iphone effect: the quality of in-person social interactions in the presence of mobile devices. Environ Behav. 2016;48(2):275–98. https://doi.org/10.1177/0013916514539755.

39. Dwyer RJ, Kushlev K, Dunn EW. Smartphone use undermines enjoyment of face-to-face social interactions. J Exp Soc Psychol. 2018;78:233–9. https://doi.org/10.1016/j.jesp.2017.10.007.

40. Kushlev K, Hunter JF, Proulx J, Pressman SD, Dunn E. Smartphones reduce smiles between strangers. Comput Hum Behav. 2019;91:12–6.

41. Panova T, Lleras A. Avoidance or boredom: negative mental health outcomes associated with use of information and communication technologies depend on users' motivations. Comput Hum Behav. 2016;58:249–58. https://doi.org/10.1016/j.chb.2015.12.062.

42. Hunter JF, Hooker ED, Rohleder N, Pressman SD. The use of smartphones as a digital security blanket: the influence of phone use and availability on psychological and physiological responses to social exclusion. Psychosom Med. 2018;80(4) https://doi.org/10.1097/PSY.0000000000000568.

43. Saunders PL, Chester A. Shyness and the internet: social problem or panacea? Comput Hum Behav. 2008;24(6):2649–58.

44. Bargh JA, McKenna KYA, Fitzsimons GM. Can you see the real me? Activation and expression of the "true self" on the internet. J Soc Issues. 2002;58(1):33–48. https://doi.org/10.1111/1540-4560.00247.
45. Pierce T. Social anxiety and technology: face-to-face communication versus technological communication among teens. Comput Hum Behav. 2009;25(6):1367–72. https://doi.org/10.1016/j.chb.2009.06.003.
46. Valkenburg PM, Peter J. Preadolescents' and adolescents' online communication and their closeness to friends. Dev Psychol. 2007;43(2):267–77. https://doi.org/10.1037/0012-1649.43.2.267.
47. Walther JB. Computer-mediated communication: impersonal, interpersonal, and hyperpersonal interaction. Commun Res. 1996;23(1):3–43. https://doi.org/10.1177/009365096023001001.
48. Longman H, O'Connor E, Obst P. The effect of social support derived from world of Warcraft on negative psychological symptoms. Cycberpsychol Behav. 2009;12(5):563–6. https://doi.org/10.1089/cpb.2009.0001.
49. Kowalski RM, Giumetti GW, Schroeder AN, Lattanner MR. Bullying in the digital age: a critical review and meta-analysis of cyberbullying research among youth. Psychol Bull. 2014;140(4):1073–137. https://doi.org/10.1037/a0035618.
50. Mehta N, Atreja A. Online social support networks. Int Rev Psychiatry. 2015;27(2):118–23. https://doi.org/10.3109/09540261.2015.1015504.
51. Zhu Y, Stephens KK. Online support group participation and social support: incorporating identification and interpersonal bonds. Small Group Res. 2019;50(5):593–622. https://doi.org/10.1177/1046496419861743.
52. Pendry LF, Salvatore J. Individual and social benefits of online discussion forums. Comput Hum Behav. 2015;50:211–20. https://doi.org/10.1016/j.chb.2015.03.067.
53. Gilmour J, Machin T, Brownlow C, Jeffries C. Facebook-based social support and health: a systematic review. Psychol Pop Media Cult. 2019; https://doi.org/10.1037/ppm0000246.
54. Indian M, Grieve R. When Facebook is easier than face-to-face: social support derived from Facebook in socially anxious individuals. Personal Individ Differ. 2014;102–106 https://doi.org/10.1016/j.paid.2013.11.016.
55. O'Leary K, Bhattacharya A, Munson SA, Wobbrock JO, Pratt W. Design opportunities for mental health peer support technologies. In: Proceedings of the 2017 ACM conference on computer supported cooperative work and social computing; 2017. p. 1470–84. https://doi.org/10.1145/2998181.2998349.
56. Rains SA, Tsetsi E. Social support and digital inequality: does internet use magnify or mitigate traditional inequities in support availability? Commun Monogr. 2017;84(1):54–74. https://doi.org/10.1080/03637751.2016.1228252.
57. Sharma E, De Choudhury M. Mental health support and its relationship to linguistic accommodation in online communities. In: Proceedings of the 2018 CHI conference on human factors in computing systems; 2018,p. 641:1–641:13. https://doi.org/10.1145/3173574.3174215
58. Maier C, Laumer S, Eckhardt A, Weitzel T. Giving too much social support: social overload on social networking sites. Eur J Inf Syst. 2015;24(5):447–64. https://doi.org/10.1057/ejis.2014.3.
59. Baglione AN, Girard MM, Price M, Clawson J, Shih PC. Modern bereavement: a model for complicated grief in the digital age. In: Proceedings of the 2018 CHI conference on human factors in computing systems; 2018. p. 416:1–416:12. https://doi.org/10.1145/3173574.3173990
60. Christopherson KM. The positive and negative implications of anonymity in Internet social interactions: "On the Internet, Nobody Knows You're a Dog". Comput Hum Behav. 2007;23(6):3038–56. https://doi.org/10.1016/j.chb.2006.09.001.
61. Kildare C, Middlemiss W. Impact of parents mobile device use on parent-child interaction: a literature review. Comput Hum Behav. 2017;75 https://doi.org/10.1016/j.chb.2017.06.003.
62. Steiner-Adair C, Barker T. The big disconnect: protecting childhood and family relationships in the digital age. Harper Business; 2013.
63. Hiniker A, Sobel K, Suh H, Sung Y, Lee C, Kientz J. Texting while parenting: how adults use mobile phones while caring for children at the playground. In: Proceedings of the 33rd

annual ACM conference on human factors in computing systems (CHI '15). Association for Computing Machinery, New York; 2015. p. 727–36. https://doi.org/10.1145/2702123.2702199.

64. Connell SL, Lauricella AR, Wartella E. Parental co-use of media technology with their young children in the USA. J Child Media. 2015;9(1):5–21. https://doi.org/10.1080/1748279 8.2015.997440.

65. Salomon G. Effects of encouraging Israeli mothers to co-observe "sesame street" with their five-year-olds. Child Dev. 1977;48(3):1146–51. https://doi.org/10.2307/1128378.

66. Nathanson AI. Identifying and explaining the relationship between parental mediation and children's aggression. Commun Res. 1999;26(2):124–43. https://doi.org/10.1177/009365099026002002.

67. Mendelsohn AL, Brockmeyer CA, Dreyer BP, Fierman AH, Berkule-Silberman SB, Tomopoulos S. Do verbal interactions with infants during electronic media exposure mitigate adverse impacts on their language development as toddlers? Infant Child Dev. 2010;19(6):577–93. https://doi.org/10.1002/icd.711.

68. Lemish D, Rice ML. Television as a talking picture book: a prop for language acquisition. J Child Lang. 1986;13(2):251–74. https://doi.org/10.1017/S0305000900008047.

69. Fish AM, Li X, McCarrick K, Butler ST, Stanton B, Brumitt GA, Partridge T. Early childhood computer experience and cognitive development among urban low-income preschoolers. J Educ Comput Res. 2008;38(1):97–113. https://doi.org/10.2190/EC.38.1.e.

70. Barron B, Martin C, Takeuchi L, Fithian R. Parents as learning partners in the development of technological fluency. Int J Learn Media. 2009;1(2) https://doi.org/10.1162/ijlm.2009.0021.

71. Eynon R, Malmberg L-E. Understanding the online information-seeking behaviours of young people: the role of networks of support. J Comput Assisted Learn. 2012;28:514–29. https://doi.org/10.1111/j.1365-2729.2011.00460.x.

72. Livingstone S, Haddon L, Görzig A, Ólafsson K. Risks and safety on the internet: the perspective of European children. Full Findings. LSE, London: EU Kids Online; 2011.

73. Coyne S, Padilla-Walker L, Stockdale L, Day R. Game on girls: associations between co-playing video games and adolescent behavioral and family outcomes. J Adolesc Health. 2011;49(3):160–5. https://doi.org/10.1016/j.jadohealth.2010.11.249.

74. ITU. Yearbook of statistics. Telecommunication/ICT Indicators 2008–2017. Geneva: ITU; 2018. http://handle.itu.int/11.1002/pub/8123c374-en

75. Lee S-J, Chae Y-G. Children's Internet use in a family context: influence on family relationships and parental mediation. Cyberpsychol Behav. 2007;10(8):640–4. https://doi.org/10.1089/cpb.2007.9975.

76. Council on Communications and Media. Children, adolescents, and the media. Pediatrics. 2013;132(8):958–61. https://doi.org/10.1542/peds.2013-2656.

77. Wac K. Quality of life technologies. In: Gellman M, editor. Encyclopedia of behavioral medicine. New York: Springer; 2020. https://doi.org/10.1007/978-1-4614-6439-6_102013-1.

78. Baumeister RF, Leary MR. The need to belong: desire for interpersonal attachments as a fundamental human motivation. Psychol Bull. 1995;117(5):497–529. https://doi.org/10.1037/0033-2909.117.3.497.

Chapter 16
Sexual Function and Quality of Life: Assessing Existing Tools and Considerations for New Technologies

Diana Barger

Introduction

Sex or sexual activity is a core human function. It refers to the way in which people experience or express their sexuality. It encompasses activities done alone (e.g. masturbation) or with another person (intercourse, oral sex, etc.) or people (group sex) [1]. Sexual activity is considered to be an important part of social functioning (also referred to as social health), which is distinct from mental and physical health and seen as an important dimension of quality of life (QoL). Echoing its definition of health, the World Health Organization's (WHO) has defined *sexual health* as a "state of physical, emotional, mental and social well-being in relation to sexuality; it is not merely the absence of disease, dysfunction or infirmity" [2]. The WHO has acknowledged its importance by including an item on sexual activity in its assessment of QoL. The item "how satisfied are you with your sex life?" seeks to capture the desire for, expression of, opportunity for and fulfillment from sex.

These definitions differ somewhat from the assessment of *sexual function*, which has been defined as "how the body reacts in the different stages of the sexual response cycle", first described by pioneering researchers Masters and Johnson in the late 1960s [3]. The *sexual response cycle* was defined as linear stages: (1) excitement, (2) plateau, (3) orgasm, and (4) resolution, all of which are experienced by both men and women [3]. This model of sexual response has since been criticized for its overt focus on the physiological rather than psychological and social components of human sexual response. Subsequent models of human sexual response have aimed to correct these shortcomings. Kaplan incorporated desire into the model [4]. Whipple and Brash-McGreer and later Basson proposed alternative, more complex, models of sexual response in women that were circular rather than linear [5, 6]. Basson's model specifically recognizes that orgasm may contribute to, but is not

D. Barger (✉)
University of Bordeaux – Inserm 1219 Bordeaux Population Health, Bordeaux, France
e-mail: diana.barger@u-bordeaux.fr

© The Author(s) 2022
K. Wac, S. Wulfovich (eds.), *Quantifying Quality of Life*, Health Informatics,
https://doi.org/10.1007/978-3-030-94212-0_16

necessary for, satisfaction, and emphasizes that relationship factors can affect one's willingness and ability to participate in sex.

As our understanding of sexual response has evolved, so have accepted diagnostic criteria for sexual dysfunctions (SD). The Diagnostic and Statistical Manual of Mental Disorders (DSM)—V, published in May 2013, sought to incorporate newer models of sexual response and to correct, expand, and clarify criteria for diagnoses [7]. Definitions of male sexual dysfunction (MSD) according to the DSM-V include erectile disorder, male hypoactive sexual desire disorder, premature (early) ejaculation, and delayed ejaculation (previously termed male orgasmic disorder). MSD involving pain was removed from the DSM-V. Definitions of female sexual dysfunction (FSD) in the DSM-IV were collapsed in the DSM-V. FSD includes female sexual interest/arousal disorder (previously female hypoactive desire disorder and female arousal disorder), female orgasmic disorder, and genito-pelvic pain/penetration disorder (previously dyspareunia and vaginismus). The DSM-V also required that frequency (dysfunction present 75–100% of the time), duration (a minimum of 6 months), and distress be investigated in order to make a diagnosis.

Although epidemiological studies are scarce, SD appear to be prevalent in the general population and varies by sex and age. In a large United States-based survey, conducted in 1992, in adults age 18–59; 43% of women and 31% of men reported experiencing SD over the past 12 months based on the DSM-IV criteria [8]. Disorders of desire/interest, arousal, orgasm and pain appear to be more common in women than in men. The most common complaint in men is premature ejaculation [8]. The causes of SD can be either physical such as common chronic conditions and/or their treatment (e.g., diabetes, heart and vascular disease, hormonal imbalance, alcoholism, etc.) or psychological (e.g., stress, depression).

The high prevalence of SD and its impact on individuals' QoL prompted the development of pharmacological interventions for addressing these disorders (e.g. erectile dysfunction). It was therefore necessary to develop valid and reliable methods for both diagnosing SD and assessing the impact of experimental interventions on sexual function and QoL. There are a number of direct physiological measures of sexual function used in men and women. For example, erectile function can be diagnosed using a Nocturnal Penile Tumescence (NPT) device [9], Intracavernous Injection of Prostaglandin E1 [10], or the penile/brachial index [11], Doppler studies [12] and sacral evoked potentials. In women, genital blood peak systolic velocity [13], vaginal pH, intravaginal compliance, and genital vibratory perception thresholds are used as direct measures of female sexual function. These direct measures appear to be correlated with indirect measures, specifically hormonal changes within the body, like estrogen, luteinizing hormone, testosterone and prolactin. However, they are useful insofar as for a diagnosis of sexual (dys)function rather than for the assessment of sexual QoL. In addition to direct and indirect objective measures, sexual function and more importantly sexual QoL can also be assessed via standardized inventories or self-administered questionnaires. These vary in their scope (comprehensiveness), measurement properties (evidence of validity, reliability and responsiveness), applicability (e.g., use in women, LGBTI populations) and context (e.g., research versus clinical practice). We hypothesized that although there are numerous instruments, covering sexual function and QoL, not all were sufficiently comprehensive nor were they applicable to all populations.

Furthermore, the use of new technologies have enabled not only the collection of self-reported measures of sexual function and sexual activity in relation to sexual and reproductive health, but may facilitate the collection of indirect measures of sexual activity via applications or personal, miniaturized mobile and wearable devices. The private sector has responded to people's interest in better understanding not only their reproductive health, for instance with the use of menstruation and/or fertility awareness applications or devices [14], but also, their sexual activity. Mobile phone applications for the assessment of sexual activity run the gamut and allow users to quantify and qualify sexual activity. These applications differ substantially from those aimed at individuals or couples trying to conceive or avoid unwanted pregnancy. Many of these applications still rely on users completing an assessment, reporting and often qualifying their sexual activity, whereas others use the smartphone's sensors or those of personal devices (Smart watches or rings or sex toys) to log sexual activity. These applications, wearables and devices might offer new opportunities for research and clinical practice. We discuss these opportunities in relation to commonly used methods, e.g., self-administered questionnaires, in this chapter.

Aims

We conducted a review of literature reviews of self-administered assessments of sexual function and QoL to identify viable instruments for the assessment of sexual function in the context of research and clinical practice. We then evaluated their breadth (dimensions assessed), summarized evidence of their psychometric properties and applications (e.g., clinical research/screening) and considered these within the context of emerging technologies.

Methods

We conducted a meta-review of literature reviews focusing on instruments assessing sexual function or sexual QoL. We identified relevant literature reviews from two sources, the COnsensus-based Standards for the selection of health Measurement INstruments (COSMIN) database[1] and the academic bibliographic databases, specifically Pubmed, PsycInfo and Scopus, using a search algorithm. The former is maintained by COSMIN, an initiative that aims to improve the choice of measurement instruments for research and clinical practice [15]. The following algorithm

---

[1] https://www.cosmin.nl/

was used ("sexual satisfaction" OR "sexual quality of life" OR "sexual well-being" OR "sexual function" OR "sexual dysfunction") AND ("questionnaire" OR "patient-reported outcome" OR "measure" OR "scale") AND ("review"). The choice of key words, ranging from "quality of sexual life" to "sexual dysfunction", reflects the lack of consensus regarding definitions of sexual QoL previously evoked. Moreover, in the first search attempts, particularly when using biomedical search engines, it was clear that the terms "sexual dysfunction" and "function" were commonly used to describe the instruments of interest. Including these terms ensured that instruments of interest, that we might have otherwise missed, were identified. We restricted the search fields to titles and abstracts and used search filters to refine the result. We restricted our search to "reviews" or "systematic reviews", available in English. No restrictions were imposed based on the date of publication and systematic reviews as well as literature reviews were considered. We choose to include non-systematic reviews as well as systematic reviews to ensure that no relevant instruments were omitted. Relevant references were selected based on a review of their title, followed by a review of the abstract by two independent reviewers. In the event of disagreement, the third reviewer provided an opinion. Full texts were then retrieved and screened for inclusion. The results from each database were imported into the Zotero,[2] free open-source reference management software. Zotero was used to identify and eliminate duplicate references.

Included literature reviews of instruments served to aid in creation of a database of instruments used to assess aspects of sexual function/dysfunction and/or sexual QoL of life in adults. Duplicate instruments were deleted. We then further selected instruments based on the following criteria. We first considered the date of development, preferring to include instruments developed after 1970 to account for changes in sexual and social norms (e.g., women's liberation and gay rights). We opted to focus this review on instruments, which were generic in nature, meaning that they had not been developed for the evaluation of sexual function/QoL in a specific patient population. The focus of this review was on self-reported measures and therefore physician or researcher administered tools were excluded from this review. We only included instruments which were associated with a peer-reviewed publication. This was re-assessed in case it had not been a requirement for inclusion in the individual reviews. We were interested in the broad assessment of sexual QoL and not merely the physical aspects of sexual function. We chose to exclude instruments, which were too narrowly focused to the physiological dimensions of sexual function (e.g., arousal, performance, orgasm). We did not distinguish between instruments intended for research and those intended for clinical practice at the selection stage; yet we did note whether the selected instrument was intended for research, clinical practice or both when evaluating the selected instruments.

The evaluation of the selected instruments covers both their content and the measurement properties. We wanted to understand whether the existing instruments were sufficiently comprehensive (number and extent of the areas covered),

---

[2] https://www.zotero.org/

reflecting the complexity of the quality of sexual life. This constitutes an assessment of their content validity, as measurement instruments must adequately reflect construct of interest, in this case sexual QoL. We coded the facets that the instruments purportedly covered according to two broad categories: sexual function and sexual QoL. Items that covered desire and the three physiological stages of sexual response: arousal, plateau (maintaining arousal) and orgasm, and those evaluating the presence of pain were considered to be part of the evaluation of physiologic sexual function. Additional items assessing satisfaction, relational-emotional aspect of sex, self-esteem or body image, and finally, the importance of distress for the individual, were considered to pertain to the subjective, often psychological, dimensions of sexual function, which together with the physiologic dimensions encompass sexual QoL. Furthermore, the anchoring of sexual activity and function with regard to its impact of QoL is important, as this is part of the current criteria for dysfunction. For the evaluation of the psychometric properties of the included instruments, we used the COSMIN taxonomy of measurement properties, based on an international consensus about the relevant properties to be evaluated for any measurement instrument, regardless of its application. We summarized evidence of the instruments' reliability, validity and responsiveness (10).

Results

Selection of Reviews of Instruments

Our search strategy identified 613 references, 133 of which were duplicates, resulting in 480 references. Based on a review of their titles, 411 references were further excluded, as they did not pertain to sexual QoL measures. Sixty-nine abstracts were reviewed independently by two reviewers. There was disagreement regarding the inclusion of 15/69 (22%) references and these were resolved by a third reviewer, ultimately resulting in the exclusion of an additional 58 references, 18 of which were excluded as they pertained to instruments used in specific patient populations. Eleven full texts were retrieved and reviewed and two additional references were excluded because they were not reviews of instruments for the evaluation of sexual QoL. The final number of "generic" reviews included was therefore nine [16–24] (Fig. 16.1). The characteristics of these reviews, i.e., date of publication, objectives, populations studied, and number of instruments evaluated, are reported in Table 16.1. Their inclusion criteria, search engines and keywords used are reported in Table 16.2. The reviews were published between 2002 and 2018, five of them were systematic reviews and the majority (7/9) targeted the general population. One review was focused on women and another on homosexual men. Most reviews included only validated instruments. In terms of the quality of the reviews, we noted that certain methodological standards for reviews of instruments were respected: the majority of reviews clearly stated their objective and their predefined inclusion criteria.

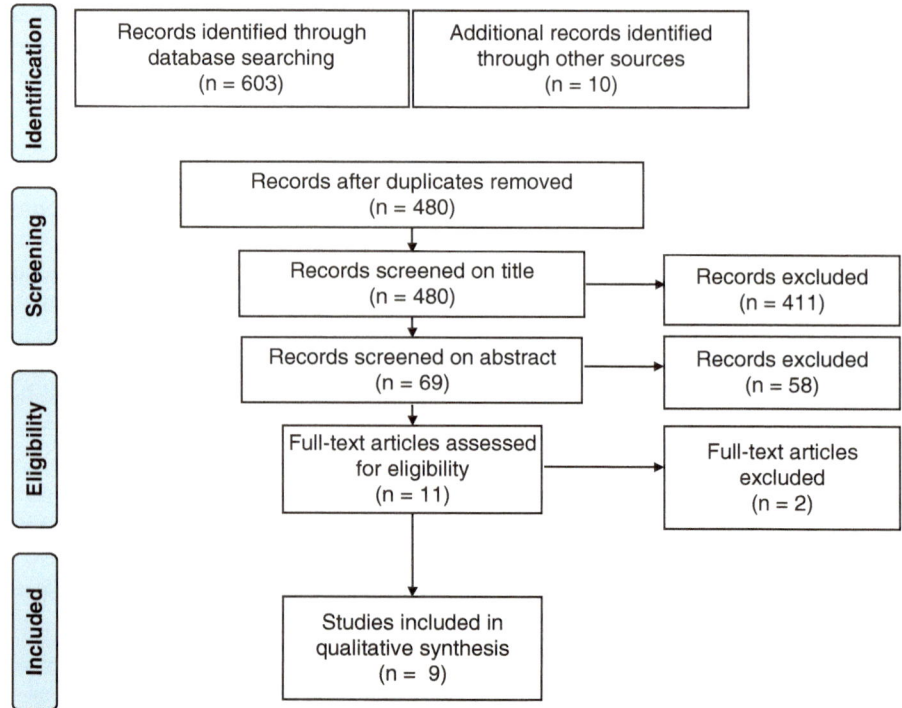

Fig. 16.1 Flow diagram illustrating the selection of reviews for inclusion

However, nearly all of the reviews did not explicitly state how they selected references, or whether they complied with a standardized method of instrument evaluation. Reviews tended to focus on instrument validation studies and the instruments themselves. They also were interested in the instruments' psychometric properties. The definition of sexual QoL appears somewhat nebulous. The number of instruments included varied significantly between reviews. For the sake of accuracy, the information summarized was extracted verbatim from the reviews themselves.

Selection of Instruments for Consideration

A total of 110 instruments were included from the nine reviews retained after the elimination of duplicate records (instruments). Subsequently, six instruments were excluded because of the date of their development (prior to 1970), four were excluded because they did not aim to measure sexual QoL, and two instruments were excluded because they seemed to lack a peer-reviewed validation study. Of the remaining 98 instruments, we retained 34 that were generic measures of sexual QoL and excluded the remainder (N = 64) that were instruments designed for use in a

Table 16.1 Characteristics of the reviews included

N°	Authors	Year	Title	Systematic review (SR) or Review (R)	Stated objective	Population	Number of instruments evaluated
1	Daker-White	2002	Reliable and Valid Self-Report Outcome Measures in Sexual (Dys)function: A Systematic Review	SR	To identify reliable and valid non-disease specific measures of sexual function, or sexual function and satisfaction, or sexual function and quality of life	General adult population	25
2	Meston et Derogatis	2002	Validated Instruments for Assessing Female Sexual Function	R	To highlight the psychometric properties of these questionnaires in an effort to assist researchers in selecting effective measurement tools for female sexual dysfunction.	Women	5
3	Arrington et al.	2004	Questionnaires to measure sexual quality of life	SR	To identify questionnaires measuring sexual function, determine the domains most commonly assessed, and examine evidence for their usefulness in different populations.	General adult population	56
4	Corona et al.	2006	Inventories for male and female sexual dysfunctions	R	To describe the main sexual inventories hitherto described and validated in different sexual areas of health and disease, and the advantages of the two main formats available to clinicians.	General adult population	30
5	DeRogatis	2008	Assessment of sexual function/dysfunction via patient reported outcomes	R	To provide an overview of the PRO instruments currently available to measure and quantify the status of an individual's sexual functioning.	General adult population	10

(continued)

Table 16.1 (continued)

N°	Authors	Year	Title	Systematic review (SR) or Review (R)	Stated objective	Population	Number of instruments evaluated
6	Giraldi et al.	2011	Questionnaires for Assessment of Female Sexual Dysfunction: A Review and Proposal for a Standardized Screener	R	To compile a comprehensive review of existing questionnaires and their psychometric characteristics with an aim to identify questions that could potentially be useful for clinicians who wish to screen for FSD in their practices.	Women	26
7	McDonagh et al.	2014	Systematic Review of Sexual Dysfunction Measures for Gay Men: How Do Current Measures Measure Up?	SR	To identify reliable and valid measures of sexual dysfunction suitable for use with gay men.	Gay men	7
8	Santos-Iglesia et al.	2018	A Systematic Review of Sexual Distress Measures	SR	To review the measures available for assessing sexual distress and to list, compare, and highlight their characteristics and psychometric properties to assist researchers and clinicians in choosing the measures that best meet their needs.	General adult population, men with sexual dysfunction, women with sexual dysfunction and patients with cancer	17
9	Cartagena-Ramos et al.	2018	Systematic review of the psychometric properties of instruments to measure sexual desire	SR	To evaluate the methodological quality of the psychometric properties and the level of evidence of the selected instruments that measure sexual desire.	General adult population	10

Table 16.2 Description of the methodology employed by the included reviews

N°	Review	Eligibility criteria	Time period	Keywords	Search Engine(s)
1	Daker-White—2002	English language. Instruments concerned primarily with the measurement of sexual behavior, attitudes, identity, or knowledge were excluded. Administered interviews, diary measures, and third-party report forms were also excluded.	1980–1999	Search terms: "Sexual function/dysfunction", "self-report", "psychometric", "instrument", "questionnaire", "scale", "reliable", "reliability", "valid" and "validity"	Embase Medline Existing reviews Reference lists
2	Meston and Derogatis—2002	Assessing various domains of sexual function in women and meet the psychometric requirements	1986–2000	NA	Medline Psychinfo
3	Arrington et al.—2004	English-language patient-reported sexual function or included sexual function as a component of a general or disease specific quality of life questionnaire.	1957–2001	MESH headings: "Sexual function", "quality of life", "sexual dysfunction", "questionnaires", "sexual satisfaction"	Medline, PubMed PsychLit, reference lists Unpublished reports
4	Corona et al.—2006	English-language. Available papers about validation inventories.	1969–2005	Search terms: "Inventories", "questionnaires", "interviews", "structured interviews" "erectile dysfunction", "impotence", "sexual health", "sexual functioning", "quality of life" "premature ejaculation" "female sexual dysfunction", "desire", "arousal", "lubrication", "orgasm", "satisfaction", "pain", "discomfort"	Medline
5	DeRogatis—2000	Instruments with at least two published validation studies.	No time restriction	NA	NA

(continued)

Table 16.2 (continued)

N°	Review	Eligibility criteria	Time period	Keywords	Search Engine(s)
6	Giraldi et al.—2011	Instruments used in public domain. Validated or in the process of being validated.	No time restriction	NA	Medline, PubMed PsychInfo Personal knowledge, books, etc.
7	McDonagh et al.—2014	Published in peer-reviewed outlets after 1980. Sufficient details available about psychometric properties. Not focused on sexual attitudes, knowledge, identity, and/or quality of life. Instruments with the minimum standards for reliability and validity as identified by Daker-White (2002).	1980–2011	Search terms: "Sexual function" or "sexual dysfunction," "self-report," "psychometric", "instrument", "questionnaire", "scale", "reliable", "reliability", "valid", and "validity"	EBSCOhost, Google Scholar PsycINFO among others
8	Santos-Iglesias et al.—2018	English-language. Definition of sexual distress measure compatible with authors. A complete psychometric validation. At least a specific separate subscale for assessing sexual distress.	1980–2011	Different relevant synonyms for sexual distress: "Sexual bother", "concern", or "worry". Limitations to make sure "sexual" and "distress" were in the same notion	Scopus (within health and social sciences journals) Pubmed
9	Cartagena-Ramos et al.—2018	Original studies. Published in Portuguese, English, and Spanish in human beings. Presenting the process of evaluating cultural validations and adaptations of sexual desire instruments, regardless of sample sex or gender.	1980–2011	Search equation: (libido) AND (psychometrics) AND (cross cultural-comparison) OR (cross-cultural AND comparison) OR (cross AND cultural AND comparison) AND (sexual desire) OR (sexual AND desire) OR (sexual AND interest) OR (sexual interest) + MESH	PubMed, Embase PsycINFO, Science Direct Web of Science databases, COSMIN

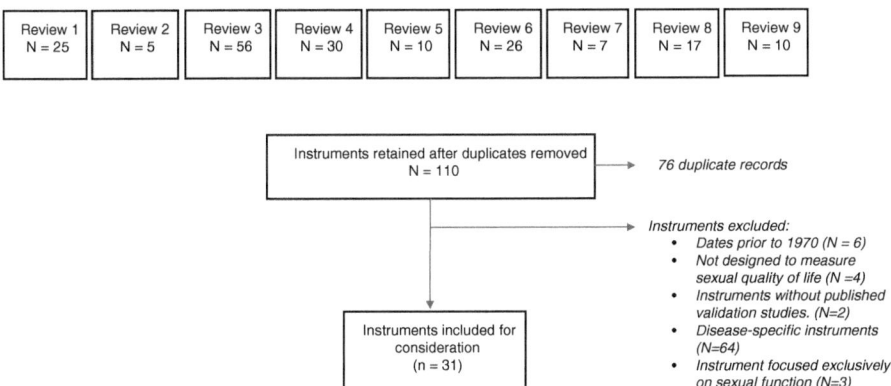

Fig. 16.2 Instrument selected from reviews for consideration

specific patient population. Finally, three of these were excluded because their focus was too narrow, exclusively on sexual function. Thirty-one instruments were ultimately selected for detailed review (Fig. 16.2).

Characteristics of Included Instruments

The characteristics of the instruments included in our study are presented in Table 16.3. The dates of instrument development and/or validation range from 1973 to 2018. Most of the instruments were developed for use in research and clinical practice (77% [24/31]) and approximately half (51%, [16/31]) targeted both males and females or heterosexual partners. The evaluation of the instruments with respect to the dimensions they consider in measuring sexual QoL is reported in Table 16.4. This evaluation allows us to judge the adequacy of an instrument in capturing the complexity of this construct. As expected, instruments often focused on physiological sexual function or phases of the sexual response cycle. On the other hand, psychosocial questions regarding relationships/emotional aspects of sex, body image and self-esteem cultural and gender expectations were only rarely taken into account in the assessment of sexual QoL. The most commonly assessed domains were arousal (N = 19), desire (N = 17), satisfaction and pleasure (N = 17), and frequency (N = 14). The most comprehensive instrument appears to be the Derogatis Sexual Functioning Inventory (DSFI), covering nine dimensions of sexual QoL. The DSFI was one of the few instruments to assess distress, relationship issues, fantasies and cultural and gender roles. The importance of sexual activity for the individual was only assessed in six of the 31 instruments reviewed. Of note, we found that the majority were not developed with the involvement of lay individuals or those coping with SD.

Table 16.3 Content of the included instruments (N = 31)

| Instruments – Year of validation study | PHYSIOLOGICAL DIMENSIONS | | | | | | SUBJECTIVE/PSYCHOLOGICAL DIMENSIONS | | | | | | |
| | Phases of sexual function | | | | | | | | | | | | |
	Desire	Arousal	Performance	Orgasm	Pain	Behavior, frequency	Importance of sexual activity	Satisfaction; Pleasure, Quality	Anxiety - Distress	Body Image - Self esteem	Relationship - Feelings and emotions	Fantasy	Cultural and gender roles
1. Sexual Orientation Methods and Anxiety (SOMA) – 1979 /2018 [25]	N	Y	N	N	N	Y	N	N	N	N	N	N	N
2. Sexual Interaction Inventory Scale (partner) (SII) – 1978 [26]	N	Y	N	N	N	Y	Y	Y	N	Y	Y	N	N
3. Questionnaire Measure of Sexual Interest (QMSI) – 1974 [27]													
4. Derogatis Sexual Functioning Inventory (DSFI) – 1979 [28]	Y	Y	N	N	N	Y	N	Y	Y	Y	Y	Y	Y
5. Female Sexual Arousability Index (SAI) – 1989 [29]	Y	Y	Y	N	N	Y	N	N	N	N	N	N	N
6. Index of Sexual Satisfaction (ISS) – 2014 [30]	Y	Y	N	N	N	N	Y	Y	N	N	N	N	N
7. Golombok Rust Inventory of Sexual Satisfaction (GRISS) – 1986 [31]	Y	Y	Y	Y	Y	Y	N	Y	Y	N	N	N	N
8. Jewish General Hospital Sexual Self-Monitoring Form – 1998 [32]	N	Y	N	N	N	N	N	Y	N	N	N	N	N
9. Derogatis Interview for Sexual Functioning - Self Report – 1997	Y	Y	N	Y	N	Y	N	N	N	N	Y	Y	N
10. Florida Sexual History Questionnaire (FSHQ) – 1991 [33]	Y	N	N	N	N	Y	Y	Y	N	N	N	N	N
11. Sexual Desire Conflict Scale for Women (SDCSW) – 1991 [34]													
12. Brief Index of Sexual Function for Women – 1994 [35]	Y	Y	N	Y	N	Y	N	Y	Y	N	Y	N	N
13. Multidimensional Sexuality Questionnaire – 1993 [36]	N	N	N	N	N	Y	N	Y	Y	Y	N	N	N
14. Sexual Function Scale (SFS) – 1988 [37]	Y	Y	Y	N	N	N	N	Y	N	N	N	N	N
15. Sexual Interaction System Scale – 1994	Y	Y	N	N	N	N	N	Y	N	N	N	N	N
16. Brief Male Sexual Function Index/Inventory (BMSFI) – 1995 [38]	Y	Y	Y	Y	N	Y	Y	Y	N	N	N	N	N
17. Sexual Activity Questionnaire (SAQ-F) – 1996 [39]	N	N	N	N	Y	Y	N	Y	N	N	N	N	N
18. Sexual Desire Inventory – 1996 [40]													
19. Sexual History Form (SHF) – 1998 [41]	Y	Y	N	N	N	Y	N	Y	N	N	N	N	N
20. Arizona Sexual Experience Scale (ASEX) – 2000 [42]	Y	Y	N	Y	N	N	N	Y	N	N	N	N	N
21. The Female Sexual Functioning Index (FSFI) – 2003 [43]	Y	Y	N	Y	Y	N	N	Y	N	N	N	N	N
22. Short scale to measure female sexuality – 2001 [44]	Y	Y	N	N	N	Y	Y	N	N	N	Y	N	N
23. Female Sexual Distress Scale (FSDS) – 2002 [45]	N	N	N	N	N	N	N	N	Y	N	N	N	N
24. The Sexual Function Questionnaire (SFQ) – 2002 [46]	Y	Y	N	Y	Y	N	N	Y	N	N	Y	N	N
25. Sexual Satisfaction Scale for Women (SSS-W) – 2005 [47]	N	N	N	N	N	N	N	Y	Y	N	Y	N	N
26. Scale for Quality of Sexual Function (QSF) – 2005 [48]	N	Y	N	N	N	N	N	N	Y	N	Y	N	N
27. Sexual Arousal and Desire Inventory (SADI) – 2006 [49]	Y	Y	N	Y	N	N	Y	N	N	N	N	N	N
28. Cues for Sexual Desire Scale (CSDS) – 2006 [50]	Y	N	N	N	N	N	Y	N	N	N	N	Y	N
29. Female Sexual Distress Scale Revised (FSDS-R) – 2008 [51]	Y	N	N	N	N	N	N	N	Y	N	N	N	N
30. Female Sexual Desire Questionnaire (FSDQ) – 2011 [52]													
31. Sexual Distress Scale (SDS) – 2018 [53] / Sexual Distress Scale Revised (SDS-R)													

Green (Y) = Yes, dimension covered by the instrument; White(N) = No, dimension omitted by the instrument,
Light grey = NA, not provided in the reviews

Green (Y) = Yes, dimension covered by the instrument; White (N) = No, dimension omitted by the instrument, Light grey = NA, not provided in the reviews

Discussion

Sexual activity is an important part of social functioning and consequently QoL. Over 100 self-administered instruments have been developed to respond to an interest in assessing sexual function/dysfunction and QoL in the context of research and clinical practice. These instruments nevertheless vary in their purpose and scope, making selecting the appropriate measurement tool potentially challenging for those wishing to assess sexual activity, function and QoL. This meta-review aimed to provide an up-to-date overview of generic rather than disease-specific instruments that can be used in men and/or women. Thirty-one self-reported measures of sexual QoL with documentation of their psychometric properties were considered. Only three, however, were developed after 2013, the year the DSM-V revised the definition of male and female SD. Furthermore, there seems to be no consensus as to what constitutes the key dimensions of sexual QoL beyond the phases in the sexual response cycle. Nevertheless, our meta-review highlights that there has been an effort to address the specific needs of women, evidenced by the 12 instruments included which were for use in women [21].

Table 16.4 Evaluation of the psychometric properties of the included instruments measuring sexual quality of life

Instruments	N items	Target population[a]	Internal Consistency (Cronbach's α)[b]	Test-retest reliability[c]	Content validity	Construct Validity[d]	Cross-cultural validity	Responsiveness	Sensitivity
1. Sexual Orientation Method and Anxiety (SOMA))	112	M/F	NI	Yes	Yes	Yes	NI	NI	NI
2. Sexual Interaction Inventory Scale (partner) (SII)	17	PART (HETERO)	0.88	0.82 (2 wks)	NI	NI	NI	NI	NI
3. Questionnaire Measure of Sexual Interest (QMSI)	140	M/F	0.83–0.95	0.68–0.92	This is the standard pattern for the questionnaire; the scale positions within each pair were then randomized. As there are four bipolar adjectives and five levels of sexual behavior to be rated, the questionnaire comprises 140 items. These are set out in random order, in the form of a questionnaire.	PCA identified five components. Factor loadings between 0.83–0.98; p < 0.001. With variances ranging from 69.2 to 92.9% for all three groups.	NI	NI	NI

(continued)

Table 16.4 (continued)

Instruments	N items	Target population[a]	Internal Consistency (Cronbach's α)[b]	Test-retest reliability[c]	Content validity	Construct Validity[d]	Cross-cultural validity	Responsiveness	Sensitivity
4. Derogatis Sexual Functioning Inventory (DSFI)	25	M/F	0.74–0.80	All > or =0.70 (1 wk) OR 0.84–0.92 (1 wk)	Y	Yes	NI	NI	NI
5. Female Sexual Arousability Index (SAI)	28	F	0.92	Yes	NI	Yes	NI	N	NI
6. Index of Sexual Satisfaction (ISS)	25	PART	0.93,0.99,0.92 (in one repeated sample)	NI	Yes	Yes	NI	NI	NI
7. Golombok Rust Inventory of Sexual Satisfaction (GRISS)	28	M/F	0.61–0.79 for men; 0.61–0.83 for women	0.52–0.84 for men; 0.47–0.82 for women	Yes	Yes	NI	N	NI
8. Jewish General Hospital Sexual Self-Monitoring Form	8	PART	NI	NI	NI	NI	NI	NI	NI
9. Derogatis Interview for Sexual Functioning—Self Report	25	M/F HETERO	0.74–0.80	0.86 (1 wk)	NI	Co-administration of a measure of well-being	NI	Yes	89%
10. Florida Sexual History Questionnaire (FSHQ)	20	M	0.90	NI	Yes	NI	NI	No	71.9%

11. Sexual Desire Conflict Scale for Women (SDCSW)	33	F	0.927	NI	To begin the process of quantitatively assessing the experience of sexual desire and its dysfunctional manifestations, a set of 43 items were written by the first author and also upon clinicians' reports of working with such patients, most typically the survivors of childhood sexual abuse	EFA and varimax rotation identified three factors. Total variance explained by 46.1%. Eigenvalues of ≥3.24. Factor loadings between 0.30 and 0.84 for the three factors	NI	NI	NI
12. Brief Index of Sexual Function for Women (BISF-W)	22	F	0.83	0.68–0.70 (1 mo)	NI	The construct validity may be criticized	NI	No	NI

(continued)

Table 16.4 (continued)

Instruments	N items	Target population[a]	Internal Consistency (Cronbach's α)[b]	Test-retest reliability[c]	Content validity	Construct Validity[d]	Cross-cultural validity	Responsiveness	Sensitivity
13. Multidimensional Sexuality Questionnaire (MSQ)	10	M/F	0.83 (anxiety) 0.92 (depression)	0.58–0.68 (3 wks) 0.68–0.71 (3 wks)	NI	1 factor (anxiety) 1 factor (depression)	NI	NI	NI
14. Sexual Function Scale	174	M/F	< 0.85	<0.99	NI	NI	NI	NI	NI
15. Sexual Interaction System Scale	48	PART	0.90	NI	Yes	The SISS distinguished between sex dysfunction couples and other couples; criterion validity: The SISS correlated with the Dyadic Adjustment Scale and an un attributed sexual satisfaction item	NI	NI	NI
16. Brief Male Sexual Function Index/inventory (BMSFI)	10	M	0.62–0.95	0.87 (1 wk)	NI	Insufficient	NI	No	NI
17. Sexual Activity Questionnaire (SAQ-F)	14	F HETRO	0.82–0.89	NI	NI	NI	NI	NI	NI

| 18. Sexual Desire Inventory | 14 | M/F | 0.87–0.88 | NI | Yes | CFA and the oblique rotation identified four factors: $x982 = 299.50$, GFI = 0.98, TLI = 0.96, RNI = 0.96, RMSEA = 0.058 at tested to two additional factors tested (fantasies and erotophilia) | Negative result: It was translated via the parallel-blind technique. This approach requires the participation of at least two people translating an inventory from the source (English) to the target language (Spanish). Then, these translators compare their individual work and collaborate on the final version. | NI | NI |

(continued)

Table 16.4 (continued)

Instruments	N items	Target population[a]	Internal Consistency (Cronbach's α)[b]	Test-retest reliability[c]	Content validity	Construct Validity[d]	Cross-cultural validity	Responsiveness	Sensitivity
19. Sexual History Form (SHF)	46	M/F	0.65 for male global scale; for the female score, item-total correlation coefficients range 0.18–0.85, majority between 0.50 and 0.70	0.92 (2 wks)	NI	NI	NI	NI	NI
20. Arizona Sexual Experience Scale (ASEX)	5	M/F HETERO/ HOMO	0.91	0.80 (1–2 wks)	Yes	The items on the ASEX did not correlate with depression scores. Significant differences on total ASEX scores between patients and controls	NI	Yes	82%
21. The Female Sexual Functioning Index (FSFI)	19	F	0.82	0.79–0.86	Yes	NI	NI	Yes	NI
22. Short scale to measure female	9	F	0.74–0.80	0.81–0.90	Yes	NI	NI	Yes	79%
23. Female Sexual Distress Scale (FSDS)	12	F	0.86–0.96	0.74–0.92 (1–4 wks)	NI	1 factor	NI	Yes	93%

24. The Sexual Function Questionnaire (SFQ)	28	F	0.79–0.91	0.42–0.70	Yes	NI	NI	Yes	NI
25. Sexual Satisfaction Scale for Women (SSS-W)	30	F	0.81–0.90	0.72–0.83 (4–5 wks)	NI	1 factor	NI	NI	NI
26. Scale for Quality of Sexual Function (QSF)	40	M/F	0.75–0.90	NI	NI	NI	NI	NI	NI
27. Sexual Arousal and Desire Inventory (SADI)	54	M/F	0.91	0.57–0.81	List of 86 English descriptors was compiled by interview in gap proximately 500 men and women	PCA and varimax rotation identified four factors. Total variance explained by 41.3%. Factor loadings ≥0.30 for the total of factors	NI	NI	NI

(continued)

Table 16.4 (continued)

Instruments	N items	Target population[a]	Internal Consistency (Cronbach's α)[b]	Test-retest reliability[c]	Content validity	Construct Validity[d]	Cross-cultural validity	Responsiveness	Sensitivity
28. Cues for Sexual Desire Scale (CSDS)	40	F	USA version: 0.78–0.92 Portuguese version: 0.87–0.90	NI	USA version: Fifty women (age range 18–67 years) were involved in the item generation stage. The 125 generated items were listed using a conventional questionnaire format	USA version: PCA and varimax rotation identified four factors. Factor loadings between 0.43 and 0.89. Portuguese version: CFA identified four factors: $x2/d.f. = 24.5$; CFI = 0.793; GFI = 0.754; RMSEA = 0.08; P[RMSEA<0.05] <0.001 PCA and varimax rotation identified six factors, but the scree plot extracted five factors. Total explained variance of 58.3% with factorial loads between 0.492–0.854, eigenvalue >1 and communalities between 0.408–0.750	The original version of the CSDS was translated into Portuguese by three independent persons fluent in Portuguese and English. The final version was back-translated by a native English speaker. The translation of the English version into Portuguese was semantically equivalent to the English original accordingly to the retro-translation.	NI	NI

	13	F	0.88–0.97	0.75–0.93 (1–4 wks)	Yes	NA	NI	NI	Yes	93%
29. Female Sexual Distress Scale Revised (FSDS-R)										
30. Female Sexual Desire Questionnaire (FSDQ)	50	F HETERO	0.80–0.92	NI	Preliminary items for the FSDQ were determined through individual interviews with 40 heterosexual partnered women. Interview data were analyzed using the principles of interpretive phenomenological analysis, and questionnaire items were developed to reflect the themes extracted from these data. Questions assessing the DSM-IV-TR criteria for HSDD. Approximately 250 candidate items were peer reviewed by a group of three researchers/clinicians for item appropriateness, relevance, redundancy, and ease of understanding	EFA and the oblimin rotation identified six factors. Total variance explained by 60%. Factor loadings >0.40	NI	NI	NI	NI

(continued)

Table 16.4 (continued)

Instruments	N items	Target population[a]	Internal Consistency (Cronbach's α)[b]	Test-retest reliability[c]	Content validity	Construct Validity[d]	Cross-cultural validity	Responsiveness	Sensitivity
31. Sexual Distress Scale (SDS)/ Sexual Distress Scale Revised (SDS-R)	13	M	0.93–0.94	0.80–0.85 (35 days)	Yes	1 factor (CFA); invariant	NI	NI	NI

[a]Target population specified by the instrument: M/F "male and female" F "female" M "male" PART "male and female partners" | HETRO "heterosexual" HOMO "homosexual"

[b]Reliability OK if Cronbach's α > 0.7

[c]ICC values less than 0.5 are indicative of poor reliability, values between 0.5 and 0.75 indicate moderate reliability, values between 0.75 and 0.9 indicate good reliability

[d]EFA Exploratory Factor Analysis, CFA Confirmatory Factor Analysis, PCA Principal Component Analysis

There is a notable lack of instruments designed for use in homosexuals. Only one of these reviews sought to explore the availability of instruments in a specific population: gay men. McDonagh's systematic review from 2014 comprised only seven instruments that were potentially viable for use in this specific population. She pointed to a number of shortcomings, specifically hetero-normative or heterosexist language [22], concluding that there was currently no psychometrically sound measure of male sexual function that can be used in gay men. The lack of involvement of sexual minorities in the development of these instruments may be at fault.

The issue of patient/individual involvement in instrument development was also raised by Arrington et al. [18] in their review from 2004. They questioned whether available instruments would be more comprehensive had items been generated using qualitative research in a diverse group of patients either representative of the general population or dealing with SD. The study population often selected for instrument validation studies is also a concern as many scales appear to have been validated in patients with SD, rather than community samples. This raises concerns about their ability to detect SD in the general population.

There is also limited evidence of instruments' responsiveness or their ability to detect change over time. Evidence of responsiveness had only been documented in seven of the 31 instruments considered. Users should be wary of this limitation as they consider these tools for use in research or clinical practice.

Surprisingly, none of the reviews addressed the issue of the electronic collection of sexual function/QoL in spite of the fact that the practicality of administering, mostly face-to-face or self-administered paper-and-pencil, patient-reported outcomes (PROs) has been documented as a major barrier to their use, particularly in clinical practice. Paper-based approaches require that a physician and/or other staff administer the questionnaire during the consultation and enter and/or analyze data manually, requiring resources to collect, analyze and utilize PROs data [54]. Computerized PRO assessments have become common in settings where PROs have been widely adopted, offering a number of advantages over pencil-and-paper assessments. For example, applying compulsory items and pre-specifying acceptable ranges or values can improve data completeness and quality. Complex skip patterns can be programmed or Computer Adaptive Testing (CAT) methods applied to ease administration [55]. Immediate data capture also reduces the burden and/or costs associated with data entry [56]. Furthermore, from a methodological point of view, there is strong evidence to suggest that electronic and paper-and-pencil administration methods are equivalent [57].

New technologies are yet another means of assessing sexual activity. We see the emergence of technologies making use of both active, memory-based data capture methods, which are subjective and socially acceptable, and passive, technology-based methods—via wearables and personal devices—which are sensory-based, momentary, non-judgmental and context-rich. Sexual activity assessment applications (also referred to as "sex trackers") take an alternative approach, seeking to help individuals improve the quality of their sex lives via various means such as educational content including erotica, personalized advice, forums, data visualization, and automated statistics (…). Wac has qualified technologies such as these as

"Quality of Life Technologies" (QoLT) or any technology designed for the assessment or improvement of individuals' QoL. These "leverage miniaturized computing, storage, and communication sensor- and actuator-based, context-rich technologies that can be embedded within various personal devices" [58]. Some technologies seek to appeal to all, whereas others focus specifically on features that appeal to sexual minorities or are deliberately gendered. Applications like Coral[3] (an intimacy app that relies on education), Rosy[4] (a platform offering women a holistic approach to sexual health and wellness, offering education, self-discovery and community), Sex Life[5] (a sex tracker, sex diary and sex calendar) are novel as their core functionalities are not geared towards sexual activity for fertility or the prevention of sexually transmitted infections (STIs). Unsurprisingly, most of these applications still rely on memory-based approaches to capture users' data, differing from now commonplace applications designed to assess and improve physical fitness. Nevertheless, they offer features that seek to maintain, prevent, or enhance sexual QoL, fitting with the aims of QoLT. Many of these application and service providers have already begun to collaborate with universities and researchers, often establishing cohorts of users and generating a wealth of real-world data.

Other mobile applications are hybrids, pairing sexual activity assessment with the promotion of sexual health (e.g., STI prevention or testing). For example, applications like Biem[6] or Safely[7] focus on convenience. The former offers telemedicine and at home or laboratory testing. The latter connects users with a network of laboratories and allows them to receive their results electronically. They have also incorporated interactive features; both allowing users to make their data available to partners or potential partners securely and privately and the functionality "Biem Connect" allows sexual partners to be notified anonymously if one of their partners tests positive for an STI. Nice[8] is a combination of a sexual activity assessment application that allows for some data on STIs to be collected. Data entered through these applications can be synched with the Apple HealthKit,[9] which allows users to record their reproductive health data, including the occurrence of sexual activity.

The use of wearables (e.g. Smartwatches, rings) and personal devices (e.g. sex toys) to measure sexual activity appears to be possible as they work by measuring whole-body motion (accelerometer), heart rate (photo plethysmography), body temperature, and respiration rate. The use of these devices, such as the Oura Ring[10]

---

[3] https://getcoral.app/

[4] https://meetrosy.com/

[5] https://play.google.com/store/apps/details?id=com.sexdiary.safeforsex

[6] http://biemteam.com/

[7] https://play.google.com/store/apps/details?id=me.safehealth&hl=fr

[8] https://apps.apple.com/us/app/sex-tracker-by-nice/id1107291612

[9] https://developer.apple.com/design/human-interface-guidelines/healthkit/overview/

[10] https://ouraring.com/

or MyMotiv Ring[11] (personal health or fitness trackers worn as rings) have been used to successfully generate detailed data corresponding to different parts of the sexual response cycle. However, most still require users to qualify increases in peaks and valleys in physiological data captured by wearable devices as sexual activity. Although, to date, the use of fitness trackers and Smartwatches to capture this highly sensitive data does not seem to be the approach most favored by QoLT developers, there is evidence, mostly from popular culture, that quantified self-ers are embracing these new technologies. They are willing to monitor their own behaviors, even their sex lives, to gain a greater understanding of their bodies, in the vein of "knowing thyself" and perhaps even in an attempt to improve their (sexual) QoL [59]. Connected personal devices offer yet another avenue for passive data collection related to sexual function or activity. For example, users of biofeedback vibrators like the Lioness[12] can, in addition to traditional functionalities, visualize their precision sensor-generated data via an application and participate in medical and academic research by opting to share their de-identified data. Another connected wearable is the Lovely 2.0,[13] a connected penis ring, which, when worn during sex learns about your "style" of having sex and provides personalized advice via a dedicated application. Some research is currently ongoing to validate these sex toys as viable research tools. There is a need to understand how data generated from applications, wearables and connected devices correlates with direct and indirect measures of sexual function and self-reported measures seeking to capture sexual QoL as a broader construct.

As it is important to get information from the best source when standardizing the assessment of sexual activity, function and QoL, we proposed the following sources of information for each of the different dimensions according to Mayo et al. taxonomy [60] (Table 16.5). These include self-reports like patient-reported outcomes (PROs), collected via the proposed self-administered instruments, which clearly vary in both length and scope and Ecological Momentary Assessment (EMA) techniques, in which individuals report states close to the time they experience them, as well as other reports like observer/proxy-reported outcomes (ObsRO/ProxyRO) and Peer-Ceived Momentary Assessments Methods (PeerMA) [61]. Both ObsRO/ProxyRO and PeerMA would rely on an "other", likely to be an intimate partner rather than a friend in the case of sexual activity reporting on an individual's perceived state and behaviors. The former doing this at fixed time points and the latter on a momentary basis, similar to the collection of EMAs. These could be useful insofar as externally observable behaviors and states are concerned, e.g. perceived desire, arousal, plateau, and orgasm as well as the frequency of sexual activity with a partner. Finally, technology-reported outcomes (TechROs) which could be collected via wearables, Smartphone behavioral metrics, or connected devices could be

---

[11] https://mymotiv.com/

[12] https://lioness.io/

[13] https://ourlovely.com/

Table 16.5 Options for measuring dimensions of sexual function/QoL according to Mayo et al. taxonomy [60]

Dimensions	Self-report	Other/ Partner-Report	Technology-reported outcomes		
	PRO and/or EMA	ObsRO, PeerMA	Wearables (psychophysiology, movement)	Smartphone behavioral metrics	Other connected devices
Desire	Overall: PRO	Overall: ObsPRO	HRV	Analytics of search	
	Momentary: EMA[a]	Momentary: PeerMA		Social network activity	
				VA	
Arousal	Overall: PRO	Overall: ObsPRO	BT	Eye movement, pupil size	BT
	Momentary: EMA[a]	Momentary: PeerMA[a]	HR		GSR
			GSR		HR
			RR		RR
			VA		
Performance	Overall: PRO	Overall: ObsPRO	BM		BM
	Momentary: EMA[a]	Momentary: PeerMA[a]	BT		BT
			GSR		GSR
			HR		HR
			RR		RR

Orgasm	Overall: PRO Momentary: EMA[a]	Overall: ObsPRO Momentary: PeerMA[a]	BM BT EMG GSR HR RR VA	Analytics of search	BM BT EMG GSR HR RR
Pain	Overall: PRO Momentary: EMA[a]	Overall: ObsPRO	BM HR RR EMG		BM HR RR EMG
Frequency	Overall: PRO Momentary EMA	Overall: ObsPRO Momentary: PeerMA	BM		Frequency, duration and context of use (weekdays, weekend, timing etc.)
Importance of sexual activity	Overall: PRO Momentary: EMA			Analytics of search, social online interactions	Use of devices for masturbation

(continued)

Table 16.5 (continued)

Dimensions	Self-report PRO and/or EMA	Other/ Partner-Report ObsRO, PeerMA	Technology-reported outcomes		
			Wearables (psychophysiology, movement)	Smartphone behavioral metrics	Other connected devices
Satisfaction Pleasure and Quality	Overall: PRO Momentary: EMA[a]	Overall: ObsPRO	BT EMG HR GSR RR VA	Analytics of search, social online interactions	Frequency, duration and context of use (weekdays, weekend, timing etc.)
Anxiety—distress	Overall: PRO Momentary: EMA	Overall: ObsPRO Momentary: PeerMA	BT[b] EMG[b] GSR[b] HR[b] RR[b]		BT EMG GSR HR RR
Body image—Self-esteem	Overall: PRO Momentary: EMA	Overall: ObsPRO		Analytics of search, social online interactions Frequency/manner of looking at a magic mirror Frequency of selfies posted	

Relationship feelings and emotions	Overall: PRO	Overall: ObsPRO	BM[b]	Analytics of social [online] interactions with partner(s)	Use of devices with others (Frequency, duration and context of use, communication with partner(s))
	Momentary: EMA	Momentary: PeerMA	BT[b]		
			GSR[b]		
			HR[b]		
			RR[b]		
			VA[b]		
Fantasy	Overall: PRO	Overall: ObsPRO		Analytics of social [online] interactions	Masturbation
	Momentary: EMA	Momentary: PeerMA			Unconventional use of devices
Cultural and gender roles	Overall: PRO	Momentary: PeerMA			
	Momentary: EMA				

BM Body motion, *BT* Body temperature, *EMG* electromyography, an electrical correlate of muscle tension, *GSR* Galvanic skin response, changes in sweat gland activity that are reflective of the intensity of our emotional state, *HR* Heart rate, *HRV* Heart rate variability, the beat-to-beat HR changes and are thought to be an appropriate indicator of a person's health and heart disorders (Camm et al., 1996), *VA* Voice stress or layered voice analysis to determine the emotional state of speakers

[a]The EMA of these dimensions would conceivably occur after actively or passively recorded sexual activity

[b]These measures could be taken in the presence of the partner(s)

used for many of the domains important to the assessment of sexual function/QoL, namely if whole-body motion and physiological parameters could be co-calibrated with valid direct, indirect, or self-reported measures of sexual function/QoL, We have summarized some options for assessing and measuring different domains below in Table 16.5.

Conclusions

There are a number of generic self-reported instruments that can ease the collection of data on sexual function and QoL in both research and clinical practice. The emerging use of sexual activity assessment applications, wearables and personal devices, may also provide another less invasive avenue for the collection of sexual activity data. There is nevertheless a need to understand the measurement properties of these novel means of data generation as well as users' willingness to submit or share their highly personal data with researchers or providers in order to facilitate their adoption in research and clinical practice. There is still a long way to go if we wish to exploit the full potential of QoLT for sexual health and QoL.

Acknowledgements I would like to thank Camille Etcheverry, Cecilia Dumar and Camille Berduras for their assistance in conducting this review and the editors for their helpful comments. FundingThis research was conducted during Diana Barger's post-doctoral research fellowship (2020–2022), funded by the Agence nationale de recherche sur le SIDA et les hépatites (ANRS), at the University of Bordeaux/Inserm Bordeaux Population Health Research Centre (U 1219).

References

1. Gebhard PH. Human sexual activity Encyclopædia Britannica 2019 June, 26 2020.
2. World Health Organization, D.o.R.H.a.R., Defining sexual health: Report of a technical consultation on sexual health, 28–31 January 2002, Geneva, WHO, Editor; 2006: Geneva.
3. Masters W, Johnson V. Human Sexual Response. Boston: Little Brown; 1966.
4. Kaplan H, Disorders of sexual desire and other new concepts and techniques in sex therapy. New York. New York: Brunner/Hazel Publications; 1979.
5. Basson R. Female sexual response: the role of drugs in the management of sexual dysfunction. Obstet Gynecol. 2001;98(2):350–3.
6. Whipple B, Brash-McGreer K. Management of female sexual dysfunction. In: Sipski M, Alexander C, editors. Sexual function in people with disability and chronic illness. A health professional's guide. Gaithersburg, MD: Aspen; 1997.
7. Sungur MZ, Gündüz A. A comparison of DSM-IV-TR and DSM-5 definitions for sexual dysfunctions: critiques and challenges. J Sex Med. 2014;11(2):364–73.
8. Laumann EO, Paik A, Rosen RC. Sexual dysfunction in the United States prevalence and predictors. JAMA. 1999;281(6):537–44.
9. Nocturnal penile tumescence study. In: Male sexual dysfunction; 2017. p. 129–32.

10. Linet OI, Neff LL. Intracavernous prostaglandin E1 in erectile dysfunction. Clin Investig. 1994;72(2):139–49.
11. Scott JR, Liu D, Mathes DW. Patient-reported outcomes and sexual function in vaginal reconstruction: a 17-year review, survey, and review of the literature. Ann Plastic Surg. 2010;64(3):311–4.
12. Varela CG, et al. Penile Doppler ultrasound for erectile dysfunction: technique and interpretation. Am J Roentgenol. 2020;214(5):1112–21.
13. Berman JR, et al. Clinical evaluation of female sexual function: effects of age and estrogen status on subjective and physiologic sexual responses. Int J Impot Res. 1999;11(Suppl 1):S31–8.
14. Symul L, et al. Assessment of menstrual health status and evolution through mobile apps for fertility awareness. npj Digital Med. 2019;2(1):64.
15. Mokkink LB, et al. The COnsensus-based standards for the selection of health measurement INstruments (COSMIN) and how to select an outcome measurement instrument. Braz J Phys Ther. 2016;20(2):105–13.
16. Daker-White G. Reliable and valid self-report outcome measures in sexual (Dys)function: a systematic review. Arch Sex Behav. 2002;31(2):197–209.
17. Meston CM, Derogatis LR. Validated instruments for assessing female sexual function. J Sex Marital Therapy. 2002;28(Suppl. 1):155–64.
18. Arrington R, Cofrancesco J, Wu AW. Questionnaires to measure sexual quality of life. Qual Life Res. 2004;13(10):1643–58.
19. Corona G, Jannini EA, Maggi M. Inventories for male and female sexual dysfunctions. Int J Impotence Res. 2006;18(3):236–50.
20. DeRogatis LR. Assessment of sexual function/dysfunction via patient reported outcomes. Int J Impotence Res. 2008;20(1):35–44.
21. Giraldi A, et al. Questionnaires for assessment of female sexual dysfunction: a review and proposal for a standardized screener. J Sex Med. 2011;8(10):2681–706.
22. McDonagh LK, et al. A systematic review of sexual dysfunction measures for gay men: how do current measures measure up? J Homosex. 2014;61(6):781–816.
23. Santos-Iglesias P, Mohamed B, Walker LM. A systematic review of sexual distress measures. J Sex Med. 2018;15(5):625–44.
24. Cartagena-Ramos D, et al. Systematic review of the psychometric properties of instruments to measure sexual desire. BMC Med Res Methodol. 2018;18(1):109.
25. Patterson DG, O'Gorman EC. The SOMA—a questionnaire measure of sexual anxiety. Br J Psychiatry. 2018;149(1):63–7.
26. LoPiccolo J, S JC. The sexual interaction inventory: a new instrument for assessment of sexual dysfunction. In: LoPiccolo J, L L, editors. Handbook of sex therapy. Perspectives in sexuality (Behavior, research, and therapy). Boston: Springer; 1978.
27. Harbison JJ, et al. A questionnaire measure of sexual interest. Arch Sex Behav. 1974;3(4):357–66.
28. Derogatis LR, Melisaratos N. The DSFI: a multidimensional measure of sexual functioning. J Sex Marital Ther. 1979;5(3):244–81.
29. Andersen BL, et al. A psychometric analysis of the sexual arousability index. J Consult Clin Psychol. 1989;57(1):123–30.
30. Mark KP, et al. A psychometric comparison of three scales and a single-item measure to assess sexual satisfaction. J Sex Res. 2014;51(2):159–69.
31. Rust J, Golombok S. The GRISS: a psychometric instrument for the assessment of sexual dysfunction. Arch Sex Behav. 1986;15(2):157–65.
32. Libman E, et al. Jewish General Hospital sexual self monitoring form. In: Davis CM, et al., editors. Handbook of sexuality related measures. Thousand Oaks: Sage; 1998. p. 272–4.
33. Geisser ME, et al. Reliability and validity of the Florida sexual history questionnaire. J Clin Psychol. 1991;47(4):519–28.
34. Kaplan L, Harder DW. The sexual desire conflict scale for women: construction, internal consistency, and two initial validity tests. Psychol Rep. 1991;68(3 Pt 2):1275–82.

35. Taylor JF, Rosen RC, Leiblum SR. Self-report assessment of female sexual function: psychometric evaluation of the brief index of sexual functioning for women. Arch Sex Behav. 1994;23(6):627–43.

36. Snell WE, Fisher TD, Walters AS. The multidimensional sexuality questionnaire: an objective self-report measure of psychological tendencies associated with human sexuality. Ann Sex Res. 1993;6(1):27–55.

37. Marita PM. Sexual function scale: history and current factors. In: Terri D, et al., editors. Handbook of sexuality-related measures. New York: Routledge; 1988.

38. O'Leary MP, et al. A brief male sexual function inventory for urology. Urology. 1995;46(5):697–706.

39. Thirlaway K, Fallowfield L, Cuzick J. The sexual activity questionnaire: a measure of women's sexual functioning. Qual Life Res. 1996;5(1):81–90.

40. Spector IP, Carey MP, Steinberg L. The sexual desire inventory: development, factor structure, and evidence of reliability. J Sex Marital Ther. 1996;22(3):175–90.

41. Creti L, et al. "Global sexual functioning:" a single summary score for Nowinksi and LoPiccolo's sexual history form (SHF). In: Davis CM, et al., editors. Handbook of sexuality-related measures. Thousand Oaks, CA: Sage; 1998.

42. McGahuey CA, et al. The Arizona sexual experience scale (ASEX): reliability and validity. J Sex Marital Ther. 2000;26(1):25–40.

43. Rosen R, et al. The female sexual function index (FSFI): a multidimensional self-report instrument for the assessment of female sexual function. J Sex Marital Ther. 2000;26(2):191–208.

44. Dennerstein L, Lehert P, Dudley E. Short scale to measure female sexuality: adapted from McCoy female sexuality questionnaire. J Sex Marital Ther. 2001;27(4):339–51.

45. Derogatis LR, et al. The female sexual distress scale (FSDS): initial validation of a standardized scale for assessment of sexually related personal distress in women. J Sex Marital Ther. 2002;28(4):317–30.

46. Quirk FH, et al. Development of a sexual function questionnaire for clinical trials of female sexual dysfunction. J Womens Health Gend Based Med. 2002;11(3):277–89.

47. Meston C, Trapnell P. Development and validation of a five-factor sexual satisfaction and distress scale for women: the sexual satisfaction scale for women (SSS-W). J Sex Med. 2005;2(1):66–81.

48. Heinemann LA, et al. Scale for quality of sexual function (QSF) as an outcome measure for both genders? J Sex Med. 2005;2(1):82–95.

49. Toledano R, Pfaus J. Original research—outcomes assessment: the sexual arousal and desire inventory (SADI): a multidimensional scale to assess subjective sexual arousal and desire. J Sex Med. 2006;3(5):853–77.

50. McCall K, Meston C. Cues resulting in desire for sexual activity in women. J Sex Med. 2006;3(5):838–52.

51. DeRogatis L, et al. Validation of the female sexual distress scale-revised for assessing distress in women with hypoactive sexual desire disorder. J Sexual Med. 2008;5(2):357–64.

52. Goldhammer DL, McCabe MP. Development and psychometric properties of the female sexual desire questionnaire (FSDQ). J Sexual Med. 2011;8(9):2512–21.

53. Santos-Iglesias P, et al. Psychometric validation of the female sexual distress scale in male samples. Arch Sex Behav. 2018;47(6):1733–43.

54. Feinstein AR. Benefits and obstacles for development of health status assessment measures in clinical settings. Med Care. 1992;30(5 Suppl):Ms50-6.

55. Sands WA, Waters BK, McBride JR, editors. Computerized adaptive testing: from inquiry to operation, vol. xvii. Washington, DC: American Psychological Association; 1997. p. 292.

56. Jones JB, Snyder CF, Wu AW. Issues in the design of Internet-based systems for collecting patient-reported outcomes. Qual Life Res. 2007;16(8):1407–17.

57. Gwaltney CJ, Shields AL, Shiffman S. Equivalence of electronic and paper-and-pencil administration of patient-reported outcome measures: a meta-analytic review. Value Health. 2008;11(2):322–33.

58. Wac K. Quality of life technologies. In: Gellman M, editor. Encyclopedia of behavioral medicine. New York: Springer; 2019. p. 1–2.
59. Wac K. From quantified self to quality of life. In: Rivas H, Wac K, editors. Digital health: scaling healthcare to the world. Cham: Springer International Publishing; 2018. p. 83–108.
60. Mayo NE, et al. Montreal Accord on patient-reported outcomes (PROs) use series – Paper 2: terminology proposed to measure what matters in health. J Clin Epidemiol. 2017;89:119–24.
61. Berrocal A, et al. Complementing human behavior assessment by leveraging personal ubiquitous devices and social links: an evaluation of the peer-ceived momentary assessment method. JMIR Mhealth Uhealth. 2020;8(8):e15947.

Chapter 17
Role of Technology-Enabled Tools for Measuring Financial Resources and Improving Quality of Life

Joan Julia Branin

Introduction

An individual's financial resources are directly related to their ability to meet current and future needs. Higher levels of financial assets and lower debt have been found to be positively associated with financial satisfaction [1, 2]. Findings from the 2012 National Financial Capability Survey report that the presence of an emergency fund has been found to be positively associated with financial satisfaction [3]. On the other hand, inadequate financial resources can lead to financial strain and financial distress. This financial burden has been associated with decreased treatment adherence, worsened mortality, increased bankruptcy. Financial strain can have a powerful impact on a person's perception of their overall well-being. At the same time, the growth of personal finance apps and technology-enabled tools have exploded over the past 20 years with the potential for objective, quantitative assessments, and self-management of financial resources. Although financial resources have been shown to have a direct effect on quality of life, few studies have addressed the impact of financial resources and financial burden on quality of life and the role of QoL technology-enabled tools for measuring and managing financial resources and improving quality of life.

Definition of Financial Resources

According to the WHOQOL theoretical model, quality of life encompasses several key domains: Physical Health, Psychological Health, Social Relationships, and Environment. One of the environmental facets of quality of life is financial resources.

J. J. Branin (✉)
Center for Health and Aging Research, Pasadena, CA, USA

© The Author(s) 2022
K. Wac, S. Wulfovich (eds.), *Quantifying Quality of Life*, Health Informatics,
https://doi.org/10.1007/978-3-030-94212-0_17

The facet of financial resources explores a person's view of how his/her financial resources and the extent to which these resources meet the needs for a healthy and comfortable lifestyle, and what the person can afford or cannot afford which might affect quality of life. This facet includes a sense of satisfaction/dissatisfaction with those things which the person's income enables them to obtain and the sense of dependence/independence provided by the person's financial resources (or exchangeable resources), and the feeling of having enough regardless of the respondent's state of health or whether or not the person is employed. It acknowledges that a person's perspective on financial resources as "enough," "meeting my needs," etc. is likely to vary greatly [4].

Originally, financial well-being was seen as synonymous with material and objective financial resources (e.g., income) and was investigated at the country level, without including individuals' perceptions. However, once Easterlin [5] recognized the importance of subjective perceptions of financial well-being, it has been mainly studied at the individual level.

Nowadays, scholars agree that financial well-being has both an objective and a subjective side, believing that objective financial well-being consists of individuals' material financial resources (e.g., income), whereas subjective financial well-being is individuals' perception and cognitive and emotional evaluation of their own financial condition [6]. This subjective side frequently has been investigated mainly in populations considered critical or atypical from a financial point of view, such as the elderly [7], unwed women [8], and divorced women [9].

Although financial resources have been shown to have a direct effect on quality of life, few studies have addressed the impact of financial resources and financial burden on quality of life and the use of personal, quantitative methods and technologies to assess the financial resources and the financial well-being of individuals.

This chapter reviews the literature about (1) the effects of financial resources and financial burden on treatment outcomes and overall quality of life; (2) the state-of-art tools for measuring financial resources by individuals and financial and health professionals; (3) the evaluation of Web-based interventions for enhancing financial resource management; and (4) the behavioral and technology-related factors for successful adoption of QoL technology-enabled methods and financial resource management tools for improving individual life satisfaction and financial well-being.

Current Research About Financial Resources and Financial Well-Being

Financial Resources and Financial Satisfaction

Researchers over the past 30 years have examined both objective and subjective measures to describe the financial condition of individuals and families. Whereas objective indicators have been used to predict one's perceptions about their financial condition, such indicators do not measure the depth of one's feelings about or reaction to their financial condition. Several researchers have examined factors contributing to psychological well-being and found economic distress to be a good predictor of lower levels of well-being [10, 11]. Mills [10] found that a key determinant of psychological well-being was the level of economic distress reported. Rettig and Danes found that people who perceived their income to be inadequate to meet even basic living expenses reported experiencing negative feelings and lower satisfaction with the perceived gap between their standard and their level of living. Such normal, negative reactions to adverse economic condition can reduce individuals' psychological well-being [12].

An individual's financial resources are directly related to their ability to meet current and future needs. One theory argues that relative income (both to others and to previous periods) rather than absolute income may be important in explaining variations in life satisfaction, i.e., this includes income relative to others, which impacts an individual's status in society, and relative to the individual's income in previous periods, which impacts one's habits and view of what is the norm [13]. Boyce [14] found that the ranked position of an individual's income predicts general life satisfaction, whereas absolute income and reference income have no effect. Similarly, the rank of a person's income or wealth within a social comparison group, rather than income or wealth themselves or their deviations from the mean within a reference group, is more strongly associated with depressive symptoms [15].

Researchers generally agree that subjective well-being (SWB) tends to be higher among people with abundant economic resources as compared to people with limited economic resources [1, 2]. However, recent research has begun to distinguish two aspects of subjective well-being: emotional well-being which refers to the emotional quality of an individual's everyday experience that make one's life pleasant or unpleasant, and life evaluation which refers to the thoughts that people have about their life when they think about it. Analyzing results from the Gallup-Heathways Well-being Index and Cantril's Self Anchoring Scale, Kahneman [16] examined whether money buys happiness separately for these two aspects of well-being. While life evaluation rises steadily with increases in income, emotional well-being also rises but there is no further progress beyond an annual income of USD75,000. High income buys life satisfaction but not happiness; low income is associated both with low life evaluation and low emotional well-being.

Although the literature on the relationship between economic standing and SWB has become substantial, only a limited number of studies, to date, have focused on this relationship in mid-life and old age [1, 2]. In general, the findings of the studies on the wealth-SWB relationship have demonstrated that greater wealth leads to higher SWB. Some studies have also demonstrated that the effect of wealth is stronger than the effect of income. Wealth refers to the stock of assets (e.g., savings, real estate, businesses) less their debt held by a person or household at a single point in time vs. income which refers to money (e.g., wages, rents on property, government assistance) received by a person or household over some period of time [2]. It should be noted, however, that the magnitude of the association between wealth and SWB may depend upon the welfare regime and the degree of social support (including type of pension) provided by the state. Similarly, those with more economic resources are more likely to cope with stressful events when they occur. Most studies on the issue were conducted within a cross-sectional research design, very few studies have addressed the wealth-SWB relationship from a longitudinal perspective taking into account the dynamics of household wealth and life transition events over time [1, 2].

Longitudinal Perspective of Financial Resources and Financial Well-Being

Prior research consistently has found that older adults, despite low incomes, are more financially satisfied than younger adults. This "satisfaction paradox" is typically attributed to elders' supposed psychological accommodation to poor financial circumstances. However, data from the first wave of the Norwegian NorLAG study shows that material circumstances are more important to the financial satisfaction of the elderly. A considerable part of the higher financial satisfaction with increasing age can thus be explained by greater assets and less debt among the elderly. Nonetheless, assets and debt do not mediate this relationship at lower incomes, because older people with little income have very little accumulated wealth. Older adults with little income and wealth have a much stronger tendency to be financially satisfied than younger, equally poor counterparts. An "aging paradox" remains in this field [1].

A life-course perspective attempts to place indicators of financial well-being in context, yet it is not widely used in the financial well-being literature. Studies that do use a life-course perspective tend to focus on the 'U' shaped gradient of financial satisfaction over time [1], arguing that lower debt and greater asset wealth explain why older people are more financially satisfied despite having lower incomes than younger people. Employment-based income was a stronger determinant of younger people's financial satisfaction whereas investment income was a stronger determinant for older adults [2].

Financial Burden and Its Effect on Treatment Outcomes and Quality of Life

The financial burden of care, especially for cancer, is slowly becoming more recognized among providers, institutions, and policymakers. The high costs of cancer care, even for patients with health insurance, can lead to poorer outcomes for patients, decrease quality of life, heighten stress, and, in some cases, increase mortality rates [17].

According to Fenn [17], increased financial burden because of cancer care costs is the strongest independent predictor of poor quality of life among cancer survivors over the age of 18. Patients who reported "a lot" of financial problems were approximately four times less likely to report a quality of life that was "good" or higher compared with patients who reported no financial problems. The magnitude of cancer-related financial difficulty was a more significant predictor of quality of life than age, education, race/ethnicity, and family income. These findings highlight the potentially powerful impact of financial strain on a patient's perception of their overall well-being after a cancer diagnosis.

Financial resources can act as a buffer for subjective well-being after the onset of a disability or disease. Using data from the Health and Retirement Study, Smith [18] found that wealth measured prior to the onset of a disability protected participants' well-being from some of the negative effects of a new disability. Participants who were above the median in total net worth reported a much smaller decline in well-being after a new disability than did participants who were below the median. There was also some evidence that the buffering effect of wealth faded with time, as below-median participants recovered some of their well-being.

Regarding frailty, financial resources, and subjective well-being in later life, Hubbard [19] found that those with greater financial resources reported better subjective well-being with evidence of a "dose–response" effect. The poorest participants in each frailty category had similar well-being to the most well-off with worse frailty status. Hence, while the association between frailty and poorer subjective well-being is not significantly impacted by higher levels of wealth and income, financial resources may provide a partial buffer against the detrimental psychological effects of frailty.

Standardized Self-Report Based Measurements of Financial Resources and Quality of Life

Two different approaches exist to measure financial resources and financial well-being. One approach utilizes a standardized instrument for self-assessment and is frequently used by physical and mental health professionals.

One standardized instrument, **The InCharge Financial Distress/Financial Well-Being Scale (IFDFW),** is an 8-item scale designed to measure a latent construct representing responses to one's financial state on a continuum ranging from overwhelming financial distress/lowest level of financial well-being to no financial distress/highest level of financial well-being. The robust Cronbach's alpha of 0.956 for the IFDFW indicates high internal consistency, and factor analysis indicates measurement of one factor, verifying that the indicators together estimate only one latent construct. Thus, the IFDFW Scale provides a high level of confidence for researchers and practitioners using the scores to indicate perceived levels of financial distress/financial well-being in individuals and groups of consumers [20].

This scale has been used to develop and deliver high quality, effective workplace financial programs [21]. However, the scale was developed to measure financial distress and well-being in general and may not include factors relevant to a specific stage of life such as emerging adulthood [20]. The IFDFW is also known as the **Personal Financial Well-Being (PFW) scale**.

Another standardized instrument, **The Multidimensional Subjective Financial Well-being Scale (MSFWB)** is a 25-item scale measuring five different aspects of subjective financial well-being (general subjective financial well-being, money management, peer comparison, having money, financial future) with acceptable psychometric properties and was developed for use with emerging adults in European countries namely Italy and Portugal. Validity of its 25 items was assessed through consultation with experts on the construct as well as the emerging adults themselves (qualitative research). Furthermore, the functioning of these items was confirmed by quantitative results showing the stability of the scale's factorial structure, its generalizability across different emerging adult subgroups, its relationship with convergent and criterion-related measures, and its internal consistency. However, the generalizability of this measure across European countries and use in the US has not been demonstrated [22].

State-of-the-Art Apps and Tools for Measuring Financial Resources and Financial Well-Being

Use of Personal Finance Apps and Financial Resources

The second approach to measure financial resources and financial well-being utilizes Artificial Intelligence (AI) and algorithms and is frequently used by financial advisors and individuals interested in the self-management of their financial resources. This approach can range from budgeting and personal financial apps to more sophisticated and complex software programs used by financial advisors and robo advisors. More recently SigFig and Jemstep offer QoL technology-enabled investment and financial management tools.

Smartphone personal finance apps have grown significantly with the growth of smartphone users. According to a recent (February 2018) Bankrate.com Report [23] about the use of finance-related apps, including those from traditional banks and fintech players, 63% of US adults who use a smartphone have at least one financial app. The average smartphone user has 2.5 financial apps. Among millennials (ages 18–37), the average is 3.6. It drops to 2.3 for Gen X (ages 38–53) and 1.4 for Baby Boomers (ages 54–72). The most common financial apps are full-service banking apps; 55% of US adults who have a smartphone have at least one full-service banking app; 23% have at least two. Peer-to-peer payment (P2P) apps (such as Venmo, PayPal and Square Cash) are the second most common. Forty-one percent of smartphone users have at least one of these and 15% have more than one. Standalone budgeting and investing apps are less common; they can be found on just 18% and 17% of smartphones users, respectively. Mint, Clarity Money, and Wally are among the best-known budgeting apps. Stash, Acorns, and Betterment are popular for investing. People with banking apps are the most engaged. Seventy percent say they use them at least once a week. Fifty-six percent of budgeting app users, 51% with investing apps, and 38% with P2P payments apps reported likewise.

Consumers who avoid mobile banking tend to fall into two camps: people who like their routine and people who are afraid of fraud. Twenty-nine percent of financial app users believe financial apps are better than non-financial apps versus just 9% who think they are worse.

Millennials are the most divided generation on this issue: 31% of them say financial apps are better, but 15% say they are worse. Regular use of mobile banking apps has potentially significant implications for the financial health of users. Using mobile banking makes the user more aware of his/her current financial status and impacts his/her financial satisfaction.

Types of Personal Financial Management Apps

According to Investopedia [24], some of the best personal finance apps for managing finance resources in 2020 are the following apps. Each of them relies on embedded sets of models and algorithms that maximize some predefined goal like track past spending and saving habits or plan a future budget or invest excess change or reminder of payment due or improve money and investment management or monitor credit score or keep track of income and expenses for multiple individual projects.

*The Best Money Management App is the **Mint.com** from Intuit.* It is a free all-in-one resource for creating a budget and tracking spending which can be linked to bank accounts giving a snapshot of one's financial situation momentarily and over time. It creates budgets based on previous spending habits and alerts users if they are over spending limits. It also offers a financial goals section that assists in identifying and tracking goals for almost any expenditure [25]. Similar websites are ***MoneyStrands.com, PearBudget.com, Wesabe.com, JustThrive.com***

*The Best Budgeting/Debt App is **You Need a Budget (YNAB)**.* YNAB is built around a simple principle that every dollar has a job in a personal budget, be it for investing, for debt repayment, or to cover living expenses and forces an individual to live within his/her actual income. If one gets off track, YNAB helps one to see what can be done differently to balance the budget. There is also a built-in accountability partner. ***EveryDollar Budgeting*** is a similar zero-based budgeting app which tracks spending and plans for purchases. ***PocketGuard***, another alternative, builds a personalized budget based on your income, bills, and goals.

*The Best Tracking Expenses App is **Wally**.* Wally helps the user to organize and track personal and professional expenses. Wally lets the user scan receipts and add notes allowing the user to see where money has been spent [26]. Wally uses machine learning and AI algorithms to adapt to user preferences and behavior.

*Best App for Easy Saving is **Acorns***. Acorns is a useful app for inexperienced investors.

It links to a user's bank account and will make investments automatically or manually [25]. It takes the user's extra change from transactions and automatically directs it to a savings account after evaluating the user's spending and income history [27]. A similar website is **Stockpile.**

*Best App for Freelancers is **Tycoon***. This app is perfect for freelancers and those who are self-employed and get paid at the completion of a contract. It standardizes the details of a contract, puts in a timetable for it, and keeps track of payments that have come in, are scheduled to come in, or that are past due. It makes it easy to see which clients have not paid yet.

App users report that these personal finance tools simplify the formerly confusing and complicated investment and financial management process into something approachable, learnable, and scalable, but the question arises if they are making the users better at managing his/her money or encouraging the user to spend more. Money apps can have a positive and a negative effect. They can make an individual more aware of how much one is spending and help one stick to a budget, but the convenience of contactless payment could also be encouraging even more spending [28].

Unlike the standardized instruments for measuring financial well-being, none of these personal finance apps have been validated, although some research in behavioral sciences provide examples of the importance of financial self-tracking on individual's spending and saving behaviors [29, 30]. A new class of applications and web sites called personal informatics is appearing that collects behavioral information about users and provides access to this information to help users become more aware of their own behaviors. Personal informatics systems support users in understanding various aspects of their life, behaviors, habits, and thoughts. They help users build self-understanding by providing a means to collect personal history, as well as tools for review or analysis. Increased self-understanding has many benefits, such as fostering insight, increasing self-control, and promoting positive behaviors such as energy conservation and financial management [30] (Figs. 17.1, 17.2 and 17.3).

Web-Based Technology Tools for Financial and Investment Analysis

Web-based tools are useful for long-term planning and financial and investment analysis. These tools measure the individual's attained level of financial resources and then estimates the future level of financial resources based on certain assumptions about investment interest rates, inflation rates, tax brackets, the amount and

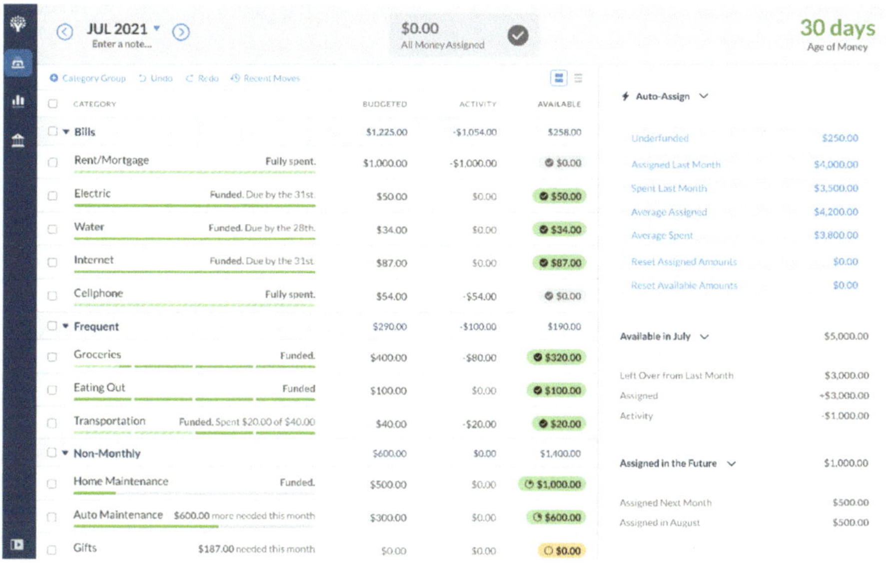

CATEGORY		BUDGETED	ACTIVITY	AVAILABLE
▾ Bills		$1,225.00	-$1,054.00	$258.00
Rent/Mortgage	Fully spent.	$1,000.00	-$1,000.00	$0.00
Electric	Funded. Due by the 31st	$50.00	$0.00	$50.00
Water	Funded. Due by the 28th.	$34.00	$0.00	$34.00
Internet	Funded. Due by the 31st.	$87.00	$0.00	$87.00
Cellphone	Fully spent.	$54.00	-$54.00	$0.00
▾ Frequent		$290.00	-$100.00	$190.00
Groceries	Funded.	$400.00	-$80.00	$320.00
Eating Out	Funded	$100.00	$0.00	$100.00
Transportation	Funded. Spent $20.00 of $40.00	$40.00	-$20.00	$20.00
▾ Non-Monthly		$600.00	$0.00	$1,400.00
Home Maintenance	Funded.	$500.00	$0.00	$1,000.00
Auto Maintenance	$600.00 more needed this month	$300.00	$0.00	$600.00
Gifts	$187.00 needed this month	$0.00	$0.00	$0.00

JUL 2021 ▾ — Enter a note… | $0.00 All Money Assigned | 30 days Age of Money

Auto-Assign ∨

Underfunded	$250.00
Assigned Last Month	$4,000.00
Spent Last Month	$3,500.00
Average Assigned	$4,200.00
Average Spent	$3,800.00
Reset Assigned Amounts	$0.00
Reset Available Amounts	$0.00

Available in July ∨ — $5,000.00

Left Over from Last Month	$3,000.00
Assigned	-$3,000.00
Activity	-$1,000.00

Assigned in the Future ∨ — $1,000.00

| Assigned Next Month | $500.00 |
| Assigned in August | $500.00 |

Fig. 17.1   Screenshot of You Need a Budget (YNAB) app

savings rate through employer-based plans plus individual savings, expected retirement age, and major funding needs such as college funding, home purchase, financial needs for incapacitated relative (such as a child or parent), major medical expenses (such as cancer or a chronic condition), or for a legacy to a charity (such as one's university). This approach attempts to provide an objective, quantitative assessment of financial resources needed to attain these financial goals and needs. Individuals can then assess the gap between their future financial needs and their estimated financial resources to develop strategies to close the gap. Often this is done with the aid of a financial advisor; hence many of the software programs and technology developed to serve this function have been developed by investment management companies.

More recently automated investing services also known as robo advisors have emerged. Robo-advisors are digital platforms that provide automated, algorithm-driven financial planning services generally with little or no human intervention used to automatically allocate, manage, and optimize a client's assets and investment portfolios.

One tool, **Personal Capital,** is an investment management service for tracking wealth and spending that combines the algorithms used by robo advisors with access to human financial advisors. While Personal Capital is primarily an investment tool, its free app includes features

helpful for budgeteers looking to track their spending. It allows one to connect and monitor checking, savings, and credit card accounts, as well as IRAs, 401(k)s, mortgages, and loans. The app provides a spending snapshot by listing recent transactions by category and displays the percentage of total monthly spending by each

Fig. 17.2 Screenshot of
PocketGuard app

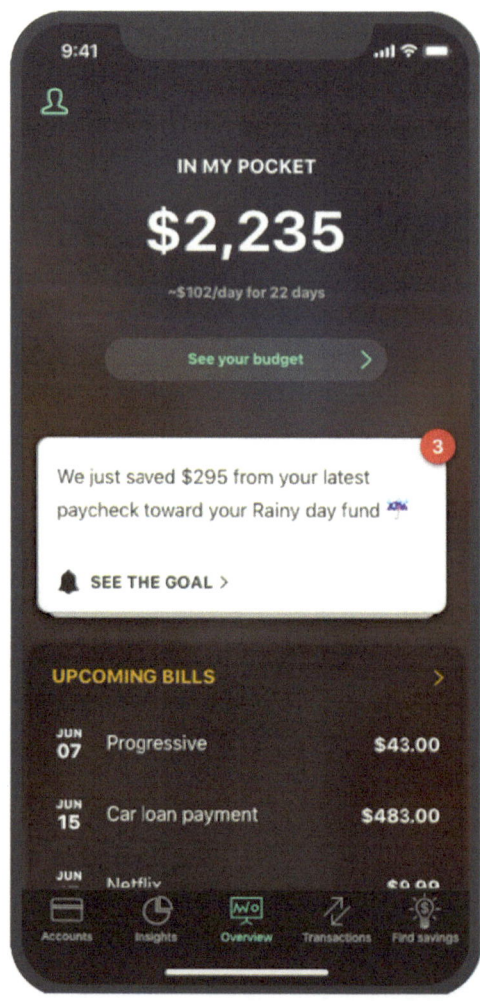

category. Personal Capital also calculates and tracks user's net worth and provides
a breakdown of one's investment portfolio [22].

Perhaps the most evident area where technology has impacted the world of
investing is in the capabilities to analyze investments and develop investment strategies. Ranging from "the basics" at Yahoo Finance and Google Finance to the more
advanced analytics like Morningstar and Bloomberg terminals, to a host of other
specialized sites for investment analysis, the capabilities for investment analysis
have come a long way.

A more recent crop of tools like **SigFig** and **Jemstep** are also making it easier for
consumers to analyze an existing portfolio and immediately get actionable advice
about what can be improved. The latest technology evolution disrupting the world
of investing uses robo advisors—services like **Betterment** and **Wealthfront**—that

Fig. 17.3 Screenshot of
Acorns Easy Saving
Roundup app

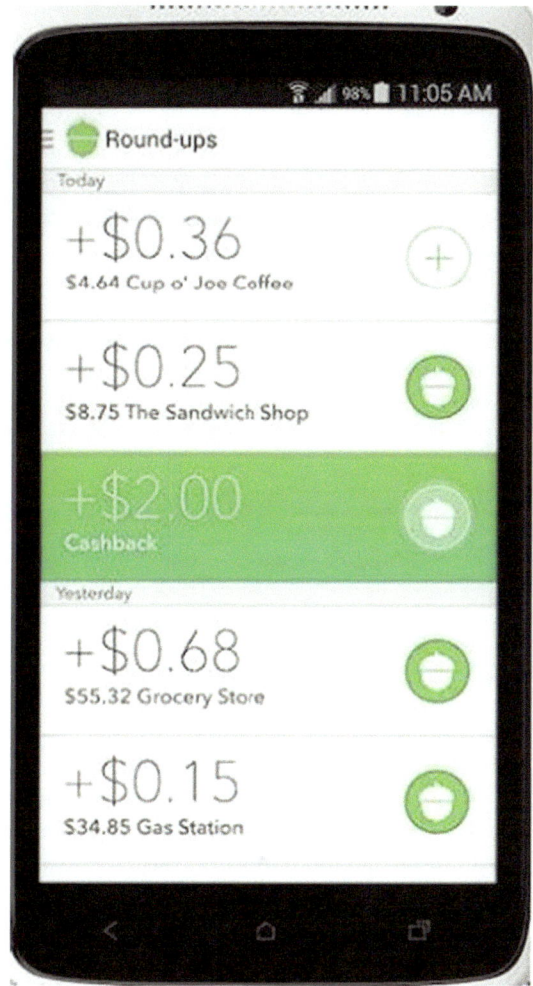

are affordable and will construct the entire asset-allocated passive strategic portfolio for the investor. The depth to the portfolio construction process also makes it feasible to leverage technology tools for additional investment value-adds that were previously done in a more arduous manual processes, from proactive tax loss (or gains) harvesting to supporting good asset location decisions. Such services can now be offered effectively directly to consumers and easily implemented via robo advisors (through advisors and their "rebalancing" and trading software tools) [22].

Robo advisors have a couple of key advantages. Because they do not have to employ many people, they typically have lower fees and lower account minimums than traditional investment advisors. Some robo advisors like Betterment, WiseBanyan, and Bloom even have *no* account minimums. Of course, robo advisors

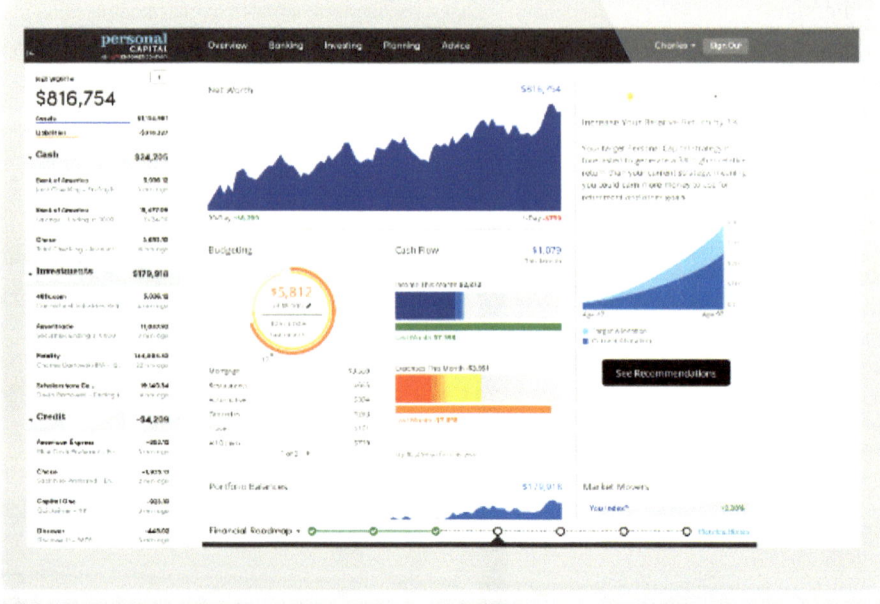

Fig. 17.4 Screenshot of Personal Capital Investment Management desktop app

have drawbacks. For one thing, the user cannot rely on a robo advisor to do complicated financial planning or to give him/her legal advice.

Moreover, a robo advisor will not be able to give the user the kind of nuanced financial advice that a human being might be able to provide.

As more and more tools become accessible directly to consumers, many of the services once provided by financial advisors to analyze, review, and construct client portfolios can now be done at a dramatically lower cost and without the advisor being involved at all. These apps and tools have continued to commoditize many financial resource management functions previously performed by stockbrokers and financial advisors. There has been specifically a rapid growth of the number of personal finance apps and app users and an increase in the amount of monies directed towards these apps, but there has been little or no empirical studies addressing the direct impact of these apps on financial resources and the financial well-being of users (Fig. 17.4).

Other Web-Based Interventions and Strategies

According to Statista, as of February 2020, Yahoo Finance, MSN Money Central, and CNN Money are the three most visited investment and finance websites in the US with 70 million monthly visits, 65 million monthly visits, and 50 million monthly visits respectively) [31].

Personal finance websites now attract one-in-four people who use the Internet, according to comScore.com which tracks Internet traffic, rivaling Facebook in popularity [32]. Currently, the financial capacity in the US is reported by FINRA Investor Education Foundation [33].

There are three categories or models of Web-based personal finance websites which can be of use in enhancing one's financial assets and financial well-being: Financial Data Aggregators, Financial Decision Aids, and Financial Communities [34].

Financial Data Aggregators are highly effective tools which use the computer's power to compile and analyze massive amounts of data harnessing computing power to perform complex analyses that few individuals would have the time, interest, or ability to perform on their own.

Aggregators are effective as personal finance tools because they continually analyze the underlying data as it changes. As a result, aggregators keep individuals coming back for fresh information, which, in turn, keeps users engaged and focused on their financial affairs.

Aggregators also encourage individuals to analyze changes, over time, in their finances. These tools apply the same format to the data each time, making it easy to compare progress over months or years. These tools, therefore, provide a framework for a richer understanding of one's financial profile.

Financial Decision Aids are effective because they calculate costs and benefits that range from simple mortgage calculators to complex online financial planning decisions. By requiring individuals to input their unique data and other personal information, individuals can generate personalized results that illuminate an important element of financial health.

Financial Communities enable individuals to meet and discuss their personal finances and provide support and advice to each other, compete in financial games, swap coupons, or attend virtual parties focused on personal finance. Even personal finance blogs are, in one way, communities and are categorized as such in this report.

Use of Web-Enabled Tools to Increase Financial Literacy and Financial Well-Being

Web-enabled tools can be used to increase financial literacy, financial well-being, and financial satisfaction in later life. The Consumer Financial Protection Bureau (CFPB) publishes financial information to increase financial literacy. They advocate that the ultimate measure of success for financial literacy efforts should be improvement in individual financial well-being [32].

According to the Consumer Financial Protection Bureau (CFPB), financial well-being can be defined as a state of being wherein individuals:

- Have control over day-to-day, month-to-month finances;
- Have the capacity to absorb a financial shock;
- Are on track to meet their financial goals; and

• Have the financial freedom to make the choices that allow them to enjoy life.

Personal finance apps can provide users with personalized financial information for establishing and monitoring budgets, spending and saving habits, progressing towards financial goals, etc.

Because individuals value different things, traditional measures such as income or net worth, while important, do not necessarily or fully capture this last aspect of financial well-being.

Joo and Grable [35] advocate for the development and testing of a framework for understanding the determinants of financial satisfaction. Direct, as well as indirect, effects on financial satisfaction were identified using a path analysis method. It was determined that financial satisfaction is related, both directly and indirectly, with diverse factors including financial behaviors, financial stress levels, income, financial knowledge, financial solvency, risk tolerance, and education. Findings support the continued and increased use of targeted education initiatives directed at improving the financial literacy and behavior of family and consumer economics constituencies. Quality of life technologies can support these initiatives.

However, although there are a tremendous number of personal finance apps and tools, relatively little has been published on a causal relationship between financial knowledge and financial behavior. Empirical studies of the effectiveness of these personalized QoL technology-enabled tools and interventions in increasing financial resources and enhancing financial well-being needs to be done.

Operationalizing Self-Report Measurements of Financial Resources and QoL

In this subsection we reflect how the above-mentioned tools could help to operationalize data collection for self-reports like IFDFW and MSFWB (defined earlier).

For the IFDFW, the data collected via these technology tools could enable the use of an analytics approach to questions ranging from "Living on paycheck-to-paycheck" basis (in case one's account is reaching low at the end of the month) to "Ability to handle $1000 financial emergency" e.g., by analyzing all the available assets and the saved amounts for at least $1000 available to an individual. Questions such as "Worry about being able to meet normal monthly living expenses" can be diagnosed and remediated through the creation and monitoring of budgets and establishment of alerts when approaching overbudget limits utilizing budgetary apps. "Ability to manage money" can be assessed through data aggregation from various banking, credit card, and investment accounts and comparison to targets utilizing money management and investment tools.

For the MSFWB which consists of five different aspects of subjective financial well-being—general subjective financial well-being (GS), money management (MM), peer comparison (PC), having money (HM), and financial future (FF), numerous questions can be operationalized using data collected from technology tools. For questions about the having money aspect of subjective well-being such as "At times I do not have the money to buy what I need," "I cannot do some things with my friends, because I do not have the money to do them," and "Sometimes I miss the cash to buy things I need," personalized tools can be used to monitor spending habits against predetermined budgets and setting alerts if going over pre-set spending limits. For questions about evaluating issues such as "Satisfaction with the way I manage my financial situation," budget and money management tools can provide spending snapshot of recent transactions by category and displays the percentage of total monthly spending by each category so that users can evaluate their money management performance against predetermined targets. Data can be individually analyzed or aggregated across multiple categories—credit card, banking, and investment management. Financial future concerns about "In the near future, having enough money to carry out my plans" can be operationalized by using apps to set up separate budgets for personal expenses from special budgets for future plans, contractual assignments, or multiple projects with different timetables and overlapping income and expenses. These tools are ideal for self-employed, gig workers or freelancers. By using robo advisors or investment management web-based tools and software with some advisory support and comparative investment data resources, a user can answer whether an individual's "Financial situation is better [or worse] than that of one's peers."

The use of quality of life technologies together with standardized self-report-based measurements of financial resources can serve as predictors of level of current and future financial well-being and financial distress and an individual's quality of life.

Success Factors for Technology Adoption

Computer self-efficacy and anxiety, ability to solve novel reasoning problems (fluid intelligence) and remembering and using previously acquired information (crystallized intelligence), attitudes and experience with technology and the World Wide Web are important predictors of the use of QOL technology-enabled tools [36]. Millennials have often led older Americans in their adoption and use of technology, and this largely holds true today. But there has been significant growth in tech adoption since 2012 among older generations—particularly Gen Xers and Baby Boomers [37, 38]. Older users may not adopt a new tool or technology merely because it is available. They must also perceive how the system is personally useful compared to existing methods [39] and that it is easy to use. As prior research has shown [40], older adult's adoption of new technologies may critically depend on whether they understand the costs and benefits of those technologies (perceived utility). Previous Pew Research Center surveys have found that the oldest adults face some unique barriers to adopting new technologies—from lack of confidence in using new technologies to physical challenges manipulating various devices [39].

On the other hand, higher evaluation of life quality was found significantly correlated with the number of new media technologies owned. Owning and using new media technology seems to have become a part of defining one's lifestyle that emphasizes enjoying oneself and enjoying life. Finally, the use of both traditional and new media, except for TV watching, was found complementary in improving people's living quality [41].

The use of tools for financial management definitely falls into the domain of use of new media technologies, where the gains may be substantial, while, at the same time, the cost of the use may be literal. Namely, the user may lose not only time to learn and operate the tool but may also lose his/her own money, e.g., because of the use of inadequately configured tools. The success for technology adoption in the financial domain is therefore multidimensional, and interdependent and shall be researched further.

Discussion and Conclusions

Around the world, the financial landscape is becoming increasingly complex. In response to this, an array of Web-based and non-Web-based technology initiatives seek to address financial participation, education, and inclusion for the broad population, as well as, for vulnerable populations. There is an increased expectation that these tools and programs will improve the financial well-being of individuals and households. But financial well-being is inadequately conceptualized and inconsistently defined, making it difficult to understand and improve financial outcomes and assess financial resources. Existing conceptualizations do not adequately account for the dynamic interplay between a person's environment and their financial well-being as well as how aspects of financial well-being can interact according to age, life stages, and culture.

The recent Quantified Self Movement (QS) and the wealth of digital data originating from wearables, applications, and self-reports is enabling QS practitioners to better address the diverse domains of daily life—physical state, psychological state, social interactions, and environmental context which contribute to an individual's Quality of Life (QoL). This systemic monitoring approach of assessing an individual's state and behavioral patterns through these different QoL domains on a continuous basis can provide real-time performance optimization suggestions [42, 43].

As the Quantified Self Movement (QS) expands and the availability of quality-of-life technologies grow, the incorporation and self-monitoring of financial resources will become an increasingly important component of an individual's overall quality of life [42, 43]. In the future, we may see wearable or digital tracking devices that encourage both fitness goals (steps taken, calories burned, hours slept) and financial goals (interest earned, fees avoided, milestones reached). This is one way that financial resources can play a role in the "quantified self"—by encouraging financial fitness. Second, inventive thinking as the intersection of fintech and healthtech can create incentives and new categories for consumers to blend sound physical and financial habits for overall well-being. The self-tracking of quantifiable financial data and qualitative data, e.g., mood, stress, depression, happiness, productivity can enable individuals to monitor changes in their financial behaviors based on changes in physiological and psychosocial factors such as the correlation between spending behaviors and emotional state. Third, self-tracking and monitoring with QS devices and related apps may lead to more proactive than reactive financial behaviors by providing users with information on their credit card and debt management and the financial performance analysis of investment portfolios. Lastly, in personal finance decisions, future wearable devices may be able to nudge the wearer with a gentle vibration, that they just should not buy the product that they are looking at and instead save the money for a more important goal thus leading to better self-management of financial resources and life satisfaction.

An important aspect of the self-tracking of financial resources is that QS financial activity fundamentally includes both the collection of objective metrics data and the subjective experience of the impact of this data. Self-trackers have an increasingly intimate relationship with QS data as it mediates the experience of reality. In self-tracking of financial resources, the cycle of experimentation, interpretation, and improvement transforms the quantified self into an improved "higher quality" self. The quantified self provides individuals with means for qualifying themselves, through which some higher level of financial performance and financial resources may be attained or exceeded leading to greater financial well-being and life satisfaction.

In the case of health impairment, the individual's financial situation and its relationship with life quality is even more complex. Despite the limited number of qualitative and quantitative studies available, likely owing to the relative infancy of the field, financial resources and the financial burden of treatment seems to play a critical role in quality of life. In particular, the use of the QS tools and methods presented here can better integrate the costs of care and available financial resources when coordinating treatment while attempting to enhance the patient's treatment outcomes and quality of life.

Thus, the use of QoL technology-enabled apps and methods can play an important role in the overall quantification and self-management of financial resources, the reduction of financial burden and distress, increased financial literacy, and the enhancement of financial well-being and financial satisfaction. Ideally, the use of these QoL technologies should be evidence-based and designed and developed to enhance the effectiveness, efficacy, and appropriateness of their use and based on an understanding of factors that influence acceptance by users. Greater development and integration of quality of life technologies can help address the need for better self-management of financial resources for short-term and long-term goal attainment, and the enhancement of the quality of life of an individual and families at large.

References

1. Hansen T, Slagsvold B, Moum T. Financial satisfaction in old age: a satisfaction paradox or a result of accumulated wealth? Soc Indic Res. 2008;89(2):323–47. https://doi.org/10.1007/s11205-007-9234-z.
2. Plagnol A. Financial satisfaction over the life course: the influence of assets and liabilities. J Eco Psy. 2011;32(1):45–64. https://doi.org/10.1016/j.joep.2010.10.006.
3. Seay MD, Preece G, Le VC. Financial literacy and the use of interest-only mortgages. J Fin Counsel Plan. 2015;28:2. https://doi.org/10.1891/1052-3073.28.2.168.
4. WHO Working Group. The World Health Organization quality of life assessment (WHOQOL): development and general psychometric properties. Soc Sci & Med. 1998;46(12):1569–85. https://doi.org/10.1016/s0277-9536(98)00009-4.
5. Easterlin RA. Does economic growth improve the human lot? Some empirical evidence. In: David PA, Reder MW, editors. Nations and households in economic growth: essays in honor of Moses Abramowitz. New York, NY: Academic Press; 1974. p. 89–125.
6. Sorgente A, Lanz M. Emerging adults' financial well-being: a scoping review. Adolesc Res Rev. 2017;2(4):255–92. https://doi.org/10.1007/s40894-016-0052-x.
7. Borg C, Hallberg I, Blomqvist K. Life satisfaction among older people (65þ) with reduced self-care capacity: the relationship to social, health and financial aspects. J Clin Nurs. 2008;15(1):607–18. https://doi.org/10.1111/j.1365-2702.2006.01375.x.
8. Lichter DT, Graefe DR, Brown JB. Is marriage a panacea? Union formation among economically disadvantaged unwed mothers. Soc Prob. 2003;50(1):60–86. https://doi.org/10.1525/sp.2003.50.1.60.
9. Smock PJ, Manning WD, Gupta S. The effect of marriage and divorce on women's economic well-being. Am Soc Rev. 1999;64(6):794–812. https://doi.org/10.2307/2657403.
10. Mills RJ, Grasmick HG, Morgan CS, Wenk D. The effects of gender, family satisfaction, and economic strain on psychological well-being. Interdiscipl J Appl Fam Stud. 1992;41(4):440–5. https://doi.org/10.2307/585588.
11. Mirowsky J, Ross CE. Social institutions and social change. In: Education, social status, and health. Aldine de Gruyter; 2003. p. 242p.
12. Rettig KD, Danes SM, Bauer JW. Family life quality: theory and assessment in economically stressed farm families. Soc Indic Res. 1991;24:269–99. https://doi.org/10.1007/BF00306083.
13. Diener E, Suh E, Lucas R, Smith H. Subjective well-being: three decades of progress. Psy Bull. 1999;1999(125):276–302. https://doi.org/10.1037/0033-2909.125.2.276.
14. Boyce C, Brown G, Moore S. Money and happiness: rank of income, not income, affects life satisfaction. Psy Sci. 2010;21(4):471–5. https://doi.org/10.1177/0956797610362671.
15. Hounkpatin HO, Wood AM, Brown GD, Dunn G. Why does income relate to depressive symptoms? Testing the income rank hypothesis longitudinally. Soc Indic Res. 2015;124(2):637–55. https://doi.org/10.1007/s11205-014-0795-3.
16. Kahneman D, Deaton A. High income improves evaluation of life but not emotional well-being. Proc Natl Acad Sci USA. 2010;107(38):16489–93. https://doi.org/10.1073/pnas.1011492107.
17. Fenn KM, Evans SB, McCorkle R, DiGiovanna MP, Pusztai L, Sanft T, Hofstatter EW, Killelea BK, Knobf MT, Lannin DR, Abu-Khalaf M, Horowitz NR, Chagpar AB. Impact of financial burden of cancer on survivors' quality of life. J Oncol Pract. 2014;10(5):332–8. https://doi.org/10.1200/JOP.2013.001322.
18. Smith DM, Langa KM, Kabeto MU, Ubel PA. Health, wealth, and happiness: financial resources buffer subjective well-being after the onset of a disability. Psy Sci. 2005;9:663–6. https://doi.org/10.1111/j.1467-9280.2005.01592.x.
19. Hubbard R, Goodwin V, Lewellyn D, Warmoth K, Lang I. Frailty, financial resources, and subjective well-being in later life. Arch Geront Geriatr. 2014;2014:364–9. https://doi.org/10.1016/j.archger.2013.12.008.

20. Prawitz AD, Garman ET, Sorhaindo B, O'Neill B, Kim J, Drentea P. Incharge financial distress/financial well-being scale: development, administration, and score interpretation. J Fin Couns Plan. 2016;17(1):34–51. https://doi.org/10.1037/t60365-000.
21. Garman TE, MacDicken B, Hunt H, Shatwell P, Haynes GC, Hanson E, Olson P, Woehler M. Progress in measuring changes in financial distress and financial well-being as a result of financial literacy programs. Consumer Interests Ann. 2007;53. https://www.consumerinterests.org/assets/docs/CIA/CIA2007/garman_progressinmeasuringchangesinfinancialdistressand-finan.pdf
22. Sorgente A, Lanz M. The multidimensional subjective financial well-being scale for emerging adults: development and validation studies. Int J Behav Dev. 2019. https://doi.org/10.1177/0165025419851859
23. Barba R. Bankrate 63% of smartphone users have at least one financial app. 2020 Feb 8. https://www.bankrate.com/personal-finance/smart-money/americans-and-financial-apps-survey-0218/. Accessed 12 Feb 2020.
24. Hong E. The best personal finance apps help you manage your money. Investopedia. 29 Aug 2019. https://www.investopedia.com/personal-finance/personal-finance-apps/. Accessed 2 Feb 2020.
25. Anders SB. Four leading money management apps. CFA J. 2015;85(9):64–5. https://www.nysscpa.org/news/publications/the-cpa-journal/article-detail?ArticleID=11699#sthash.Hr8XeFTr.dpbs. Accessed 12 Feb 2020.
26. Cooney S. 5 apps to help you get your finances in check. Bus Insider. 2016 June 25. https://www.businessinsider.com/personal-finance/5-best-personal-finance-apps-2016-6. Accessed 2 Feb 2020.
27. Mitchell DS. Shortfall savings: the all-important financial buffer against volatility; 2017. Retrieved Feb 1, 2020, from The Aspen Institute epic: http://www.aspenepic.org/wp-content/uploads/2017/06/06-2017_ASPEN_EPIC_SHORTFALL_WEB.pdf. Accessed 2 Feb 2020.
28. Belton P. Do money apps make us better or worse with our finances? BBC News. 2019. https://www.bbc.com/news/business-47075429. Accessed 12 Feb 2020.
29. Becker BW. The quantified self: balancing privacy and personal metrics. Behav Soc Sci Librarian. 2014;33(4):212–5. https://doi.org/10.1080/01639269.2014.964595.
30. Li l. Designing Personal Informatics Applications and Tools that Facilitate Monitoring of Behaviors Ian Li Human Computer Interaction Institute, Carnegie Mellon University 5000 Forbes Avenue, Pittsburgh, PA 15213 ianli@cmu.edu
31. Statista Leading investment and finance websites in the U.S. 2020, by monthly visits Published by M. Szmigiera, Jun 3, 2020. https://www.statista.com/statistics/203953/us-market-shares-of-selected-print-imedia-websites/. Accessed 12 Feb 2020
32. Consumer Financial Protection Bureau Financial. Well-being: the goal of financial education. 2015. Report, Iowa City, IA: Consumer Financial Protection Bureau. Accessed from https://files.consumerfinance.gov/f/201501_cfpb_report_financial-well-being.pdf. Accessed 12 Feb 2020
33. FINRA Investor Education Foundation. Financial capability in the United States: report of findings from the 2012 National Financial Capability Study; 2013. Available at http://usfinancialcapability.org/ and, OECD. PISA 2012 Results: Students and Money: Financial Literacy Skills for the 21st Century (Volume VI). PISA, OECD Publishing. http://www.oecd.org/pisa/keyfindings/PISA-2012-results-volume-vi.pdf (2014) Accessed 12 Feb 2020
34. Blanton K. Financial literacy on the Web. Spring 2011. http://crr.bc.edu/wp-content/uploads/2012/04/FL-on-Web-GUIDE.pdf. Accessed 2 Feb 2020.
35. Joo S, Grable JE. An exploratory framework of the determinants of financial satisfaction. J Family Eco Iss. 2004;25:25–50. https://doi.org/10.1023/B:JEEI.0000016722.37994.9f.
36. Pak R, Stronge A. Health maintenance, older adults, and the Internet. Psy Sci Agenda | November 2008. https://www.apa.org/science/about/psa/2008/11/pak. Accessed 2 Feb 2020.
37. Vogels E. Millennials stand out for their technology use, but older generations also embrace digital life. Fact Tank News in the Numbers September 9, 2019, Pew Research Center. https://

www.pewresearch.org/fact-tank/2019/09/09/us-generations-technology-use/. Accessed 9 June 2020.

38. Bixter M, Blocker K, Rogers W. Enhancing social engagement of older adults through technology. In: Pak R, McLaughlin CA, editors. Aging, technology, and health. New York: Academic; 2018. p. 179–214.

39. Davis FD, Bagozzi RP, Warshaw PR. User acceptance of computer technology: a comparison of two theoretical models. Manag Sci. 1989;35(8):982–1003. https://doi.org/10.1287/mnsc.35.8.982.

40. Melenhorst A-S, Rogers WA, Bouwhuis DG. Older adults' motivated choice for technological innovation: evidence for benefit-driven selectivity. Psy Ag. 2006;21:190–5. https://doi.org/10.1037/0882-7974.21.1.190.

41. Wei R, Leung L. Owning and using new media technology as predictors of quality of life. Telematics Informatics. 1998;15(4):237–51. https://doi.org/10.1016/S0736-5853(98)00008-2.

42. Wac K. From quantified self to quality of life. In: Rivas H, Wac K, editors. Digital health. Health informatics. Springer; 2018. https://doi.org/10.1007/978-3-319-61446-5_7.

43. Wac K. Quality of life technologies. In: Gellman M, editor. Encyclopedia of behavioral medicine. New York: Springer; 2020. https://doi.org/10.1007/978-1-4614-6439-6_102013-1.

Part V
Environment

Chapter 18
Artificial Intelligence and Quality of Life: Four Scenarios for Personal Security and Safety in the Future

Sylvaine Mercuri Chapuis

Introduction

At the beginning of the 2000s, international institutions such as the European Union invested heavily in artificial intelligence (AI; [1]). Academic studies of AI started appearing within the decade. Later on, researchers began studying the implications of AI for quality of life (QoL). They focused on individuals' attitudes and motivations regarding the use of AI, as well as the behaviors, and practices of AI use (or non-use) they engaged in. Studies such as those carried out by Wright [1] and Wright et al. [2] quickly drew attention to the potential benefits and drawbacks of AI use, discussing its potential for fostering economic growth, convenience, security, and individual and social safety, as well as growing concerns over reduced privacy, profiling, surveillance, spamming, and identity theft or fraud, among other issues. In relation to freedom, a key factor in QoL, the emergence of AI has coincided with the rapid development of the Internet and associated digital technologies, effectively erasing the physical borders between individuals and granting them new freedoms. During this time, various questions around the issue of AI have emerged, including ones related to the quantification of "thought" and the protection of the individual online.

Non-academic literature has often discussed AI and its implications for QoL, since technological developments over the past decades have created new needs for the average individual, while researchers have continued to develop new

The original version of this chapter was revised. The correction to this chapter can be found at https://doi.org/10.1007/978-3-030-94212-0_26

S. Mercuri Chapuis (✉)
Institute of Sustainable Business and Organizations, Sciences and Humanities Confluence Research Center - UCLY, ESDES, Lyon, France
e-mail: smercurichapuis@univ-catholyon.fr

applications of AI to fill those needs. For example, Pew Research Center,[1] a non-partisan fact tank, has revealed several benefits and risks of AI use while stimulating open discussions about AI, the Internet, and the future of social relations and humanity, including the visions of the millennial generation, and the disruption of established business models.

At present, there are still gaps in the research concerning the social impact of AI's emergence, particularly regarding individuals' attitudes and beliefs about AI and its actual use. There are vast differences between the way AI use is popularly imagined and how it is really used, and these gaps only widen when one considers future trends. This chapter aims to bridge this gap by presenting the results of a survey of 1000 university students from Western Switzerland regarding how they imagine the future of AI use and its implications for technologies and QoL, especially in relation to personal security and safety.

After a brief outline of the state of the art of AI, we present a tool called "Futurescaper" designed to animate collective reflection and foresight. We then present the methodology that was used to collect data in this study, along with the four scenarios for the future developed using the Futurescaper tool according to students' responses, which are called, respectively, "AI for the best," "AI down," "AI for business," and "AI freeze." These scenarios may serve as starting points to enrich discussions on AI and QoL. It should be noted that in this study, we do not consider questions concerning wealth, employment, the environment, physical and mental health, education, recreation and leisure time, social belonging, or religious beliefs that factor into QoL. Instead, the study focuses on questions on safety and security in relation to QoL.

State of the Art of AI and Questions for the Future

Artificial intelligence (AI) refers to a quality of systems that rely on datasets typically originating from a range of connected objects (e.g., computers, tablets, smart-watches, and connected wristbands) and that are individually and collectively capable of quickly analyzing data implementing search, pattern-recognition, learning, planning, and induction processes. Artificial intelligence has considerable power [3] and enables individuals to access information more simply and efficiently, as "the proactivity of the environment lightens the cognitive load that the user currently has to deal with to access information via computers" [4, p. 482].

The development of AI is paving the way for a large number of innovative, highly responsive, personalized applications and services. However, it is also a significant risk driver, particular in the areas of personal security and safety. Artificial intelligence is a factor in QoL, due in large part to the emotional responses it provokes

---

[1] https://www.pewresearch.org/?s=ambient%20intelligence, visited 1 December 2020, keywords for the research engine: AI.

(e.g., safety/lack of safety, security/insecurity). Given the threats presented by hackers and other malicious actors, the protection of sensitive personal data in an environment of widespread AI use is essential and must be an object of constant attention for both service providers and the users themselves. Recent scandals indicate the kinds of AI security and safety threats that may become increasingly prevalent in the future. For example, the cyber-attack launched in May 2017 via the "Wannacry" virus infected more than 200,000 computers belonging to individuals in over 150 countries [5]. Given these risks, the consequences of AI are a major concern for individuals that is likely to affect QoL.

In protecting against threats to private life and overall safety in the context of cybertechnologies, individual users must assume primary responsibility. As the domain of personal AI technologies continues to grow, individuals must continually increase their digital literacy and vigilance against risk. In 2017, in its biannual report on information security, the Swiss Confederation stated in an inventory carried out at the international level that it expected the number of connected objects to grow to 20 billion globally in 2030 (compared the 6 billion connected objects that existed when the report was issued; [6]). The report also indicated that individuals' vulnerability was "due to the inappropriate safety culture" that predominated in information security practices [6, p. 8], including the behaviors of service providers. The report demonstrates that cybersecurity ultimately depends on the actions of and interactions between several cultures of data protection. The aim should be to define appropriate individual and collective practices that would allow the development of a "good" safety culture [7]. Developing shared values, norms, and symbols with respect to security requires mobilizing not only experts but citizens at large as well, the latter of which are the first vectors of risk and have a genuine role to play in developing cultures of safety.

To properly mobilize individual citizens, an analysis of their actual behaviors and practices with regards to AI is highly needed as an initial step. Once these current behaviors and practices are understood, developing awareness and vigilance should be key areas of focus to allow individuals to ensure their safety and security and achieve better QoL outcomes as a result. An individual's perception of security and safety is a central point that must be understood in order to transform behaviors and develop prevention practices. The use of *foresight processes* is a promising technique for facilitating such transformation, because it helps individuals and collectives consider systems of action [8] as a whole (i.e., political, economic, sociocultural, technological, ecological, and legal [PESTEL] megatrends). Additionally, foresight processes consider weak signals (i.e., areas where information is incomplete or lacking) and, while encouraging exploration of alternative future scenarios, potentially improve participants' decision-making processes and actions. Rather than anticipating foreseeable and unalterable changes and better prepare for them (by being *pre-active*), foresight processes encourage individuals to act in order to bring on a desirable future (by being *pro-active*; [9]). This distinction between the (pre-)acceptance of changes (after the fact) and proactive contribution to change is important. When proactive actions are identified and perceived by a collective, individuals feel more motivated and responsible because they take the future upon themselves instead of simply dealing with the changes. In addition, in the case of AI, individuals are likely to be more involved in developing preventive actions to the extent that they feel safe and secure in their own use of AI.

Considering the need for research regarding AI use and attitudes towards AI among individuals, as well as the potential for collective foresight processes as a tool to drive individuals' future actions on AI, this chapter seeks to answer the following questions: What attitudes and motivations characterize individuals' current AI use, and what behavior and practices of AI use (or non-use) do they engage in? What future scenarios for AI are possible in Switzerland? How will companies, administrations, and the public sector in Switzerland be affected by developments in AI? How can individuals in Switzerland seize opportunities to improve security and safety and respond to AI-related threats? To answer these questions, we carried out a study between September 2016 and January 2017 involving 1000 university students in Western Switzerland. This chapter reports the results of the study and discusses the implications of these results for efforts to ensure individual users' cyber-security and safety in the midst of technological developments and improve QoL in the long term.

Methods: Futurescaper, Animating Collective Reflection and Foresight

According to some researchers, foresight consists of developing plausible scenarios for the future that shed light on imminent decisions and actions [10]. The development of such scenarios facilitates individuals' strategic actions in an uncertain environment [11]. Change is a central concern of foresight, and the scenarios developed through foresight explore new opportunities resulting from changes in a systemic environment [12].

In this study, we used an online platform called "Futurescaper" (Futurescaper. com) to encourage forward-looking reflections and operationalize foresight processes. Futurescaper is a collaborative tool that supports the imagining of future solutions by a group of participants, each of whom can provide significant contributions and insights into specific situations. Tools for foresight processes like Futurescaper helps individual users describe and analyze their prospective actions in the context of a collectively generated future. Several similar tools also exist, such as Scenario Management International (ScMI.de), 4strat (4strat.com), FIBRES (fibresonline.com), and the Futures Platform (futuresplatform.com).[2] Since we had the opportunity to use Futurescaper for in-class learning in a course on foresight with bachelor's- and master's-level students in Western Switzerland, this chapter focuses on this tool in particular.

---

[2] https://rossdawson.com/futurist/futurist-software-and-tools/, visited 1 December 2020.

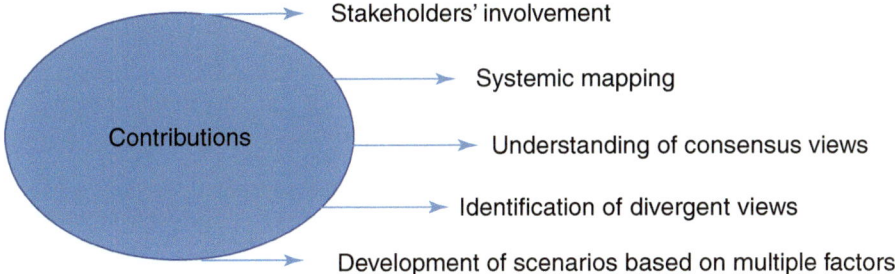

Fig. 18.1 Contributions of Futurescaper and similar future scenario platforms

Futurescaper's contributions are qualitative (Fig. 18.1), as it helps groups of individuals collectively imagine the change factors influencing potential futures, the consequences of these change factors, and the consequences of these consequences. Futurescaper facilitates collective reflection by all the individuals within a group, allowing participants to propose new combinations of change factors and their corresponding future scenarios. In this study, the new scenarios included opportunities to be seized relating to the development of AI and security and safety threats to be avoided. Given these opportunities and threats, participants can articulate the attitudes they should adopt in order to change or pursue the imagined future scenario. Participants verbalize courses of action, taking into account change factors that other individuals in the group might not have identified. In this way, Futurescaper is a future scenario-building tool that encourages proactive action.

The course on foresight in which the present study was conducted took place between September 2016 and January 2017 in Western Switzerland. During this course, 1000 bachelor's and master's students participated in a crowdsourced foresight exercise. The students came from the fields of economics and management, engineering, art, and design. A facilitator oversaw the overall process of the exercise, including configuring the platform and encouraging the participants to brainstorm the scenarios with multiple change factors.

Implementing Futurescaper involves the following four-step process (as shown in Table 18.1): project preparation (step 1), stakeholder engagement (step 2), interpretation and analysis (step 3), and the restitution and presentation of the results (step 4). As considerations concerning the general environment (e.g., PESTEL factors) are the starting point of any definition of plausible future scenarios, such considerations are a key factor in the project preparation step [10]. Table 18.1 presents a brief description of the actions carried out for each step of the process in the present study, as well as the results of each step.

To prepare the project (step 1), the facilitator predefined 52 change factors that were derived through exploratory research and collection of data concerning AI. The facilitator used newspaper articles, scientific and professional studies, blog posts, and other documents to provide quality starting content. The facilitator then held workshops with students (step 2), each of which began with students viewing an introductory video on the current state of AI in Switzerland. In total, 25 workshops

Table 18.1 Steps of Futurescaper approach to future scenario building

Step 1: Project preparation	Step 2: Stakeholder engagement	Step 3: Interpretation and analysis (from change factors to future scenarios)	Step 4: Presentation of the results
Desk research conducted by researchers (authors, domain experts)	25 workshops (average 40 students per workshop)	Interpretation and analysis of crowdsourced change factors and future scenario development carried out by researchers (authors)	An academic paper in a book Academic article in a peer-reviewed journal (to be submitted)
52 initial change factors (see Annex A)	2193 crowdsourced change factors	337 merged crowdsourced change factors 4 contrasted, plausible scenarios along 2 axes	

Table 18.2 Survey questions

Question 1: What are the change factors that you would intuitively associate with AI? Change factors can be political, economic, social, technological, environmental, or legal (PESTEL). Answer type: you may select up to 3 pre-registered items or create new ones.
Question 2: What could the impact of those change factors (answers to question 1) be? Answer type: you may select up to 3 pre-registered items or create new ones.
Question 3: What could the impact of those change factors (answers to question 2) be? Answer type: you may select up to 3 pre-registered items or create new ones.
Question 4: What is the most significant opportunity or threat that you see emerging from those change factors (any change factor)? Answer type: you may describe that opportunity/threat.
Question 5: From now on, which actions might you recommend to companies, public administrations, and individuals?

(with an average of 40 students per workshop) were conducted, all of them structured the same way, starting with an introduction to AI, then a presentation of the survey (see Table 18.2 for survey questions), followed by the students taking the survey and a wrap-up. To stimulate students' initial thoughts and provoke discussion, the facilitator presented three recent examples of AI in practice (autonomous vehicles, smartwatches, and drones) at each workshop and compiled students' reactions. At each workshop, the facilitator also led a discussion among the students to elicit insights regarding their knowledge of AI, their attitudes and motivations regarding it, and the behaviors and practices of AI use (or non-use) they engage in. The process of engaging the students in step 2 was highly beneficial, leading to the identification of 2193 crowdsourced change factors.

Step 3 of implementing Futurescaper involves interpreting the data collected. In this case, the 2193 crowdsourced change factors were reduced to 337 merged change factors, which we then analyzed in order to identify interrelationships and possible axes along which to consider distinct scenarios. The identification and analysis of the merged crowdsourced change factors during step 3 supported the generation of four contrasted and plausible future scenarios for AI with respect to personal security and safety. We defined two prime axes (see Fig. 18.2) to represent

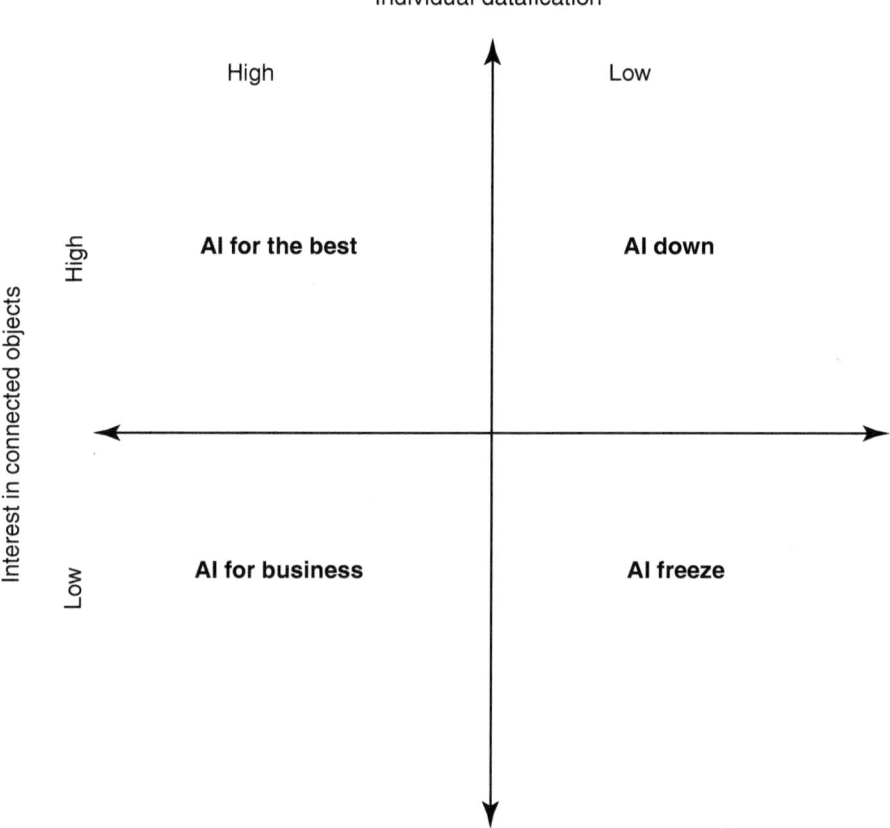

Fig. 18.2 Possible scenarios for the futures of AI in Switzerland

these four future scenarios. We then developed 564 proposals directed towards businesses, governments, and individuals regarding actions that could be taken to seize opportunities and protect against threats related to the use of AI. By outlining this study in the present chapter, we carry out step 4, the final step of the process, which is to present the results to the international academic community. We hope to also present these results in a peer-reviewed article in the near future.

Results

In this section, we present the results of the study described above regarding possible futures for AI in Switzerland. In the presentation of these future scenarios, we consider the implications of future technologies for the security and safety of users and the possibility of such technologies leading to improvements to QoL in the long term. As mentioned in the previous section, our crowdsourced foresight approach led to the identification of 337 change factors. These allowed for the elaboration of the four scenarios discussed below (see Fig. 18.2): "AI for the best," "AI down," "AI for business," and "AI freeze." We named each scenario according to the descriptions provided by the participants of the study. Each of these four scenarios presents opportunities related to AI that businesses and governments might seize in the present to improve long-term QoL, as well as security and safety threats that individuals can proactively address. As mentioned above, the scenarios were defined along two main axes, which were based on two composite change factors whose future outcomes were deemed the least certain due to their high degree of dependence on social and technological developments. These two change factors are *individual datafication* and *interest in connected objects*. Datafication consists in moving from data to useful information: "insurance companies can, for example, use data relating to the movements of the vehicles of their policyholders in order to establish contracts as close as possible to the risks actually presented by their customers (and no longer contracts taking into account their age, their sex, and their driving history)" [13, p. 95]. Datafication refers to the quantification of various aspects of our daily lives in the form of data and information that can be used for diverse purposes [14]. High levels of individual datafication means that large amounts of information are extracted from individuals' data via everyday devices for potential use by companies, governments, or the individual themselves. In a society with low levels of individual datafication, the amount of data extracted from individuals, or the amount of useful information derived regarding individuals' behaviors, will be lower than it is today, either as a result of individuals' rejection of data-collecting devices or because of an improved culture of security around digital devices. High and low interest in connected objects/AI are expressed through what we analyzed from individuals' attitudes and motivations regarding AI and the behaviors and practices of use they engage in. High and low interest in connected objects/AI refer to the degrees to which individuals, businesses, and governments take interest in AI technologies and assess the benefits and drawbacks of its various applications. In a society with high AI interest, individuals are knowledgeable about developments in AI and their implications; in a society with low AI interest, developments in AI may be widely used, but there will be little awareness of or interest in AI from the culture at large. In what follows, we offer a brief description of the scenarios developed based on these and associated change factors.

"AI for the Best" (A Utopian Scenario)

This scenario is characterized by strong interest in AI combined with a high level of datafication of individuals. In this scenario, individuals view new technologies with enthusiasm. They are expert users of connected objects and are eager to exploit their full potential, treating them as instruments to facilitate community-building and individual growth. Individuals in this scenario are proactive and willing to share—in real time, if possible—the data they generate with the manufacturers of connected objects and service providers. They feel safe and secure and are convinced that this is the best way to contribute to the continuous improvement of user experiences. They are confident and feel free, and they view the impact of AI on their QoL as positive.

In the "Al for the best" scenario, AI becomes a natural extension of the human body. At the same time, individuals develop genuine connection with AI (made possible by the widespread deployment and generalization of connected objects). Artificial intelligence reassures users; it follows and supports individuals' needs in real time. In this scenario, one assumes that there is no possibility of AI harming the individual. Secure in their use of AI, individuals are able to use it to maximize their QoL in the long term.

In this scenario, the design of AI technologies is informed by the high level of overall datafication of individuals, allowing various highly personalized services to be offered to the users. In the "Al for the best" scenario, safety and security experts face a surprising question: What is left for them to do? What does society expect from them?

"AI Down"

The "AI down" scenario is characterized by strong interest in AI combined with a low level of individual datafication, where vast data processing capabilities are nevertheless available to individuals. In this scenario, while connected objects continue to accumulate, they do not offer many new opportunities for users nor new security and safety risks. The reason for this is that entities and individuals are unable to transform the data collected into actionable information or relevant knowledge in their particular context. Individuals' QoL is partially impacted, since interest in AI does not result in significant improvements in digital technology or widespread cultural developments, leading to frustration and a sense of stagnation.

In the "AI down" scenario, the design of AI technologies is affected by weak datafication, and there is a lack of highly personalized services available to users. In this scenario, safety and security experts are mainly concerned with the future of AI and its potential uses, asking what the potential of AI is and when it will be fully realized. It is unlikely that AI affects QoL, since it seems impossible for technological developments to produce changes in culture or daily life.

Consequently, said security experts must react to new developments in the same manner as users to predict as early as possible potential security and safety threats and opportunities produced by new services, objects, and classes of connected objects.

"AI for Business"

The "AI for business" scenario is characterized by low interest in AI combined with high levels of datafication of individuals. In this scenario, individuals have only slight interest in AI, although they individually and collectively generate ever-increasing amounts of data through their numerous daily transactions. While the datasets they produce provide information and high value-added knowledge to businesses and government entities, the safety and security of users are also well ensured.

In the "AI for business" scenario, the data that are produced form a "knowledge pool" (i.e., a commons) that is made available to innovators to help them accelerate the development of prototypes and shorten the time required for them to reach the market. Individuals contribute to open innovation dynamics [15], which most actors in the AI field, including computer security experts, have widely embraced. Individuals' QoL is significantly impacted because AI users, whether they approve or not, contribute knowledge that leads to innovations in everyday life technologies.

In this scenario, the design of AI technologies benefits from high levels of datafication, but there is a lack of highly personalized services offered to users, as the users do not express a need or desire for these services.

In this version of the future, the large amount of user data available to safety and IT security experts allows continuous improvements to data and infrastructure security and greater awareness among experts of the challenges associated with cybertechnology. This ensures that businesses do not simply take advantage of individuals' data for their own benefit. Experts are concerned about establishing and maintaining high standards of cybersecurity to ensure the safety and security of individuals and maximize their QoL in the long term.

"AI Freeze"

The fourth and final scenario, "AI freeze," is characterized by low interest in AI combined with low levels of datafication of individuals. In this scenario, AI consists of a variety of gadgets. However, most people are afraid that, sooner or later, a disaster will occur that compromises their safety and security. There is a lack of confidence in the manufacturers of connected objects and service providers, which ultimately impacts users' QoL. The latter ask themselves questions such as "When is it going to happen?" "Will that connected, autonomous vehicle be hacked and cause the death of one or more bystanders?"

At the same time, people distrust the large companies that collect, analyze, and perhaps even resell their data. From the perspective of safety and security experts, the "AI freeze" scenario is not a "surveillance" scenario. However, no one questions the importance and necessity of the continuous monitoring of data traffic. Meanwhile, no one assumes responsibility for detecting attacks and other fraudulent behavior, which can be potentially catastrophic, putting individuals' safety, security, and long-term QoL at risk.

In this scenario, the design of AI technologies is limited by low levels of datafication, and highly personalized services for users are lacking.

Discussion

There are numerous consequences of the development of AI for individuals, public agencies, companies, and communities. Studies that analyze the implications of AI must therefore consider the issue at both the individual and collective levels, including those focusing on AI's impact on QoL. Trends at one level of analysis can have chain effects with consequences for others, such as changes in collective employment that impact individuals' health. However, in considering these facets, research on AI's social impact should ultimately assess whether AI helps to meet humanity's needs or, on the contrary, creates additional problems. To predict AI's future influence on a society, researchers must specifically consider the perceptions of the individuals in that society regarding AI, since cultures and value systems around the world vary widely and are subject to continuous evolution and change. Expectations regarding standards of security and safety, for instance, may differ from one population to another. Furthermore, it is important to consider the perceptions of individuals in their particular context, because this ultimately determines individual attitudes and drives behaviors. These perceptions, which influence social action, can also vary from one individual to another. To predict the potential of AI on QoL in the future, individuals' perceptions of AI must therefore be a key consideration.

Our study, which surveyed 1000 university students in Western Switzerland regarding future AI scenarios in the context of workshops on this topic, increased these students' level of familiarity with AI, allowing them to gain experience thinking about this topic while addressing additional issues, such as personal security and safety, that they may not have considered elsewhere. The study also contributed to individual datafication, since Futurescaper stores the responses used to develop its plausible future scenarios in a large server. Our study corroborates the idea associated with the quantified self-movement that practices of 'measuring yourself' using technological innovations can be highly beneficial to the individual. The quantified self-movement, first described in 2007, encourages self-knowledge through the collection and analysis of data relating to the body and its activities. Recent studies [16] have argued that "quantified selfers" are the individuals with the highest levels of datafication and interest in connected objects. Unfortunately, it was beyond the scope of this study to consider the degree to which participants identify with this movement.

In different research settings, researchers have also defined and evaluated the technology of wearables [17]. It would be worthwhile, in future research, to imagine or ask participants to imagine future individuals' attitudes and motivations regarding wearables that incorporate AI and their behaviors and practices of use (or non-use). Based on the findings of studies like that of Estrada-Galiñanes and Wac [17], which analyzed 438 off-the-shelf wearables, new questions along the following lines could be introduced to the survey given to students (see Table 18.1): What impact could the use of small personal devices have on individuals? Could these devices be considered a new organ of the body? Since these devices can be worn on one's head, wrist, legs, torso, arms, and ears, do the quantified selfers who wear them perceive themselves to be advanced people, cyborgs, or humanoids?

Our study has certain methodological limitations. One of these is the fact that the merged crowdsourced change factors used to develop the future scenarios described above were derived from only one particular population or category of stakeholders [18], namely students from Western Switzerland. Although these students may correctly identify future trends related to personal security and safety in connection with AI, they are not representative of the entire population of Switzerland. Furthermore, the facilitator in this study played a determining role in how students engaged in the foresight process. The participants also may have been more motivated to participate because they knew each other and belonged to the same school. One can imagine participants having very different attitudes regarding the future of AI in other study contexts (e.g., alone at home or in a country at war such as Syria). The results could also be affected by participants' level of technical mastery or their motivation to use an electronic tool like Futurescaper, or scenarios in which participants are not allowed or able to use such tools.

Despite these limitations, our study contributes to research on QoL technologies [19] and their future implications. As greater social interaction in environments such as the university classroom improves students' learning and affects future behavior, our study demonstrates how QoL technologies like Futurescaper can enhance learning by allowing experimentation with new forms of social interactions and intelligence that individuals ordinarily would not directly or spontaneously engage in, while promoting collective learning rather than solely individual learning. By inspiring proactive attitudes towards the future, Futurescaper and similar QoL technologies encourage participants to renew their commitments to particular courses of actions or, on the contrary, revise them. In the future, QoL technologies may be a common good that any individual (or other form of intelligence) in the world has the right to freely access and use.

Concluding Remarks

In this chapter, we have examined perceptions of personal safety and security in the context of today's cybertechnology—specifically, the development of AI solutions and services. We have also discussed the implications of AI in daily life for future QoL. In particular, based on the survey responses of 1000 bachelor's- and master's-level students in Western Switzerland, we have generated four scenarios for the future of AI using Futurescaper, a platform that facilitates foresight processes.

Ultimately, these scenarios may serve as a test bed for experts' future research. We have also indicated several strategic options and policies that can improve safety and security in the future. Based on the lessons these scenarios present, readers of this chapter may adjust their decisions regarding the use of AI or even consider alternatives that they had not thought of previously. By representing the potential implications of AI in different plausible scenarios, this study also demonstrates the need for future research focusing on the ethics of AI use, citizen's vulnerability to AI-related threats, and the relationship between AI and QoL.

Building plausible forward-looking scenarios and assessing them in relation to various strategies or policy options gives governments, businesses, and individuals the opportunity, at a relatively low cost, to anticipate the direct and indirect long-term impacts of particular decisions on personal security, safety, and QoL. Foresight processes like those promoted by Futurescaper can benefit organizations, communities, and societies around the world. In the domain of public governance from the local to national levels—where decision-making is often plagued by instability, uncertainty, and contradiction—the possibility arises of institutionalizing foresight practices, which are remarkably inexpensive and accessible in comparison to the strategic gains they provide.

Acknowledgments We want to thank Thomas Gauthier, Professor at Emlyon Business School, and Professor Katarzyna Wac (UNIGE) for their support in this research (especially TG) and for co-writing the initial draft of this chapter (especially KW).

Appendix

The list of 52 initial change factors

Political	1. Unemployment up
	2. Unemployment down
Economical	3. Information flow up
	4. Telecommuting up
	5. Jobs up
	6. Jobs down
	7. Machine-to-machine interactions up
	8. Logistic performance up
	9. Automation up
	10. Decisions made directly by machines and objects up
	11. Quality of work up
	12. Quality of work down
Social	13. Human interactions up
	14. Free time available up
	15. Home support and follow-up for older people
	16. Free time available down
	17. Decreased human interactions

Technological	18. Hyperconnectivity up
	19. Hyperconnectivity down
	20. Remote monitoring on the rise
	21. Geolocation of objects and people up
	22. Comfort up
	23. Rising security
	24. Traceability up
	25. Identification of objects rising
	26. Rising environmental observation
	27. Remote action capability up
	28. Assistance to the human rising
	29. Patents up
	30. Autonomy of machines and objects up
	31. Safely down
	32. Decreased human assistance
	33. Comfort down
Ecological	34. Pollution down
	35. Increased traffic flow
	36. Lower energy consumption
	37. Electromagnetic waves up
	38. Pollution rising
	39. Rising energy consumption
	40. Traffic flow down
Legal	41. Cyberattacks on the rise
	42. Data protection up
	43. Privacy protection on the rise
	44. Information exchange up
	45. Rising safety standards
	46. Data collection up
	47. Falling safety standards
	48. Privacy protection down
	49. Data protection down
	50. Cyberattacks down
	51. Transparency on the rise
	52. Transparency down

References

1. Wright D. The dark side of ambient intelligence. J Policy Regulation Strat Telecommun Inf Media. 2005;7(6):33–51.
2. Wright D, Gutwirth S, Friedewald M. Shining light on the dark side of ambient intelligence. Foresight. 2007;9(2):46–59.
3. Russell S, Norvig P. Artificial intelligence: a modern approach. 2002.
4. Entretien (2009), Distances et savoirs, vol. 7, 3, p. 479–500.
5. Le Temps. Le rançongiciel Wannacry a peu touché la Suisse, 15 mai; 2017.
6. Melani. Sureté de l'information, situation en suisse et sur le plan international, Centrale d'enregistrement et d'analyse pour la sûreté de l'information, Confédération Suisse; 2017. https://www.newsd.admin.ch/newsd/message/attachments/47967.pdf

7. Chevreau F-R, Wybo J-L. Approche pratique de la culture de sécurité. Pour une maîtrise des risques industriels plus efficace. Rev Fr Gest. 2007;5(174):171–89.
8. Von Bertalanffy L. (2010). General Systems Theory. The Science of Synthesis: Exploring the Social Implications of General Systems Theory, 103.
9. Godet M, Durance P. La prospective stratégique pour les entreprises et les territoires. Paris: Dunod; 2011.
10. Godet M. Manuel de prospective stratégique, tome 2: L'art et la méthode. Paris: Dunod; 2004.
11. Johnson G, Whittington R, Scholes K, Angwin D, Regnér P. (2011). Exploring strategy. Financial Times Prentice Hall.
12. Mercuri Chapuis, S. & Gauthier, T. (2016). L'économie collaborative en Suisse romande à l'horizon 2030 : les MOOCs à l'heure de Coursera. Gestion 2000, 33, 55-73. https://doi.org/10.3917/g2000.334.0055.
13. Chamaret C. La révolution big data. Annales des Mines – Gérer et comprendre. 2014;116(2):94–6.
14. Cukier, K., Mayer-Schoenberger, V. (2013), "The rise of big data: how it's changing the way we think about the world", Foreign Aff, vol. 92, n°3, p. 28–40.
15. Rogers E. Diffusion of innovations. 5th ed. Simon and Schuster; 2003.
16. Wac K. From quantified self to quality of life. In: Digital health. Cham: Springer; 2018. p. 83–108. https://doi.org/10.1007/978-3-319-61446-5_7.
17. Estrada-Galiñanes V, Wac K. Collecting, exploring and sharing personal data: why, how and where. Data Science, IOS Press; 2019. https://content.iospress.com/articles/data-science/ds190025
18. Freeman RE. Strategic management: a stakeholder approach. Boston: Pitman series in business and public policy; 1984.
19. Wac K. Quality of life technologies. In: Gellman M, editor. Encyclopedia of behavioral medicine. New York: Springer; 2020. https://doi.org/10.1007/978-1-4614-6439-6_102013-1.

Chapter 19
A Matter of Distance? A Qualitative Study of Data-Driven Early Lifestyle Assessment in Preventive Healthcare

Troels Mønsted

Introduction

The importance of lifestyle-related diseases (LRD) such as type 2 diabetes (T2DM), COPD, cardiovascular disease (CVD) [1, 2] and the consequent importance of assessing citizens' risk for developing LRDs at an early stage is widely acknowledged, but has proven to be notoriously difficult to achieve. The reasons for this are many: Most importantly, health risk behaviors are deeply ingrained in the daily life of people or part of valued habits and social interactions, and hence difficult to change, as well as many individuals are either unable or unwilling to articulate these. In effect, it requires a massive effort to effectively screen a population through conventional means such as health checks and interviews performed at regular intervals by e.g. general practitioners (GPs). Hence, too many citizens remain effectively at a distance from preventive healthcare services such as smoke cessations courses, dietary consultancy, physical exercise or even sleep.

From a public health perspective, decreasing the distance between citizens and preventive care services is essential in the sense that such health services must be available, accessible, affordable, and acceptable for the citizens. Currently, LRDs such as T2DM, asthma, COPD, CVD, and certain types of cancers account for 50–60% of all hospital admissions [3], and have a significant impact on the life expectancy and quality of life of the individual by affecting both the physical well-being and mental health [4–7]. LRDs are often, although not exclusively, incurred by health-risk behaviors such as smoking, harmful use of alcohol, poor diet, and a sedentary lifestyle [1, 2], and it is predicted that it will lead to a surge in LRDs if rates of obesity and physical inactivity continue to increase [8, 9].

Health-risk behaviors, early symptoms, and socio-demographic conditions can in principle be used for identification of individuals at risk [10–12]. This requires a

T. Mønsted (✉)
Roskilde University, Roskilde, Denmark
e-mail: monsted@ruc.dk

proactive approach in which providers reach out to people at risk while their condition is still revocable, rather than merely react and provide treatment when people eventually reach providers and get diagnosed because the damage is beyond repair [13].

Current initiatives in preventive healthcare in developed countries have primarily focused on improving the ability of healthcare services to take responsibility by reaching out to citizens, for instance through information campaigns and systematic health checks performed by general practitioners (GPs) [14, 15]. Experience, however, shows that these approaches are either ineffective or very resource intensive for the healthcare systems.

While the preventive healthcare system struggles to reach out to citizens, recent trends suggest that digital technologies hold great potential for enabling people to assess and monitor their own health condition autonomously or in collaboration with health professionals. Under the umbrella term *quantified-self*, this entails various types of devices enabling production of data representations of the behavioral patterns and health condition of an individual, which in turn can support health management and lifestyle change, leading to better life quality in the long term. The purpose of this chapter is to investigate how quantified-self technologies can contribute to decreasing the *distance* between preventive healthcare services and citizens by addressing the following two research questions: **RQ1:** *In what ways are citizens at a distance from the preventive healthcare system?* **RQ2:** *And how can quantified-self technologies enable citizens to decrease this distance?*

Based on a qualitative case study of the pilot implementation of a preventive healthcare intervention in Denmark, this chapter first analyses the challenges faced by citizens for engaging in early assessment of risk for LRD and then proceeds to discuss the potential of quantified-self technologies to increase the accessibility to preventive healthcare.

Background

Towards Data-Driven Approaches to Early Detection

In Denmark, the responsibility for supporting citizens (5.8 million) through lifestyle change is shared among municipal health centers and GPs. These health services are tax funded, free to use for all citizens, and developed to support citizens through all stages of lifestyle change. However, while citizens can get referred to lifestyle offers, for instance by their GP, it is their own responsibility to reach out in case they want support. The municipalities (98) are responsible for offering consultancy and support for lifestyle change through services such as smoke cessation courses and dietary consulting. The GPs (app. 3500) are responsible for identifying citizens with risk behaviors and for monitoring their health status over time. Preventive healthcare first and foremost involves assisting citizens in converting to a healthier lifestyle, and in retaining this for a lasting period of time [16–20]. Conversion not only involves supporting behavior change through, for instance, improved diet, physical exercise, or smoking cessation [11], but also generates motivation in the citizens for engaging in behavior change, and ensure that the preventive care offers match the person's stage of readiness. Likewise, retention is challenging as health-risk behaviors are often deeply ingrained habits and preventive care must therefore extend over sufficient time to allow new habits to form [21–23]. Without effective assessment of the lifestyle of citizens at risk of developing LRD, and without the means to attract these to the right lifestyle offers, preventive healthcare will, however, remain at a distance to the citizens.

Recently, the manner of which analysis of big data in healthcare can assist early detection by enabling analysis of more comprehensive, diverse, and timely health data has attracted significant interest. This trend stems from the fact that a substantial amount of relevant data is already accumulated in, for instance, electronic medical records, as well as quantified-self technologies which facilitate the production of new data representations of the citizens. Hence, relevant data can originate from a combination of sources, including clinically-based electronic medical records (EMRs) and other health information systems used to routinely collect patients' data, as well as from sensors and mobile phone apps, and crowdsourced information [24]. In addition, advancements within artificial intelligence and machine learning provide new opportunities for more efficient analysis of big data and for seeing new patterns in these data [25]. This approach, however, comes with its own set of challenges. First of all, it requires a digital infrastructure that is capable of handling the high volume, velocity, and variety of data [25] and often extracting data from otherwise non-integrated and incompatible health information systems [24]. Furthermore, it requires that the clinical models, workflows and organizational procedures for offering preventive care are adjusted to accommodate this potentially more precise way of detecting risk in a larger population of citizens. Lastly, this trend may raise concern about the risk of mental stress incurred by false positives produced by information systems, privacy of citizens and attention towards the potential mental response that increased awareness of risk may bring [24].

Even more importantly, this may not solve the fundamental challenge of bringing citizens closer to the preventive healthcare services, as conventional data-driven approaches assume an interest and willingness from the citizens to participate. One example is the Danish research and development initiative, TOF—Danish abbreviation of 'Early Detection and Prevention'.[1] The purpose of this project was to develop an intervention that enabled detection of health-risk behaviors in the broad population of citizens, and in particular among those not in frequent contact with the healthcare system. The core of the intervention was an information system that automatically stratified citizens into risk groups in order to connect them to a suited lifestyle offer either in municipal healthcare or general practice.

---

[1] The details of the TOF intervention will be introduced in the findings.

The TOF intervention aimed at ensuring 'right to health' [26] by (a) securing *availability* of prevention care services for all citizens, (b) creating *accessibility* by inviting citizens through eBoks (a secure, public electronic mailbox), (c) *economic affordability*, as participation was free for all citizens, (d), *acceptability*, as the intervention was non-discriminatory and inclusive for all citizens, (e) and *quality*, by basing the intervention on existing, well-established care services.

The TOF intervention has been under development since 2009, and in 2012 the clinical precision of the stratification algorithm was confirmed through a feasibility study [12]. In addition, the intervention was tested through two full-scale pilot implementation projects in 2016 and 2019, each lasting 3 months, and including a total of 4201 citizens (2016: N = 2661; 2019: N = 1540) [27, 28]. The research evaluations of the pilot implementation projects proved the ability of the TOF intervention to attract a large number of citizens and facilitate automatic stratification based on citizen-reported health information. However, the research evaluations also highlighted that while this type of data-driven approach could in fact identify citizen in moderate to high risk of developing LRD, it mainly succeeded in retaining those with fairly moderate health risk behaviors and with a relatively high level of health literacy [28]. Hence, there is a strong indication that this type of data-driven approach to detection of citizens in need of preventive healthcare alone cannot decrease the distance between preventive healthcare services and the broad population of citizens, and especially those who may need it the most.

Quantified Self

An issue of existing data-driven approaches to early detection, such as TOF, may be that they maintain a traditional division of responsibility between the citizen and the preventive health providers and hereby assume that the distance between citizens and healthcare services can be decreased by enabling the healthcare system to *just reach out* to the citizens in improved ways (in the case of TOF via eBoks and the software platform 'the Health Folder'). This alone will, however, not solve the fundamental challenge that citizens must willingly accept and follow up on the offers provided in order for disease prevention to be successful in the long term. At the other end of the spectrum, a long-standing interest exists in terms of how technologies can make citizens engage autonomously in their own health and well-being. It is well known that digital technologies can assist people in assessing and monitoring their own health condition on their own or in collaboration with health professionals. Under the umbrella term, *quantified-self*, this entails individuals engaging in self-tracking of biological, physical, behavioral, and environmental information using various types of technologies readily available on the consumer market, including exercise wristbands and mobile phone applications, as well as equipment developed to support specific types of health monitoring [29]. Recent numbers from the US show that 69% of the adult population track key health indicators such as bodyweight, exercise routines or symptoms. While many use pen and paper, digital technologies such as wearables (for instance fitness wristbands or other specialized sensor technologies) or smartphone applications are becoming increasingly common tools to track physical indicators such as fitness, sleep, nutrition, vital signs, as well information mental factors such as mood and stress level, or even interactions with one's social network [30, 31]. In fact, it has been documented that 27% of all Internet users in the US of age 18 or older track health data such as body weight and exercise routines online, and that 29% of all adults have downloaded health management applications [24].

Quantified-self technologies are primarily aimed at supporting individual, autonomous health management and intended to make citizens more qualified to identify relevant health concerns that require professional assistance and seek healthcare at their own initiative. Therefore, they also hold the potential for individuals to track indicators of their quality of life and consequently, they are increasingly referred to as *quality of life technologies* [32]. This, however, entails a number of ethical and societal concerns related to the movement towards transferring, not only autonomy, but also responsibility towards potentially vulnerable individuals [33]. However, certain types of care can benefit greatly from using technology to create a stronger link between the patient and the health provider. This is not least the case in treatment of chronic conditions, where telemedicine applications have shown to be valuable, for instance by enabling the patient to take an active role in monitoring his/her condition and detecting exacerbations at an earlier stage than in conventional care [34, 35]. In contrast to telemedicine, which refer to situations where measurements are sent to a healthcare provider as part of a mandated type of care [36], the purpose of quantified-self technologies is to enable citizens to monitor health and well-being autonomously [32, 36], and hence drive individuals to engage in the pre-diagnostic work involved in identifying potential health concerns. This has not yet, however, provided the required evidence of efficacy and effectiveness maturity to become a medical device and reliably support the citizen in important health decisions [37]. Additionally, quantified-self technologies do not by definition bring the citizens any closer to the preventive care providers when needed. Hence, the purpose of this chapter is to explore how quantified-self technologies can help decrease the distance between citizens and care providers in preventive healthcare.

Methods

Setting

The study has been conducted within the preventive healthcare system in the Region of Southern Denmark (RSD). RSD is a rural area of Denmark comprised of 22 municipalities with a total population of approximately 1.2 million citizens. Compared to the four other regions in Denmark, the population of RSD has a relatively low socio-economic standing, and the prevalence of lifestyle-related diseases is relatively high. Like the rest of the country, the responsibility for offering preventive healthcare services is distributed across two sectors in RSD. First, the municipalities (N = 22) are responsible for offering support for lifestyle change, e.g. through dietary consultancy and smoke cessation courses, as well as for reaching out and informing the population, for instance through information campaigns. To enable this, many municipalities have created *physical health centers* over the past decade where these services are integrated. Second, general practitioners are responsible for responding to citizens they identify as having significant health-risk behaviors as part of their normal routine in the clinic, and offer support for lifestyle change, for instance by examining, advising and monitoring the citizen, or by referring the citizen to one of the municipal offers. Currently, these preventive care services are, however, largely uncoordinated, and, hence, it is of high priority for the Region of Southern Denmark to increase the cohesiveness of the system.

The case of the study is the research and development project TOF, which is a partnership between a research unit on general practice, the RSD, the Organization of General Practitioners, and ten municipalities, and aims to develop and implement a data-driven approach to early detection that integrates all preventive care services in the region.[2] The TOF intervention has been under development since 2009. Since then, two 3-month pilot implementation projects have been conducted in 2016 and 2019 involving a total of three municipalities and 4201 citizens [27, 28].

Data Collection and Analysis

The study presented has been conducted as an *ethnographic study* of the practical use of the TOF intervention, which has been followed by the author since 2013. In order to identify the challenges for citizens in engaging in preventive healthcare, a qualitative field study was conducted during two 3-month pilot implementation projects in 2016 and 2019. During each round of data collection, citizens and health professionals (GPs, nurses, and municipal health professionals) were interviewed about their experiences. All study participants have given informed consent for the study. The TOF pilot project, part of which this study was conducted, has been approved by the Research & Innovation Organization, University of Southern Denmark (SDU RIO). The registration numbers of the approvals are 15/60562 (2016) and 18/32742 (2019). The main focus has been on the experience of the 26 citizens included in the study (12 males, 14 females, aged 32–60). The **citizens** were followed during their one to three-month participation in either 2016 (N:13) or 2019 (N:13). 13 of the interviewees opted out of the TOF intervention before the health interview with the GP. 21 of the respondents were interviewed over the phone and five interviews were conducted either at the care facilities or in the citizens' homes. Ten respondents were interviewed twice during the pilot implementation in 2016. In

---

[2] Details about the intervention are introduced in the findings.

total, 36 interviews (duration of 15–45 min) have been included in this analysis. The interview guide revolved around (a) the interviewees' general perception of health and previous experiences with lifestyle change, (b) their general experience of participation in TOF, and (c) their experience in using a digital system to prepare and assess their own risk profile and connect to relevant preventive healthcare providers. All interviews were audio recorded.

For the purpose of the analysis, the interviews were partially transcribed by the author. That is, the author listened through all interviews and transcribed the parts of specific relevance to the analysis. In line with the ethnographic approach applied in this study, the analysis of the empirical data was developed through three stages of writing with the aim of building theory through an inductive and interpretative process [38]. This explorative process allows for patterns, occurrences, phenomena and theorising to emerge and connections between themes that otherwise seem disparate and later generalised into a coherent argument [39].

The first stage consisted of writing the field notes (notes taken during field visits and after interviews). The second stage consisted of writing out data (listening through interviews, exploring emerging themes in these and sorting them into categories). From this analytical process, the general theme of 'distance' gradually emerged as a concept encompassing both the physical and psychological challenges involved in connecting citizens to preventive healthcare services through the data-driven approach. The third stage consisted of writing up the findings in order to capture the multiple facets of distance experienced by the citizens [38].

Assessing Risk at a Distance

A Data-Driven Approach to Creating Risk Awareness

The fundamental goal of the TOF intervention was to design a data-based workflow that would relieve health professionals from the labor-intensive tasks of detection and enrolment of citizens, and hereby allow them to focus on supporting the conversion and retention of improved lifestyles among citizens. To achieve this, the task of early detection was distributed to a stratification algorithm design to automatically assess the risk profile of potentially all citizens in the region. More specifically, this algorithm would assess the risk of developing hypertension, hyperlipidemia, and COPD (chronic obstructive pulmonary disease), and, based on this, refer the citizen to a relevant preventive healthcare service in either municipal healthcare or general practice. To ensure clinical validity, the algorithm was based on four guidelines: The Swedish National Guidelines for Disease Prevention [40], COPD-PS screener [41], the Danish Diabetes Risk Model [42], and the Heartscore BMI score [43]. To enable the stratification, two sources of data were needed. The first was pre-existing clinical data extracted from the GPs' electronic medical records. These consisted of structured data, including prescription codes, National Health Service disbursement codes, and International Classification of Primary Care (ICPC-2) codes. The second source was citizen-reported information on risk behaviors. To facilitate this, a 15-item web-based questionnaire was developed covering behaviors such as alcohol consumption, smoking, amount of physical exercise, diet (intake of sweets, fruit, vegetables, and fish), observable symptoms (shortness of breath, coughing), and own experience of general health.

Hereby, the TOF intervention took a great deal of responsibility for decreasing the distance between the preventive healthcare services and the citizens with moderate to high risk of developing LRD, by providing an intentionally efficient channel to convey the required data and the result of the risk assessment.

Dealing with Risk at a Distance

A general theme in the interviews was that citizens found the TOF intervention to be an interesting approach that facilitated renewed awareness of the potential consequences of their current lifestyle. While three citizens missed the opportunity to report the health information directly to a health professional and get direct feedback, a more general theme was that it was more meaningful for the citizens to complete the initial risk assessment in the comfort of their own home. The analysis shows that the citizens generally felt reflective and honest when answering the questionnaire, although one citizen stated that a health professional would be judgmental about the person's way of living, indicating a possible social acceptability bias. Thus, to a certain extent, the format of the TOF intervention resulted in some citizens feeling more comfortable with relating to their lifestyle, and illustratively speaking, TOF brought them closer to the possible risk of this. Yet, the quantitative evaluation of the TOF pilot implementation in 2019 showed a relatively high drop-out rate of citizens. Specifically, 134 out of 358 citizens identified with moderate risk completed the health interview with a municipal health professional, and 144 out of 321 identified with high risk completed the health check at their GP [28].

As observed during the pilot implementation of the TOF intervention, the digital system was useful for the citizens as a tool for reporting and receiving information. Hereby, this system supported the overall intentions of improving the connection between citizens and preventive healthcare services by developing an efficient method of *informing* citizens about the potential risks of their lifestyle, assuming that this would incite many of them to engage in behavior change, either supported by a GP or other healthcare services, or on their own. As expressed by the notion of health literacy, this, however, presumes not only the ability of the citizen to obtain this information, but also to understand and act upon it [44].

The information provided through the automatic stratification did not come as a surprise to any of the citizens participating in the study indicating a good preexisting understanding of their health condition, the degree of risk, and the causes of this:

> I am fully aware of what my issue is and that is smoking. To my knowledge I have no dis-
> eases, so it was <u>not</u> unpleasant for me to participate (in the intervention). (Citizen 5)

In spite of knowing about the correlation between lifestyle and risk, approximately half of the citizens participating in the study did not complete the TOF intervention by attending either the health interview with a municipal health professional or the health check at their GP. Some citizens altered their lifestyle at their own initiative, for instance by introducing moderate changes in their diet or exercise habits (citizens 18, 20, 22, and 23). Among those who did not change their lifestyle following their participation in TOF, the dominant theme emerging from the analysis was that they found it difficult to relate to the risk assessment performed by the stratification algorithm, or in other words; they felt distanced to their own risk of developing LRD rather than to the preventive healthcare services.

First, the analysis showed that that citizens often found the assessment performed by the algorithm to be un-nuanced and not fully reflect their lifestyle the way they perceived it, or not respectful of their priorities in life. Part of the reason for this was that the questionnaire reflected a very structured clinical logic for assessing health, which inhibited their ability to report what they found to be relevant nuances about their behaviors. For instance, the questionnaire asked "do you eat sweets every day" without specifying the amount. Some citizens who had a small but steady consumption of, e.g., dark chocolate experienced that they, by answering yes to this question, would be placed in the same category as people with a significantly less healthy lifestyle, yet faced the dilemma that they had to answer yes in order to respond honestly to the questionnaire:

> There were a lot of questions where I felt that I couldn't be precise in my answers, so I had
> to choose the lesser of two evils. (Citizen 2).

From a clinical logic, this question was meaningful, as people who respond positively to this statistically have a higher risk of developing LRD without considering the actual nutritional value of the diet. To the citizens, however, this made it difficult to relate to the assessment, in some cases causing them to become frustrated about the assessment or even reject it all together. Adding to this, the system was based purely on one-way communication, hereby relying on the ability of the citizens to ask questions to a health professional at a later stage of the intervention:

> Actually, I was slightly appalled by the answer. Of course, the system should do what it
> does, but the way it was communicated was like, well okay, is it really that bad? With a
> human I could have asked questions about how this conclusion was reached (Citizen 3)

Second, the long temporal span through which lifestyle may cause LRD obscured the importance of changing behavior in the short term. In contrast to acute disease that typically manifests itself through symptoms that are immediately present (e.g. the flu), the consequence of LRDs appears over a much longer timespan, often years or even decades. Coupled with the fact that risk behaviors such as diet or alcohol consumption are often deeply ingrained in social habits and cultures and hence also contribute positively to the quality of life for many people, a common theme of the interviews was that long-term risk had lower priority than short-term benefit and the pleasure, it would be to maintain the current lifestyle. Out of context of the specific data-driven format of the TOF intervention, citizens therefore showed a temporal distance to the concept of risk, which formed a barrier for acting upon this knowledge.

Certain factors, however, provided a push towards engaging in lifestyle changes. Most notably, it created awareness when citizens looked at themselves and considered the consequences of LRDs in the light of someone they held dear, for instance family or friends, who suffered or had passed away due to avoidable disease:

> It made me look at myself in a new perspective. There were questions (in the TOF questionnaire) about my social network and family. I realized that there have been heart conditions on my father's side. Compared to his age when he developed his heart disease, that will be me in 7 years. So perhaps I should start to eat healthier (Citizen 12).

An even more significant trigger was the prospect of losing the ability to fulfill their moral social responsibility, for instance by being a good parent, spouse or friend in the years ahead. However, this was not directly supported by the TOF intervention and, hence, it remained a challenge for the citizens in the study to make long-term risk a present, short-term concern.

Summary of Findings

In response to **RQ1** *"In what ways are citizens at a distance from the preventive healthcare system?"*, this study has found that citizens can both be physically and psychologically distanced from preventive healthcare services. While the physical distance did not impose a barrier for any of the citizens participating in this study, the psychological distance was prominent. More specifically, these citizens felt distanced to their own risk of developing LRD. This was either because they did not recognize themselves in the assessment performed by the algorithm, or because the concept of risk to many was difficult to grasp because of the long time span until it affects one's health and well-being.

Decreasing Distance with Quantified Self

Design Implications

As argued so far, data-driven approaches to early detection have the potential to decrease the burden of the preventive healthcare services when performing stratification of a large population. Furthermore, it creates a space where the citizens in the comfort of their own home can engage in their own health without feeling exposed to moral scrutiny of health professionals. While such an approach in one sense decreases the distance between preventive healthcare services and the citizens by enabling healthcare professionals to actively reach out to a larger number of individuals, it does not solve the fundamental challenge of making the concept of risk more palpable for the citizens. Rather, as the analysis showed, the citizens often felt distanced to their own risk of developing LRD, either because they did not recognize themselves in the assessment performed by the algorithm, or because the concept of risk to many was difficult to grasp because of the long time span until it affects one's health and well-being.

In response to **RQ2** *"How can quantified-self technologies enable citizens to decrease this distance?"*, two implications can be elicited to guide design of systems for data-driven early detection aimed at decreasing the distance between citizens and healthcare providers in preventive healthcare:

Design implication 1: The risk assessment provided by the system must be experienced as relevant by the citizen by accounting for individual variances with regard to physical, psychological, and social characteristics.

Design implication 2: To generate motivation for lifestyle change, a system must not only ensure that citizens realize the consequences of the lifestyle for their physical well-being and lifespan, but also for their psychological well-being and ability to fulfill social roles and moral obligations for family and friends.

Discussion

The findings of this study provide insights into what requirements quantified-self technologies must fulfill in order to contribute to decreasing distance between preventive healthcare services and the citizens in need of these. The most important implication is that design should not only focus on creating a connection between existing preventive healthcare services and the citizens in order to facilitate efficient early detection, but also create fertile ground for citizens to gain an interest in and motivation for engaging in the activities related to first detecting potential risk and later for converting to a healthier lifestyle. Quantified-self technologies first off hold the potential for creating risk profiles that are accurate, relevant, timely and relatable to the citizens. This can be achieved by combining multiple sources of data, for instance different types of wearables, such as activity trackers, and self-reported behavioral and health information, which is collected over time. While this approach may require technical skills and interest from the citizens as well as alignment of the underlying risk assessment model with clinical guidelines, it will address the most important issue related to existing data-driven detection models, namely that citizens gain influence on shaping a representation of their risk profile that takes the specific details and conditions of their lifestyle into account.

Another potential of quantified-self technologies is that they often include a social dimension by connecting the user to a community of other users in comparable situations and life conditions, often for the purpose of creating an incentive to perform well, e.g. in relation to physical exercise and weight loss. In the context of preventive healthcare, this can enable the citizens to compare their own lifestyles, not only to population statistics, but also to how different lifestyles affect the abilities and quality of life of other people, especially other individuals like them (for instance with regard to age, gender, socio-demographics, family context, etc.). While this approach requires conscientious considerations towards the potential ethical and privacy-related issues that it may entail to induce citizens to expose very private information about their lifestyle to others, it will address an important barrier for engaging in early detection and lifestyle change, that is not breached by existing data-driven approaches to early health assessment: as found in this study, it is very challenging for citizens to relate to the very long-term effects of an unhealthy lifestyle. Also, many citizens are not as concerned about how an unhealthy lifestyle will affect their own health and wellbeing as they are about how LRDs will influence their ability to fulfill their social roles and obligations for people they hold dear. Connecting citizens to a social network that can expose them to knowledge about how the consequences of an unhealthy lifestyle can also lead to the loss of ability to fulfill these obligations is therefore likely to incite more to engage in lifestyle change.

The study hereby shows that the matter of distance in early health assessment in preventive healthcare goes much beyond the challenge of connecting preventive healthcare services to relevant citizens, and assisting citizens in reaching out to relevant health care services. From the perspective of the citizens, the issue is not as much *where* to go but *why* to go, and the barrier for creating motivation for this is a psychological distance between the citizens and their own risk.

In a recent study, Wulfovich et al. [37] found that, in order to offer improved support for health self-management, quantified-self technologies must be further refined by addressing various human aspects of their use. These include the usability of devices and applications, personalization and context-awareness towards the user's life conditions, routines, and lifestyle, as well as the lifestyle advise provided by the devices must be timely and non-judgmental. In line with this, the results of this study suggest that data-driven approaches in preventive healthcare must be further developed to decrease the distance towards the citizens in order to ensure their 'right to health' [26]. While the physical distance, as demonstrated by the TOF intervention, can in fact successfully be bridged by existing digital means, and hereby ensure availability and accessibility for citizens, the psychological distance remains a greater challenge. From the perspective of the citizens, this first and foremost will require that the early health assessment provided by data-driven preventive healthcare interventions becomes increasingly nuanced to be immediately recognizable and reliable for the citizens and hereby become mentally and cognitively affordable for the user. Furthermore, the early assessments provided must be adaptable to personal differences between citizens to accommodate for the varying physiological preconditions as well as circumstances in life that characterize the heterogeneous populations to whom such services are offered. Lastly and most fundamentally, the health assessments must take into account that citizens do not only experience risk as personal loss of health and capability, but also as the possible loss of ability to fulfill social roles and moral responsibilities, in order to produce the intended motivation for lifestyle change. This speaks for, to a greater extent, focusing on such factors that evidently increase quality of life in the design of such technologies, rather than on developing increasingly sophisticated ways to quantify health.

Conclusion

As lifestyle-related diseases continue to have a significant impact on the life expectancy and quality of life of individuals, it is becoming increasingly critical to develop efficient approaches to assist citizens through lifestyle change. An essential aspect of preventive healthcare is to detect citizens in need of preventive care services at an early stage, when the risk is still reversible. Current approaches to early detection tend to focus on how the large number of citizens as well as the physical distance between these and the preventive care services prevent health professionals in screening for risk at a population level. Hence, development of data-driven approaches that support citizens in receiving a risk assessment and support for conversion of their lifestyle in the comfort of their own home is attracting significant interest. Based on a qualitative study of the practical use of data-driven early detection in Denmark, this chapter concludes that this approach, while showing great promise, does not succeed in decreasing the distance between citizens and preventive healthcare services. This is that many citizens feel distanced from this type of risk assessment due to a lack of personalization as well as they find it difficult to relate to how an unhealthy lifestyle will affect their health and quality of life in the long term. Quantified-self technologies may help address these barriers by providing an opportunity for producing a more personalized representation of users' lifestyles and helping them gain autonomy and influence in creating life changes and a better understanding of how their quality of life is *actually* affected by an unhealthy lifestyle and what are *actual* behavior changes that would fit their social context. This will entail the psychological distance between citizens and preventive healthcare being shortened from a citizen perspective, rather than a system perspective, leveraging the human intrinsic motivation.

References

1. Booth FW, Roberts CK, Layne MJ. Lack of exercise is a major cause of chronic diseases. Compr Physiol. 2014;2(2):1143–211.
2. Knight JA. Physical inactivity: associated diseases and disorders. Ann Clin Lab Sci. 2012;42(3):320–37.
3. Glümer C, Hilding-Nørkjær H, Jensen H, et al. Sundhedsprofil for region og kommuner 2008. Glostrup: Region Hovedstaden; 2008.
4. Juel K, Sørensen J, Brønnum-Hansen H. Supplement: risk factors and public health in Denmark. Scand J Public Health. 2008;136(Suppl 1):11–227.
5. WHO Working Group. The World Health Organization quality of life assessment (WHOQOL): development and general psychometric properties. Soc Sci Med. 1998;46(12):1569–85.

6. World Health Organization. Diet, nutrition, and the prevention of chronic diseases. Geneva: WHO; 2003.
7. World Health Organization. Global status on noncommunicable diseases. Geneva: WHO; 2010.
8. Shaw JE, Sicree RA, Zimmet PZ. Global estimates of the prevalence of diabetes for 2010 and 2030. Diabetes Res Clin Pract. 2010;7(1):4–14.
9. Yusuf S, Reddy S, Ounpuu S, Anand S. Global burden of cardiovascular diseases: part I: general considerations, the epidemiologic transition, risk factors, and impact of urbanization. Circulation. 2001;104(22):2746–53.
10. Armstrong D. Screening: mapping medicine's temporal spaces. Sociol Health Illness. 2012;34(2):177–93.
11. Bauer UE, Briss PA, Goodman RA, Bowman BA. Prevention of chronic disease in the 21st century: elimination of the leading preventable causes of premature death and disability in the USA. Lancet. 2014;384(9937):45–52.
12. Larsen LB, Soendergaard J, Halling A, Thilsing T, Thomsen JL. A novel approach to population-based risk stratification, comprising individualized lifestyle intervention in Danish general practice to prevent chronic diseases: results from a feasibility study. Health Informatics J. 2017;23(4):249–59.
13. Prochaska JO, Velicer WF. The transtheoretical model of health behavior change. Am J Health Promot. 1997;12(1):38–48.
14. Engelsen CD, Koekkoek PS, Godefrooij MB, Spigt MG, Rutten GE. Screening for increased cardiometabolic risk in primary care: a systematic review. Br J Gen Pract. 2014;64(627):616–26.
15. Si S, Moss JR, Sullivan TR, Newton SS, Stocks NP. Effectiveness of general practice-based health checks: a systematic review and meta-analysis. Br J Gen Pract. 2014;64(618):47–53.
16. Bouton ME. Why behavior change is difficult to sustain. Prev Med. 2014;68:29–36.
17. Gaziano TA, Galea G, 1 Reddy, K.S. Scaling up interventions for chronic disease prevention: the evidence. Lancet. 2007;370(9603):1939–46.
18. Huang TT-K, Glass TA. Transforming research strategies for understanding and preventing obesity. J Am Med Assoc. 2008;300(15):1811–3.
19. Landon BE, Hicks L, O'Malley AJ, Lieu TA, Keegan T, McNeil BJ, Guadagnoli E. Improving chronic disease management at community health centers. N Engl J Med. 2007;356(9):921–34.
20. Pollard RQ, Betts WR, Carroll JK, Waxmonsky JA, Barnett S, deGruy FV, Pickler LL, Kellar-Guenther Y. Integrating primary care and behavioral health with four special populations: children with special needs, people with serious mental illness, refugees, and deaf people. Am Psychol. 2014;69(4):377–87.
21. Consolvo S, McDonald DW, Landay JA. Theory-driven design strategies for technologies that support behavior change in everyday life. In: Proceedings of the CHI2009 conference on human factors in computing systems. New York: ACM Press; 2009. p. 405–14.
22. Fjeldsoe B, Neuhaus M, Winkler E, Eakin E. Systematic review of behavior change following physical activity and dietary interventions. Health Psychol. 2011;30(1):99–109.
23. Mlinac M, Lees F, Stamm K, Saint J, Mulligan J. Maintaining late life health behaviors: comparing clinician rating and self-reported resilience. Topics Geriatric Rehabilit. 2014;30(3):188–94.
24. Barrett MA, Humblet O, Hiatt RA, Adler NE. Big data and disease prevention: from quantified self to quantified communities. Big Data. 2013;1(3):168–75.
25. Raghupati W, Raghupati V. Big data analytics in healthcare: promise and potential. Health Inf Sci Syst. 2014;2:3.
26. United Nations. Committee on Economic, Social and Cultural Rights, Twenty-second session, Geneva, 25 April-12 May 2000. http://docstore.ohchr.org/SelfServices/FilesHandler.ashx?enc=4slQ6QSmlBEDzFEovLCuW1AVC1NkPsgUedPlF1vfPMJ2c7ey6PAz2qaojTzDJmC0y%2B9t%2BsAtGDNzdEqA6SuP2r0w%2F6sVBGTpvTSCbiOr4XVFTqhQY65auTFbQRPWNDxL. Accessed 18 June 2020.
27. RUGP (Research Unit on General Practice). Evalueringsrapport for TOF pilotprojektet: Resultater af den kvalitative evaluering. University of Southern Denmark; 2016.

28. Thilsing T, Svensson N, Søndergaard J, Larsen LB. Resultater af den kvantitative evaluering. Forskningsenheden for Almen Praksis: University of Southern Denmark; 2020.
29. Swan M. The quantified self: fundamental disruption in big data science and biological discovery. Big Data. 2013;1(2):85–99.
30. Estrada-Galiñanes V, Wac K. Collecting, exploring and sharing personal data: why, how and where. Data Science. 2020:1–28.
31. Fox S, Duggan M. Tracking for health. Pew Research Center's Internet and American Life Project; 2013.
32. Wac K. Quality of life technologies. In: Gellman M, editor. Encyclopedia of behavioral medicine. New York: Springer; 2020.
33. Sharon T. Self-tracking for health and the quantified self: re-articulating autonomy, solidarity, and authenticity in an age of personalized healthcare. Philos Technol. 2017;30:93–121.
34. Andersen T, Bansler JP, Kensing F, Moll J, Mønsted T, Nielsen KD, Nielsen OW, Petersen HH, Svendsen JH. Aligning concerns in telecare: three concepts to guide the design of patient-centred e-health. J Comput Supp Cooperative Work. 2018;27(3–6):1181–214.
35. Robbins R, Krebs P, Jagannathan R, Jean-Louis G, Duncan DT. Health app use among US mobile phone users: analysis of trends by chronic disease status. JMIR Mhealth Uhealth. 2017;17(5)
36. Verdezoto N, Grövall E. On preventive blood pressure self-monitoring at home. Cogn Tech Work. 2016;18:267–85.
37. Wulfovich S, Fiordelli M, Rivas H, Conception W, Wac K. "I must try harder": design implications for mobile apps and wearables contributing to self-efficacy of patients with chronic conditions. Front Psychol. 2019;23.
38. Madden R. (2010). Analysis to interpretation: Writing "out" data. In R. Madden (Ed.), Being ethnographic: A guide to the theory and practice of ethnography (pp. 136–151). Los Angeles, CA: SAGE Publications Ltd.
39. O'Reilly, K. (2012). Ethnographic Methods. 2nd ed. Routledge.
40. Svenska Socialstyrelsen. Sjukdomsförebyggande metoder. Vetenskabeligt underlag för nationella riktlinjer. Svenska Socialstyrelsen; 2011.
41. Martinez FJ, Raczek AE, Seifer FD, Conoscenti CS, Curtice TG, D'Eletto T, et al. Development and initial validation of a self-scored COPD population screener questionnaire (COPD-PS). COPD: J Chron Obstruct Pulmon Dis. 2008;5(2):85–95.
42. Christensen JO, Sandbaek A, Lauritzen T, Borch-Johnsen K. Population-based stepwise screening for unrecognised type 2 diabetes is ineffective in general practice despite reliable algorithms. Diabetologia. 2004;47(9):1566–73.
43. European Society of Cardiology. Heartscore (webpage). http://www.heartscore.org/en_GB/. Accessed 13 June 2020.
44. Nutbeam D. The evolving concept of health literacy. Soc Sci Med. 2008;67:2072–8.

Chapter 20
Assessment of Activities of Daily Living Via a Smart Home Environment

Qing Zhang and Mohan Karunanithi

Introduction

Globally, the population is aging. It is expected that by 2050, 1 in 6 people will be over the age of 65. It is a significant increase from the current rate of 1 in 11 [1]. In Australia, the population aged 65-and-over is 1 in 7, and it is expected to rise to 1 in 5 by 2050 [2]. This increase in the aging population, accompanied by the prevalence of age-related chronic diseases, presents significant challenges to public healthcare policymakers, societies at large, and individuals themselves [3]. An increasing shortage of care staff exacerbates it due to the aged care and nursing industry's inability to attract the workforce and the departure of existing staff from the workplace as they get older themselves. As a result, this will not only reduce access to the already limited residential care placements, but families will subsequently be expected to meet the health, social, safety, and other daily needs of their older parents. Hence, the future of aged care will increasingly move to the community or home [4]. As advances in technology and the Internet of Things (IoT) are becoming rapid and part of everyday life, it is inevitable that smart home developments will pave the way for future technologies supporting older people in the community and enable their quality of life as they age.

To ensure the health and well-being of older people in the community or home, an accurate assessment of the functional independence measures is crucial. Activities of Daily Living (ADL) is a good predictor of a wide range of health-related behaviors [5]. Hence, it is used routinely in the clinical setting and before hospital discharge, and patients return to home. Over the past 40 years, more than 43 indexes of ADL have been published to determine the fundamental functional disability status of both patients and the general, undiagnosed population [5]. The Katz ADL scale

Q. Zhang (✉) · M. Karunanithi
Australian eHealth Research Centre, CSIRO, Brisbane, QLD, Australia
e-mail: qing.zhang@csiro.au; mohan.karunanithi@csiro.au

© The Author(s) 2022
K. Wac, S. Wulfovich (eds.), *Quantifying Quality of Life*, Health Informatics,
https://doi.org/10.1007/978-3-030-94212-0_20

is arguably the first such instrument used in the clinical framework [6]. The Katz ADL assessment requires evaluating activities pertaining to bathing, dressing, toileting, transferring, continence and feeding. This instrument scores each activity with 1 if an older person can achieve it independently and 0 if it is dependent on assistance. Hence, the Katz ADL index scores will range from 0 to 6, indicating an older person's ability to function as being dependent (0) to independent (6), respectively. Barthel's ADL score [7] is very similar to Katz's ADL score. It is used to measure the performance of ten daily living activities, including feeding, bathing, grooming, dressing, bowel control, bladder control, toilet use (hygiene), transfers, mobility, and use of stairs. The possible scores range from 0 to 20, with lower scores indicating increased disability. There are also other ADL scales to measure more sophisticated functional independence, such as the full range of activities necessary for independent living in the community [8], for stroke patients receiving in-patient rehabilitation [9], and for patients with cognitive impairment such as Alzheimer's disease or dementia [10].

The challenge is that ADL assessment approaches in the clinical setting are still based on subjective assessments from a clinical assessor [11]. There are several drawbacks to this approach. First, it is neither representative of the individual older people's living environment nor reflective of their daily living routines. Second, it is a subjective measurement because it relies on a clinical staff's observation and corresponding responses from the individual older people being assessed. Hence, this approach's reliability is neither objective nor consistent for it to be reliable [12]. There is also evidence of varied ADL-based assessments, which are likely to be skewed by different individual responses to questions. These responses could vary from routine- or memory-based answers to socially desirable ones, which could be different for each individual depending on their culture, language, and educational backgrounds [13]. Furthermore, communication barriers from those with cognitive impairment could have significant implications on achieving reliable ADL assessment. Thirdly, current assessments are clinically resource-intensive, particularly when applied in a home setting, making them impractical for long-term care of the older people or disabled populations.

To address subjective bias and reduce human resource time on home-based functional assessments, the 'Smart Home' concept was proposed in the 1980s and applied to support independent-living older people's health and aging [14]. Along with the emergence of new technology in mobile computing, smart sensors, and the Internet of Things, smart home is becoming topical and relevant in home automation and assistance for health and well-being [15–18]. A few smart home-like products are emerging in the marketplace that employs motion, and movement sensors to detect daily activity patterns and the possibility of detecting falls [19, 20]. However, the wide adoption and deployment of smart homes in the senior community are still elusive. We believe there are two main reasons which have limited the smart home initiatives. First, the cumbersome use of technology stems from the lack of adequate design choices influencing their ease of use by the older community and influencing the users' perception that technology could compromise their privacy and security. Second, the lack of utility—in the form of a personalized and objective measure of functional independence that determines the individual's activities in their home setting reflects health and well-being. Current functional independence measures through subjective assessment, such as Katz ADL and Barthel ADL, are population-based one-fits-all models and not personalized to individual profiles. Given this state of the art, this chapter presents a design, implementation, and evaluation of a novel daily activity assessment tool, the objective ADL (OADL) aimed at older populations and achieved through fusing data from simple, non-intrusive, always-on, wireless sensors placed in a home environment called Smarter Safer Home (SSH).

This chapter is structured as follows. Section "A Smarter Safer Homes Approach" describes our SSH platform and the OADL index computed from the SSH platform's sensor data. Section "Clinical Trial and Discussion" introduces a small observational trial of the SSH platform and demonstrates the OADL index's effectiveness. Section "Conclusive Remarks" concludes this chapter.

A Smarter Safer Homes Approach

The CSIRO Smarter Safer Home (SSH) platform was designed as a home-based passive activity monitoring system without the need for residents' intervention to capture their health state and potential needs for care support and services. The platform was designed to be interoperable with commercially available sensors and devices. Furthermore, the design included privacy and security considerations, ensuring informed consent of all monitoring and data collection and processing processes.

System Requirements and High-Level Architecture

To enable an appealing and acceptable platform that seamlessly integrates with the home of an older person, the SSH platform's design challenge was to use only non-wearable, environmentally build sensors. From the consultation with the network of older individuals and their family members, it was clear that wearables could be intrusive, burdensome, and introduce anxiety. In general, the criterion for sensor inclusion in the SSH platform was for individuals to attend to their normal lifestyle while enabling the SSH platform to monitor their behavioral patterns silently and derive their activities towards health status assessment. It was also made clear that to maximize the protection of smart home residents' privacy, the SSH platform can not use any video/audio sensors. Instead, it should use advanced machine learning techniques to analyze raw environmental sensor data and infer, for example, individual's indoor activities of the state.

The SSH platform was designed as the outcome of multiple iteration cycles, takes advantage of wireless sensor technology and health home monitoring to provide a smart home integration that aligns with the older individual needs, and enables an engagement of informal (e.g., family) support and formal care services. SSH employs machine learning techniques that capture an individual's profile of Activity of Daily Living from sensor data. Based on that, personal level functional independence can be determined. The platform includes a sensor-based in-home monitoring system (data collection), a cloud computing server (data analyses), and a client module (data presentation) with a tablet app, a family/care-taker portal, and a formal caregiver service provider portal. Figure 20.1 shows the architecture of the SSH platform.

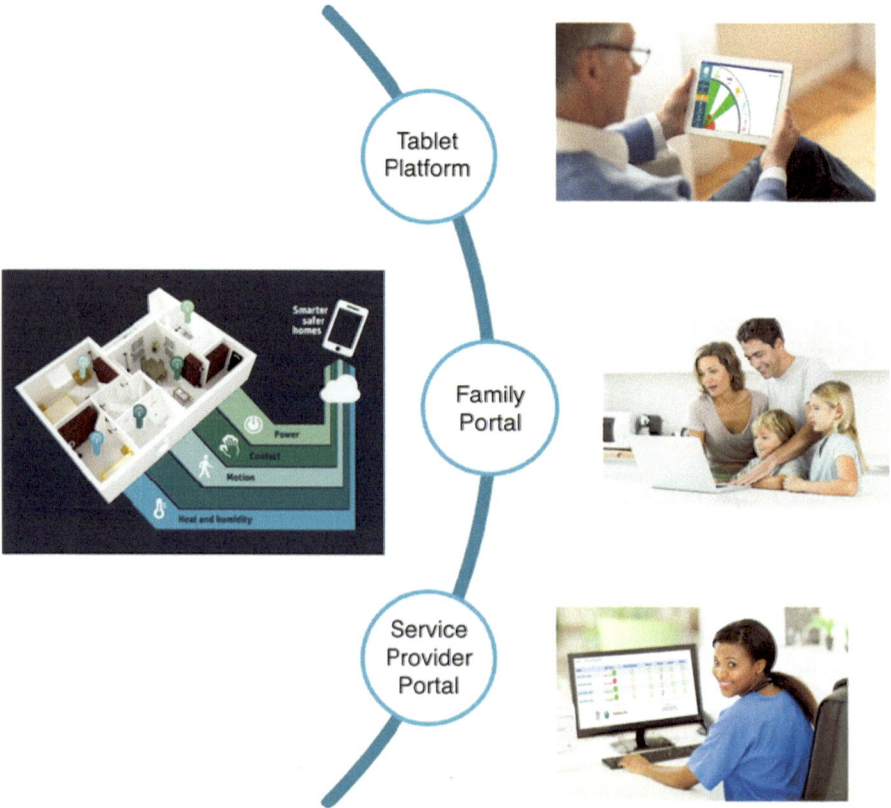

Fig. 20.1 Smarter Safer Homes platform and its three modules for client, family, and care service providers

The Wireless Sensor Network of SSH

The SSH platform includes wireless sensors that can monitor the physical environment and daily human activities within the home. In each home, environmental sensors are deployed with minimal footprints, i.e., they are deployed seamlessly 'invisible' in key areas of the older residents' home environment. In an observational trial conducted in 2013 [21], trial participants reported that they completely forgot those sensors 1 week after installation. These sensors communicate in-home activity data with a local server through the Zigbee protocol, enabling low-power, secure, and reliable data transmission [22].

The main types of environmental sensors deployed by SSH include *motion, power and circuit meter, ambient temperature* and *humidity, contact* and *sleep* sensors. Motion sensors detect the presence of people in its vicinity. They are passive infrared sensors installed in every room's corner (one motion sensor per room) to monitor the location and transition within the home. Power plug sensors and circuit meter sensors are either directly plugged into power outlets or connected in the meter box to measure home appliances, ovens, and cooktops' electrical power consumption. Ambient temperature and humidity sensors evaluate the indoor temperature and humidity periodically (one temperature and humidity sensor per room); usually, they report readings every 5 min. The contact sensor records open/close doors (including pantry, fridge, and wardrobe) and windows. Finally, sleep sensors are installed under the bed mattress to monitor sleep quality, length, and different sleep stages of individual residents. Table 20.1 shows a full list of sensors, and Fig. 20.2 illustrates a typical SSH sensor installation in a two-bedroom unit (approximately 70 m^2). Sensors used in the SSH platform are all commercially available from Aeotec® [23], Fibaro® [24], and EmFit® [25].

All sensors (except the power meter and circuit meter sensors) are running on batteries and can last around 8 months in general in a domestic home environment. It makes sensor installations flexible and easy because the sensors are untethered

Table 20.1 SSH sensors details

Sensors	Position	Sampling freq.	Value
Motion	Corners in all rooms	When status changes	Binary status
Power	Power plugs used by appliances	Every minute	Kwh
Circuit meter	Meter box	Every minute	Kwh
Temperature	Corners in all rooms	Every 5 min	Degree
Humidity	Corners in all rooms	Every 5 min	Percentage
Contact	Doors of room and fridge	When status change	Binary status
Sleep	Under mattress	Every day	Duration, efficiency, sleep stages, heart rate, respiration rate

Fig. 20.2 An example SSH in-home sensor installation (area of around 70 m²)

and can be positioned less intrusively and close to the activity being assessed, independent of a power source. It also benefits easy sensor maintenance. Furthermore, sensor communication generally requires little bandwidth and is relatively insensitive to latency. Energy-efficient communication protocols and event-based communication strategies can be applied, i.e., only uploading sensor data whenever an event change has been detected. In this way, the sensor's battery life can be greatly extended.

Activity Recognition from Ambient Sensors

Raw sensor data collected at indoor environments are first uploaded to the CSIRO secure cloud sever, to be initially processed to extract activities of daily living. Figure 20.3 illustrates an example of data samples collected at home by the SSH platform, where the ID represents a unique sensor ID (Table 20.2).

From the raw sensor data, activities are recognized by applying machine learning and pattern recognition algorithms. Specifically, the SSH platform evaluated five types of ADL, including Mobility, Hygiene, Dressing, Postural Transferring, and Meal preparation, which are included in many ADL instruments such as Katz ADL Barthel ADL [6, 7]. Specifically, Barthel ADL assessments' mappings compared to the types of ADLs evaluated by the SSH platform are listed in Table 20.3.

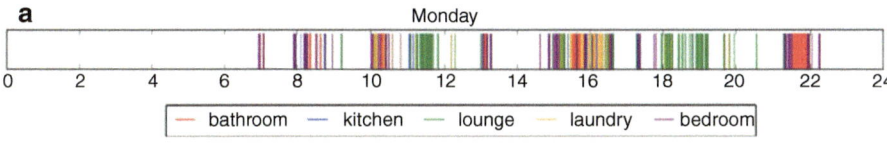

One day's motion sensor data in a home, (00:00:00 ~ 23:59:59)

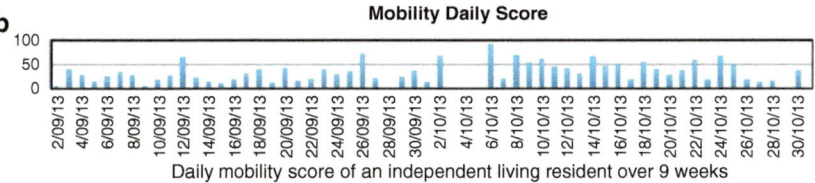

Daily mobility score of an independent living resident over 9 weeks

Fig. 20.3 An example of motion sensor data and mobility scores of one clinical trial participant. (**a**) One day's motion sensor data in a home (**b**) Daily mobility score of an indepenent living resident over 9 weeks

Table 20.2 Examples of sensor data collected in-home

ID	Type	Location	Timestamp	Value
17	Motion sensor	Bathroom	2016-08-14T07:24:00	1
31	Power sensor	Kettle	2016-08-14T08:00:00	0.9 kw
18	Humidity sensor	Bedroom	2016-08-14T08:10:00	58%
19	Temperature sensor	Bedroom	2016-08-14T08:22:00	25.6 °C
23	Contact sensor	Front door	2016-08-14T08:15:00	1

Table 20.3 Mapping Barthel ADL to SSH ADL

Barthel ADL	SSH ADL
Mobility/stairs	Mobility
Bowels/bladder/grooming/toilet use/bathing	Hygiene
Dressing	Dressing
Transfers	Postural transferring
Eating	Meal preparation

It is important to note that the Barthel ADL instrument evaluates whether an older individual can eat meals independently. However, due to limitations of technology and privacy violation concerns, we only monitored meal preparation instead, assuming the meal consumption was the natural next step. Table 20.4 lists types of ADL recognized by SSH, the assessment criteria, and the connected sensors. Every day for each recognized activity, we compute its own objective ADL scores, including *S_mobility, S_hygiene, S_dressing, S_transferring, and S_meal,* representing the individual's daily performance in a given area. To compute an objective ADL score, we utilize the assessment criteria for a given domain (e.g., hygiene) and compare it against normal performance, i.e., performance during the baseline period,

Table 20.4 ADLs monitored by the SSH platform

ADL	Assessment criteria	Sensors implied
Mobility ($S\_mobility$)	Indoor movements	Motion in all rooms
Hygiene ($S_{hygiene}$)	Grooming, showering	Bathroom motion, humidity, and temperature
Dressing ($S_{dressing}$)	Access wardrobe/cloth room	Wardrobe motion and contact
Postural transferring ($S_{transferring}$)	Sit/stand from chairs/beds	Motion in all rooms
Meal preparation (S_meal)	Cooking, accessing fridge	Kitchen motion, contact, power, and circuit

defined as 28 days after the SSH platform installation. Specifically, any objective ADL score ranging from 0 to 100 represents the likelihood of the assessed activity to be normal, with 0 meaning absolutely abnormal, i.e., very low, and unlikely activity performance such as no mobility for a whole day; and 100 meaning normal daily activity. In the subsections below, we explain the calculation of individual objective scores and the overall objective ADL (OADL) score.

Mobility, Dressing, Postural Transferring ($S\_mobility$, $S_{dressing}$, $S_{transferring}$)

These three ADLs are mainly evaluated through motion and contact sensors. From motion sensors deployed in all rooms, the in-home motion patterns can be inferred. With details from contact sensors, SSH can recognize the indoor movement activity and compute its performance scores, i.e., S_mobility. Figure 20.3a shows an example of motion sensor data from different rooms in one participant's single day. Figure 20.3b shows 9 weeks mobility scores within a home when the home resident was in a rehabilitation period after hip replacement of the same participant. We can see some of the days in this 9-week, the participant's mobility performance is abnormal, i.e., performance scores are less than 50, which is much lower when compared to the performance score during the baseline period participant.

Hygiene ($S_{hygiene}$)

This activity represents how well the resident maintains hygiene status, inferred through bathroom access and water usage. Hygiene-related activities can usually be recognized through significant changes in the humidity sensor readings in the bathroom. As illustrated in Fig. 20.4a, around 19:00, SSH assessed a shower activity when applying peak detection techniques in humidity sensor time-series data. Similarly, the hygiene scores can be calculated from humidity readings and bathroom access occurrences, as illustrated in Fig. 20.4b. It is interesting to notice that the hygiene was a bit low from September to October when the weather was cold but was higher towards the end of October when the weather became warmer in the trial location.

Meal Preparation (*S_meal*)

Extracting meal activity implies the use of multiple types of sensors deployed in the kitchen area. We assume that when all kitchen sensors report data within a short period, meal preparation is happening with high probability during that period. Specifically, the SSH platform applies multi-scale pattern recognition and unsupervised learning techniques to evaluate the probability of a meal preparation activity and the intensity of kitchen appliances usages as its corresponding score. Figure 20.5a shows one day's kitchen sensor events. SSH computed a meal preparation between 11:07 and 12:55 with a 98% confidence and a meal preparation between 20:01 and 22:10 with 30% confidence. Figure 20.5b shows daily meal preparation probability scores over 9 weeks.

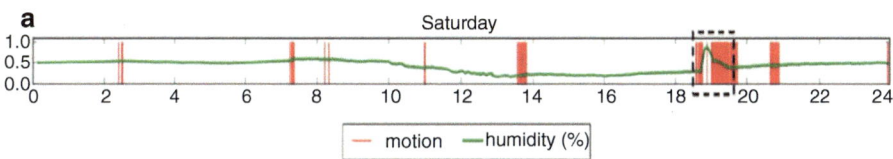

Bathroom motion and humidity sensor readings. The box indicates a showering event

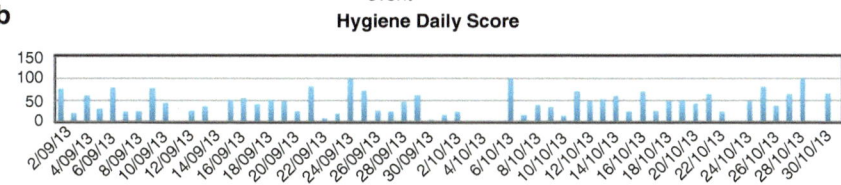

Daily hygiene score of an independent living resident over 9 weeks

Fig. 20.4 An example of humidity sensor data and hygiene scores of a clinical trial participant

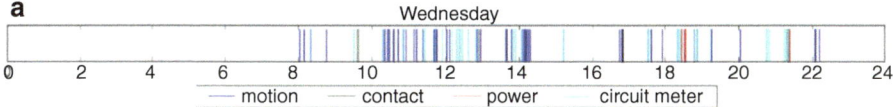

Kitchen sensor data. SSH computed a possible lunch preparation 11:07 ~ 12:55.

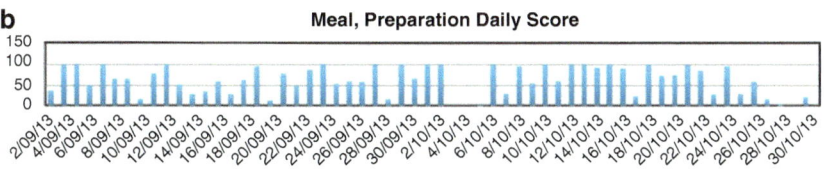

Daily meal preparation score of an independent living resident over 9 weeks.

Fig. 20.5 An example of kitchen sensors data and meal preparation scores of a clinical trial participant

Computing Objective ADL in a Smart Home (OADL)

Having the daily ADL value $S_{mobility}$, $S_{hygiene}$, $S_{dressing}$, $S_{transferring}$, and S_{meal} computed leveraging all the environmental sensors in a smart home, the objective ADL score, i.e., OADL, can be calculated through their aggregation. The OADL represents the total health and well-being status of home residents in a day:

$$OADL = F\left(S_{-mobility} + S_{-hygiene} + S_{-dressing} + S_{-transferring} + S_{-meal}\right)$$

Like the individual ADL scores (S_i), the OADL score ranges from 0 to 100, indicating the likelihood of the OADL being normal, with 0 implying very unlikely and 100 implying a normal day. Note that S_i represents the activity i, and F is a function learned from 28 days baseline data (assuming representative data for normal days). In our study, we asked participants to do a self-evaluation of their ADL during the baseline period via Barthel ADL [15]. We then apply Gaussian Process Regression to learn F with minimum regression error.

Clinical Trial and Discussions

Since 2013, the SSH platform has been deployed and trialed in Armidale, a local town of New South Wales in Australia. The study was supported by the Australian Centre for Broadband Innovation (ACBI) and CSIRO and recruited participants from a cohort of aged care facility residents who live in independent units. The study was conducted following Health and Medical Research Human Research Ethics (HREC# 12/17). The participants in the study agreed to the installation of in-home sensors and data analytics via SSH. The participants consented to 12 months of participation in the trial.

To be eligible to participate in the pilot, participants had to be aged over 70 years and have no home care arrangements. Participants with cognitive difficulties were excluded. Of those who self-selected (N = 23), 17 signed consent forms; however, three residents withdrew before the sensors were installed. Participants' retention in longitudinal trials with the older population was problematic for reasons including morbidity, mortality, relocation, or others. Over the course of the 12 months of the study (9/2013 to 9/2014), there were further withdrawals. Eventually, we collected five complete sets of data (sensor and interview data at 3Overse 5 eligible participants were aged between 79 and 88 years (Mean 83.6 ± 3.6). There were more female (n = 4) than male participants (n = 1). Two participants listed primary or secondary school as their highest level of education, 2 had non-university certificates or diplomas, and only 1 had a university education. Three interviews of participants' general daily routine were conducted education level after sensor installation [14]. Interviews were recorded and transcribed. Relatives or friends were present at most interviews and contributed to the discussion.

SSH App

To access the progress and summarized information derived from the SSH platform, participants are provided with a tablet and the preinstalled 'Smarter Safer Homes' app. The app interface was designed with independent living older people and a professional graphic design company during an earlier study [26]. The app displays the progress status of their daily activities of living and physical and social activity, vital signs (measured through clinical devices). Residents can connect to their family or care services via video conferencing services within the app. An example of the app's dashboard reflecting the daily status of well-being is represented by the colored rays (Fig. 20.6). A full-extension, green-ray, indicates the individual's well-being measures are within the expected range for that individual.

In contrast, two-thirds of an amber ray (not presented within Fig. 20.6) means a decline to the unexpected range. Furthermore, a one-third red-ray implies a very unexpected well-being measure that should trigger an intervention by the caretakers, to clarify the individual's current well-being. There are five main measurement modules in the SSH app: *Health Check*, *Sleep*, *Social*, *Walking*, and *Daily Activity*. The core module—Daily Activity, reflects the individual's OADL. Besides this module, SSH also provides other modules to perform remote physiological data measurements, sleep efficiency and sleep stage monitor, social connectivity promotion, and indoor and outdoor step counting.

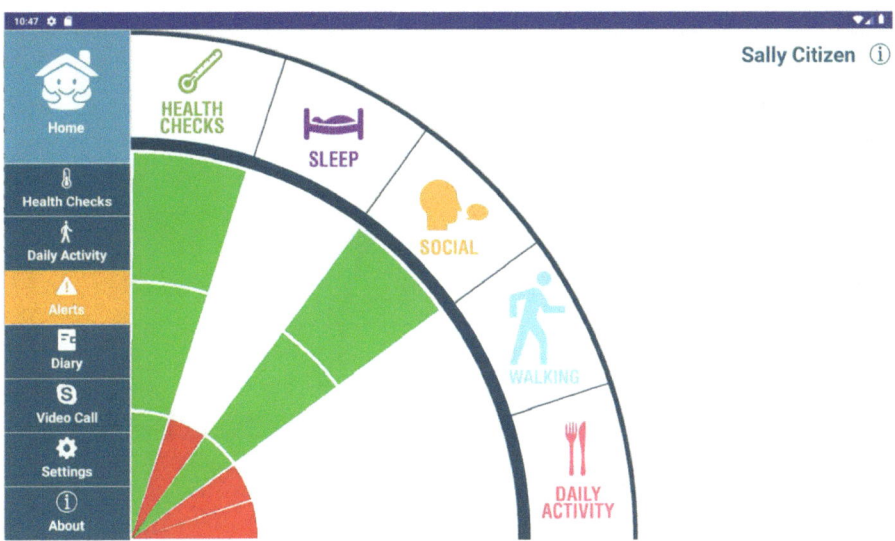

Fig. 20.6 An example screen for the SSH tablet app

Family Portal

Family members and friends of older people living alone often are anxious about their welfare. The family portal allows others to gain an insight into the lives of the older resident by communicating some of the information about their everyday lives via a website. There are four levels of access that the resident can make available to family members or nominated contacts. These levels are *No access*, *Limited*, *Standard*, and *Full details*. Figure 20.7 shows the front page of a family member with full access to the smart home data.

Care Service Provider Portal

The care service provider portal provides access to the SSH platform for formal caregivers, such as aged care service providers, to monitor the participant's profile and OADL scores. The Service Provider Portal can present an individual's OADLs

Fig. 20.7 Family portal available through an internet browser

over various periods (weekly or monthly). It can also display OADLs of previous days on the dashboard (Fig. 20.8) for trend checking and comparison purposes. The portal can be accessed by multidisciplinary healthcare teams engaged in an individual's care.

Trial Result

We measure the similarity between daily OADL and self-reported Barthel ADL of each participant by computing their Pearson correlation coefficients. In our pilot study, the OADL aligns well with the self-reported ADL with an average Pearson coefficient (75%), indicating this novel instrument's great potential as an effective tool for accurate state assessment of the individual. Figure 20.9 shows an example of the calculated OALD score and self-reported Barthel ADL from one trial participant.

The results for five trial participants who participated in the 10 months are summarized in Table 20.5.

Discussion

ADLs and Home Environment

In the SSH platform that we present in this chapter, non-wearable, non-intrusive sensors are deployed within the home to monitor older people's behavior patterns. The SSH design choices assume that a modern home's basic infrastructure is available, such as running water, electricity, private toilet, internet connectivity, to facilitate the data collection and processing. The home environment examines the principal place where a person lives and contributes to its quality. As shown above,

Fig. 20.8 An example screen for the SSH service provider portal

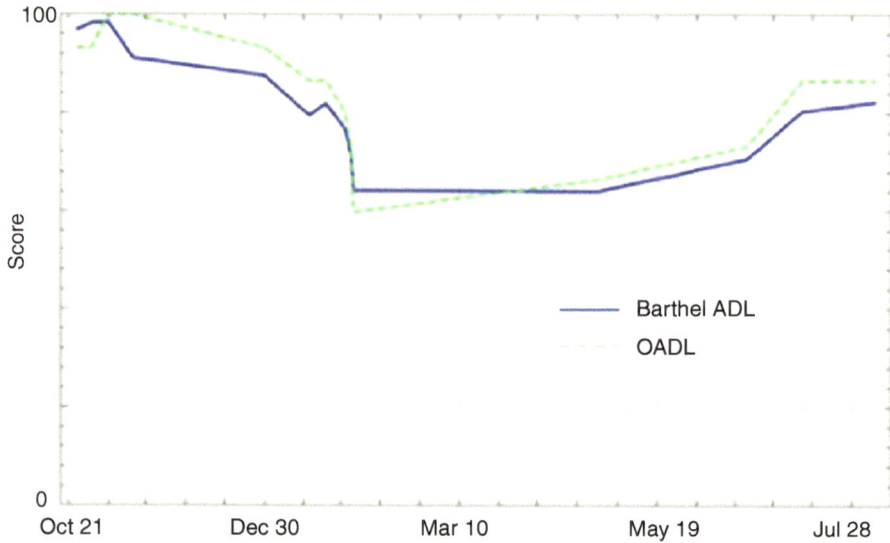

Fig. 20.9 An example of OADL vs. Self-reported Barthel ADL of a clinical trial participant

Table 20.5 Similarity between OADL and self-reported ADL of all five trial participants

Trial participant code	Age	Gender	The similarity between OADL and self-reported ADL (Pearson Coefficients) (%)
SC01	79	Male	66
SC02	81	Female	80
SC03	84	Female	58
SC04	86	Female	89
SC05	88	Female	80

in Figs. 20.3, 20.4 and 20.5, we can demonstrate a technology-supported assessment that objectively evaluates its habitant daily behavioral patterns.

Additional quality measures of the home itself could be derived for a wider assessment scope. At first, some existing SSH sensor data such as motion, temperature, and humidity could help quantify the quality of different parts of the home environment, such as temperature and humidity of the living room vs. bedroom. Additionally, the comfort quality of the home environment for the older residents could be assessed by deploying additional environmental sensors, such as a light sensor for visual comfort, a noise sensor for acoustic comfort, or air quality sensors for optimal air quality. Furthermore, given the recent climate change, SSH datasets could be leveraged to understand the external temperature changes on the residents' home activities. It has the potential to support the design of new homes for heat mitigation in hot areas.

Additional home environment quality measures, not considered within SSH yet, may include self-reported safety or intimacy measures, i.e., whether the residents appreciate the amount of space available for them and if they have opportunities for privacy if they do not live alone. Further home environment considerations, like the quality of the building's construction (such as roof leaking and damp), as well as the quality of the immediate neighborhood around the home, could also be self-reported. However, they are not likely to influence the algorithmic power on the OADL computation of the SSH itself. The overall assessment of the home environment's quality, including additional metrics mentioned above, can be another future research direction of extending SSH towards a generic smart home platform.

Human Factors Influencing the Quality of Data

In the first clinical trial that demonstrated the SSH platform's feasibility for independent living [21], tablets were provided to older participants to observe their activity profile through the SSH app towards potentially motivating and encouraging them for better self-management. Although tablets' interactive touch screen is a natural and friendly communication channel for most people, some participants found it somewhat difficult to use, especially when required to tap the screen with their fingers. Along with the recent development in voice and facial recognition through deep learning techniques, one possible future research direction of the SSH platform could include developing and integrating social robots as a virtual assistant, leveraging voice and gesture recognition as input.

Although the SSH platform uses only environmental sensors to enable convenience and reduce the potential intrusion for residents attending to their daily life, the major limitation is that this platform is currently reliable only for a single independent person in a single home. This is the case until a reliable environmental sensor technology distinguishes multiple residents in the home. Occasionally, this requirement could be relaxed for a dual occupancy home because there may be few joint activities between the two residents. For example, in our study, we recruited two multi-residential homes: one was a mother and an adult child, the other was a couple. Indeed, we noticed a decline in the SSH data quality of these two homes.

Nevertheless, these changes did not drastically affect the OADL score computation as the mother and adult child differed in their activities and schedules. Examples of this include cooking at different times. In the case of the couple, where one has a full-time job and is away from home during the day, the other can be assumed as living independently during the daytime.

Furthermore, to allay individual concern about privacy violation and personal information disclosure in a smart home, the SSH platform has strictly been designed not to use video/audio sensors to monitor in-home activities. Sensors are located in unobtrusive positions, such as ceilings, corners, and so on. Blinking indication lights of sensors, especially those deployed in the bedroom, are masked to avoid disruption. The SSH platform only records raw sensor data identifiable only by sensor serial numbers, and no mapping details between the sensors and the home or the resident are made available. The individual personal information is stored separately on a dedicated secure server. In our study, most participants forgot that the SSH platform was present after 1 week of the installation [21].

Smart Home and Quantified Self

The current SSH platform focuses mainly on evaluating five domains that are closely related to the ability of older people to live independently. This platform can also be expanded into a self-tracking tool that supports Quantified Self and ultimately contributes to improving the Quality of Life of individuals [27]. This could be achieved by adding additional sensors (e.g., wearable sensors [28]) to the SSH platform to extend its capabilities to measure various home and outdoor activities, including intensity of physical activity (via accelerometer and heart rate) or stress levels (via heart rate). Specifically, the SSH platform could develop a digital profile of an individuals' daily activities. By tracking and analyzing changes and trends in the digital profiles, the SSH system could facilitate individual's self-tracking, self-experimentation, and ultimately self-management, thereby promoting their health and well-being.

Quality of Life Technologies

The SSH platform is designed to support the independent living of older people. It satisfies the definition of a Quality of Life (QoL) technology [29]. It provides an object Activity of Daily Living index as a tool to assess the health-related QoL, i.e., ADL, of older people. Additionally, it connects older people to their caregivers through the family and service provider portals and thus enables timely interventions when necessary to maintain/prevent the decline of QoL. Furthermore, it also provides information to older people to support their goal-oriented self-management, in turn facilitating the enhancement of their QoL.

Conclusive Remarks

In this chapter, we focused on quantifying the home environment as contributing to the individual's life quality and introduced a novel objective activity of daily life (ADL) assessment through a smart home platform. This platform enables the aging individuals to engage their family/care-takers and/or aged care providers to access information about their state in real-time and support services that maintain or enhance their health and well-being. Through rich and up-to-date information about the resident's progress of functional measures of independence and health status via the SSH platform, more informed and timely care and support intervention can be provided following individual needs.

The novelty of the SSH platform lies in its features of providing an objective and personalized measure of ADL components and scoring through a set of miniaturized non-intrusive and non-wearable sensors in the home environment; and the ability to correlate this measure with the self-reported or care-reported status of the individual's health and well-being. The SSH platform allows for dignified aging for all older people in the community by facilitating efficient and effective aged care delivery. Consequently, the SSH platform can deliver enhancement of the quality of life to the older people and peace of mind to their families.

References

1. United Nations, Department of Economic and Social Affairs, Population Division; 2019. World Population Ageing 2019: Highlights (ST/ESA/SER.A/430).
2. AIHW. Australia's health; 2018. http://www.aihw.gov.au/reports/australias-health.
3. WHO. Global health and aging; 2011. www.who.int/ageing/publications/global_health.pdf.
4. Wiles JL, Leibing A, Guberman N, Reeve J, Allen RES. The meaning of "Aging in Place" to older people. Gerontologist. 2011;52(3):357–66. https://doi.org/10.1093/geront/gnr098.
5. Feinstein AR, Josephy BR, Wells CK. Scientific and clinical problems in indexes of functional disability. Ann Intern Med. 1986;105(3):413–20.
6. Katz S. Assessing self-maintenance: activities of daily living, mobility, and instrumental activities of daily living. J Am Geriatr Soc. 1983;31(12):721–7.
7. Mahoney F, Barthel DW. Functional evaluation: the Barthel index: a simple index of independence useful in scoring improvement in the rehabilitation of the chronically ill; 1965.
8. Lawton MP, Brody EM. Assessment of older people: self-maintaining and instrumental activities of daily living. Gerontologist. 1969;9(3):179–86.
9. Sulter G, Steen C, De Keyser J. Use of the Barthel index and modified Rankin scale in acute stroke trials. Stroke. 1999;30(8):1538–41.
10. Bucks RS, Ashworth DL, Wilcock GK, Siegfried K. Assessment of activities of daily living in dementia: development of the Bristol activities of daily living scale. Age Ageing. 1996;25(2):113–20.
11. Wallace M, Shelkey M. Monitoring functional status in hospitalized older adults. Am J Nurs. 2008;108(4):64–71. quiz 71–2
12. Guralnik JM, Branch LG, Cummings SR, Curb JD. Physical performance measures in aging research. J Gerontol. 1989;44(5):M141–6.
13. Linn MW, Hunter KI, Linn BS. Self-assessed health, impairment and disability in Anglo, Black and Cuban elderly. Med Care. 1980;18(3):282–8.
14. Ayuningtyas C, Leitner G, Hitz M, Funk M, Hu J, Rauterberg M. Activity monitoring for multi-inhabitant smart homes. SPIE Newsroom; 2014.
15. Smith G, Lunde A, Hathaway J, Vickers K. Telehealth home monitoring of solitary persons with mild dementia. Am J Alzheimers Dis Other Dement. 2007;22(1):20–6.
16. Tomita M, Mann W, Stanton K, Tomita A, Sundar V. Use of currently available smart home technology by frail elders: process and outcomes. Top Geriatric Rehabilit. 2007;23(1):24–34.
17. Wild K, Boise L, Lundell J, Fousek A. Unobtrusive in-home monitoring of cognitive and physical health: reactions and perceptions of older adults. J Appl Gerontol. 2008;27(2):181–200.
18. Doyle J, Bailey C, Dromey B. Experiences of in-home evaluation of independent living technologies for older adults. Proceedings of I-HCI; 2009.
19. Just checking. https://www.justchecking.com.au. Accessed 7 Mar 2021.
20. Quiet care. https://www.careinnovations.com/quietcare/. Accessed 7 Mar 2021.

21. van Kasteren Y, Bradford D, Zhang Q, Karunanithi M, Hang D. Understanding smart home sensor data for ageing in place through everyday household routines: a mixed method case study. JMIR Mhealth Uhealth. 2017;6(5):e52.
22. ZigBee applications. In: Gislason D, editor. Zigbee wireless networking. Burlington: Newnes; 2008. p. 111–206.
23. Aeotec Multisensor 6. https://aeotec.com/z-wave-sensor/. Accessed 12 Feb 2021.
24. Fibaro Motion sensor. https://www.fibaro.com/en/products/motion-sensor/. Accessed 12 Feb 2021.
25. EMFIT Sleep Tracker. https://shop-eu.emfit.com/products/emfit-qs. Accessed 12 Feb 2021.
26. Dana B, Jill F, M Karunanithi. Sensors on my bed: the ups and downs of in-home-monitoring. International Conference on Smart Homes and Health Telematics, 2013.
27. Wac K. From quantified self to quality of life. In: Digital health. Cham: Springer; 2018. p. 83–108. https://doi.org/10.1007/978-3-319-61446-5_7.
28. Estrada-Galiñanes V, Wac K. Collecting, exploring and sharing personal data: why, how and where. Data Science, IOS Press. https://content.iospress.com/articles/data-science/ds19002.
29. Wac K. Quality of life technologies. In: Gellman M, editor. Encyclopedia of behavioral medicine. New York: Springer; 2020. https://doi.org/10.1007/978-1-4614-6439-6_102013-1.

Chapter 21
Granting Access to Information Is Not Enough: Towards an Integrated Concept of Health Information Acquisition

Maddalena Fiordelli and Nicola Diviani

Introduction

The overarching objective of this book is to explore the potential of technology-enabled methods and tools for objective, quantitative assessment and improvement of Quality of Life. This chapter aims at exploring possible ways to enhance both the conceptualization and the measurement of the subdomain of quality of life labeled *opportunities for acquiring new information and skills*. After a brief overview on the definition of the subdomain under investigation and its original measurement, this chapter will present a summary of current studies aiming at the assessment and/or improvement of the variable, making the point for the urgency to find novel ways to conceptualize and measure it. The core of the chapter will be dedicated to the discussion of how research around the concept of *health literacy*, which is conceptually very close to the subdomain of interest and has received major attention within the academic community in the last decades, might inform developments from the point of view of the contents. On the other hand, we will show how current practices in the fields of marketing and computer science could inspire possible

M. Fiordelli (✉)
Institute of Public Health, Faculty of Biomedical Sciences, Università della Svizzera italiana, Lugano, Switzerland

Department of Health Sciences and Medicine, University of Lucerne, Lucerne, Switzerland

Swiss Paraplegic Research, Nottwil, Switzerland
e-mail: maddalena.fiordelli@usi.ch

N. Diviani
Swiss Paraplegic Research, Nottwil, Switzerland

Department of Health Sciences and Medicine, University of Lucerne, Lucerne, Switzerland
e-mail: nicola.diviani@paraplegie.ch

© The Author(s) 2022
K. Wac, S. Wulfovich (eds.), *Quantifying Quality of Life*, Health Informatics,
https://doi.org/10.1007/978-3-030-94212-0_21

advancements as regards measurement. The chapter will conclude with the discussion of some of the challenges and opportunities for future research on the topic.

Definition of the Variable "Opportunities for Acquiring New Information and Skills"

The subdomain of quality of life labeled *opportunities for acquiring new information and skills* has been defined by WHO as "a person's opportunity and desire to learn new skills, acquire new knowledge, and feel in touch with what is going on [...] through formal education programs, or through adult education classes or through recreational activities, either in groups or alone (e.g. reading)". The subdomain is included in the environmental domain and refers to the individuals' feeling of being in touch with, and having news of, what is going on around them. The focus is on a person's chances to fulfill a need for information and knowledge, whether this refers to knowledge in an educational sense, or to local, national or international news, that has some relevance to the person's quality of life. Depending on one's specific circumstances, this can be interpreted either broadly (e.g., being up-to-date with "world news") or in a more limited way (e.g., knowing what is going on in the local community).

The construct is complex, because it comprises both an objective and a subjective dimension. The objective dimension refers to the *possibility* to acquire information in terms, for example, of accessibility of sources of information. These include formal education sources, such as the school system, but also informal ones, for instance family and friends, which in turn can be accessed through different channels and in different formats. The subjective dimension of the subdomain, instead, refers to the individual's *ability* to satisfy the need of accessing new information and developing new skills.

Current Studies Aiming at the Assessment of the Variable

The questions included in the original WHOQOL-100 instrument are deemed to cover both dimensions of the subdomain. Three questions are used for each dimension, as the two are deemed equally important. Questions are phrased in order to be able to capture all relevant aspects of acquiring new information and skills ranging from world news and local gossip to formal educational programs and vocational training. It is assumed that questions will be interpreted by respondents in ways that are meaningful and relevant to their position in life [1].

Studies observing different population subgroups and cultures used the classical WHOQOL-100, WHOQOL-BREF [2] or other widely spread measures of quality of life such as the Health-Related Quality of Life score (HRQOL) [3]. Findings related to the measurement of acquiring new information and skills are consistent as they show a positive correlation between this subdomain and the educational level of the individuals in the sample [4]. Findings are mixed in describing the relationship between financial resources and opportunities for acquiring new information and skills, as also very poor subgroups of the population have a positive perception of their environmental quality of life [5].

The studies presented used the classical measurement tools, whose psychometric properties have been consistently proven across cultures, conditions and against other measures [6–8]. A measure of the environmental domain, though, has to keep up with the historical changes, therefore, to reflect what the current environment actually is in terms of offering opportunities to acquire new information and skills. These studies highlight connections with other constructs, and these connections point to another very relevant construct that will be presented in the next section.

Changes in the Information Landscape and the Need to Update the Subdomain

As outlined above, current studies aiming at assessing *opportunities for acquiring new information and skills* still largely rely on the questions included in the WHOQOL-BREF. The instrument, however, was developed based on the original definition of the subdomain, which dates back to 1994 [9]. In the almost 30 years after the development of the instrument, however, a major societal change has occurred: the advent and the global diffusion of the Internet and affordable personal Internet-enabled technologies and its consequences. The magnitude of this change, moreover, makes it something that cannot be neglected by researchers interested in studying this phenomenon and urges them to reflect on possible ways to update both the conceptualization and the measurement of the subdomain to better reflect today's reality. First, the Internet has allowed people worldwide to have access to an unprecedented number of sources of health-related information on virtually every possible topic [10]. Second, the possibility offered to everyone by the new media, independently from background or qualifications, to contribute to the discussion online, has contributed to the "mushrooming" of websites, blogs and social media posts providing unverified information of varying quality [11].

How does this societal change affect the subdomain *opportunities for acquiring new information and skills*? On the one hand, it makes the issue of access to diverse sources and types of information, i.e., the objective dimension of the subdomain, no longer a major problem. If, during its infancy, accessing the Web required the availability of technologies which were not for everyone, with the relatively recent widespread diffusion of smartphones information can be truly considered ubiquitous [12]. On the other hand, however, the widespread diffusion of online communication has created new disparities [13]. One recent concrete example is the declaration of Tedros Adhanom Ghebreyesus, Director-General of WHO at a gathering of foreign policy and security experts in Munich, Germany, in mid-February during the COVID-19 pandemic. The term *"infodemic"* refers to an excessive amount of information about a problem that is viewed as being a detriment to its solution [14]. This example clearly shows how, over the last years, we have witnessed a shift of the problem from the *availability* of health related information to the more and more essential *ability* of accessing this information [15]. This is not restricted only to the ability to use technology but refers more in general to all the competences needed to make good use of the opportunities the technology has to offer to maximize quality of life. We argue that in such an information landscape, the subjective dimension of our subdomain, namely one's ability to deal with information collected, should become the conceptually predominant dimension, and therefore the most important to assess and potentially improve, if necessary.

Suggested Approach for Quantitative Assessment/ Improvement of the Variable

In order to explore possible ways to improve the conceptualization of the subdomain *opportunities for acquiring new information and skills* we undertook a critical review of the literature in the field of psychology, educational sciences, health communication, technology, and marketing. This effort serves to explore some constructs that in our view are closely related to the subdomain at stake, by giving particular attention to the evolution in the conceptualization and measurement of their main dimensions over the last years, but also to innovative ways offered by technology to measure them and adapt to them. As regards the conceptualization, we will in particular discuss the concept of health literacy and its evolution, also because health literacy has a direct link to empowerment and health behavior and, in the long term, to improved health outcomes and reduced healthcare costs [16, 17]. Research undertaken in the field of marketing and technology will instead be used as a starting point to suggest possible innovative future directions in measurement.

Learning from Health Literacy Research

The individuals' ability to deal with information has been at the center of research in the field of health literacy [18]. We therefore believe that the advancements in this field could provide precious insights on possible future developments of the subdomain, both from a conceptual and a measurement point of view. The concept of health literacy was originally introduced in the 1970s in the context of school education and was initially understood as a set of basic literacy skills (i.e., reading and writing) in the health domain [19]. Following the societal changes outlined above, researchers in the field started to realize that being health literate entailed more than merely being able to access and read health-related information. Already in the early 2000s, Nutbeam proposed a new definition of health literacy, which has three main dimensions. The first dimension is *basic/functional* health literacy and entails having basic skills in reading and writing to be able to function effectively in everyday situations. A second dimension is labeled *communicative/interactive* health literacy and refers to more advance cognitive and literacy skills, which, together with social skills, can be used to participate in everyday activities, to extract information and derive meaning from different forms of communication, and to apply new information to changing circumstances. The last dimension, critical literacy, entails more advanced cognitive skills that, together with social skills, can be applied to critically analyze information, and to use this information to exert greater control over life events and situations [20]. From here, also following the growing interest in the concept related to the increasing evidence of a link with health outcomes [16], among researchers in the fields of medicine, public health, and health communication, several authors have contributed to expand the breadth of the concept. As a result, all the most recent definitions of health literacy recognize the multi-faceted nature of the concept and the need to include, besides functional skills, the more advanced skills needed to make sense and evaluate the increasingly complex information that is available to the public, including media literacy skills [21].

We believe that the evolution in the concept of health literacy presented above could be useful to inform the refinement of the contents of the subdomain *opportunities for acquiring new information and skills* and to shift the focus from its functional dimension to a more communicative and, what is even more important, a critical one. Besides the considerable efforts that have been devoted to the conceptualization of health literacy however, a significant amount of scholarly attention has also been devoted to the refinement of existing measurement tools and to the development of new ones [22]. In the following, we will describe some of the mostly used instruments in an evolutionary perspective and briefly discuss the current trends and future directions as they have been described in the numerous reviews that have been conducted recently both in the field of health literacy. The most commonly used measures of health literacy, to date, are the Rapid Estimate of Adult Literacy in Medicine (REALM) [23] and the Test of Functional Health Literacy in Adults (TOFHLA) [24]. The first tool measures a patient's ability to pronounce 66 common medical words and lay terms for body parts and illnesses, while the was developed using actual hospital materials and consists of a 50-item reading comprehension and 17-item numerical ability test. Both measures were developed in the early years of health literacy research. It has now been recognized by experts in the field that these measures do not fully capture the complexity and richness of the concept of health literacy, but are limited to its functional dimension, i.e., the ability to read and understand health-related information [25]. Starting from this consideration, many research groups around the world have started to develop new measuring tools with a broader scope. Examples of such measures are the All Aspects of Health Literacy (AAHLS) [26], the European Health Literacy Survey (HLS-EU) [27], or the Swiss Health Literacy Survey (HLS-CH) [28]. In contrast with the REALM and the TOFHLA, which are commonly considered objective measures as they ask individuals to perform a concrete task, the new measures are mostly subjective. This means that they ask individuals to rate their ability to perform a task. Whereas this evolution has substantially improved the content validity of the measurement, it has been argued that this type of tools do not actually measure actual ability but rather confidence or self-efficacy [21]. Moreover, several authors have suggested that new tools need to be developed to overcome the limitations of existing health literacy measurement [29]. Overall, despite the advancements in measurement, tools to assess health literacy are still very traditional and do not seem to take advantage, if not in some rare cases, of the possibilities offered by new technologies.

Advancing Measurement Using Insights from Marketing Research

Whereas, from a content perspective, the field of health literacy and its evolution might be a suitable example to learn from, it does not seem to provide useful insights as regards advancing the measurement of the subdomain under investigation.

The field of marketing is a perfect example of how it is now possible both to acquire precious information about the individuals (e.g., by tracking consumers' behaviors) and to tailor information to their needs, preferences, momentary context and abilities. Online Behavioral Advertising (OBA) is also called "online profiling" and "behavioral targeting" [30] and its definitions are multiple in the literature. One of them is the following from the Federal Trade Commission: 'the tracking of a consumer's activities online—including the searches the consumer has conducted, the web pages visited, and the content viewed—in order to deliver advertising targeted to the individual consumer's interest'. This is just one example of the many definitions; however, they all have in common two distinguished components: the monitoring of users' online behavior and the use of the monitoring data to target future advertising. Behavioral monitoring happens through use of software elements called *cookies*, or simply through the information that we give to our social media. In our online activity, everything can be tracked in principle, but also, we are giving out much information on specific channels. On the ground of the data collected the system make predictions of our behaviors and attitude. As a result, we receive advertising that is tailored to the research we made, or even in a more subtle way, we are exposed to contents because we interacted with a post or we just spent more time on it. Because of our actions, be them conscious or not, our network, and our history, we are timely tailored with the contents that are more prone to trigger an intention or even a behavior of ours. Behavioral data are therefore used to predict new behaviors, or even to arouse behavioral change (which usually results in some kind of financial gain for service provider).

This algorithm-driven approach to marketing and advertising is a novelty compared to the classical "one size fit all" mass media advertising, but also compared to the simple targeted advertising made possible by the Internet so far [30–32]. Based on a large amount of data, the algorithm can also become more refined, and be informed by persuasion and communication techniques, that make our behavioral change more likely to happen [31]. OBA can simply be based on our online activity through the more classical devices such as computers, tablets or smartphones, but it can also be using data derived from wearables and other more sophisticated devices. Whatever is able to collect and track data about our daily routine, our device usage or content consumption, can inform the algorithm for tailoring the content. The ethical and legal considerations about this practice have accompanied the development of the field since its infancy. The regulatory frames of data protection have been developed worldwide also in consideration of this, and the perception of the users towards his data privacy can strongly influence the persuasive effect of OBA practice. However, if this practice is disputable because of its ultimate aims being directed to profit, there is a chance that the mechanism can be exploited for higher purposes such as the ones related to the health and the wellbeing of individuals.

It was already some years ago when scientists were envisioning technologies able to adapt to the health literacy level of an individual [33]. When technologies able to improve user knowledge in specific chronic conditions were already a reality, researchers advocated for intelligent systems able to improve skill deficits in health care and basic literacy skills, such as numeracy through coaching. Beyond the provision of knowledge, they said, technologies could influence other constructs closely related to health literacy, like for instance self-efficacy and motivation for behavioral change using persuasion techniques and counseling agents. Information technologies could also serve to activate low literate individuals during doctor patient encounters by offering a list of questions and issues at hand. Wac's definition of Quality of life technologies goes in this direction when describing its aims [34]. Technologies able to respond to the needs of the user, and particularly at enhancing his/her quality of life are the ones that prove effective in ameliorating health literacy and related constructs.

Despite some first endeavors in this direction, this is not (yet) happening in health, at least on a large scale. Mobile health has exploited behavioral assessment for content tailoring in specific interventions or for self-management of chronic condition [35], but online (neither offline) behavior is not tracked and used in practice to deliver a more understandable health content. It would thus be essential to follow this line within the health domain. This means to keep developing and improving systems that are able to measure needs, preferences, and abilities through the individual actions (e.g., measuring health literacy level through Natural Language Processing or through real world actions) and to automatically adapt the information provided based on this data and the individual's context [36, 37].

Open Challenges and Future Directions

The goal of measuring the entire construct of Quality of Life, the way it is conceptualized by the WHO, is an ambitious one. Every single subdomain of the construct would deserve a separate scale covering all its dimensions, and this is true also when it comes to "opportunities for acquiring new information and skills". Based on our critical review, we conclude that, in the current information landscape, the measurement of this specific subdomain of the environmental domain (opportunities for acquiring new information and skills) should prioritize the subjective component. Indeed, individuals must be able not only to access information but also to appraise it critically. Only that way the new information and the new skills will contribute to enhance quality of life.

Health literacy research has shown that taking into account—and working towards the improvement of—citizens' and patients' ability to critically appraise information has several tangible benefits, making it a valuable investment for governments and health institutions. First, it would enable citizens to practice their "right to health", making healthcare services more available and equitable [38]. Second, but not less important, it would contribute to the containment of healthcare costs, for instance by reducing utilization of non-necessary health services, increasing participation rates to preventive services, or improving compliance with and adherence to treatment plans [16].

While developing systems that are able to assess and collect essential data in order to adapt information to the individual, we should take into account the ethical challenge related to a "tracking" on the one hand, and wrong adaptation effort, on the other hand, which would contribute to an exacerbation of disparities. A system collecting the wrong measures or interpreting one single measurement as an absolute indicator would offer information platforms that are too restricted, in terms of content, to the "predicted" need and preference of the user. Measurement would need to be comprehensive (and valid) not just in terms of constructs and data collected but also in temporal terms. We need to take a longitudinal perspective in order to work on the effective tailoring approach. Beyond that, we can leverage on what the Quantified Self movement has supported so far [39]. By getting to know more and become more aware about ourselves through technologies, we could contribuite to develop a self-determined and an highly democratic process.

Conclusion

Our personal digital devices are always with us, are able to track our actions, to collect contextual information, and even to ask us direct questions. We envision a system able to unobtrusively measure important characteristics of an individual (e.g., educational background, emotional state, beliefs, self-efficacy and health literacy level, health behaviors in daily life) in the long run together with environmental information. This way, we could build an highly tailored system, always at hand, that is able to offer information and recommendations that are not just timely but, hopefully, more useful and persuasive, thus effectively and safely contributing to behavior change, better health outcomes and the long term Quality of Life of the individuals.

References

1. The Whoqol Group. The World Health Organization quality of life assessment (WHOQOL): development and general psychometric properties. Soc Sci Med. 1998;46:1569–85. https://doi.org/10.1016/S0277-9536(98)00009-4.
2. W. Group. Development of the World Health Organization WHOQOL-BREF quality of life assessment. Psychol Med. 1998;28:551–8.

3. Horner-Johnson W, Krahn G, Andresen E, Hall T. Developing summary scores of health-related quality of life for a population-based survey. Public Health Rep. 2009;124:103–10.
4. Chang C-Y, Hung C-K, Chang Y-Y, Tai C-M, Lin J-T, Wang J-D. Health-related quality of life in adult patients with morbid obesity coming for bariatric surgery. Obes Surg. 2010;20:1121–7. https://doi.org/10.1007/s11695-008-9513-z.
5. Nayak A, Pradhan J. A comparative analysis of the quality of life between the poor and non-poor: a study of Rourkela city, India, (n.d.) 6.
6. Den Oudsten BL, Van Heck GL, Van der Steeg AFW, Roukema JA, De Vries J. The WHOQOL-100 has good psychometric properties in breast cancer patients. J Clin Epidemiol. 2009;62:195–205. https://doi.org/10.1016/j.jclinepi.2008.03.006.
7. Leung KF, Wong WW, Tay MSM, Chu MML, Ng SSW. Development and validation of the interview version of the Hong Kong Chinese WHOQOL-BREF. Qual Life Res. 2005;14:1413–9. https://doi.org/10.1007/s11136-004-4772-1.
8. Huang I-C, Wu AW, Frangakis C. Do the SF-36 and WHOQOL-BREF measure the same constructs? Evidence from the Taiwan population*. Qual Life Res. 2006;15:15–24. https://doi.org/10.1007/s11136-005-8486-9.
9. Whoq. Group. The development of the World Health Organization quality of life assessment instrument (the WHOQOL). In: Quality of life assessment: international perspectives. Springer; 1994. p. 41–57.
10. Dutta-Bergman MJ. Media use theory and internet use for health care. Internet Health Care: Theory Res Pract. 2006:83–103.
11. Eysenbach G, Powell J, Kuss O, Sa E-R. Empirical studies assessing the quality of health information for consumers on the world wide web: a systematic review. JAMA. 2002;287:2691–700.
12. Pandey A, Hasan S, Dubey D, Sarangi S. Smartphone apps as a source of cancer information: changing trends in health information-seeking behavior. J Cancer Educ. 2013;28:138–42.
13. Viswanath K, Kreuter MW. Health disparities, communication inequalities, and eHealth. Am J Prev Med. 2007;32:S131–3.
14. WHO Director-General's opening remarks at the media briefing on COVID-19 – 5 March 2020 (n.d.). https://www.who.int/dg/speeches/detail/who-director-general-s-opening-remarks-at-the-media-briefing-on-covid-19%2D%2D-5-march-2020. Accessed 24 Mar 2020.
15. Diviani N. On the centrality of information appraisal in health literacy research. Health Literacy Res Pract. 2019;3:e21–4.
16. Berkman ND, Sheridan SL, Donahue KE, Halpern DJ, Crotty K. Low health literacy and health outcomes: an updated systematic review. Ann Intern Med. 2011;155:97–107.
17. Crondahl K, Eklund Karlsson L. The nexus between health literacy and empowerment: a scoping review. SAGE Open. 2016;6:2158244016646410.
18. World Health Organization, Health Literacy. The Solid Facts; 2013. http://publichealthwell.ie/search-results/health-literacy-solid-facts?&content=resource&member=572160&catalogue=none&collection=none&tokens_complete=true. Accessed 16 Jan 2019.
19. Simonds SK. Health education as social policy. Health Educ Behav. 1974;2:1–10.
20. Nutbeam D. Health literacy as a public health goal: a challenge for contemporary health education and communication strategies into the 21st century. Health Promot Int. 2000;15:259–67. https://doi.org/10.1093/heapro/15.3.259.
21. Frisch A-L, Camerini L, Diviani N, Schulz PJ. Defining and measuring health literacy: how can we profit from other literacy domains? Health Promot Int. 2012;27:117–26.
22. Haun JN, Valerio MA, McCormack LA, Sørensen K, Paasche-Orlow MK. Health literacy measurement: an inventory and descriptive summary of 51 instruments. J Health Commun. 2014;19:302–33.
23. Davis TC, Long SW, Jackson RH, Mayeaux E, George RB, Murphy PW, Crouch MA. Rapid estimate of adult literacy in medicine: a shortened screening instrument. Fam Med. 1993;25:391–5.
24. Parker RM, Baker DW, Williams MV, Nurss JR. The test of functional health literacy in adults. J Gen Intern Med. 1995;10:537–41.

25. Pleasant A. Advancing health literacy measurement: a pathway to better health and health system performance. J Health Commun. 2014;19:1481–96.
26. Chinn D, McCarthy C. All aspects of health literacy scale (AAHLS): developing a tool to measure functional, communicative and critical health literacy in primary healthcare settings. Patient Educ Couns. 2013;90:247–53. https://doi.org/10.1016/j.pec.2012.10.019.
27. Sørensen K, Pelikan JM, Röthlin F, Ganahl K, Slonska Z, Doyle G, Fullam J, Kondilis B, Agrafiotis D, Uiters E. Health literacy in Europe: comparative results of the European health literacy survey (HLS-EU). Eur J Pub Health. 2015;25:1053–8.
28. Wang J, Thombs BD, Schmid MR. The Swiss health literacy survey: development and psychometric properties of a multidimensional instrument to assess competencies for health. Health Expect. 2014;17:396–417.
29. Pleasant A, McKinney J, Rikard RV. Health literacy measurement: a proposed research agenda. J Health Commun. 2011;16:11–21.
30. Boerman SC, Kruikemeier S, Zuiderveen Borgesius FJ. Online behavioral advertising: a literature review and research agenda. J Advert. 2017;46:363–76.
31. Yang K. Online behavioral advertising: why and how online customers respond to it?: an experimental study into the effects of personalized levels, rewards on click-through intentions towards ads between Chinese and Dutch. Master's Thesis, University of Twente, 2020.
32. Varnali K. Online behavioral advertising: an integrative review. J Mark Commun. 2019;1–22
33. Bickmore TW, Paasche-Orlow MK. The role of information technology in health literacy research. J Health Commun. 2012;17:23–9.
34. Wac K. Quality of life technologies. In: Gellman M, editor. Encyclopedia of behavioral medicine. New York: Springer; 2019. p. 1–2. https://doi.org/10.1007/978-1-4614-6439-6_102013-1.
35. Chandler J, Sox L, Kellam K, Feder L, Nemeth L, Treiber F. Impact of a culturally tailored mHealth medication regimen self-management program upon blood pressure among hypertensive Hispanic adults. Int J Environ Res Public Health. 2019;16:1226. https://doi.org/10.3390/ijerph16071226.
36. Allen LK, Snow EL, McNamara DS. Are you reading my mind? Modeling students' reading comprehension skills with natural language processing techniques. In: Proceedings of the fifth international conference on learning analytics and knowledge; 2015. p. 246–54.
37. Fuchs K, Barattin T, Haldimann M, Ilic A. Towards tailoring digital food labels: insights of a smart-RCT on user-specific interpretation of food composition data. In: Proceedings of the 5th International Workshop on Multimedia Assisted Dietary Management. Nice: Association for Computing Machinery; 2019. p. 67–75. https://doi.org/10.1145/3347448.3357171.
38. Logan RA, Wong WF, Villaire M, Daus G, Parnell TA, Willis E, Paasche-Orlow MK. Health literacy: a necessary element for achieving health equity, NAM perspectives; 2015.
39. Wac K. From quantified self to quality of life. In: Rivas H, Wac K, editors. Digital health: scaling healthcare to the world. Cham: Springer International Publishing; 2018. p. 83–108. https://doi.org/10.1007/978-3-319-61446-5_7.

Chapter 22
Using Technology to Predict Leisure Activities and Quality of Life

Andrijana Mušura Gabor and Igor Mikloušić

Measuring Quality of Life (QoL)

WHO defines QoL as an "individual's perception of their position in life in the context of the culture and value systems in which they live and in relation to their goals, expectations, standards and concerns" [1]. Four areas of life that QoL describes includes physical health, psychological state, social relationships and relationship with the environment. QoL is often mentioned with related term "subjective well-being" (SWB), recently defined as overall evaluation of the quality of a person's life from her or his own perspective [2]. Since concepts that relate to quality of life have been defined in numerous ways [3], many instruments and tools are available for measuring quality of life. For example, French MAPI Research Institute offers access to more than 1000 QoL instruments available through a database [4]. Linton et al. [5] did a review of 99 self-report measures for assessing well-being in adults, and concluded with a warning about major variability between instruments and the need to pay close attention to what is being assessed under the concept of "well-being".

Although many of these instruments show strong psychometric properties, there is much debate over self-report as a method to measure quality of life concepts. People are, consciously or unconsciously, deceiving themselves or others about what really is the truth regarding their own well-being. On the other side, some? research show that reaching a decision about someone overall quality of life is not based on careful and systematic analysis of all personal experiences, but on emotionally intensive peak-end moments in a person's life [6]. In contrast to many

A. M. Gabor (✉)
Zagreb School of Economics and Management, Zagreb, Croatia
e-mail: andrijana.musura@zsem.hr

I. Mikloušić
Ivo Pilar Institute of Social Sciences, Zagreb, Croatia
e-mail: igor.miklousic@pilar.hr

© The Author(s) 2022
K. Wac, S. Wulfovich (eds.), *Quantifying Quality of Life*, Health Informatics,
https://doi.org/10.1007/978-3-030-94212-0_22

self-report measures of QoL concepts, there are some measures emerging that exclude self-report and rely on patterns of behavior such as intensity of smiling in Facebook photos [7] or the content of tweets posted on Twitter [8]. By analyzing the words and topics from collection of tweets, researchers were able to improve accuracy in predicting life satisfaction over and above standard demographic and socio-economic controls such as age, income or education.

This brings us to the trend of "quantified self (QS)", developed due to increased availability of wearables and tracking applications. According to Lee [9], QS involves extended tracking and analysis of personally relevant data. For example, results of Gfk's global study [10] on health and fitness self-tracking reported that one in three Internet users track their fitness health via mobile apps or other wearable technology. Many of these self-trackers report that tracking changed how they approach to maintaining their health. Some researchers suggest that mobile technologies and wearables accompanied by Internet of Things (IoT) will revolutionize the way we understand ourselves and live our lives [11]. Besides the value of QS movement in learning sciences [9], a line of research that has been recently developed proposes a new term—"QoL Technologies (QoLT)", and gives rich insight into the ways QS can improve quality of life [12]. QoIT collects data from hardware and/or software technologies, provides objective and minimally intrusive assessments of QoL, and via feedback mechanisms aims at improving individual's QoL [13]. As such, QoLT gives great promise of providing greater self-awareness and opportunities for better quality of life.

Leisure Engagement and Quality of Life

WHO defined leisure within QoL model as "Participation in and opportunities for recreation/leisure activities as "a person's ability, opportunities and inclination to participate in leisure, pastimes and relaxation. The questions include all forms of pastimes, relaxation and recreation." [14, 15, p. 66]. Leisure engagement, more objectively, is defined by the amount of time, diversity, or frequency of person's participation in leisure activities [16]. Defining features of leisure activities is intrinsic motivation and freedom to engaging in them, and as such, they predict subjective well-being [17, 18]. There are many studies that report on significant relationship between engaging in leisure activities and improving quality of life. Recent meta-analysis of 37 effect sizes, with more than 11,000 participants, reported on strong evidence for the moderately positive association between leisure engagement and SWB [19]. Moreover, the relationship was mediated by leisure satisfaction, while measures of the frequency and diversity of leisure engagement were more strongly associated with SWB than measures of time spent in leisure.

Usually, leisure activities are classified into two categories: relaxed leisure activities (more passive, e.g., sedentary activities) and serious leisure activities (more active, e.g., physical activities) [20]. While Passmore [21] talks about three types of leisure activities—active, social, and time-out, Silverstein and Parker [22] propose 6 domains: culture-entertainment, productive-personal growth, outdoor-physical, recreation-expressive, friendship, and formal-group. Not all leisure activities lead to positive outcomes. Studies have shown that relaxed and passive leisure activities correlate to less satisfaction and well-being, compared to more active or physical leisure activities [23–26]. Kahneman and Kruger [27] offer a list of activities with accompanying self-reported measures of positive emotions and proportion of time with negative affect. Three activities that lead to most positive and least negative affect are intimate relations, socializing after work and relaxing. At the bottom of the list, with up to 30% of time feeling negative emotions, are activities of commuting and working. Oishi, Diener and Lucas [28] report that happiest people are most successful in terms of close relationships and volunteer work. On the other side, most commonly chosen leisure activity, watching TV, offers only limited enjoyment and needs satisfaction [29]. Overall, it seems that physical leisure activities and social leisure time rank high in predicting quality of life aspects.

Results of research on leisure activities and quality of life measures have important implications for developing alternative and less intrusive measures of QoL, backed by modern technologies. Although research on this topic is extremely limited, there are some research paving the way through the emerging field of QoL technologies.

Quality of Life Technologies

If we assume that there is a strong relationship between engaging in leisure activities and benefits for individual quality of life, especially in the long term, then it is also legitimate to assume that alternative measures for leisure activities could point to outcomes related with quality of life. This leads us back to quality of life technologies (QoLT). Widely available and affordable tools such as personal smartphones or wearable technology provide us with ample of data that can be correlated with how individuals spend their leisure activities. For example, smartphones can be seen both as tools that provide valuable information about overall activities individuals engage in, as well as tools that influence usage of leisure time.

Because smartphones have become quite pervasive [30], they are associated with many different behaviors and behavioral patterns such as social interactions, daily activities, and mobility patterns. Smartphone devices are becoming behavioral data-collection tools, and thanks to their computational power, logs and sensors, they provide us with unprecedented access to people's social interactions, daily activities, and mobility patterns [31, 32]. Research on this topic has only started to accumulate. Although there are number of studies that explore the relationship between smartphone usage, leisure activities and affective states [33–36], they all rely on self-reports. One interesting study, that also used self-reports, asked a representative sample of participants to report on the type of communication (text message of voice call) and physical proximity from individuals' whom the participants was in contact with [37]. Authors reported that geographic proximity of individuals was related with mobile communication patterns and social leisure activities between people who communicate. Let us now image how this study would look like if all the data was coming from smartphone devices. If backlog communication and geolocation data would be available for participants and their contacts from this study, we could determine the type and length of mobile communication (including recently popular use of messaging apps, beyond standard texting or calls), as well as geolocation of, both, caller and receiver. Social leisure activity could be, also, identified by geolocation data (e.g. restaurants, movies, theaters, concerts, sport games, bars and clubs). Thus, relying only on backlog smartphone data, we could learn about communication patterns and make assumptions about quality of life of these individuals, as well. These assumptions could take form of key predictors of important life outcomes, such as subjective well-being.

To test whether smartphone data can be predictive of subjective assessments, de Masi and Wac [38] used smartphone logger data to predict quality of experience assessed by in-situ quality of experience survey. Authors reported that predictive model for "good" and "bad" quality of experience can be build using quality of service information, mobile application name, user task (e.g., consuming or producing content) data within an app and physical activity of user. By combining self-reports and quantitative data, authors were able to determine alternative measure of quality of experience. Using similar study design, alternative measures for leisure activities can be built, as well. For each alternative leisure activity measure, self-reported satisfaction and positive/negative affect should be obtained. In Table 22.1, we offer a possible list of leisure activities and potential objective indicators, without affective indication, i.e., if the leisure activity indeed was enjoyable for the individual (which would require more physiological assessment with respect to the lower levels of stress, more calmness and happiness).

Having identified some of the objective sources of data for measuring leisure activities, we need to mention that future alternative measures of QoL will need to include not only the type of activity, but also its frequency, duration and diversity, as well as resulting affective state of the individual after the activity is finished (e.g., lower stress levels). This calls for different index measures that will have greater validity in assessing QoL. Once relevant alternative measures are identified, QoLT could help improve quality of life via feedback mechanisms such as notifications, reminders or motivational messages, leading the individuals to managing their leisure activity nest matching their momentary needs and context (e.g., if there are resources or opportunities for leisure in location where they are, with whom they could meet). With this heightened self-awareness, individuals will be able to change their behavioral patterns by overcoming self-regulation barriers that arise due to lack of planning or lack of goal progress. This way, individuals will be able to have direct impact into their quality of life.

Table 22.1 List of leisure activities and potential objective measures

Leisure category	Leisure activity	Type of objective data	Source of objective data
Culture and entertainment	Movies/cinema (offline)	Geolocation data	Smartphone
	Movies (online)	Media-service providers	Smart TV, personal computer
	Cultural events	Geolocation data	Smartphone
	Restaurants	Geolocation data	Smartphone
	Games (offline)	Geolocation data	Smartphone
	Games (online)	Gaming activity, gaming apps	Gaming consoles, Smartphone
	Apps	Smartphone data	Smartphone
	TV	TV tracking	Smart TV
	Internet surfing	Internet usage data, smartphone data	Personal computer, smartphone, tablet
	Going dancing	Geolocation data, wearable data	Smartphone, wearable
Personal growth	Books (online)	E-book readers	Tablet
	Courses (online)	E-mail enrollment notifications	Smartphone, personal computer
Outdoor and physical	Walks	Geolocation data, blood pressure and heart rate, steps, stress level	Smartphone, smartwatch, wearable
	Sports	Geolocation data, pressure and heart rate, steps, stress level	Smartphone, smartwatch, wearable
Passive	Sleeping	Blood pressure and heart rate, sleep apps	Smartwatch, smartphone
	Relaxing (indoor, sedentary)	Blood pressure and heart rate, movement sensors	Smartwatch, smartphone
Socializing	Phone call chatting	Communication logs	Smartphone
	Meeting with friends	Geolocation data, communication logs, calendar data, stress level, talking-to-listening ratio, tone of voice	Smartphone, wearable, sociometric badge
	Internet interaction/ chatting	Internet usage data, social media activity, messaging apps data	Smartphone, personal computer
	Club activity (church, hobby group, etc.)	Geolocation data	Smartphone

Possible Negative Impact of Screen-Time
on the Quality of Life

Since 2008, daily hours spent with digital media per adult user have risen from 2.7 h to 5.9 h in 2018, according to Meeker [39]. In fact, we rely more on devices for leisure time but not all of us (hence it's important to collect diversity of data to quantify leisure) [40, 41].

A notable rise in technology use in the American youth from in recent years, according to Pew research center (2018) is related to the increase in the smartphone ownership. 95% of teens reporting owning a smartphone, with the majority using some online social networks such as YouTube, Instagram, Snapchat or Facebook. Although the mobile phone technology and instant Internet connectivity is a somewhat new phenomenon, it has opened up new opportunities for improvements on various aspects of our lives and the quality of life. Also, it created new and superior ways of tracking and studying human behaviour [31, 42] there are already some research point out possible negative aspects of this hyper connectivity. Most of the studies done so far have been focusing on possible negative impact information communication technologies and social networks such as Facebook have on children and teenagers while some research that linked recently observed decrease in well-being and happiness in adolescent populations to the increase in screen activities facilitated mostly by widespread use of smartphones [43]. In many ways, focus on the negative impacts of new technologies has been a rule in psychological research as fears of detrimental impact of television as well as video games spreads through the population. For the most part our intuitions on the catastrophic effects of for instance video games and their link to violence have for the most part been dispelled [44, 45], however it seems that that there is reason for justifiable concern regarding this new technological trend.

As Pew research study suggested, although the participation in social media is almost ubiquitous in the teen population, the impact of it is not straightforward [46]. About a third of the teens report mostly positive effect of social media such helping them connect with friends and family, find information and meet people with whom they share interest, as much as a quarter report mostly negative effects of social media, with bullying, reducing in person contact, imposing unrealizing views of others' lives and distraction being poised as biggest issues. In a recent review of paper dealing with the impact of online social media on adolescent mental health Keles, McCrae and Grealish [47] found a positive relation between the use of social media and mental health problems. More precisely, the exposure to social media in the context of the time spent on the social media networks, type of activity (i.e. frequency of checking the profile, number of "selfies"), participant investment and addiction to social media were implied to be risk factors for the development of depression, anxiety and general psychological distress. High frequency users are, especially, under great risk from decreasing their quality of life due to less physical activity and poorer physical condition, and having greater connection to their smartphones in spending leisure time [48]. High frequency users report that smartphones

make leisure more enjoyable, increase personal freedom, are intrinsically reward-
ing, make it easier to engage in and experience leisure [49].

There are some caveats in these findings. Amongst others, the main issues with
studies conducted so far include small samples and reliance on self-report measures
[47]. Also, the majority of these studies are correlational and have a hard time establishing the direction of a causal relationship as it could be also likely that people
with pre-existing psychological disorders spend more time on social networks.
However, there are some experimental studies that do imply how abstaining from
social networks can have a positive impact on our well-being. For instance, Tromholt
[50] conducted an experiment on more than one thousand participants where the
experimental group was made to abstain from using Facebook for a week. The
results showed abstaining from Facebook had a positive effect on life satisfaction
and emotional life, especially for heavy Facebook users and people prone to experiencing envious feelings while using Facebook. Another similar study conducted in
organizational setting [51] showed that Facebook use and effects happiness in a way
that it promotes comparison which has negative effect on happiness and that this
effect was stronger for younger users. Another problem would also be the depth of
analysis. The quality of interactions that people have on these networks also seems
to influence the outcomes. In an extensive review of 70 studies dealing with social
network sites and well-being depression and anxiety showed that the impact of
social network use marked with positive interactions, social support and connected-
ness is related to lower levels of both depression and anxiety, while negative interactions, social comparison and addictive behavior has the opposite effect [52]. Also,
the positive interactions on social networks might benefit those that otherwise struggle with face to face communication.

So far, only a limited number of researchers reflected on the relationship between
mobile technologies use and leisure activities, drawing from similarly limited number studies on the use of mobile phones and their impact on our behavior and emotions. Lepp [34, 35] sees two major domains of overlap between mobile
communicating technologies and leisure. First one is facilitation of leisure activities
through enhancing communication and coordination. Access to this technology can
help in planning efforts for various outdoor activities but also in creating opportunities for spending leisure time in sedentary behaviors for individuals who are less
prone to face to face interactions. The second is related to the depth of our experiences whilst engaged in leisure activities. In one hand taking away from our ability
to isolate ourselves from outside influences and being in the moment, but also
enhancing our outdoor experiences trough access to information providing naviga-
tional aid. A creative experiment by Kushlev et al. [53] showed exactly how mobile
technology can take away meaningful experiences from our everyday activities.
Participants were asked to navigate through campus and find a particular building
either with or without using smartphones, and whilst individuals using smartphones
were more efficient at completing the task the ones that were left to their own
devices—looking at signs, asking for direction, etc.—felt more socially connected.
Dwyer, Kushlev and Dunn [54] reported similar findings in a study where participants either had their smartphones on the table or put away during a meal. The

results pointed toward phone use taking away the enjoyment experienced in real world interactions, and in a follow up study the negative effect of smartphone presence was found in other types of face-to-face interactions. Mostly, the phone was seen as a distraction, preventing people to fully engage with their environments. A negative relationship between smartphone addiction and productivity has also been reported, with spending time on smartphones taking considerable amount of both work and leisure—off time that could have been used on more meaningful pursuits [55]. This hyperconnected, interruptive and addiction forming quality of smartphones seems to be a cause of psychological distress. However, there is some proof that some of the content available on smartphones, such as health apps aimed at promoting health lifestyles and increased physical activity can produce positive outcomes (i.e. [56]).

Conclusive Remarks

This chapter focuses on the possibilities of combining behavioral patterns collected through use of smartphone and other technologies in predicting leisure time and quality of life. With the constant rise of internet users and ever-developing technology, the abundance of data gives new opportunities for behavioral research that goes beyond traditional methods and its deficiencies. One such opportunity includes using behavioral lifestyle data to recognize leisure activities and outcomes of engaging in such activities. Research has shown that life outcomes, such as quality of life and related constructs (i.e. SWB), are highly correlated with engaging in leisure activities. Throughout this chapter we offered numerous ways of using technology and data proxies for assessing leisure activities. We, also, imply that QoLT has a major opportunity to impact individual lives through feedback mechanisms that they offer to its users.

There are great advantages to using personal mobile technologies in research on leisure activities, and plethora of data it collects is in many ways more reliable than standard self-report measures we relied on so far. Especially since there is reason to believe that much of our self-report data on our mobile technology use is flawed, as recent studies comparing actual smartphone use to self-reports demonstrated people grossly underestimate the time spent using our smartphones (i.e. [57]). These trends will not only affect individual lives and their quality of life, but will also strengthen interdisciplinary research, and possibly transform field of psychology and its research methods [42, 58].

However, just the extent to which information and communication technologies should be fully incorporated in leisure activities is left to be determined. A recent interesting step away from digitalization, and toward the trend of so called "Digital detox" was reported by MIT Technology review [59]. The so-called Google Paper Phone, a product of Google creative lab is basically a piece of paper where a person would print out the relevant information for the day—telephone numbers, to do list, shopping list, a map, and use it to go about the day whilst leaving phone at home. That proves that maybe QoLT is not the only answer, and the person's ability, opportunities and inclination to participate in leisure, pastimes and relaxation may be when technology doesn't reach.

References

1. WHO. The World Health Organization quality of life assessment (WHOQOL): position paper from the World Health Organization. Soc Sci Med. 1995;41(10):1403–9. https://doi.org/10.1016/0277-9536(95)00112-K.
2. Diener E, Lucas RE, Oishi S. Advances and open questions in the science of subjective well-being. Collabra Psychol. 2018;4(1):15. https://doi.org/10.1525/collabra.115.
3. Camfield L, Skevington SM. On subjective well-being and quality of life. J Health Psychol. 2008;13:764–75. https://doi.org/10.1177/1359105308093860.
4. Emery MP, Perrier LL, Acquadro C. Patient-reported outcome and quality of life instruments database (PROQOLID): frequently asked questions. Health Qual Life Outcomes. 2005;3(1):12. https://doi.org/10.1186/1477-7525-3-12.
5. Linton M, Dieppe P, Medina-Lara A. Review of 99 self-report measures for assessing well-being in adults: exploring dimensions of well-being and developments over time. BMJ Open. 2016;6(7):e010641. https://doi.org/10.1136/bmjopen-2015-010641.
6. Kahneman D, Riis J. Living, and thinking about it: two perspectives on life. In: Huppert FA, Baylis N, Keverne B, editors. The science of well-being. Oxford University Press; 2005. p. 285–304. https://doi.org/10.1093/acprof:oso/9780198567523.003.0011.
7. Seder JP, Oishi S. Intensity of smiling in Facebook photos predicts future life satisfaction. Soc Psychol Personal Sci. 2012;3(4):407–13. https://doi.org/10.1177/1948550611424968.
8. Schwartz HA, Eichstaedt JC, Kern ML, Dziurzynski L, Lucas RE, Agrawal M, Park GJ, Lakshmikanth SK, Jha S, Seligman MEP, Ungar LH. Characterizing geographic variation in well-being using tweets. In: Proceedings of the Seventh International AAAI Conference on Weblogs and Social Media (ICWSM). Boston, MA; 2013.
9. Lee VR. What's happening in the quantified self movement? In: Polman JL, Kyza EA, O'Neill DK, Tabak I, Penuel WR, Jurow AS, O'Connor K, Lee T, D'Amico L, editors. Learning and becoming in practice: the international conference of the learning sciences (ICLS) 2014, vol. 2. Boulder, CO: ISLS; 2014. p. 1032–6.
10. Gfk.com. Global studies – fitness tracking; 2020. [online] Available at: https://www.gfk.com/global-studies/global-studies-fitness-tracking/. Accessed 28 Feb 2020.
11. Shehab A, Ismail A, Osman L, Elhoseny M, El-Henawy IM. Quantified self using IoT wearable devices. In: Hassanien A, Shaalan K, Caber T, Tolba M, editors. Proceedings of the international conference on advanced intelligent systems and informatics 2017, AISI 2017. Advances in intelligent systems and computing, vol. 639. Cham: Springer; 2018. https://doi.org/10.1007/978-3-319-64861-3.
12. Wac K. From quantified self to quality of life. In: Rivas H, Wac K, editors. Digital health: scaling healthcare to the world. Springer. Health Informatics Series; 2018. p. 83–108. https://doi.org/10.1007/978-3-319-61446-5_7.
13. Wac K. Quality of life technologies. In: Gellman M, editor. Encyclopedia of behavioral medicine. New York: Springer; 2020.
14. The WHOQOL Group. The World Health Organization quality of life assessment (WHOQOL): development and general psychometric properties. Soc Sci Med. 1998;46(12):1569–85. https://doi.org/10.1016/s0277-9536(98)00009-4

15. WHO. Programme on mental health: WHOQOL user manual, 2012 revision; 1998. Available at: https://apps.who.int/iris/handle/10665/77932/. Accessed 6 June 2020.
16. Kleiber DA, Walker GJ, Mannell RC. A Social Psychology of Leisure. Venture; Andover, MA; 2011.
17. Deci EL, Ryan RM. The support of autonomy and the control of behavior. J Pers Soc Psychol. 1987;53:1024–37. https://doi.org/10.1037/0022-3514.53.6.1024.
18. Ryan RM, Deci EL. Self-determination theory and the facilitation of intrinsic motivation, social development, and well-being. Am Psychol. 2000;55:68–78. https://doi.org/10.1037/0003-066X.55.1.68.
19. Kuykendall L, Tay L, Ng V. Leisure engagement and subjective wellbeing: A meta-analysis. Psychol Bull. 2015;141(2):364–403. http://doi.org/10.1037/a0038508.
20. Kleiber DA, Larson R, Csikszentmihalyi M. The experience of leisure in adolescence. J Leis Res. 1986;18:169–76. https://doi.org/10.1080/00222216.1986.11969655.
21. Passmore A. The occupation of leisure: three typologies and their influence on mental health in adolescence. Occup Participat Health. 2003;23:76.
22. Silverstein M, Parker MG. Leisure activities and quality of life among the oldest old in Sweden. Res Aging. 2002;24:528–47. https://doi.org/10.1177/0164027502245003.
23. Csikszentmihalyi M, Hunter J. Happiness in everyday life: the uses of experience sampling. J Happiness Stud. 2003;4:185–99. https://doi.org/10.1023/A:1024409732742.
24. Holder MD, Coleman B, Sehn ZL. The contributions of active and passive leisure to children's well-being. J Health Psychol. 2009;14(3):378–86. https://doi.org/10.1177/1359105308101676.
25. Ussher MH, Owen CG, Cook DG, Whincup PH. The relationship between physical activity, sedentary behavior and psychological wellbeing among adolescents. Soc Psychiatry Psychiatr Epidemiol. 2007;42(10):851–6. https://doi.org/10.1007/s00127-007-0232-x.
26. Wiese C, Kuykendall L, Tay L. Get active? A meta-analysis of leisure-time physical activity and subjective well-being. J Posit Psychol. 2017;13(1):57–66. https://doi.org/10.1080/17439760.2017.1374436.
27. Kahneman D, Krueger AB. Developments in the measurement of subjective well-being. J Econ Perspect. 2006;20:3–24. https://doi.org/10.1257/089533006776526030.
28. Oishi S, Diener E, Lucas RE. The optimum level of well-being can people be too happy? Perspect Psychol Sci. 2007;2(4):346–60. https://doi.org/10.1111/j.1745-6916.2007.00048.x.
29. Kubey RW, Csikszentmihalyi M. Television and the quality of life: how viewing shapes everyday experience. Hillsdale, NJ: Erlbaum; 1990.
30. Dey AK, Wac K, Ferreira D, Tassini K, Hong J-H, Ramos J. Getting closer. In: Proceedings of the 13th International Conference on Ubiquitous Computing – UbiComp '11; 2011. https://doi.org/10.1145/2030112.2030135
31. Harari GM, Lane ND, Wang R, Crosier BS, Campbell AT, Gosling SD. Using smartphones to collect behavioral data in psychological science: opportunities, practical considerations, and challenges. Perspect Psychol Sci. 2016;11(6):838–54. https://doi.org/10.1177/1745691616650285.
32. Lane N, Miluzzo E, Lu H, Peebles D, Choudhury T, Campbell A. A survey of mobile phone sensing. IEEE Commun Mag. 2010;48(9):140–50. https://doi.org/10.1109/mcom.2010.5560598.
33. Jankovic B, Nikolic M, Vukonjanski J, Terek E. The impact of Facebook and smartphone usage on the leisure activities and college adjustment of students in Serbia. Comput Hum Behav. 2016;55:354–63. https://doi.org/10.1016/j.chb.2015.09.022.
34. Lepp A. Exploring the relationship between smartphone use and leisure: an empirical analysis and implications for management. Manag Leis. 2014;19 https://doi.org/10.1080/13606719.2014.909998.
35. Lepp A. The intersection of smartphone use and leisure: a call for research. J Leis Res. 2014;46(2):218–25. https://doi.org/10.1080/00222216.2014.11950321.
36. Lepp A, Barkley JE, Li J. Motivations and experiential outcomes associated with leisure time smartphone use: results from two independent studies. Leis Sci. 2017;39(2):144–62. https://doi.org/10.1080/01490400.2016.1160807.

37. Campbell SW, Kwak N. Mobile communication and social capital: an analysis of geographically differentiated usage patterns. New Media Soc. 2010;12(3):435–51. https://doi.org/10.1177/1461444809343307.
38. Masi AD, Wac K. Predicting quality of experience of popular mobile applications from a living lab study. In: 2019 Eleventh International Conference on Quality of Multimedia Experience (QoMEX). IEEE; 2019. https://doi.org/10.1109/QoMEX.2019.8743306
39. Meeker M. Internet trends 2018 [Power point slides]; 2018. Available at: https://www.kleinerperkins.com/files/INTERNET_TRENDS_REPORT_2018.pdf. Accessed 22 Jan 2020.
40. Livingston G. Americans 60 and older are spending more time in front of their screens than a decade ago; 2019. Available at: https://www.pewresearch.org/fact-tank/2019/06/18/americans-60-and-older-are-spending-more-time-in-front-of-their-screens-than-a-decade-ago/. Accessed 6 June 2020.
41. Wang W. The 'leisure gap' between mothers and fathers; 2013. Available at: https://www.pewresearch.org/fact-tank/2013/10/17/the-leisure-gap-between-mothers-and-fathers/. Accessed 6 June 2020.
42. Miller G. The smartphone psychology manifesto. Perspect Psychol Sci. 2012;7(3):221–37. https://doi.org/10.1177/1745691612441215.
43. Twenge JM, Martin GN, Campbell WK. Decreases in psychological well-being among American adolescents after 2012 and links to screen time during the rise of smartphone technology. Emotion. 2018;18(6):765. https://doi.org/10.1037/emo0000403.
44. Ferguson CJ. The good, the bad and the ugly: a meta-analytic review of positive and negative effects of violent video games. Psychiatry Q. 2007;78(4):309–16. https://doi.org/10.1007/s11126-007-9056-9.
45. Ferguson, C. J., Colwell, J., Mlačić, B., Milas, G., & Mikloušić, I. (2011). Personality and media influences on violence and depression in a cross-national sample of young adults: Data from Mexican–Americans, English and Croatians. Comput Hum Behav, 27(3), 1195–1200. doi:https://doi.org/10.1016/j.chb.2010.12.015.
46. Anderson M, Jiang J. Teens, social media & technology; 2018. Available at: https://www.pewresearch.org/internet/2018/05/31/teens-social-media-technology-2018/. Accessed 22 Jan 2020.
47. Keles B, McCrae N, Grealish A. A systematic review: the influence of social media on depression, anxiety and psychological distress in adolescents. Int J Adolesc Youth. 2020;25(1):79–93. https://doi.org/10.1080/02673843.2019.1590851.
48. Lepp A, Barkley J, Sanders G, Rebold M, Gates P. The relationship between smartphone use, physical and sedentary activity, and cardiorespiratory fitness in a sample of U.S. college students. Int J Behav Nutr Phys Act. 2013;10(1):79. https://doi.org/10.1186/1479-5868-10-79.
49. Lepp A. Smartphones and leisure: an exploratory study. Paper presented at National Recreation and Park Association's Leisure Research Symposium, Anaheim, California; 2012.
50. Tromholt M. The Facebook experiment: quitting Facebook leads to higher levels of Well-being. Cyberpsychol Behav Soc Netw. 2016;19(11):661–6. https://doi.org/10.1089/cyber.2016.0259.
51. Arad A, Barzilay O, Perchick M. The impact of Facebook on social comparison and happiness: evidence from a natural experiment; 2017. Available at SSRN 2916158. https://doi.org/10.2139/ssrn.2916158
52. Seabrook EM, Kern ML, Rickard NS. Social networking sites, depression, and anxiety: a systematic review. JMIR Mental Health. 2016;3(4):e50. https://doi.org/10.2196/mental.5842.
53. Kushlev K, Proulx JD, Dunn EW. Digitally connected, socially disconnected: the effects of relying on technology rather than other people. Comput Hum Behav. 2017;76:68–74. https://doi.org/10.1016/j.chb.2017.07.001.
54. Dwyer RJ, Kushlev K, Dunn EW. Smartphone use undermines enjoyment of face-to-face social interactions. J Exp Soc Psychol. 2018;78:233–9. https://doi.org/10.1016/j.jesp.2017.10.007.
55. Duke É, Montag C. Smartphone addiction, daily interruptions and self-reported productivity. Addict Behav Rep. 2017;6:90–5. https://doi.org/10.1016/j.abrep.2017.07.002.
56. King AC, Hekler EB, Grieco LA, Winter SJ, Sheats JL, Buman MP, Barnejee B, Robinson TN, Cirimele J. Effects of three motivationally targeted mobile device applications on initial

physical activity and sedentary behavior change in midlife and older adults: a randomized trial. PLoS One. 2016;11(6) https://doi.org/10.1371/journal.pone.0156370.

57. Andrews S, Ellis DA, Shaw H, Piwek L. Beyond self-report: tools to compare estimated and real-world smartphone use. PLoS One. 2015;10(10) https://doi.org/10.1371/journal.pone.0139004.

58. Gosling SD, Mason W. Internet research in psychology. Annu Rev Psychol. 2015;66(1):877–902. https://doi.org/10.1146/annurev-psych-010814-015321.

59. MIT Technology Review. Google's big plan to fight tech addiction: a piece of paper; 2019. [online] Available at: https://www.technologyreview.com/s/614669/googles-big-plan-to-fight-tech-addiction-a-piece-of-paper/. Accessed 22 Jan 2020.

Chapter 23
The Importance of Smartphone Connectivity in Quality of Life

Alexandre De Masi and Katarzyna Wac

Introduction

The World Health Organization (WHO) has defined Quality of Life (QoL) as an "individual's perception of their position in life in the context of the culture and value systems in which they live and in relation to their goals, expectations, standards, and concerns." The WHO expands this definition across several domains, namely physical and psychological health, social relationships, and the environment. In this chapter, we focus on one facet of the environmental domain: the physical environment. We explore the availability of mobile network connectivity in one's environment without considering other variables that contribute to this environment, such as noise, pollution, climate, and the general aesthetic. Determining the impacts of connectivity on an individual's QoL is important for considering improvements or adverse effects on their day-to-day life.

Wireless networks have been present in our physical environment since the invention of over-the-air transmission of information (ALOHAnet [1]) in 1970. Recent developments in communication technology have now made it affordable to own a powerful, ubiquitous, network-enabled device. Today, wireless networks are present throughout the shared physical environment, especially in the developed world and in areas with high population density. Indeed, the accelerated

A. De Masi (✉)
Geneva School of Economics and Management, Center for Informatics, Quality of Life Technologies Lab, University of Geneva, Geneva, Switzerland
e-mail: alexandre.demasi@unige.ch

K. Wac
Geneva School of Economics and Management, Center for Informatics, Quality of Life Technologies Lab, University of Geneva, Geneva, Switzerland

Quality of Life Technologies Lab, Department of Computer Science, University of Copenhagen, Copenhagen, Denmark
e-mail: katarzyna.wac@unige.ch

© The Author(s) 2022   523
K. Wac, S. Wulfovich (eds.), *Quantifying Quality of Life*, Health Informatics,
https://doi.org/10.1007/978-3-030-94212-0_23

digitalization of the population can be attributed to the global adoption of smartphones. The number of smartphone users reached 3.2 billion worldwide in 2019 and will continue to grow [2]. Likewise, the networks that support them have been deployed at a similar pace and are continuously updated to cover larger areas and upgraded to utilize new technologies (e.g., from 3G to 5G).

The majority of mobile applications require an Internet connection, and in this study we focus on connectivity to mobile networks, whereby human-to-human interaction is enabled by computer-based networks. Networks support instant information transfer in various formats, including text, image, and video, and enable the necessary interaction between nodes (i.e., people or machines). Furthermore, they provide access to a number of services that can be used to improve an individual's decision-making capabilities and ultimately their QoL. A 2018 study by Chan et al. [3] found that smartphone use predicts relationship quality and subjective well-being, while Kim et al. [4] suggested that the use of information and communication technology, such as smartphones, in old age generally plays a positive role in enhancing the psychological, mental, and social aspects of one's QoL.

This chapter presents features of mobile network connectivity derived from smartphone use data collected from different cohorts in the Geneva area (Switzerland) between 2015 and 2020. We explore four connectivity features and examine the evolution of connectivity during the last 5 years as derived from data gathered unobtrusively from the consented mQoL (mobile QoL) Living Lab participants.

This chapter is structured as follows. We present the literature review in section "Related Work". In section "Mobile Network Connectivity Study: Methods", we provide the study parameters, describe the collected data, and outline the studied connectivity features. In section "Mobile Network Connectivity: Results", we report the results obtained from the analysis of the features. In section "Discussion", we discuss the limitations of the study and different approaches to connectivity quantification. Finally, in section "Conclusion", we describe the lessons learned and provide recommendations for future areas of work, especially the quantification of the impact of mobile connectivity on QoL.

Related Work

Mobile Network Connectivity and QoL

Previous work has shown the benefits of deploying mobile networks in rural and developing areas (e.g., Ghana, Nigeria, Kenya, and Tanzania) [5]. Researchers have found that it facilitates improved communication between the local population and distant services such as health and governance. The same authors have documented income growth in the Southeast Asia region in the last 10 years due to the rising usage of mobile applications and voice calls as the population gained access to new services and information relating to the weather, agriculture, finance, and music, for example. The income growth has only been reported in low-income countries, but surprisingly, in 2018, the GSM Association [6] found that the top reason to use mobile instant messaging was the same for low, middle, and high-income countries. This indicates that the benefits of messaging applications are not the prerogative of high-income countries. In recent years, messaging applications have created new markets and services that are available on their platforms. For instance, WeChat (est. 2011), Facebook Messenger (est. 2011), and WhatsApp (est. 2009) have all integrated payment functions into their applications in selected countries including China (WeChat Pay), Brazil, and the USA. Before the prevalent use of smartphones, the development of mobile payment solutions using a fast and reliable network was stagnant. Today, mobile networks are a critical gateway to the digital economy, as these solutions have been widely adopted to simplify the exchange of money and goods. Overall, 90% of Chinese tourists claim that they would use WeChat Pay overseas if given the opportunity [7].

The direct impact of broadband network access on GDP per capita has also been studied; one investigation found that a 10% increase in broadband penetration has a notable impact on GDP per capita, increasing it from 0.9 to 1.5 percentage points on average for OECD economies. Furthermore, the authors explained that if digital services are established alongside a reliable infrastructure, new services will be created [2].

In recent years, the Asia-Pacific region has been improving its environmental QoL through connectivity and will continue to do so particularly by way of smart city initiatives [8]. Such initiatives are described as cross-sector endeavors that link people to public and private infrastructures. Connectivity is crucial for smart city services, from the use of Internet of Things devices and a cloud-based platform to monitor and analyze air quality at street level, to the publishing of open data by public authorities to enable faster development of online-based services. In summary, a link between mobile connectivity and QoL around the world has been proven to exist—to such an extent that connectivity has a direct impact on a country's GDP.

Smartphone Apps and Their Impacts on QoL

The revolution in mobile devices, which have evolved from basic cell phones to smartphones, has created a new market for mobile applications. New application types were created for those devices, and as of November 2020, the Google Play Store hosted 2.56 million different applications across 32 application categories and 17 game categories [9]. Two application categories that may have a direct impact on users' health are (1) health and fitness, including personal fitness, workout tracking, dieting and nutritional tips, health, and safety applications, and (2) medical, including drug and clinical references, calculators, medical journals, news, and handbooks for healthcare providers.

Health and fitness applications such as food diaries allow users to track their food intake for multiple purposes. These applications connect to a central database that contains nutritional information about various foods (e.g., calories, carbohydrates, fat, and vitamin content). Users have to scan the barcode on a food item or use the search box to find and manually add the specific food and item weight, and the application computes its total nutritional value. Chen et al. [10] reported that users' quality of experience is much higher with smartphone application diaries than with pen and paper diaries. They also found that diabetic patients using application diaries reported a better food intake control than those using pen and paper diaries. Furthermore, a recent study by Bracken et al. [11] demonstrated that non-patient users wishing to lose weight (e.g., managing pre-obesity) and others wishing to gain weight (e.g., building muscle) utilize diary applications to attain their nutritional goals.

Medical applications are oriented towards health workers and healthcare practitioners. These professionals can use these applications as a productivity tool in their work, which enables them to automate necessary tasks [12]. Recent work [13] has indicated the advantages of medical applications: they increase access to point-of-care tools, thus improving patient outcomes that stem from better clinical decision-making. Wattanapisit et al. [14] investigated whether a medical smartphone-based application can replace a general practitioner. They praised the use of an application for tasks such as recording medical history, making diagnoses, promoting health, performing some physical examinations, and assisting in urgent, long-term, and disease-specific care. However, the application was unable to support clinicians in performing medical procedures, appropriately utilizing other professionals, or coordinating a team-based approach. A recent literature review by Wattanapisit et al. [15] focused on medical counseling for physical activity and returned mixed findings regarding the usability and utility of medical applications. The review suggested that technical issues and the complexity of programs were barriers to usability, thereby implying the possibility of unfavorable patient outcomes such as inaccurate advice and diagnoses.

Mobile network connectivity plays a significant role in always-online smartphone applications. These applications may help to enhance an individual's decision-making and thus result in an improved QoL through connectivity to the Internet. However, such applications can also lead to the reverse effects. One example is smartphone addiction. According to the observations of Kwon et al. [16], "the overuse of smartphones can be easily seen in today's society." The examples provided in the study include physical impacts (e.g., car accidents caused by smartphone use) and mental impacts that create issues for smartphone-addicted children (e.g., a loss of concentration in class). The authors proposed the Smartphone Addiction Scale (SAS) to quantify this addiction. The SAS consists of 48 items relating to smartphone usage in distinct contexts, such as taking the smartphone to the toilet or feeling stressed when the smartphone is not connected to a network. Also derived from this scale is the Smartphone Addiction Scale for Adolescents (SAS-SV) [17], evaluated by the same authors. The SAS-SV was used by Haug et al. [18] in a study on young people in Switzerland, which found that social networking applications were the applications most closely associated with smartphone addiction.

Smartphone addiction has also been attributed as a source of loneliness, poor bonding, and lack of integration, as shown by Bian et al. [19]. Samaha et al. [20] observed the relationships between smartphone addiction, stress, academic performance, and satisfaction with life. Through the use of multiple surveys, the SAS-SV, the Perceived Stress Scale, and the Satisfaction with Life Scale, they found addiction risk to be positively related to perceived stress. Finally, a large study by Carbonell et al. [21] demonstrated a substantial overlap between smartphone use, Internet addiction, and social media use in a student population. Smartphone addiction also has physical effects. For instance, Akodu et al. [22] described higher scapular dysfunction found in a population of students who are addicted to their devices.

A considerable amount of literature has been published on the influence of smartphones and has found that smartphone applications may influence users' QoL. Applications can contribute to users' well-being both positively and negatively, depending on the applications used and the user profile.

Smartphones as Sensors of Daily Life

Research by Dey et al. [23] established that smartphones are within arm's reach of their users an average of 88% of the time. Therefore, they are a beacon of one's presence. Indeed, smartphones have been used during the COVID-19 pandemic as a proximity sensor for contact tracing [24]. In recent years, smartphones have become a critical tool for researchers in all fields, as one of the greatest challenges to conducting a study is collecting participants' data. To solve this problem, a set of applications and software libraries have been developed to collect raw sensor data from smartphones as proxies for their users. These libraries collect similar data in different ways, although iOS devices are more restricted than Android devices.

Smartphone data can be collected from the following onboard sensors: accelerometer, location, proximity, barometer, gravity, light, magnetometer, audio, and temperature. Communication data can also be recorded from Bluetooth, SMS, telephony, and social applications. Tools such as AWARE exist to simplify the data collection process [25]. However, AWARE is often unable to integrate with other software platforms, while other tools such as Sensus [26] have customization issues. Meanwhile, libraries such as SensingKit [27] cannot support data collection alone. Furthermore, other software platforms like the CARP Mobile Sensing framework [28] propose a multi-platform approach (Android and iOS) with a reusable UI (Flutter) and support sensing for numerous features, but they lack low-level, hardware-based, detailed information.

Smartphone data collected with such tools have been successfully used in human studies [29]. For example, Ciman et al. [30] and De Ridder et al. [31] leveraged data collected from smartphone sensors to propose a stress assessment method. The first study used the data generated by finger swipes on the screen to detect stress, while the second paper showed through a meta-analysis that a tailored smartphone application can directly extract the heart rate variability (HRV), which is a stress indicator, from images of the subject's finger as it touches the smartphone's camera under illumination from the smartphone's flashlight. This process is called photoplethysmography. Smartphones are also used as sleep duration sensors, which was explored by Ciman et al. [32], and can predict users' intimacy, as claimed by Gustarini et al. [33].

In summary, smartphones are a proven source of daily-life data in multiple research domains, and their output has been validated experimentally.

Mobile Network Connectivity Study: Methods

QoL Lab was established in 2010, and since 2011, our research group has collected smartphone-based datasets for various human-based research studies and has used its own logging software for research into human activity recognition [34], mobility [35], and intimacy [33], among the other areas of study. The goal of this prior research was to quantify those aspects of human behavior with the use of smartphone sensors (i.e., gathering data using accelerometers, gyroscopes, and networking information, for example) and participants' self-reported inputs. We now focus on human subject studies "in the wild" and the practical aspects of smartphone data collection [36] through various research topics such as Quality of Service (QoS) [37], Quality of Experience (QoE) [38], and behaviors such as sleep [32] or stress assessment [39]. Smartphone data is collected in these different studies using the same framework (mQoL-Log), and it is tailored for each study. The mQoL-Lab application [40] enables background data collection through the mQoL-Log framework and implements surveys and remote notification to support human and smartphone-based research studies. Updates are necessary as the target system (Android OS) is always evolving. This section presents the tools used to acquire the data as well as their characteristics and discusses the selection of the derived features that are important for modeling individuals' day-to-day mobile network connectivity. Furthermore, we detail the processes used for feature engineering and data filtering.

Data Collection Periods and Overall Summary of the Collected Data

We investigated participant connectivity with the use of mQoL-Log data records. We focused on the networking data collected through different studies conducted in Geneva over three time periods, which is presented in Table 23.1. Each participant was only present during their period. P1 was aggregated from a mQoL-Lab Living

Table 23.1 Data collection periods

Period ID	Period Years	Study focus	References	Number of participants (N) Pre/post-filtering	
P1	2015–2017	QoS, mQoL LLab	[41]	53	50
P2	2018–2019	QoE, PeerMA	[42, 38, 39]	63	55
P3	2020	QoE	[forthcoming]	5	5
Total				121	110

Table 23.2 Participation statistics for the filtered datasets in each data collection period

Period ID	Avg number of measurement days/ participant	Standard error	Min	Q1 days (25%)	Q2 days (50%)	Q3 days (75%)	Max	Missing days avg ± std. err
P1	168.6	15.4	6	79	187	270	322	59 ± 10
P2	30.4	2.2	4	24	29	32	98	22.8 ± 1
P3	32.6	1.6	30	31	31	32	39	0 ± 0
Total	93.3	9.66	13.33	27	32.5	170	153	27 ± 3

(mQoL LLab) observational study that focused on people's smartphone usage. P2 studies were also "observational" and focused on quantifying the QoE of smartphone applications. They also focused on stress assessment via peers (PeerMA [39]). The P3 study was the first "interventional study of smartphone application category recommendations made based on the QoE model", where the intervention aimed to maximize user QoE in any context.

The presented meta-study focuses on participants' mobile connectivity throughout their days. The 121 participants collected a total of 69,761,823 samples. A *sample* is a piece of timestamped network-related information that was collected automatically via the mQoL-Log either when mQoL-Log requested information (i.e. by pulling the network state every 60 s for P1) or when an event occurred, such as a handover between different network connection types (e.g. 4G to WiFi, network connection or disconnection for P2 and P3) being *pushed* to the logger. The different ways of collecting the networking data (push/pull) were dictated by the Google API changes over the years. A *"day of the collection"* is a calendar day (midnight-to-midnight) for which at least one sample exists. On average, each participant collected data for 85 days (± std. err 9), 21 days for the 25th percentile (Q1), 31 days for the 50th percentile (Q2), and 128 days for the 75th percentile (Q3). We observed outliers in the aggregated dataset: one participant recorded 322 days of collection (max), while another only submitted one day of collection (min). Filtering was applied to the dataset following two exclusion criteria: (1) a participant collected less than ten samples, or (2) a participant collected less than three consecutive days of recording. The filtered dataset contained 110 participants; the filter removed 11 participants and 18,550,170 randomly distributed samples. The remaining 51,211,653 samples were retained for further analysis. Table 23.2 presents the participation statistics for the filtered datasets collected in each period. A *"day of measurement"* is defined as any sample collected in a 24 h period during the collection period; this is valid for P1, P2, and P3. Contrary to a *"day of the collection"*, this new metric is not based on a calendar day but on the availability of samples in a 24 h period (defined as a moving window or 24 h from a previous sample).

For example, for the P1 participant who recorded 322 days of collection (max), we have defined 322 days of measurement, meaning that the time difference between any two samples was less than 24 hours and that at least one sample per calendar day (Monday, Tuesday, …) was available. On average, each participant collected data for 93.3 days of measurements (± std. err 9.66), 27 days for the 25th percentile (Q1), 32.5 days for the 50th percentile (Q2), and 170 days for the 75th percentile (Q3).

Given that a *sample* is a piece of timestamped network-related information, if $n >= 1$ samples are generated at a specific *minute* (hh:mm), we classified this as *one minute of data collected*. Table 23.3 details the total number of minutes of data collected per period. We computed the mean rate of minutes acquired to understand how much data was collected per collection period overall. This rate differs from the days of data collection and the days of measurement, as it is minute based. We compared each sample acquired at a minute level to the possible number of data collection minutes during the collection period, assuming zero data loss, i.e., with data for all the minutes available. The last column of the table shows the overall acquired minute rate over the three data collection periods. Compared to P2 and P3, as explained above, the data collected in P1 was acquired more frequently.

Measurement Framework: mQoL-Log

In 2011, within the context of the mQoL Living Lab, we developed the first version of a smartphone logger for the Android operating system, and we implemented a cloud-based infrastructure to collect smartphone data. The smartphone application was composed of two modules: the data logger (mQoL-Log) and the user interface (mQoL-Lab). The user interface contained the participant's communication medium to complete the study and provide the possibility to contact the study's principal

Table 23.3 Number of minutes of data collected for the three periods

Period ID	Avg [min]	Standard error	Min	Q1 (25%)	Q2 (50%)	Q3 (75%)	Max	Mean acquired minute rate ± std.err [%]
P1	86,148	9387.8	855	15,769	89,008	138,676	224,684	34 ± 2
P2	1499	151	189	634	1202	2021	5559	4 ± 0.2
P3	1370	286	546	1017	1232	1992	2045	3 ± 0.5
Total	39,970	5858	536	1088	2471	75,339	77,429	17 ± 1.8

investigator. A cloud-based (our university-hosted) component was able to trigger surveys remotely and control the quality of the data collected on the smartphone, for integrity purposes.

mQoL-Log collected the data from the smartphone as mentioned previously (see section "Smartphones as Sensors of Daily Life"). Table 23.4 presents data collected from the smartphone's sensors through mQoL-Log. Table 23.5 presents in detail the

Table 23.4 Data collected by mQoL-Log

Variable name	Definition	Study period	Trigger and frequency of collection
Screen activity	The status of the smartphone screen and the user interaction.	P1, P2, P3	Changes in screen events (on, off, user presence, rotation) (push)
Touches	Number and duration of user touches on the screen during a usage session.	P1, P2, P3	Screen event-based: Each smartphone session (push)
Active app name	Application name on the user screen	P1, P2, P3	Changes in the application on-screen (push)
Background app	Application services running in the background (list)	P1	Every 60 s (pull)
Connectivity and network	WiFi level, WiFi BSSID, WiFi SSID, WiFi interface speed, cell ID, cell operator, cell strength, cell radio access technology (RAT), cell network code, Internet connection status, cell bandwidth up and down stream, number of packets and bytes sent and received on wireless interfaces	P1	Every 60 s (pull)
		P2, P3	Changes in network connection state and during user app usage (push)
Round Trip-Time [ms]	The RTT is the time needed for a ping to be sent by a smartphone to a server, plus the amount of time taken for an acknowledgment to be received.	P1 (always unige.ch server)	Every 60 s (pull)
	A ping is an active probing connection to a specific server via its address. A ping is executed six times; the first is discarded to remove any noise from DNS resolution time. We derived statistics (mean, stdev, and variance) from five executions.	P2 (app server)	When the app usage session starts (pull)
Battery	Battery status (e.g., charging, full, discharging), battery level, battery temperature	P1, P2, P3	Changes in battery state (push)
Physical activity	Physical activity of the user from Google Play Services activity (still, tilting: between two states, in-vehicle, on a bicycle, on foot, running).	P1, P2, P3	Changes in the user activity (push)

Table 23.5 mQoL-log network data

Name	Description
Network type	Type of cellular or WiFi network (RAT).
Signal strength	The signal strength is defined as the received power present in the WiFi and cellular radio in dBm (RSSI). dBms were transformed into the representation used in the Android OS, i.e. bars, as the participant would see this information on-screen.
Operator	The name of the cellular network operator.
Unique identifier (ID)	Cellular network tower ID (cell ID) or WiFi basic service set identifier (BSSID).
Network name	WiFi network name.
Handover	Flag indicating a change in network type, cell ID, or BSSID.
Total downloaded data	Cumulative sum in bytes of downloaded data since the last smartphone reboot.
Total uploaded data	Cumulative sum in bytes of uploaded data since the last smartphone reboot.

connectivity and network data collected. The logger included an energy policy to preserve the participant's smartphone battery life by stopping all data collection at a threshold of 30% battery capacity. Collection resumed once the smartphone was charging or when the battery capacity was above the threshold.

Final Dataset

As we wished to compare the connectivity of participants for the given data collection periods, we resampled the acquired P1, P2, and P3 datasets to one sample per minute, and completed the missing data points with the last known connectivity value. This method interpolates the missing points between two samples (upsampling), thus enabling a minute-based analysis of the smartphone's connectivity. The following assumption was made to validate this dimension change (i.e., to discretize it to 1 min frequency): if no data is present between two samples, this means that no event occurred. With this, we propagate the last known value to the next minute until a different event-generated sample is found. However, we are fully aware that this process does not allow us to make generalizations about a representative sample of the population.

Theoretically, P1 should have been sampled at one-minute frequency, since the pull method was leveraged for collecting the data every minute. However, we observed a skew in the pulling time, due to the Android OS giving lower priority to the collection process; the mean acquired pull rate was not 100% at 1 min period. Following the resampling process, the P1 was hence resampled to a one-minute frequency. As for P2 and P3, the resampling process generated a time series from

Table 23.6 Average measurement minutes collected post resampling, per participant in a period

Period ID	Avg [min]	Standard error	Min [min]	Q1 (25%)	Q2 (50%)	Q3 (75%)	Max [min]	Mean acquired minute rate ± std.err [%]
P1	241,832	22,261	7245	113,091	267,415	388,762	462,466	100 ± 0
P2	42,171	3144	3931	33,402	39,637	45,089	139,726	100 ± 0
P3	45,122	2473	41,448	42,447	42,930	43,902	54,885	100 ± 0
Total	109,708	9292	17,541	62,980	116,660	159,251	219,025	100 ± 0

the discrete events collected by the push method. The total size of the resampled dataset is 234 million samples as presented in Table 23.6.

Features Derived from Mobile Network Connectivity

In this subsection, we describe the four features derived from the raw dataset: (1) network access technology, (2) signal strength, (3) data consumption, and (4) user's physical mobility.

Network access technology or radio access technology (RAT) is defined as the physical connection system for a radio-based communication network. Smartphones support several RATs, such as WiFi, Bluetooth, GSM, UMTS, LTE, or 5G NR (New Radio). The focus of this analysis lies on RATs that enable Internet connection, so the Bluetooth standard is out of scope.

The *signal strength* is defined as the received power present in the WiFi and cellular radio signal. The signal strength feature directly impacts a user's network context and provides an insight into the connectivity level at that moment to the current Internet provider (i.e., a cell tower or WiFi access point).

Data consumption is defined as the amount of data (bytes) transferred from and to the smartphone through upload and download. The amount of data transferred during a specific time window provides information about the immediate network bandwidth. Some types of smartphone applications consume more data than others; for example, a video call application sends and receives more bytes than a text-based chat application.

The fourth feature is the *user's physical mobility*. Smartphones are used on the move, and their small size allows users to keep them in their pockets. In this way, they are a proxy for the user's mobility. Mobile connectivity is dependent on the physical network infrastructure around the user. Therefore, we analyzed the mobility aspect registered in the dataset. Mobility is defined as the number of cell towers and/or WiFi access points with unique identifiers that a participant passes through during a specific time window.

Network Access Technology

Wireless network access technology on a smartphone consists of two Internet-enabled subtypes: WiFi and cellular. WiFi allows smartphones to connect to a wireless local area network (WLAN). Often these local networks are also routed to provide Internet access. A smartphone's WiFi interface connects to an access point (AP) to provide an Internet connection, which has a network name and a unique identifier. In contrast to a cellular connection, WiFi enables a smaller coverage range depending on the generation used (on the scale of meters rather than the kilometers of a cellular connection). For this reason, WiFi is primarily used to connect to the Internet from home, work, or university. Various generations of cellular networks have been developed (e.g., 3G, 4G, 5G) with the evolution of access technology (see Table 23.7).

A cell tower offering Internet connectivity also has a unique identifier, but the main differences between cell-based technologies generation are the speed of the connection and their coverage range from the antenna. A smartphone's baseband processor is the chip on its motherboard, which manages all radio functions. This processor is separated from the main smartphone processor for three reasons: (1) radio performance: the main processor is too slow to handle the type of work done by the baseband processor, such as encoding and modulation; (2) legal: authorities require the software that manages radio transmission to be certified; and (3)

Table 23.7 Generation of cellular network access technologies

Generation	Acronym	Full name	Max download speed	Estimated download time for a 3 min 1080p YouTube video (75 MB)
2G	GPRS	General Packet Radio Service	0.0125 Mbit/s	800 min
	EDGE	Enhanced Data Rates for GSM Evolution	0.0375 Mbit/s	27 min
3G	UMTS	Universal Mobile Telecommunications System	0.0375 Mbit/s	27 min
	HSPA	High Speed Packet Access	0.9 Mbit/s	11 min
	HSDPA	High Speed Downlink Packet Access	14 Mbit/s	1.1 min
	HSUPA	High Speed Uplink Packet Access	14 Mbit/s	1.1 min
	HSPA+	Evolved High Speed Packet Access	42 Mbit/s	13.8 s
4G	LTE (Cat4)	Long-Term Evolution	150 Mbit/s	0.001 s
5G	NR	New Radio	400 Mbit/s (sub-6Ghz) 1.8 Gbits/s (mmWave)	1.5 s

reliability: the OS or new application versions should not interfere with the baseband processor functions. The baseband processor is the component that manages the handover between network access technologies. When a tower is located too far from a smartphone, the signal may drop and end the user's connectivity. The baseband processor then automatically connects to a closer antenna to provide network access. If an antenna is not available in the same RAT, the baseband processor, selects a lower technology RAT, as older RAT often provide a larger range of coverage. For instance, if a 4G signal is unavailable because the user is on the move, and no other 4G link can be established, the smartphone will attempt to connect to a 3G antenna.

The type of network access technology is important because it is directly linked to the quality of connectivity. As Table 23.7 shows, an EDGE-based connection theoretically has a maximum download speed of 0.0375 Mbit/s, which is not enough to watch a YouTube video [43]. WiFi technologies have also undergone several stages of evolution with different maximum download speeds, i.e., WiFi type (e.g., a, b, g, n, ac). However, this information was not available during dataset collection, so information regarding WiFi speed is not included in this analysis. Connection to a WiFi network is not automatic, as the user must enter credentials to connect to the network. These credentials ensure the encryption of the communication between the smartphone and the wireless AP. The credentials exchange is transparent on a cellular connection, in which case the baseband processor communicates with the Security Information Management (SIM) card and the operator network to authenticate the smartphone on the network.

Signal Strength

We examined the overall network connectivity signal strength over the collection periods. Signal strength is always presented on the smartphone screen and is located in the upper right-hand corner on Android and iOS. Icons represent the signal strength sensed by the onboard antennas for both WiFi and cellular networks in a human-readable format. The mQoL-Log application was able to collect that information in decibel-milliwatts (dBm). To utilize this information, we determined how the Android OS presented this data to the end-user, and mapped the dBm to the number of bars (0 to 4) shown on-screen. The signal strength represents the power present in the received radio signal. For smartphones, this directly impacts the QoE of smartphone services such as video streaming and online games. The minimum signal strength needed to achieve a "good" experience when watching an online video on the move depends on the network access technology and the video format (e.g. HD or 4 K). The signal strength plays a significant role during handovers. The baseband processors collect the signal strength continuously and choose whether to switch between RATs (i.e. conduct a vertical handover for the same RAT or a horizontal handover if RATs change) or between cell antennas. Connectivity-wise, the smartphone user sees the signal strength as an overall health indicator of the network connection. Thus, a user may decide not to start a video call if the smartphone reports low signal strength, instead preferring to communicate via audio call only.

Data Consumption

Data consumption is a significant feature in the context of connectivity. The RAT limits the amount of data that can be transmitted, measured in seconds. Accordingly, the amount of data consumed is bound to the current network access technology. The data consumption depends on the type of services utilized by the smartphone user. Video applications consume a large amount of data, e.g. by downloading video, while a video calling application simultaneously generates and consumes a large amount of data by uploading and downloading a video. The overall data consumption also provides insight into the network traffic state. If the network encounters a large amount of traffic, this impacts the bandwidth available for use in a live video or other application by the user, and the user connectivity is affected. The amount of data downloaded and uploaded also indicates the user profile type, as some users consume less data than others. This may be due to the nature of their subscription to their operator (financial), the services used on their smartphones (behavior), and the quality of the link connecting them to the Internet (structural) over time [44].

User's Physical Mobility

User mobility is essential, as discussed previously. Indeed, connectivity and mobility are crucial to understanding participants' smartphone usage and connectivity changes. We explored participants' mobility per hour and the number of times each participant connected to the same tower or the same AP for multiple periods (days to weeks). A large number of unique identifiers (ID) is an indication of high mobility for a participant. One cell tower covers a few kilometers of land in a densely populated area (e.g. a 4G tower has a 16 km range), while a WiFi AP covers only a few meters (e.g. a WiFi ac reaches 12–35 m inside and up to 300 m outside).

Mobile Network Connectivity: Results

We analyzed results for the four features that quantify the connectivity level of an individual relying on the connection and usage of their smartphone network: the network access technology (section "Network Access Technology"), its signal strength (section "Signal Strength"), overall data consumption (section "Data Consumption"), and mobility (section "Users' Physical Mobility"). For each feature, we present the overall statistics (post-filtering) of the 110 participants organized by their respective collection period.

Network Access Technology

Table 23.8 presents the overall average of RAT distribution per measurement period. Figures 23.1, 23.2, and 23.3 illustrate the distribution of network access technology for P1, P2, and P3 participants, respectively. The figures clearly show the adoption of LTE (4G). In the P1 distribution, we observe a high presence of HSPA, while the P2 distribution suggests that some participants (particularly P2S98 and P2S64) were not connected (NOCO) for the majority of the study. Overall, we see lower access to the Internet in P2 than in P1 and P3. The most recent data demonstrate the rise of LTE and WiFi over the RAT. Furthermore, during P3 the participants had the most stable connection to the Internet (low NOCO), as presented in Table 23.8.

Figure 23.4 presents the overall average distribution over the three periods. We observe that LTE is more present than WiFi in P3.

The data imply that overall, on average for all periods, any connection to the Internet is present $93 \pm 0.8\%$ of the time (averaging $104,540 \pm 64.36$ min across all periods). This information is computed from the RAT distribution. Table 23.9 presents the distribution of the connectivity and the average minutes of connection for each period and reveals that P2 connectivity is lower than that of P1 and P3.

Table 23.8 Overall average RAT distribution (%) per data collection period

RAT/period [%]	P1	P2	P3
WiFi	52.1 ± 3.7	51.0 ± 3.2	47.7 ± 7.1
LTE	29.8 ± 4.1	28.7 ± 2.9	49.8 ± 0.5
HSPA+	3.1 ± 0.8	3.3 ± 1.2	0.8 ± 0.5
HSUPA	0.1 ± 0.1	0.0 ± 0	0.0 ± 0
HSDPA	2.2 ± 1.2	0.0 ± 0	0.0 ± 0
HSPA	3.9 ± 1.1	0.4 ± 0.1	0.1 ± 0.1
UMTS	3.1 ± 1	0.7 ± 0.3	0.3 ± 0.2
EDGE	2.0 ± 0.8	0.6 ± 0.2	0.2 ± 0.1
GPRS	0.0 ± 0	0.0 ± 0	0.0 ± 0
GSM	0.0 ± 0	0.0 ± 0	0.0 ± 0
UNKNOWN	2.6 ± 0.9	0.5 ± 0.3	0.3 ± 2
NOCO	1.0 ± 0.2	14.7 ± 2.4	0.7 ± 3
Download speed on cell network in Mbit/s Avg ± std.err	6.9 ± 3.3	6.4 ± 3	9.4 ± 4.6

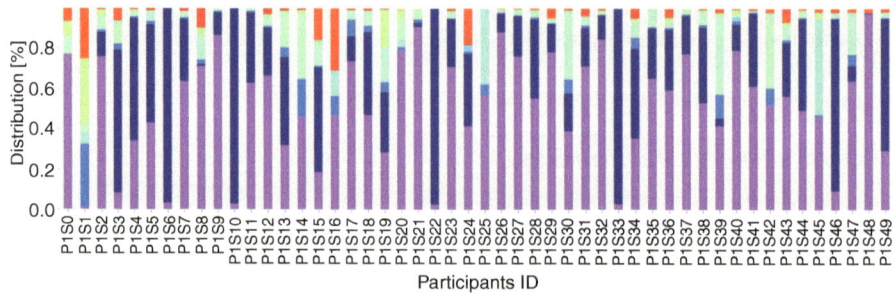

Fig. 23.1 RAT distribution of participants in P1 (N = 50)

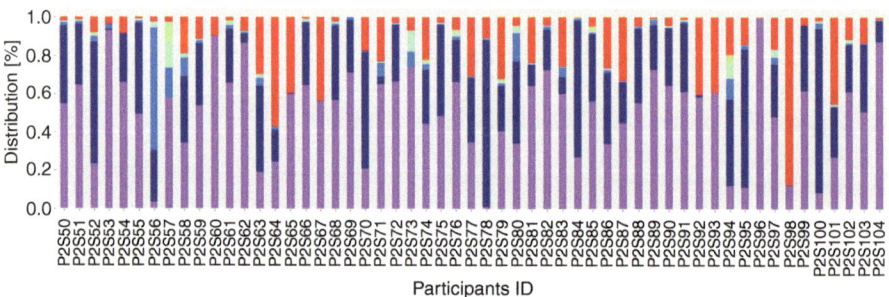

Fig. 23.2 RAT distribution of participants in P2 (N = 55)

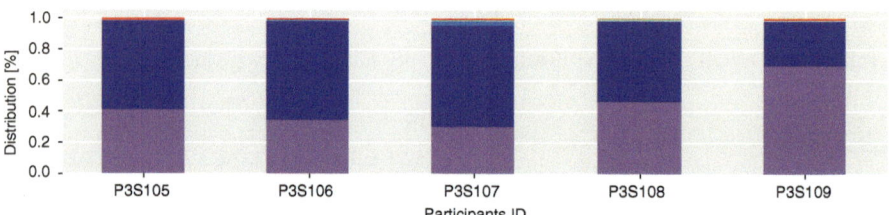

Fig. 23.3 RAT distribution of participants in P3 (N = 5)

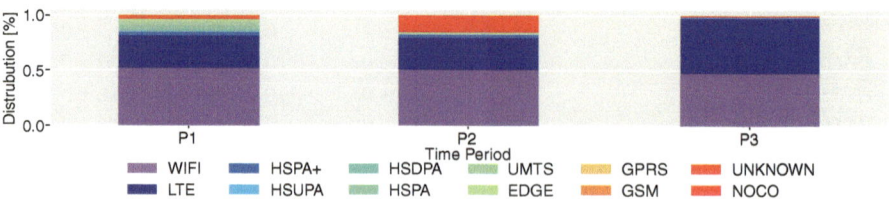

Fig. 23.4 Overall average RAT distribution over P1, P2, and P3

Table 23.9 Percentage of connectivity to internet distribution per data collection period

Connectivity (%) per period	P1	P2	P3
Mean	0.96	0.85	0.99
Std	0.06	0.18	0.00
Min	0.69	0.12	0.99
25%	0.97	0.77	0.99
50%	0.99	0.92	0.99
75%	0.99	0.97	0.99
Mean in Minutes/total time/in period ± std.err	233,162.51 ± 21,371	35,775.89 ± 2672	44,682.18 ± 2448

Signal Strength

The temporality of signal strength for each group is presented in Fig. 23.5. Signal strength increased with time for each group. P1 and P2 feature homogenous signal strength, in contrast to P3, which exhibits a higher signal strength at weekends and during mornings.

Figure 23.6 presents the overall signal strength distribution per period. The resampling process explains the high prevalence of the 0 bar.

Figure 23.7 presents the correlation between the signal quality and the connection type over all three periods. We note a high degree of correlation between WiFi

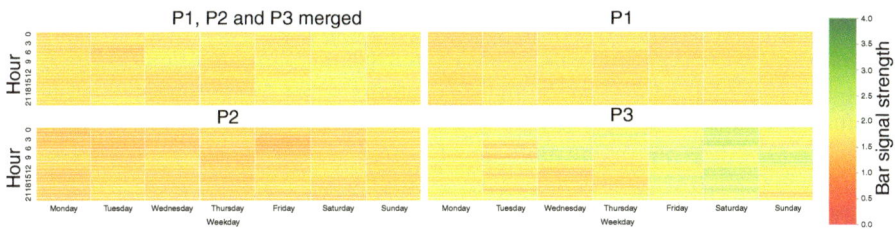

Fig. 23.5 Mean signal strength per data collection period

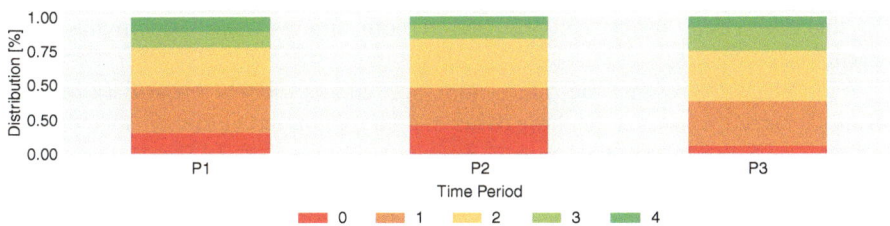

Fig. 23.6 Overall signal strength distribution per data collection period

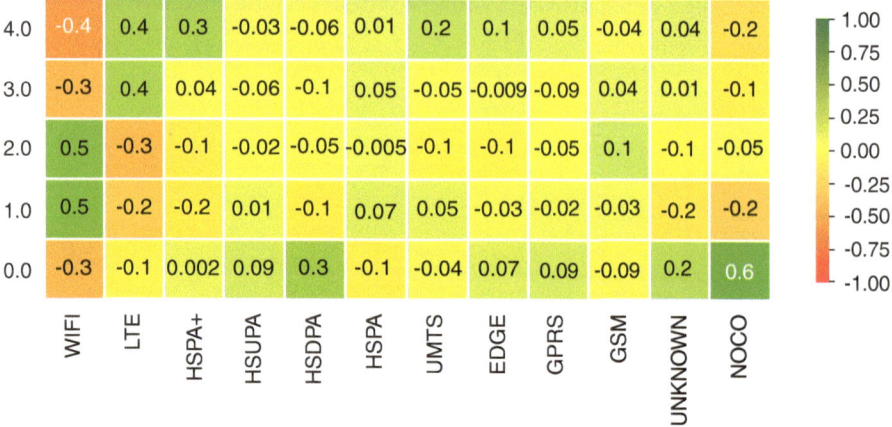

Fig. 23.7 Pearson Correlation Between Signal Strength and Network Access Technology Type

and signal strengths of 1 and 2 bars, while LTE network technology and signal strengths of 3 and 4 bars display a moderate correlation.

Data Consumption

During the analysis, we observed high data consumption by particular participants, as depicted in Fig. 23.8 with the cumulative distribution function (CDF) for monthly data usage and in Fig. 23.9 with the daily data usage for each participant (each data point on one of the lines corresponds to a participant). In both figures, each data point represents the average monthly data consumed by one study participant in terms of (1) rx (received, downlink) and (2) tx (transmitted, uplink), overall and for cell-based networking. The majority of the participants display similar data-consuming behavior, regarding both data receiving and transmitting. In both temporal modalities, the amount of data transmitted from the smartphone to the cellular network is lower than the amount of data received. The monthly and daily data usage follows the same pattern (Fig. 23.8), while we observed faster consumption in the daily data usage (Fig. 23.9), in both figures each sign represents a participant.

Figure 23.10 presents the min − max-normalized weekly mean data received from all participants over the three periods. A larger amount of downloaded data can be observed during the weekends compared to the rest of the week. Participants consumed more data during mornings and evenings, and downloaded more data on weekends. We observed clusters of spikes during afternoons and evenings. The P3 participants received fewer data during the weekend. As expected, a low volume of data was received by smartphones during the night.

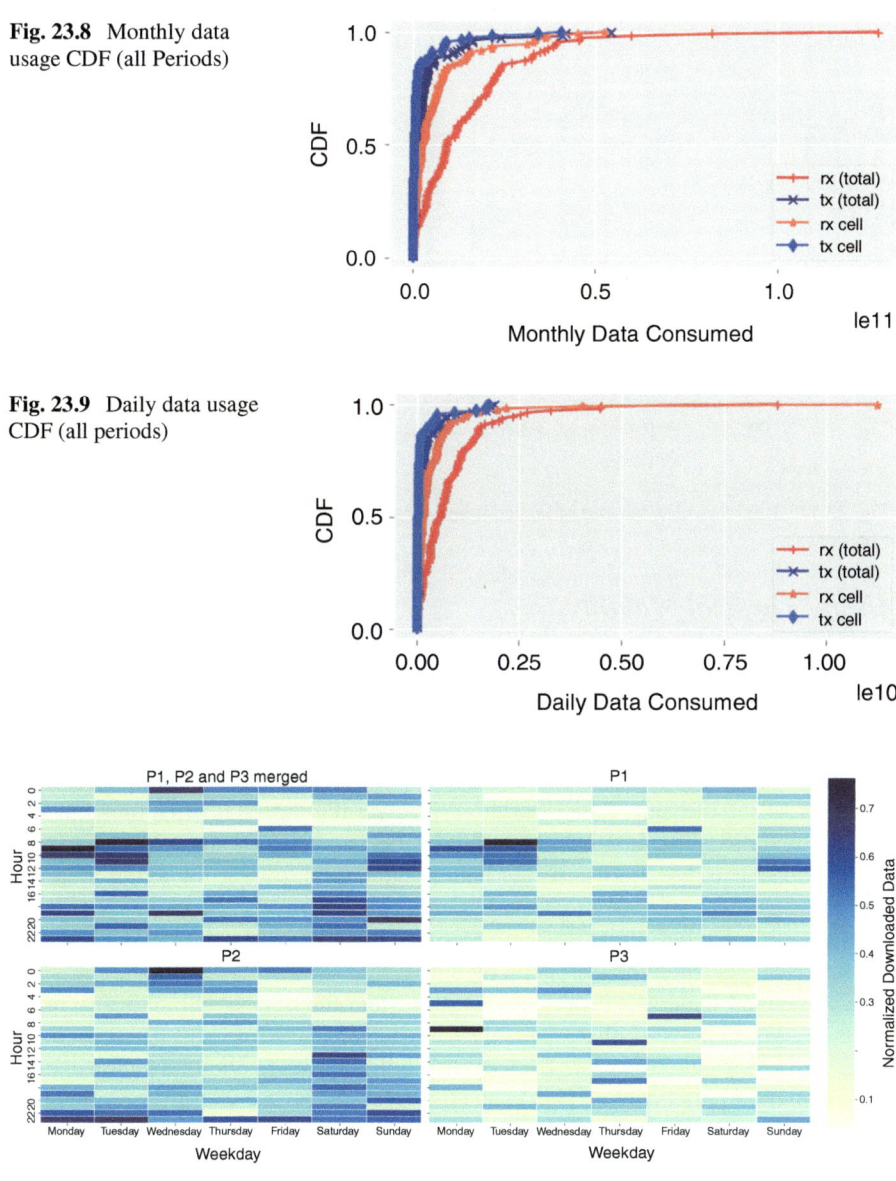

Fig. 23.8 Monthly data usage CDF (all Periods)

Fig. 23.9 Daily data usage CDF (all periods)

Fig. 23.10 Normalized weekly mean amount of data received per data collection per period

We found that participants in P2 consumed more data than the other cohorts, as shown in Fig. 23.11. P3 data consumption is less sparse, likely due to the number of participants in this cohort. In all three periods, we observed some outliers that consumed more data than other participants.

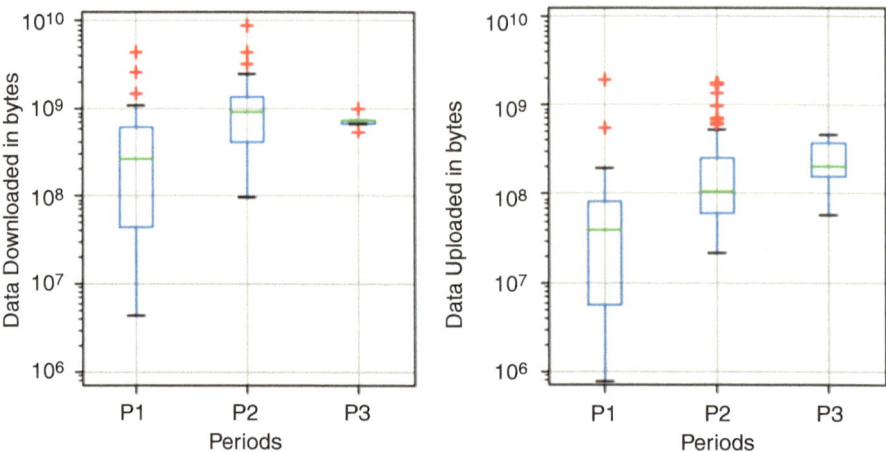

Fig. 23.11 Total downloaded and uploaded bytes per period normalized by number of data collection days

Users' Physical Mobility

We focused the analysis on the number of individual cells and AP IDs. The full dataset contains 59,602 unique cell IDs and AP IDs combined. It is important to note that the same Wi-Fi network can be accessed via different APs, in which case the ID is different, but the network is the same. This enables roaming between the different APs in the same domains. This type of configuration is often found in large networks, for example in companies, universities, and large houses. In such cases, a Wi-Fi repeater is installed to obtain better signal quality over the entire area. The repeater has the same Wi-Fi network name as the main AP, but it has a different ID. Like smartphones, these devices reconnect to another AP when they lose a connection, such as when the user is on the move.

The vertical handover process is seamless, and the device automatically reconnects to a Wi-Fi network that shares the same name as the previous network. In this case, the device already knows the security configuration to obtain a secure connection, namely a previously established authentication. Figure 23.12 shows the mean cumulative cell tower and Wi-Fi ID changes per hour and per day of the week, normalized from 0 to 1. In Fig. 23.12, we observe a lower number of unique IDs on Sundays for all periods. Other patterns are present; for instance, on Friday evenings participants were highly mobile, and the reverse is found during the night. P3 demonstrates lower mobility on Saturday evenings than P1 and P2. One possible explanation is the data collection time; P3 was recorded after the end of the first partial-lockdown in Switzerland during the COVID-19 pandemic. At this time, participants would have been less inclined to participate in external social gatherings on two consecutive nights.

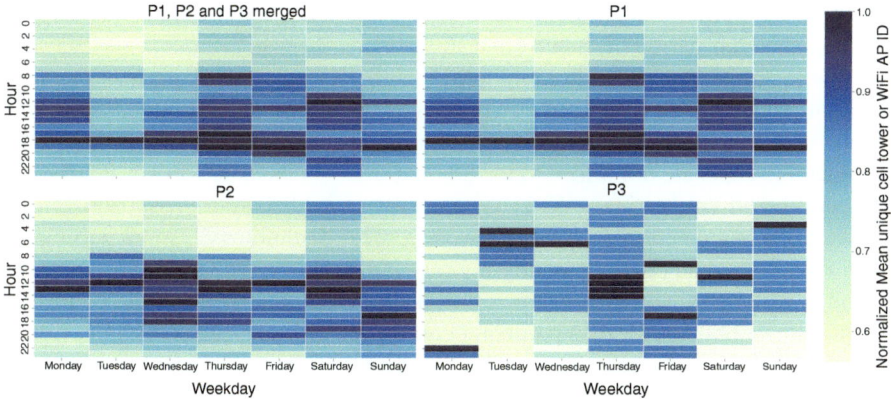

Fig. 23.12   Mean cumulative cell Tower and Wi-Fi AP ID Changes per Data Collection per Period

Discussion

The results confirm our hypothesis that network connectivity and consequently the mobile Internet is widely available in today's developed world. The results indicate that the participants' smartphones were connected to the Internet for 93% of their day (±0.8%) on average. Their devices were always either connected or searching for new network access via Wi-Fi APs and cellular towers. The quality of the connection was high overall, and we found a strong correlation between LTE and high signal strength. Furthermore, as data quantity is directly connected to the services used on the smartphone and the available network bandwidth, we observed multiple data consumption patterns that could be used to profile the users. Taken together, these findings provide important insights into the four features that impact users' connectivity and may influence an individual's decision-making and consequently their QoL. In this section, we discuss the results and their limitations before recommending other data sources for modeling environmental QoL via connected services.

Discussion of Overall Results

Over the data collection periods (2015–2020), the adoption of 4G (LTE) network access technology was close to complete in the Geneva area. The low presence of network access technologies other than LTE and Wi-Fi in P3 can be attributed to the continuous efforts of the mobile operators in updating network infrastructure (i.e., new antenna deployment), an update in performance of the smartphones' baseband processor (i.e., which leads to a faster handover), and the low mobility of the participants. A participant would have a higher number of connections if they were more mobile. Furthermore, contrary to the data in P1 and P2, data from P3 was acquired during a shorter period of time from a smaller sample size.

We found a strong prevalence of Wi-Fi usage during all three periods. As Wi-Fi is commonly used at home and at work, we made the assumption that Wi-Fi usage occurs mostly indoors, where participants are located. Furthermore, while Wi-Fi connection costs are not linked to the amount of data consumed, this is not the case for cell-based connections. Wi-Fi is generally provided by a broadband Internet connection. As noted in section "Network Access Technology", some participants use Wi-Fi less than others, possibly for cost and quality reasons. The cost of broadband is high in Switzerland, and it is cheaper to obtain an unlimited 4G connection than to have both a (Wi-Fi) broadband connection at home and a 4G subscription. The broadband connection quality also plays a role; if an area has a low population density, broadband operators will not invest in high-throughput infrastructures. As a result, using a smartphone's 4G connection to provide home Internet may become convenient and financially attractive.

Our results introduce an additional reflection with respect to the cellular and Wi-Fi connectivity, and especially the handover between the two. Autonomous handover between cell-based networks and Wi-Fi has not always been possible in smartphones. However, smartphone OSs have evolved and can now automatically switch between a Wi-Fi and a cell-based connection. In fact, the switching between the two types of connection is common in everyday smartphone usage. For example, after entering a home, a smartphone will automatically connect to the home's Wi-Fi router. A smartphone will switch to cell-based connectivity if the Wi-Fi connection is of low quality. This so-called *smart assist* feature is totally transparent to the user and does not require interaction with the smartphone. However, this process only operates in one direction (i.e., Wi-Fi to 4G); the smartphone does not subsequently test the Wi-Fi network to attempt to revert to the Internet's connection source. Given our results, we would recommend that the smart assist feature operate both ways.

Additionally, connection and disconnection events to a cell tower are important data for a network operator. Notably, operators ultimately use the data collected from their core network, particularly the number of smartphones connected to an antenna, to generate connectivity maps and understand how to improve their services. Indeed, the services can also be improved by the network operator by enabling better connection during times of increased demand for connectivity in a given area. Conversely, the network operator could also enable a low-power mode of their system during low data consumption hours, decreasing their standby energy consumption. As shown in section "Data Consumption", we found the same pattern as Walelgne et al. [45]: low data consumption during the night and higher consumption during the evening and early morning. These patterns reflect how people use their devices and connectivity. The observed higher throughput could originate from video consumption (leisure) or video conferencing with loved ones. This information can be used by a network operator to rent more bandwidth from its network provider, thus enabling a high-quality video conferencing experience at a specific time.

Study Limitations

This study has several limitations. The populations of participants in the three periods are not identical, so we are unable to comment on the evolution of the individual populations. Additionally, the two main OSs for smartphones are Android (Google) and iOS (Apple), but data was only collected from Android users in this study. As a result, information and insights about the population of iOS users is missing from this study. Additionally, the number of participants and the duration of P3 is lower than that of P1 and P2, so the generalization of the results between the cohorts is limited. We encountered another limitation during data logging due to the shortcomings of the OS de-prioritizing our logger. With the P1 dataset, we found that it was impossible to collect at least 50% of minute-based samples, even by sampling with the minute-based pulling method. Future studies shall be designed such that they address these limitations.

Quantified Self Movement

The Quantified Self (QS) movement brings together individuals from different backgrounds who wish to learn about themselves. The QS practitioners use tools, principles, and methods that are mostly enabled by smartphone applications and services and allow them to measure, analyze, and share their data [46]. The QS tools can include medical test results or well-being-oriented connected objects (e.g., fitness trackers, smartwatches), mobile applications, and web applications. Those sources of information can also contribute to collecting a high-dimensional connectivity dataset and data to quantify individuals' behaviors, health, and QoL [47]. Currently, the QS practitioners are mostly interested in their habits and health. They collect large amounts of data that they usually openly share on online platforms (e.g., quantifiedself.com, openhumans.org) for others to experiment with. In doing so, they expect to learn about themselves through their own analyses and through others'.

In the QS movement, smartphones are the main collection devices (c.f., Chap. 1). For instance, diary and reminder applications are often deployed to collect one's day-to-day emotions, mental states, social interactions, and other aspects of human life currently unquantifiable via autonomous, connected devices. Those devices and applications depend on mobile network connectivity to function. However, the collection of network connectivity by the followers of the QS movement is often neglected. At the same time, smartphone data loggers that collect smartphone user habits, such as mQoL-Log, are uncommon in QS. In the future, it will be important to explore the potential use of additional data sources in the QS movement such as smartphone connectivity levels and their influence on the daily life of the individuals. Knowledge and anecdotal data (i.e., a study with one participant) obtained by QS's followers could prompt further investigation by researchers into the links between mobile network connectivity, physical health, social iterations, individuals' overall decision-making, and QoL, for example.

QoL Technologies

The evolution of QoL technologies (QoLT), defined as technologies that enable assessment and assurance of life quality for individuals [48], is deeply linked to the development of individuals' connectivity. The possibility to improve one's life with QoLT would likely involve a component of communication to the Internet (e.g., a cloud) or edge network devices. The large amounts of data produced by personal wearable health sensors and smartphones, for example, would be processed for immediate use (in emergency situations) or for later use. The degree of QoS offered by QoLT would depend on the supported mobile network connectivity level. Therefore, the four features described in this chapter are important, as they are fundamental aspects that define the individual's connectivity. Without connectivity, there may be no QoLT. To elaborate on this point, we discuss QoL aspects defined according to the WHO and connectivity-dependent services.

The domain of physical health includes many important facets, including daily living activities. Some of these activities rely on indoor connectivity being provided in the home or at work, school, or other frequent locations. The activities may require a low-latency, high-throughput network connection to operate. For instance, smartphone applications can provide medication schedule reminders and notifications to a patient and their family. Energy, fatigue, and mobility are factors that can be quantified by smartphone and wearable data, and adequate real-time personalized care services can be provided to the individual, depending on their needs. The applications can also help a population with substance dependence issues; for example, some applications can put at-risk individuals in real-time communication with medical professionals. In the case of assisted living, connectivity can enable support services like remote healthcare and, in the future, robot care. Overall, many day-to-day physical health services provided to an individual in a given context can be supported by connectivity.

The psychological health domain of QoL may be influenced positively or negatively by smartphone applications. Connectivity to services through smartphone applications can contribute to improving this domain. Services that influence this field include entertainment (e.g., watching a video), which can influence feelings, and information services (e.g., reading news on social media), which can influence thinking processes.

In the social relationships domain, services enabled through an Internet connection can range from simple text-based messaging to smartphone-based video conferencing. More generally, opportunities for social relationships provided by connected services are extensive and are evolving. These services may range from interactive entertainment services (e.g., joint use of online games, which influences feelings) and social networks (i.e., communication and exchange of information, thus influencing the quality of the relationship). The sex industry understood this potential market and created multiple devices for remote sexual interaction through the Internet, providing intimacy for long-distance couples [49]. In the social relationships domain, the specific challenge is to ensure sufficient mobile network connectivity for both receivers to enable content exchange with sufficient user experience during the interaction.

The features of the environmental domain of QoL may be difficult to quantify, as it contains the most facets of any QoL domain and is influenced by contextual variables that may not yet be understood. For example, opportunities for leisure or education may involve the possession of interactive entertainment (and a joint use of devices such as smart TVs, for example, thus influencing feelings), the use of social networks, or the use of online education services (e.g., services for peer communication and the exchange of information). Because of the high interactivity of these examples of online leisure and education opportunities rely on connectivity to succeed. Overall, there are many services in the users' environments that may enable a better QoL and rely on mobile connectivity to be provided. However, the challenge is that a unified, well-understood model of these services and their connectivity does not exist yet.

In conclusion, QoLT may impact all the QoL domains in beneficial and detrimental manners, all depending on the implementation of the services it supports.

Conclusion

This chapter quantifies the mobile network connectivity of individuals in the Geneva area during three data collection periods between 2015 and 2020. Our results demonstrate that connectivity is ubiquitous in the day-to-day life of the participants of this study, as they could access their online services anytime and from any location. We also observed a time-based evolution of the participants' Internet connection throughout the day. Overall, our results suggest that connectivity in the same geographic location improves over time. The explored features (signal access technology, signal strength, data consumption, and users' physical mobility) offer some insights into the participants' connectivity.

We observed a high correlation between signal strength and several network access technologies. According to our data, on average, a better signal strength is available on LTE than on Wi-Fi. Furthermore, knowing the individual data consumption patterns (amount of data received and transmitted) permits the profiling of study participants. Users who consume more data during a short period (spike) may use services that other users may not access because of their low connectivity. Additionally, we considered the amount of data received and transmitted by the smartphones at different times of the day. Although we found peaks during the evenings for P1 and P2, P3 did not exhibit this pattern. It is possible that a large amount of data consumption was taking place on other devices for a better experience during the evening (e.g., watching YouTube videos on a television screen instead of a smartphone screen). In addition, we also observed less mobility on Sundays across all periods. We compared the overall mobility in all periods and noticed a lower mobility in P3, which was possibly due to the COVID-19 situation in Switzerland at the time of the study.

We discuss the results in the context of emerging QoLT, which, embedded in personal devices including wearables and smartphones, enable the collection of health information, which may support an individual's progress towards better health behaviors and, consequently, a better QoL. Overall, an increase in the use of QoLT may contribute to a better life. The range of services provided by QoLT rely on network connectivity, so future research work is needed to ensure that this connectivity matches the requirements of the technologies anywhere the user may be at any time.

Acknowledgments This work was supported by AAL Guardian (AAL-2019-6-120-CP, 2019-2022), SNSF MIQModel (157003, 2015-2019), H2020 WellCo (769765, 2018-2020), AGE-INT (2021-2024), QoL@hip2neck, and QoL@GVA.

References

1. Abramson N. THE ALOHA SYSTEM: another alternative for computer communications. Fall Joint Comput Conf. 1977;37:281–5. https://doi.org/10.1145/1478462.1478502.
2. Katz and Callord. The economic contribution of broadband digitization and ICT regulation. 2018. https://www.itu.int/en/ITU-D/Regulatory-Market/Documents/FINAL_1d_18-00513_ Broadband-and-Digital-Transformation-E.pdf
3. Chan M. Mobile-mediated multimodal communications, relationship quality and subjective well-being: an analysis of smartphone use from a life course perspective. Comput Hum Behav. 2018;87 https://doi.org/10.1016/j.chb.2018.05.027.
4. Kim M-Y. The effects of smartphone use on life satisfaction, depression, social activity and social support of older adults. J Korea Acad-Indus Cooperation Soc. 2018;19(11):264–77. https://doi.org/10.5762/KAIS.2018.19.11.264.
5. Benefits of mobile communication in rural and developing areas. Ericsson.com, Sep. 20, 2010. https://www.ericsson.com/en/press-releases/2010/9/benefits-of-mobile-communication-in-rural-and-developing-areas. Accessed 20 Oct 2020.
6. Connected Society, The State of Mobile Internet Connectivity 2019. GSM Association. https://www.gsma.com/mobilefordevelopment/wp-content/uploads/2019/07/GSMA-State-of-Mobile-Internet-Connectivity-Report-2019.pdf. Accessed 20 Oct 2020.
7. "Over 90% Chinese Tourists Would Use Mobile Payment Overseas Given the Option." https://www.nielsen.com/cn/en/insights/report/2018/nielsen-over-90-percent-chinese-tourists-would-use-mobile-payment-overseas-given-the-option. Accessed 25 Nov 2020.
8. Connectivity and QoL: how digital consumer habits and ubiquitous technology are driving smart city development in Asia Pacific. MIT Technology Review Insights. 2017.
9. Select a category and tags for your app or game – play console help. https://support.google.com/googleplay/android-developer/answer/113475?hl=en. Accessed 23 Oct 2020.
10. Chen J, Lieffers J, Bauman A, Hanning R, Allman-Farinelli M. The use of smartphone health apps and other mobile health (mHealth) technologies in dietetic practice: a three country study. J Hum Nutr Diet. 2017;30 https://doi.org/10.1111/jhn.12446.

11. M. L. Bracken and B. M. Waite, "Self-efficacy and nutrition-related goal achievement of MyFitnessPal users," Health Educ Behav, vol. 47, no. 5, pp. 677–681, Oct. 2020, doi: https://doi.org/10.1177/1090198120936261.
12. White A, Thomas DSK, Ezeanochie N, Bull S. Health worker mHealth utilization: a systematic review. Comput Inform Nurs. 2016;34(5):206–13. https://doi.org/10.1097/CIN.0000000000000231.
13. Ventola CL. Mobile devices and apps for health care professionals: uses and benefits. P T. 2014;39(5):356–64.
14. Wattanapisit A, Teo CH, Wattanapisit S, et al. Can mobile health apps replace GPs? A scoping review of comparisons between mobile apps and GP tasks. BMC Med Inform Decis Mak. 2020;20:5. https://doi.org/10.1186/s12911-019-1016-4.
15. Wattanapisit A, Tuangratananon T, Wattanapisit S. Usability and utility of eHealth for physical activity counselling in primary health care: a scoping review. BMC Fam Pract. 2020;21(1):229. https://doi.org/10.1186/s12875-020-01304-9.
16. Kwon M, et al. Development and validation of a smartphone addiction scale (SAS). PLoS One. 2013;8(2):e56936. https://doi.org/10.1371/journal.pone.0056936.
17. Kwon M, Kim D-J, Cho H, Yang S. The smartphone addiction scale: development and validation of a short version for adolescents. PLoS One. 2013;8(12):e83558. https://doi.org/10.1371/journal.pone.0083558.
18. Haug S, Castro RP, Kwon M, Filler A, Kowatsch T, Schaub MP. Smartphone use and smartphone addiction among young people in Switzerland. J Behav Addict. 2015;4(4):299–307. https://doi.org/10.1556/2006.4.2015.037.
19. Bian M, Leung L. Linking loneliness, shyness, smartphone addiction symptoms, and patterns of smartphone use to social capital. Soc Sci Comput Rev. 2015;33(1):61–79. https://doi.org/10.1177/0894439314528779.
20. Samaha M, Hawi NS. Relationships among smartphone addiction, stress, academic performance, and satisfaction with life. Comput Hum Behav. 2016;57:321–5. https://doi.org/10.1016/j.chb.2015.12.045.
21. Carbonell X, Chamarro A, Oberst U, Rodrigo B, Prades M. Problematic use of the internet and smartphones in university students: 2006–2017. Int J Environ Res Public Health. 2018;15(3) https://doi.org/10.3390/ijerph15030475.
22. Akodu AK, Akinbo SR, Young QO. Correlation among smartphone addiction, craniovertebral angle, scapular dyskinesis, and selected anthropometric variables in physiotherapy undergraduates. J Taibah Univ Med Sci. 2018;13(6):528–34. https://doi.org/10.1016/j.jtumed.2018.09.001.
23. Dey AK, Wac K, Ferreira D, Tassini K, Hong J-H, Ramos J. Getting closer: an empirical investigation of the proximity of user to their smart phones. In: Proceedings of the 13th international conference on Ubiquitous computing. New York; 2011. p. 163–72. https://doi.org/10.1145/2030112.2030135.
24. Cencetti G, et al. Digital proximity tracing in the COVID-19 pandemic on empirical contact networks. medRxiv, p. 2020.05.29.20115915, 2020. https://doi.org/10.1101/2020.05.29.20115915.
25. Ferreira D, Kostakos V, Dey AK. AWARE: mobile context instrumentation framework. Front ICT. 2015;2 https://doi.org/10.3389/fict.2015.00006.
26. Lathia N, Rachuri K, Mascolo C, Roussos G. Open source smartphone libraries for computational social science. In: Proceedings of the 2013 ACM conference on Pervasive and ubiquitous computing adjunct publication. New York; 2013. p. 911–20. https://doi.org/10.1145/2494091.2497345.
27. Katevas K, Haddadi H, Tokarchuk L. SensingKit: evaluating the sensor power consumption in iOS devices. In: 2016 12th International conference on intelligent environments (IE); 2016. p. 222–5. https://doi.org/10.1109/IE.2016.50.

28. Bardram JE. The CARP mobile sensing framework – a cross-platform, reactive, programming framework and runtime environment for digital phenotyping. arXiv:2006.11904 [cs], 2020. Accessed 16 Dec 2020. [Online]. Available: http://arxiv.org/abs/2006.11904

29. Opoku Asare K, Visuri A, Ferreira DST. Towards early detection of depression through smartphone sensing. In: Adjunct proceedings of the 2019 ACM international joint conference on pervasive and ubiquitous computing and proceedings of the 2019 ACM international symposium on wearable computers. New York; 2019. p. 1158–61. https://doi.org/10.1145/3341162.3347075.

30. Ciman M, Wac K. Individuals' stress assessment using human-smartphone interaction analysis. IEEE Trans Affect Comput. 2018;9(1):51–65. https://doi.org/10.1109/TAFFC.2016.2592504.

31. De Ridder B, Van Rompaey B, Kampen JK, Haine S, Dilles T. Smartphone apps using photoplethysmography for heart rate monitoring: meta-analysis. JMIR Cardio. 2018;2(1):e4. https://doi.org/10.2196/cardio.8802.

32. Ciman M, Wac K. Smartphones as sleep duration sensors: validation of the iSenseSleep algorithm. JMIR Mhealth Uhealth. 2019;7(5) https://doi.org/10.2196/11930.

33. M. Gustarini, M. P. Scipioni, M. Fanourakis, and K. Wac, "Differences in smartphone usage: validating, evaluating, and predicting mobile user intimacy," Pervas Mob Comput, vol. 33, pp. 50–72, Dec. 2016, doi: https://doi.org/10.1016/j.pmcj.2016.06.003.

34. Hausmann J, Wac K. Activity level estimator on a commercial mobile phone: a feasibility study. Proc Int Workshop Front Act Recognition Using Pervasive Sens. 2011:42–7.

35. Wac K, Pinar G, Gustarini M, Marchanoff J. More mobile & not so well-connected yet: users' mobility inference model and 6 month field study. In: 2015 7th International Congress on Ultra Modern Telecommunications and Control Systems and Workshops (ICUMT), 2015, p. 91–9. https://doi.org/10.1109/ICUMT.2015.7382411

36. Gustarini M, Ickin S, Wac K. Evaluation of challenges in human subject studies 'in-the-wild' using subjects' personal smartphones. In: Proceedings of the 2013 ACM conference on pervasive and ubiquitous computing adjunct publication – UbiComp '13 Adjunct; 2013. p. 1447–56. https://doi.org/10.1145/2494091.2496041.

37. K. Wac, A. van Halteren, and D. Konstantas, QoS-predictions service: infrastructural support for proactive QoS- and context-aware mobile services (position paper). In: On the move to meaningful internet systems 2006: OTM 2006 workshops, Berlin, 2006, pp. 1924–1933, doi: https://doi.org/10.1007/11915072_100.

38. De Masi A, Wac K. Towards accurate models for predicting smartphone applications' QoE with data from a living lab study. Qual User Exp. 2020;5(1):10. https://doi.org/10.1007/s41233-020-00039-w.

39. Berrocal A, Concepcion W, Dominicis SD, Wac K. Complementing human behavior assessment by leveraging personal ubiquitous devices and social links: an evaluation of the peer-Ceived momentary assessment method. JMIR Mhealth Uhealth. 2020;8(8):e15947. https://doi.org/10.2196/15947.

40. Berrocal A, Manea V, Masi AD, Wac K. mQoL lab: step-by-step creation of a flexible platform to conduct studies using interactive, mobile, wearable and ubiquitous devices. Procedia Comput Sci. 2020;175:221–9. https://doi.org/10.1016/j.procs.2020.07.033.

41. De Masi A, Ciman M, Gustarini M, Wac K. mQoL smart lab: quality of life living lab for interdisciplinary experiments. In: Proceedings of the 2016 ACM international joint conference on pervasive and ubiquitous computing: adjunct. New York; 2016. p. 635–40. https://doi.org/10.1145/2968219.2971593.

42. De Masi A, Wac K. You're using this app for what? A mQoL living lab study. In: Proceedings of the 2018 ACM International Joint Conference and 2018 International Symposium on Pervasive and Ubiquitous Computing and Wearable Computers. New York; 2018. p. 612–7. https://doi.org/10.1145/3267305.3267544.

43. Wamser F, Seufert M, Casas P, Irmer R, Tran-Gia P, Schatz R. YoMoApp: a tool for analyzing QoE of YouTube HTTP adaptive streaming in mobile networks. In: 2015 European

Conference on Networks and Communications, EuCNC 2015; 2015. p. 239–43. https://doi.org/10.1109/EuCNC.2015.7194076.

44. Fiedler M, Hossfeld T, Tran-Gia P. A generic quantitative relationship between quality of experience and quality of service. Blekinge Tekniska hogskola. 2010;24:36–41. https://doi.org/10.1109/MNET.2010.5430142.

45. Walelgne EA, Asrese AS, Manner J, Bajpai V, Ott J. Understanding data usage patterns of geographically diverse mobile users. IEEE Trans Netw Service Manage. https://doi.org/10.1109/TNSM.2020.3037503.

46. Bode M, Kristensen DB. The digital doppelgänger within: a study on self-tracking and the quantified self movement. Assembl Consump. 2016:119–35.

47. Wac K. From quantified self to quality of life. In: Rivas H, Wac K, editors. Digital health: scaling healthcare to the world. Cham: Springer International; 2018. p. 83–108.

48. Wac K. Quality of life technologies. In: Gellman M, editor. Encyclopedia of behavioral medicine. New York: Springer; 2019. p. 1–2. https://doi.org/10.1007/978-1-4614-6439-6_102013-1.

49. N. Liberati, "Teledildonics and new ways of 'being in touch': a phenomenological analysis of the use of haptic devices for intimate relations," Sci Eng Ethics, vol. 23, no. 3, pp. 801–823, Jun. 2017, doi: https://doi.org/10.1007/s11948-016-9827-5.

Chapter 24
TRAWEL: A Transportation and Wellbeing Conceptual Framework for Broadening the Understanding of Quality of Life

Bhuvanachithra Chidambaram

Introduction

Transportation is an essential functionality. Individuals perform routine activities to meet their specific daily demands, and transportation provides the opportunities to perform these, as well as to attend to a variety of commercial and social activities, amongst others. For instance, working individuals commute to/from work, children get to school or take part in extracurricular activities, families participate in leisure activities, and the elderly population engage in social or voluntary gatherings. Taking part in these activities not only enhances social interaction but also contributes to the physical and emotional wellbeing of individuals [1, 2]. Studies show that the perceived quality of public transportation [3, 4], physical mobility [5], participation in recreational travel [6], residential relocation [7] and active travel (walkability: [8] and cycling: [9]) affect quality of life.

Conversely, negative impacts of transportation causes individual quality of life (QoL) to deteriorate. For instance, an increase in vehicle kilometers traveled (VKT) by private motorized vehicles negatively affects the environment through vehicle emissions and traffic congestion [10]. These traffic problems cause increased costs, travel delays, health impacts, air pollution and a reduction in individual *subjective wellbeing* (SWB) [11]. A similar study reported that long commuting to work induces stress and impacts psychological wellbeing [12].

In transportation-based QoL, most studies have mainly focused on SWB measures to improve the transport mobility of specific groups. For instance, many studies address SWB for transport disadvantaged groups with a special emphasis on the aging of society [13–18]. To date, the holistic approach to understanding the association between various aspects of transportation and the SWB measures has not

B. Chidambaram (✉)
Transport Research Group, Department of Spatial Planning, TU Dortmund University, Dortmund, Germany
e-mail: bhuvanachithra.chidambaram@tu-dortmund.de

© The Author(s) 2022
K. Wac, S. Wulfovich (eds.), *Quantifying Quality of Life*, Health Informatics,
https://doi.org/10.1007/978-3-030-94212-0_24

received enough attention. Against this background, the central question addressed in this paper is *how do various transportation aspects affect QoL and how can these aspects be quantified in the individual's daily life? What are the relevant transport policies and practices that can enhance wellbeing at the individual and societal level?*

The remainder of this chapter is organized as follows. Section "Transportation Related QoL" presents the findings of the scoping and expert consensus and identifies the extant literature that explores transport-related indicators in the context of QoL. Following this, the conceptual framework (TRAWEL) is then presented in section "TRAWEL: A Conceptual Framework for Transportation Based QoL", explaining the link between five broader dimensions of transport (transport infrastructure and services, built environment, transport externalities, travel-based time use and travel satisfaction) and measures of SWB (community, social, economic, physical and psychological). The indicators, measures and methods used for the continuous, longitudinal and quantitative assessment of these five aspects in a wellbeing context are discussed in section "Quantifying Transportation: Method and Measures". Finally, the chapter discusses its contribution in section "Discussion" and concludes by surveying policy implications towards improving wellbeing and future research directions in section "Conclusive Remarks".

Transportation Related QoL

The World Health Organization [19] defined individual QoL as *"individuals' perception of their position in life in the context of the culture and value systems in which they live and in relation to their goals, expectations, standards, and concerns"*. Individual QoL in the context of transport mobility is defined by the WHO as *"the person's view of his/her ability to get from one place to another, to move around the home, move around the workplace, or to and from transportation services"*. Additionally, Myers [20] state that *"a community's QoL is constructed of the shared characteristics of the residents' experience in places (for example air and water quality, traffic or recreational opportunities) and the subjective evaluations residents make of those conditions"*.

The central focus of this chapter is how transportation alters the QoL of an individual. To this end, a number of studies were critically examined to understand the association between transportation aspects and wellbeing. Methods leveraged in this paper include the scoping review method and expert opinion consensus. Articles and grey literature published from January 2005 to June 2020 were identified, 55 of which met the inclusion criteria and are presented in this paper. The inclusion criteria stated that a study had to focus on one of the seven performance indicators of transportation in the wellbeing context: mobility, affordability, accessibility, connectivity, externality, travel needs and attitudes. In the following paragraphs, we define and discuss these separately.

Transport mobility is defined as the ability to move from one place to another using different types of movement such as walking, cycling, transit and driving [21, 22]. Studies on transport mobility broadly focus on measuring the impact of the transport sector on the QoL of the elderly population, especially in developed economies [2, 23]. Several studies highlight that out-of-home mobility positively influences the QoL of aging individuals. Older people (>65 years) with a driving license are likely to enjoy better mobility benefits in terms of out-of-home activities compared to non-drivers [23]. Social exclusion occurs when the transport mobility needs of elderly people are not adequately addressed. For instance, Jalenques et al. [24] found that older people who do not drive tend to be adversely affected with low QoL due to their dependence on others. Besides older people, Jones et al. [25] explored the impacts of micro mobility i.e., e-bikes, on individual health and wellbeing. Based on qualitative assessments, the purchase and usage of e-bikes positively correlate with personal wellbeing. Moreover, pedestrian-oriented development plans such as walkability and bike ability services also positively impact individual QoL [26, 27].

Transport affordability as defined by Litman [28] is the ability by all households to make journeys and access services while devoting less than 20% of household budgets to transport. In this manner, studies have analyzed the economic aspect of transportation using QoL indicators to understand individual wellbeing. For instance, De Groot and Steg [29] examined the impact of transportation pricing policy on car use using 22 QoL indicators (e.g., comfort, money, income and environmental quality). The study found that cost of car use negatively affects certain wellbeing indicators such as environmental quality, money, change and work, while it positively impacts indicators such as comfort and safety. Using the Delphi method, Zelinková [30] assessed the effect of road pricing on QoL measures such as safety and stress and found that successful transportation implementation depends on political will and public acceptance. Schwarzlose et al. [31] found that users are willing to pay for a flexible transport route service to improve QoL for the rural elderly.

Transport accessibility is defined as "the extent to which land-use and transport systems enable individuals to reach activities or destinations by means of a (combination of) transport mode(s)" [32, p. 128]. The accessibility of various facilities around the neighbourhoods is measured by the urban density, diversity of neighbourhoods, land use mix, green space, open spaces, walkability and connectivity. In particular, Ritsema et al. [33] suggested that living in high-density environments enables greater transport accessibility than in low-density suburbs. Other studies have conceptualized QoL dimensions to enable transport accessibility in transport planning. Lee and Sener [34] developed the Transportation QoL (TQoL) framework encompassing four dimensions of physical, mental, social and economic wellbeing and three components of the transportation system, namely accessibility, the built environment and vehicle traffic. In another study, Nakamura et al. [35] categorized transportation access, amenity and safety as QoL dimensions for understanding residential choices focusing on Transit-Oriented Development (TOD) for different socioeconomic groups.

Transport connectivity focuses on the links of the entire system that represent the interaction between multimodal transport modes and the ease of access to them [36]. Haslauer et al. [37] explored how the proximity and connectivity of public transport services within a community enhance QoL. Other studies have suggested that public infrastructure, connectivity, public space and green space positively influence the wellbeing of residents and thus enhance the neighborhood QoL [38, 39]. In addition, a series of studies on the built environment has confirmed that land-use heterogeneity characterized by mixed land use, walkability and park density enhances social connectivity and thus improves community wellbeing, while population density and proximity to mass-transit stations negatively impact health-related QoL [40–44].

Traffic safety and air quality are major concerns of road transport planning that directly and indirectly affect health [45]. A series of studies on traffic safety demonstrates that overall QoL decreases in cases of injury and illness [46–49]. In addition, Putra and Juwita [50] analyzed how a public transport service with low safety standards increases stress for passengers. Regarding personal safety, studies have found that perceived safety on urban streets affects QoL. In this context, Deegan and Baker [51] assessed residents' perceptions on road infrastructure and find that dark and narrow streets, illegally parked cars and low social cohesion affect personal safety. Also, Pánek et al. [52] reported that residents fear crime in underpasses, on train stations and on dark and narrow roads. Besides safety, traffic-related pollution directly affects SWB [53]. It has been found that exposure to traffic-related air pollution impacts physical wellbeing, while subjection to noise annoyance affects psychological wellbeing [54].

Travel needs are derived from different types of activities and vary not only from person to person but also with different life stages [55]. Many travel time-use studies have found that travel serving social interaction or recreation enhances both physical and mental wellbeing [18, 56–61]. Job and job-related travel causes stress and negatively impacts wellbeing due to the congestion, crowding and unpredictability of peak time travel [62–65].

Travel related attitudes refer to the psychological evaluation of transport systems and daily travel elements (e.g., travel modes, trips and travel time), conveying some degree of favor or disfavor [66]. To understand the impact of changing travel behaviour on wellbeing, studies evaluate the association between travel attitudes towards travel choices (mode use, commute time, route destinations) and emotional wellbeing. For instance, studies have quantitatively investigated how attitudes towards commute trips by different modes or travel time influence hedonic and eudemonic wellbeing [67–71]. These studies suggest that shorter travel times and active travel modes positively affect hedonic wellbeing in the short-term and induce an eudemonic state in the long run.

Summarizing the research results so far, the following research gaps are observed in the quantification of QoL dimensions in transportation. First, in addressing the impact of transportation infrastructure and services, it may be noted that relatively little research has examined the impact of mobility and affordability on QoL for economically disadvantaged social groups. For instance, differences in household income levels affect individual and household wellbeing, also affecting mobility and the affordability of transport services, as pointed out by Hernández [72]. Existing studies fail to address the impact of reducing transport costs in terms of vehicle purchase, fuel prices, transit fares, etc. on wellbeing. An interesting research direction with reference to QoL could be to investigate the effect on community wellbeing of supply-based factors like physical functions including road networks, cross-sectional design and vehicular traffic. Second, within literature on the built environment, studies on accessibility and connectivity suggest that high-density regions and proximity to public transport improve an individual's QoL. However, the existing research lacks empirical evidence on how residential differences, spatial-temporal constraints or a lack of adequate connectivity influences individuals' QoL.

Third, existing studies on transport externalities mostly focus on the effect of air pollution or traffic safety on health-related burdens, ignoring the socio-economic aspects. For instance, the impact of the socio-economic burden of transport related emissions and safety issues on QoL is still unknown.

Fourth, regarding individual travel time use, travel needs are highly interconnected with work, family and social life. However, existing studies on QoL mostly focus on the health impacts of leisure and commuting trips. Daily travel activities also involve maintenance trips such as shopping and childcare (organized activities, help with homework, overseeing playtime and escorting), which involve car use and related stress. However, little is yet known about such negative impacts on health over time. Finally, travel attitudes and their association with QoL have mostly been analyzed in the context of travel satisfaction e.g., mode usage, travel time and their effect on hedonic wellbeing. However, travel motivations also involve emotion, attitudes and preferences that are bound by other long-term choices (destination, residential occupation), as pointed out by De Vos et al. [68]. In this context, less is known in the literature about the impact of travel satisfaction on eudemonic wellbeing in the long run.

Based on the key findings synthesized from the literature review, it can be seen that existing studies in transportation research have not completely addressed the QoL dimensions in transportation. Some conceptual models have been developed [16, 73–76]; however, these frameworks are firmly rooted in SWB understandings and focus on the mobility of older people, out-of-home or leisure-based activities, the built environment and travel choices. Overall, a complete picture of how transportation hinders or facilitates wellbeing is still missing from the literature, as transportation factors such as transport infrastructure and services, the built environment, transport externalities, travel time use and travel satisfaction and wellbeing dimensions such as social, community, economic, physical and psychological are interdependent and influence each other at various levels. This additionally calls for analysis of how each aspect of transportation impacts wellbeing not just at the individual level but also at the community level at large.

TRAWEL: A Conceptual Framework for Transportation Based QoL

The study presented here proposes a conceptual framework (Fig. 24.1) that has been derived via an expert consensus process to holistically synthesize the impact of transportation on wellbeing at two levels. First, the elements of a transport system (infrastructure and services, built environment, and externalities) and its impact on wellbeing are discussed at the societal level (sections "Transport Infrastructure and Services (Mobility and Affordability)", "Built Environment (Accessibility and Connectivity)" and "Transport Externalities (Safety and Air Quality)"). Second, individual travel needs and travel satisfaction with regard to travel elements and effects on their wellbeing are explained at the individual level (sections "Travel and Time Use (Travel Needs)" and "Travel Satisfaction (Travel Related Attitudes)"). The societal level is discussed first, as it is a context within which the individual behaviors occur. The conceptual framework TRAWEL will add to the existing literature as it explains the association between five broader aspects of transportation, as derived from the literature review: (1) transport infrastructure and services, (2) the built environment, (3) transport externalities, (4) travel and time use and v) travel satisfaction and wellbeing measures.

Transport Infrastructure and Services (Mobility and Affordability)

Transport infrastructure and services form an integral part of the transport system as they enable people to cater for their travel needs. So far, as described in the state-of-the-art section above, the mobility dimensions of older people and the affordability of various pricing strategies have been discussed in assessments of QoL. Here we discuss how the underlying factors such as the road network and transport services affect individual QoL.

The capacity of road infrastructure and the availability of transport services support individuals' mobility and capacity to take part in a range of socio-economic activities. A recent study found a positive association between road and infrastructure development and community satisfaction [77]. Meanwhile, the development of road infrastructure that cannot cope with increased mobility demands may lead to induced traffic [78, 79]. In such a case, the congestion is not appreciably reduced, but rather causes a strain on other mobility infrastructure services such as the availability of public transportation, pedestrian facilities, Non-Motorized Traffic (NMT), or parking facilities. This is because cities follow car centric policies and most services are available far from residential neighbourhoods. For these reasons, public transport services become unavoidable for individuals with low car access.

A recent paradigm shift in urban transport is the application of Intelligent Transport Systems (ITS) that cover a range of innovative demand-oriented services

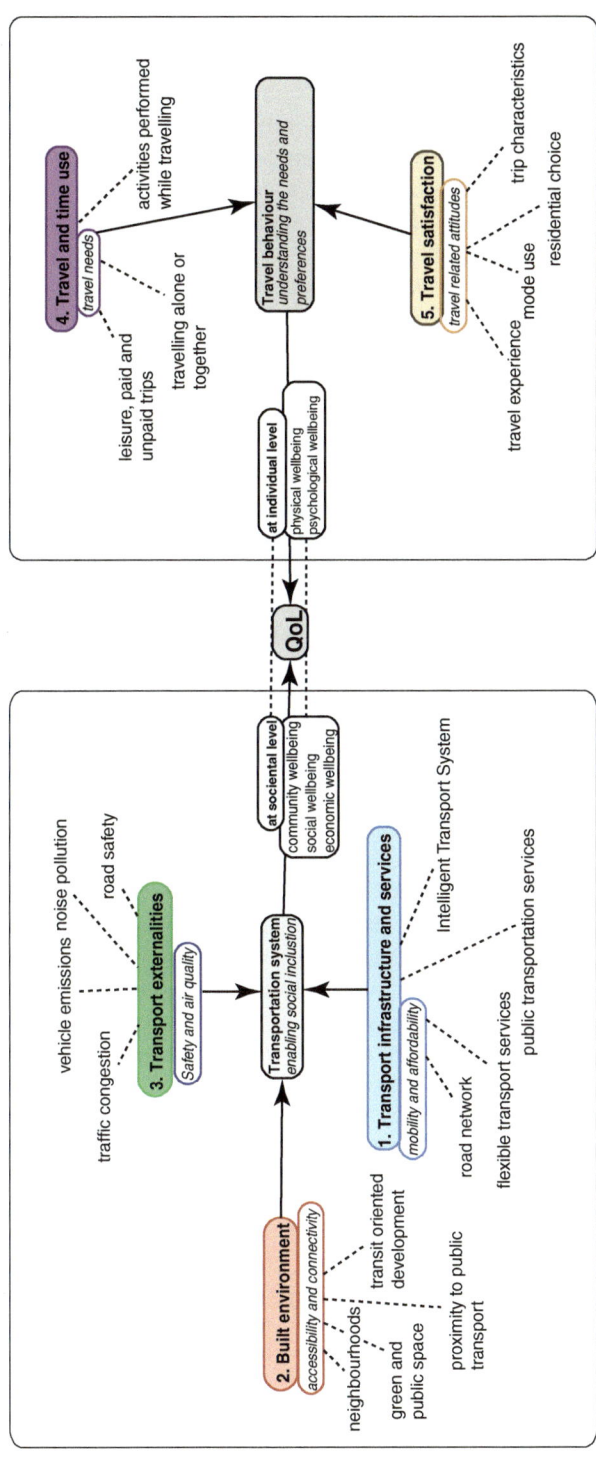

Fig. 24.1 Conceptual framework of TRAWEL explaining the interconnectedness between transportation aspects and QoL
Source: Author's compilation

in transportation. They include Mobility as a Service (MaaS), driver monitoring systems, vehicle technologies, automated driving, travel time prediction and congestion management. Each of these services aims to cut car dependency, curb traffic congestion, enhance autonomy, improve flexibility and safety, provide cost effective services and promote cleaner transport. Jones et al. [25] found positive attitudes towards technology interventions like e-bikes and QoL.

Digital advancement in ITS plays a prominent role in shaping individual QoL, especially for transport disadvantaged groups who are economically and physically constrained. First, transition in existing mobility services comes at the price and disturbs individuals who are economically stretched. Evidence shows that there is a negative association between transport affordability for individuals and SWB [80, 81]. This is because affordability for individuals largely depends on the resources they possess: physical capital (income), human capital (educational level) and social capital (participation in reciprocity networks), as pointed out by Hernández [72]. Second, a study confirms that older people are less adaptable to ITS in their daily lives, contributing to unintended social exclusion [16].

Alternatively, the mobility options could be enhanced by on-demand mobility services, known as Flexible Transport System. Some flexible transport services such as shared taxicabs, shuttle vans, dial-a-ride services and para-transit services are offered as stand-alone services. These services generally cater for a specific group of the population or fill a specific need. Shared or collective flexible transport services include services such as ridesharing, car sharing and car-hailing services and bring together public and private transport services and volunteer transport services to provide cost-efficient connectivity for the entire community [82]. In this respect, the system supports the mobility needs of diverse groups such as women with young children, individuals living in rural communities, older people and disabled people.

Besides a flexible transport service, well-kept pavements (sidewalks), lighting and crossings play a significant role in catering for essential mobility demands such as walkability and bike-ability. Both cycling and pedestrian infrastructure play a crucial role in enhancing the QoL, especially in the community where pedestrian mobility is predominant [83]. A study of older people revealed that problems with standing and a lack of walking infrastructure for at least 400 m decreases QoL [5, 84]. Overall, the successful implementation of new mobility services depends on the pricing policy that constitutes the subsidies, differentiated pricing and different land use mixes. In this way, both mobility and affordability are ensured so that the transportation system enhances community wellbeing.

Built Environment (Accessibility and Connectivity)

The built environment encompasses a range of neighborhood services such as healthcare and education services and amenities such as green space, parks or shopping malls. So far, factors like urban density or proximity to public transport have been predominantly discussed in the literature, ignoring the effect of constraints on QoL such as differences in residential location or a lack of connectivity.

Access to services and amenities is ensured by the connectivity of road infrastructure and proximity of public transportation in neighborhoods. The ability to access essential services like health care facilities, grocery shops or public transit services enhances active travel behaviour (walking and biking), leads to significant health benefits [39, 85, 86]. Besides physical health, the built environment provides opportunities for social participation and a sense of belonging to the place in question. Studies find that proximity to public transport services and access by bicycle or on foot to green spaces enhance the possibility of social interactions among different population groups, which positively influences QoL [87–92].

Built environments with well-connected street network accessibility and connectivity encourage individuals to take part in value-related out-of-home activities such as commerce, employment or recreation. These opportunities attribute a sense of belonging to a community and contribute to social wellbeing. However, if the built environment fails to provide neighbourhood accessibility, then it becomes a burden for the residents to travel and develop social connections. For instance, in residential suburbs on the outskirts of urban areas or in newly developed areas, the road network may lack continuity and provides limited access to services and amenities, which poses a threat to individuals with physical disabilities. Also, the economically constrained population live in the suburbs, while the high-income groups mostly live in the downtown region. Such spatial mismatches lead to unemployment as the skills possessed by the workers do not match the opportunities, as pointed out by Levinson and Krizek [93, p. 132]. It is proven that low-income and unemployed people mostly depend on public transport [94].

An absence of connectivity to public transit also reduces employment opportunities. In such a situation, access to transit use may positively impact the QoL of low-income individuals, while the growth of car-reliant communities along the transit-oriented development (TOD) corridors negatively affects the QoL of low-income groups [35]. Individuals who have access to private cars can travel further in search of opportunities, which affects groups with no or limited car access. For instance, Cao et al. [95] found that living in suburban areas positively affects car dependence compared to living in urban areas. Altogether, a multidisciplinary approach to policy planning is warranted to move towards creating more activity-friendly communities and catering for individuals who are spatially or economically constrained.

Transport Externalities (Safety and Air Quality)

Transport externalities are the negative impacts of transport on the environment (e.g., congestion, air pollution, noise) and on society (e.g., road safety and public safety). So far, as described in section "Transportation Related QoL", city level planning adopts three different strategies to address traffic externalities. First, road safety is ensured through the physical attributes of road space (e.g., width and the cross-sectional attributes of pavements) and physical functions (e.g., zebra crossings, lighting), and by implementing road safety measures (e.g., creating awareness for road users, reducing drink driving and the strict enforcement of speed limits) [96]. Second, air quality is maintained by adopting vehicle emission control measures, alternative fuel usage and enforcing regulations for older vehicles [97]. Third, congestion is addressed by reducing car usage through push-pull demand management measures such as parking, pricing and subsidies [98]. From the QoL perspective, these strategies concern the physical and technical aspects of road infrastructure but ignore the social aspects.

Studies across the global south (Latin America, Sub-Saharan Africa) and global north (Europe, United Kingdom and the United States) have documented how transport-related externalities such as congestion, vehicle emissions, noise pollution and road safety affect the livelihoods of various social groups [97, 99–108]. These studies together emphasize that externalities become a social problem when non-transport users are affected by such problems. Drawing on these studies, the following contributing factors are identified. First, individuals living in the downtown area or along the highways are predominantly exposed to greenhouse gas (GHG) emissions, which directly impacts their health and wellbeing. Second, individuals who reside in the outskirts with poor transportation suffer from congestion-related stress. Third, individuals with a vulnerable mode of transport (walking or cycling) combined with poor quality transport are highly exposed to traffic accidents, potentially leading to permanent illness and injury. Fourth, deficient road infrastructure hampers the livelihood of residential communities as it cuts off the community, hampers social interaction and disturbs social wellbeing. Finally, the physical functioning of road infrastructure pose serious hazards to the social security of women, children and older people.

Overall, it can be seen that transport externalities widely affect social groups and their wellbeing in the long run. To enable a healthy, safe, and communal way of life, it is recommended that policies addressing traffic externalities should consider social aspects such as income level differences, gender dimensions and age-related barriers besides the technical aspects of road infrastructure and vehicle design.

Travel and Time Use (Travel Needs)

Individuals' daily travel needs are driven by their participation in out-of-home activities related to work, education, errands, sports, leisure and recreation. So far, as described in the section "Transportation Related QoL", travel needs linked to leisure and commuting have been discussed. Such studies observe that travelling for recreational needs such as socializing with friends, visiting recreational places, sports and exercise involves physical movement and enhances enjoyment and amusement, which improves the physical and mental wellbeing of individuals [1, 109–112]. Studies on job-related travel suggest that job or business travel causes commuting stress, starting from the pre-trip planning (work, family and travel arrangements), including the travel itself (delay and safety issues) and ending with post-trip workloads (e.g., family commitments); however, such trips may have a positive impact on economic wellbeing such as income, productivity and personal development [113–119].

The following contributing factors have been identified in this context. First, the gender aspect recognizes that women carry a disproportionate share of unpaid work such as errands, shopping and childcare. These responsibilities leave them with no choice but to combine trips to perform work and family obligations. As suggested by many travel time-use studies within households, it is women who undertake shorter commutes, have less choice on the labour market, and have less access to a car in case of one-car households and traditional household patterns [120–123]. Second, research shows that individuals who travel with household members (e.g., joint travel for pickup and drop off, shopping activities, leisure trips) tend to travel more happily than when travelling alone [122, 124]. Third, studies on travel-based multitasking reportedly claim that women do more socializing and accompanying of children, while men do more work-related tasks [125, 126].

Aside from the gender differences and the travel time use perspective, overall, studies on wellbeing mostly focus on the professional sphere and recreational aspects but exclude the domestic sphere. However, such trips could affect daily moods and emotions. For instance, a recent study on travel time use finds that activity during travel determines hedonic wellbeing or short-term happiness [127]. However, less is known in the literature about the impacts of unpaid trips, joint trips and travel-based multitasking on individual wellbeing in the long run. Henceforth QoL studies on travel behavior should jointly consider the various dimensions of travel needs at individual level and intrahousehold dynamics at household level, as pointed out by Sweet and Kanaroglou [128].

Travel Satisfaction (Travel Related Attitudes)

Adopting an individual view on transportation and wellbeing also includes travel satisfaction. Studies have found a positive association between travel satisfaction and SWB (e.g., [5, 68, 75, 129–131]). SWB is commonly approached from two perspectives: travel hedonic or affect and travel eudemonia. So far, as described in section "Transportation Related QoL", the association between travel satisfaction and hedonic wellbeing has been given primary importance. Some of the contributing factors identified in the proposed framework are travel experience, mode use, residential and destination choice.

Travel experience is characterized by hedonic feelings such as emotional state or mood and cognitive evaluations of an individual's context. Some common hedonic feelings include commuting stress in public transport, feelings of pleasure and happiness when riding a bicycle or driving a car. Such hedonic feelings influence travel decisions. For instance, individuals who enjoy cycling may reduce their car commuting. In other cases, individuals who experience feelings of stress during congestion ought to seek alternate routes or modes. Ettema et al. [69] found that mode switching to sustainable modes triggers higher levels of SWB.

In the long term, hedonic wellbeing contributes to personal growth, finding purpose or meaning, and self-actualization, or achieving one's full potential, termed eudemonic wellbeing [68]. Travel eudemonia relates to intrinsic motivations for traveling such as personal development (self-confidence, physical or mental health), competence (e.g., driving, riding), autonomy (freedom), and relatedness (e.g., social connections) [132]. For instance, commuting by car is connected to freedom, comfort and self-identity, while riding a cycle may yield health benefits.

Regarding mode use, numerous studies explore the association between active travel and hedonic-eudemonic wellbeing. For instance, Singleton [133] found that active transport such as walking and biking positively influences SWB [133]. In another study, Woodcock et al. [134] examined the impact of encouraging active transport on individuals' health and found that active transport reduced heart disease among men and reduced depression among women. In addition, a recent study found that active travel certainly enhances eudemonic wellbeing more than public transit [135]. Public transport users have needs and preferences, including reliability, convenience, safety, comfort, accessibility and affordability, that affect their satisfaction with the services provided. Some studies have shown that commuting by public transport reduces QoL compared to driving by car [136].

Regarding residential choice, people who live in their preferred neighborhood based on travel preferences (e.g., car lovers living in suburban types of neighborhoods) are more satisfied than people who do not. Participation in travel activities in larger cities is more weakly associated with life satisfaction than in smaller cities [1]. In addition, travel satisfaction might also affect performance of and satisfaction with activities at the trip destination. For instance, experiencing frequent positive emotions during travel may depend on the type of activity at the destination [74, 137]. The studies addressing SWB have so far focused on traditional mobility options, as pointed out by Ettema et al. [69]. From a policy perspective, travel satisfaction with respect to new mobility options such as technology-based modes (electric, autonomous) and shared mobility (car sharing, ride sharing, para transit) could provide further insights to understanding individual perceptions and preferences.

Quantifying Transportation: Method and Measures

In transport planning, a variety of methods have been used to identify potential indicators at micro and macro-level. Existing studies employ both qualitative and quantitative approaches to evaluate SWB in travel-based QoL. Building on the previous literature, a series of indicators for each aspect of the conceptual framework are listed in Table 24.1. Discussion focuses on the various survey-based and automatic, sensor-based approaches to collecting data related to measures and recommended possible subgroups for the analysis; the aim is to understand group differences in the association between transportation aspects and wellbeing.

Table 24.1 Measures and methods for assessment of Travel Based QoL

Dimensions	Elements	Measures	Wellbeing	Methods: Survey and Sensory-based procedures	Methods: Possible sub-groups
Transport infrastructure and services	Road network Flexible transport services Public transportation services Intelligent Transport System (ITS)	1. Traffic measures (annual average daily traffic) and transport mobility (road mobility and transit mobility) 2. Individual usage of alternate travel options (car ownership, alternate travel mode options, trip characteristics, vehicle kilometres traveled) 3. Public transport service (Level of Service, frequency and connectivity) 4. ITS options (digital enhancement and technology advancement) 5. Transport costs	Community wellbeing Social wellbeing Economic wellbeing	Infrastructural data from municipalities, transportation providers GIS data model Automated Fare Collection Automatic Passenger Counts GPS Logger Travel diary app Face to face interviews, Social platforms	Road type Mode availability Residential type Social dimensions (age, gender, income, mode access)
Built environment	Neighborhoods Proximity to public transport Green or public space Transit oriented development	1. Urban density 2. Diverse neighborhoods 3. Land use mix 4. Accessibility 5. Greenspace 6. Open spaces 7. Safety 8. Walkability 9. Connectivity	Community wellbeing Social wellbeing Economic wellbeing	Geospatial data Land use and land cover data Travel survey data Transit network data	Road network Neighborhood Residential type Social dimensions (age, gender, income, mode access)

(continued)

Table 24.1 (continued)

Dimensions	Elements	Measures	Wellbeing	Methods: Survey and Sensory-based procedures	Methods: Possible sub-groups
Transport externalities	Traffic congestion Air pollution Noise pollution Traffic safety	1. Exposure to traffic related pollution (air and noise pollution) 2. Transport resources (car ownership, public transport availability) 3. Trip characteristics (vehicle kilometers travelled) 4. Noise decibels	Community wellbeing Social wellbeing Economic wellbeing	Household travel survey Traffic count survey Face to face interviews Air quality data from air quality monitoring stations Portable Air Quality Monitoring sensors (PAMs) Open-source National Household Travel Surveys National Health Surveys	Vehicle type Road type Neighborhood Social dimensions (age, gender, income, mode access)
Travel and time use	Leisure/paid/Unpaid trips Traveling alone or together Activities performed while traveling	1. Time spent on leisure/paid work/unpaid work trips 2. Time spent on joint trips 3. Time spent on travel-based multitasking 4. Percentage share of Active commuting: cycling, walking (percentage share of total modes) 5. Percentage share of public transport usage 6. Percentage share of car usage	Physical and psychological wellbeing	Time use surveys National Survey of Parents Face-to-face online survey Travel diary Harmonized European Time use Surveys	Marital status Social status Presence of children Gender Income Age Ethnicity Trip purpose Mode availability

| Travel satisfaction | Travel experience Mode use Trip characteristics Residential choice | 1. Travel-related attitudes (mode choice, preferred time of travel) choice of destination) 2. Traffic congestion and road infrastructure related attributes in connection to driving stress | Psychological wellbeing (hedonic and eudemonic) | Virtual passive data Smartphone applications Online web-based interviews Focus group Semi-structured interviews Narrative interviews Citizen satisfaction data National Household Health Surveys World Health Surveys | Age Gender Income Travel attributes (mode use, travel time) Destination type |

Transport Infrastructure and Services

The data for transport infrastructure and services can be obtained using five measures: (1) the physical attributes of the transport network such as road capacity, road geographical features and traffic composition can be collected using the Geographical Information System (GIS) data model, for example managed by the local city or regional authorities or other stakeholders, as pointed out by Croce et al. [138]; mobility-relevant data such as frequency of trips, travel distance and duration [139]; (2) information about individuals' use of different travel modes such as flexible transport services, ride sharing services, paratransit, car availability, walking and cycling may be obtained from national or regional household travel surveys, as in the case of most recent studies [140, 141]; (3) data on public transportation services such as level of service, mobility and accessibility factors can be collected using GIS tools complemented by vehicle location system and smart card fare data [142, 143]; (4) automated data collection systems in ITS can provide data on automated fare collection, automatic passenger counts including information about passengers and vehicle spatial attributes such as the GPS positioning of vehicles, time of the day, inbound/outbound details of actual arrival and departure times at the beginning of the route, end of the route and at every bus stop along the route [144]; and (5) data attributed to transport costs (taxes, accident costs, medical expenditure) can be obtained from respective transit agencies, insurance agencies and medical care facilities.

Built Environment

The geospatial method can be adopted to examine the association between an individual's participation in an activity (e.g., going to a library, going to a supermarket, etc.) and geographic accessibility [86]. Available land-use and land-cover data from local city or regional authorities or other stakeholders and from open sources like European Union open data portals can be used to understand the spatial distribution of different land uses. Travel survey data of residents living in traditional neighborhoods, suburbia [95] and other regions can be used to compare different neighborhoods' mobility characteristics. Both the geospatial data and travel survey data can also be obtained from secondary sources like government organizations and certified private agencies. The data attributed to the transport network must include walking trails and bicycle networks [145] and the traditional road network data. Extending these data sources, geospatial techniques like those leveraged by D'Orso and Migliore [83] could be used to assess walkability to and from transport services based on practicability, pleasantness and safety. In addition to these data techniques, transit network data could be collected from local transport agencies to analyze the connectivity and proximity of transit services [91].

Transport Externalities

The data for traffic congestion such as traffic volume, mode classification and level of service can be obtained from household travel surveys and traffic count data [146]. In addition, data on safety in daily travel can be obtained through face-to-face interviews and online feedback forms or can be incorporated as a component in household survey data. Air quality data from air quality monitoring stations can help model exposure level to gaseous pollutants and particulate matter. Recent advances in technology interventions have led to portable instruments for measuring vehicular emissions on-board [147, 148] that could also be employed in data collection. As an individual's exposure level also varies with different mobility patterns, it can be measured using the Portable Air Quality Monitoring sensors (PAMs) over a period [149]. These data can form a basis for modelling the health-related QoL of the public and vulnerable groups. Additionally, a noise emission model can capture the noise generated by the impact of road traffic on vehicle kinematics [100].

To measure the social impact of transport externalities on social groups, data on individual attributes (gender, age, employment status and personal income, etc.) and household attributes (household size, number of children and dwelling type, etc.) are required. These data are mostly derived from structured self-reports (e.g., household travel surveys) where the respondents (mainly the household head) self-report individual, household and travel attributes. Health data can be collected through national health surveys [150].

Travel and Time Use

Travel and time use related data can be obtained from the cross-sectional time use surveys. A typical time-use survey comprises 24 h activity patterns of each respondent for three random, representative days. The data derived from these surveys (1) includes in-home and out-of-home activities classified in four broader groups: each activity (paid work, unpaid work, leisure) and associated travel; (2) includes primary and secondary activities that allow analysis of travel-based multitasking; and (3) provides details about whether the individual performs the activity alone or with other household members, which could be used for joint household travel analysis. Besides time use surveys, the National Survey of Parents, the Family study including surveys and the Experience Sampling Method can provide information about primary and secondary activities [151]. Time spent on activities could additionally be obtained from self-reported longitudinal or cross-sectional epidemiological investigations conducted either online or using face-to-face surveys. Also, travel diary applications can be utilized to obtain the mobility patterns adopted for different in-home and out-of-home activities. In addition, health behavior associated with active travel can be obtained using the Harmonised European Time use Survey [152].

Travel Satisfaction

The travel-related attitudes on various scales such as satisfaction, liking, happiness, enjoyment and subjective valuation of time over travel modes, destination and activity can be qualitatively derived from virtual passive data (including location data) and from smartphone applications (mobile phone-based surveys and locations captured via WiFi, GPS or GSM data). For instance, Susilo and Liotopoulos [153] used various data collection formats, including an online web-based, real time questionnaire using an Android application, and a focus group interviewing method to collect data related to travel satisfaction. In another study, Kwon and Lee [154] adopted a qualitative and quantitative approach to measure the travel satisfaction. Quantitative data is collected using face-to-face interviews, semi-structured interviews and narrative interviews, and qualitative data is derived from longitudinal measurements with 15-day intervals before and after a given trip, leveraging the qualitative dataset collection via an online portal. The data on individual satisfaction about mode usage reflect attitudes towards traffic congestion, road conditions and availability of alternate or public transport services. Such information may be collected by using face-to-face interview surveys and questionnaires [17], leveraging online social platforms [155], analyzing citizen satisfaction data from annual surveys [156], or administering perceived quality surveys [157]. Apart from the readily available data, perceived benefits and burdens can be acquired via administration of validated QoL questionnaires.

Residential choices can be determined by subjective measures, such as satisfaction with neighbourhood attractiveness, comfort, convenience and safety. These data can be obtained from perception and attitudinal surveys administered in face-to-face interviews, via mobile apps or online forms. Factors contributing to eudemonic wellbeing [132] can be modelled across different socioeconomic and sociodemographic groups. For instance, the overall physical health of participants and the daily frequency of feeling rushed could be collected using self-reported questionnaires [158]. Example studies in this area incorporate physical and mental health data using National Household Health Surveys [159] or World Health Surveys [160, 161]. In addition, the Questionnaire for Eudemonic Wellbeing (QEWB) and the Flourishing Scale (FS) [162] can be leveraged.

Discussion

In this chapter we discuss transportation aspects such as transport infrastructure and services, the built environment, transport externalities, travel time use and travel satisfaction, and familiar wellbeing dimensions such as social, community, economic, physical and psychological, which are interdependent and influence each other at various levels. Overall, the transportation aspects affect the wellbeing of diverse social groups in the following ways.

First, poor transport infrastructure and limited mobility services affect mobility and increase transportation costs. This widely affects the community wellbeing of social groups, especially impacting groups with low car access, the older population and poor income groups. Second, built environments that lack local and regional accessibility, transport connectivity, green or public spaces and are subject to disconnected services disrupt social integration and isolate the community. As the social support of the community is threatened, the social wellbeing of vulnerable groups is affected. This further influences social conditions and potentially leads to social problems such as unemployment and crime. Third, transport-related air and noise pollution negatively impacts the health of residents who are constrained by income, mode and residential location, while lack of traffic safety leads to road accidents and affects the overall wellbeing (economic, social, mental and physical health) of residents. Fourth, feelings of being rushed during paid work trips, additional trips to cater for unpaid work demands such as errands, escort and shopping, and the lack of (pure) leisure related trips jointly affect the mental and physical wellbeing of individuals, in particular working women with traditional gender role orientations. Finally, low travel satisfaction due to commuting stress, traffic congestion or delays, a low quality of public transport services, and a lack of interest in cycling or walking affect the physical health and psychological wellbeing of individuals.

The functions of the transportation aspects (transport infrastructure and services, built environment, transport externalities, travel time use and travel satisfaction) overlap one another, while wellbeing, on the other hand, is contextual and multifaceted and varies according to social groups. For instance, if transportation services are not available to all segments of the population equitably, then the problem of social injustice emerges in a society where the vulnerable groups who contribute less to congestion and pollution are those that are most exposed to these problems.

At the **societal level**, transport policy planning requires the consideration of horizontal equity and vertical equity in transportation, as suggested by many studies [163–166]. Horizontal equity enables the equal distribution of transport services among groups with the same transport needs, while vertical equity accounts for social differences between groups with different transport needs. For instance, on the horizontal level, the availability and proximity of public transport services improve the transport mobility of all residents in the community, while on the vertical level, reasonable ticket fares in public transport services enhance affordability for low-income groups, seat availability improves accessibility for individuals with physical constraints (wheelchair users, those with impaired sight or older people) and safety enhances women's mobility.

At the **individual level**, travel demand measures should consider the intrinsic motivations of individuals to travel. For instance, the first-and-last mile (F&LM) problem and the lack of proximity to services (employment, health and education) may push individuals to increase car trips and long-distance travel. In such cases, transport pricing policies are crucial as they significantly impact travel choices and help in regulating vehicle traffic and managing infrastructure and natural resources (e.g., vehicle taxation in France, Germany and Sweden) [167]. Alternatively, land use policies like transit-oriented development can enhance life satisfaction by enabling ease of access to essential services and promoting safe out-of-home activities. Likewise, strategies for integrating multi-modal trips provide better connectivity and improve individuals' wellbeing. In addition, policy experimentation to enhance subjective measures (comfort, convenience, safety, reduced stress levels) can encourage usage of public transport. For example, the tele-bus on-demand transport service in Australia and Germany [168, 169] aims to provide a bus service on demand in less populated areas; in this way both individual wellbeing and community wellbeing can be improved.

The study has **limitations**. First, both wellbeing and transportation are multifaceted and complex, as well as unfolding and evolving over time, and this discussion addresses only part of this complex relationship, captured momentarily. Second, the internal and the external validity of the wellbeing measures extracted from the conceptual framework with respect to travel context remain unknown. Third, although the study conceptually untangles the well-being and travel attributes, the question of a suitable methodological approach and deployment of statistical methods remains. Further, adopting QoL as the primary goal in assessing transport systems and understanding travel behaviors requires more research on the framework, measures and methods, and clear demonstration with case studies and empirical evidence. Finally, the study does not address the effect of recent technological changes in transportation that aim to improve individual QoL; the study focuses on the assessment aspects of transportation's contribution to individual QoL [170]. Some of the additional technological innovations not considered here include the Quantified Self (QS) technologies including wearables to self-track individual behavioral patterns that may influence their behavioral choices and lifestyles [171, 172]; smart city innovations e.g., automated/electric vehicles, or advanced traffic management systems and urban mobility apps that aim to improve transport mobility [173]. Future research could extend the current study to examine whether these technological innovations in transportation improve wellbeing.

Conclusive Remarks

This chapter puts forward and discusses the hypothesis that transport systems affect wellbeing on the societal level, while travel behavior influences wellbeing on the individual level. The extensive literature method identifies the relevant transportation aspects that impact wellbeing at the individual and societal levels. Based on the extant literature, the conceptual framework TRAWEL is put forward to understand the association between transportation aspects and wellbeing. The determinants of the transport system such as the mobility, affordability, accessibility, connectivity, safety and air quality influence community or social wellbeing, while travel needs and travel attitudes influence the psychological and physical wellbeing of individuals. Such interconnectedness between transportation aspects and wellbeing has not been fully explored elsewhere in transport research to date.

Following the conceptual framework TRAWEL, the review of existing measures and survey procedures suggests the scope for quantifying the impact of transportation aspects on QoL. Finally, the discussion on policy relevance at the societal level suggests an agenda of addressing horizontal and vertical equity in transport system planning, to enable community, social and economic wellbeing. Additionally, it suggests the need for further understanding of the intrinsic motivations of travel needs and travel satisfaction with transport at the individual level, which may be leveraged to enhance individuals' physical and mental wellbeing. To advance understanding of the impact of transportation on individual wellbeing, future research should broaden the intersectionality between transportation aspects and wellbeing dimensions.

Acknowledgments Thanks to Ashish Verma at Indian Institute of Science for his feedback and Hemanthini Alli rani for her assistance in the draft version of the paper. Thanks also to anonymous peer reviewers and the book editors whose comments significantly improved the entire paper.

References

1. Morris EA. Should we all just stay home? Travel, out-of-home activities, and life satisfaction. Transport Res Part A: Policy Pract. 2015;78:519–36. https://doi.org/10.1016/j.tra.2015.06.009.
2. Rantakokko M, Portegijs E, Viljanen A, Iwarsson S, Kauppinen M, Rantanen T. Changes in life-space mobility and quality of life among community-dwelling older people: a 2-year follow-up study. Qual Life Res. 2016;25(5):1189–97. https://doi.org/10.1007/s11136-015-1137-x.
3. Fresher-Samways K, Roush S, Choi K, et al. Perceived quality of life of adults with developmental and other significant disabilities. Disability Rehab. 2003;25:1097–105. https://doi.org/10.1080/0963828031000148638.
4. Kim J, Schmöcker J-D, Nakamura T, et al. Integrated impacts of public transport travel and travel satisfaction on quality of life of older people. Transport Res Part A: Policy Pract. 2020;138:15–27. https://doi.org/10.1016/j.tra.2020.04.019.
5. Delbosc A. The role of well-being in transport policy. Transport Policy. 2012;23:25–33. https://doi.org/10.1016/j.tranpol.2012.06.005.
6. Andereck KL, Nyaupane GP. Exploring the nature of tourism and quality of life perceptions among residents. J Travel Res. 2011;50:248–60. https://doi.org/10.1177/0047287510362918.
7. Gerber P, Ma T-Y, Klein O, et al. Cross-border residential mobility, quality of life and modal shift: a Luxembourg case study. Transport Res Part A: Policy Pract. 2017;104:238–54. https://doi.org/10.1016/j.tra.2017.06.015.
8. Friedman D, Parikh NS, Giunta N, et al. The influence of neighborhood factors on the quality of life of older adults attending New York City senior centers: results from the Health Indicators Project. Qual Life Res. 2012;21:123–31. https://doi.org/10.1007/s11136-011-9923-6.
9. Crane M, Rissel C, Standen C, Greaves S. Associations between the frequency of cycling and domains of quality of life. Health Promot J Aust. 2015;25:182–5. https://doi.org/10.1071/HE14053.
10. Chidambaram B. A comprehensive integrated framework linking vehicle emissions and traffic simulation complemented with social-institutional analysis. Int J Energy Environ. 2011;5:733–43.
11. Higgins CD, Sweet MN, Kanaroglou PS. All minutes are not equal: travel time and the effects of congestion on commute satisfaction in Canadian cities. Transportation. 2018;45:1249–68. https://doi.org/10.1007/s11116-017-9766-2.
12. Olsson LE, Gärling T, Ettema D, Friman M, Fujii S. Happiness and satisfaction with work commute. Soc Indicators Res. 2013;111:255–63.
13. Friman M, Olsson LE. Daily travel and wellbeing among the elderly. Multidisciplinary Digital Publishing Institute; 2020.
14. Lättman K, Olsson LE, Friman M, Fujii S. Perceived accessibility, satisfaction with daily travel, and life satisfaction among the elderly. Int J Environ Res Public Health. 2019;16:4498.
15. Metz DH. Mobility of older people and their quality of life. Transport Policy. 2000;7:149–52. https://doi.org/10.1016/S0967-070X(00)00004-4.
16. Pangbourne K, Aditjandra PT, Nelson JD. New technology and quality of life for older people: exploring health and transport dimensions in the UK context. IET Intell Transport Syst. 2010;4:318–27. https://doi.org/10.1049/iet-its.2009.0106.

17. Wong RCP, Szeto WY, Yang L, Li YC, Wong SC. Elderly users' level of satisfaction with public transport services in a high-density and transit-oriented city. J Transport Health. 2017;7:209–17. https://doi.org/10.1016/j.jth.2017.10.004.

18. Woo E, Kim H, Uysal M. A measure of quality of life in elderly tourists. Appl Res Qual Life. 2016;11(1):65–82.

19. World Health Organization (WHO). The World Health Organization Quality of Life assessment (WHOQOL): position paper from the World Health Organization. Soc Sci Med. 1995; https://doi.org/10.1016/0277-9536(95)00112-K.

20. Myers D. Community-relevant measurement of quality of life: a focus on local trends. Urban Affairs Rev. 1987;23(1):108–25. https://doi.org/10.1177/004208168702300107.

21. Spinney JEL, Scott DM, Newbold KB. Transport mobility benefits and quality of life: a time-use perspective of elderly Canadians. Transport Policy. 2009;16:1–11. https://doi.org/10.1016/j.tranpol.2009.01.002.

22. Vasconcellos EA. Urban transport environment and equity: the case for developing countries. Routledge; 2014.

23. Spinney JEL, Newbold KB, Scott DM, Vrkljan B, Grenier A. The impact of driving status on out-of-home and social activity engagement among older Canadians. J Transport Geogr. 2020;85(April):102698. https://doi.org/10.1016/j.jtrangeo.2020.102698.

24. Jalenques I, Rondepierre F, Rachez C, et al. Health-related quality of life among community-dwelling people aged 80 years and over: a cross-sectional study in France. Health Qual Life Outcomes. 2020;18:126. https://doi.org/10.1186/s12955-020-01376-2.

25. Jones T, Harms L, Heinen E. Motives, perceptions and experiences of electric bicycle owners and implications for health, wellbeing and mobility. J Transport Geogr. 2016;53:41–9. https://doi.org/10.1016/j.jtrangeo.2016.04.006.

26. Khan MA. Impact of public transit and walkability on quality of life and equity analysis in terms of access to non-work amenities in the United States. PhD Thesis, North Dakota State University; 2020.

27. Talmage CA, Frederick C. Quality of life, multimodality, and the demise of the autocentric metropolis: a multivariate analysis of 148 mid-size US cities. Soc Indicators Res. 2019;141(1):365–90.

28. Litman T. Transportation affordability: evaluation and improvement strategies; 2013. http://www.vtpi.org/affordability.pdf

29. De Groot J, Steg L. Impact of transport pricing on quality of life, acceptability, and intentions to reduce car use: an exploratory study in five European countries. J Transport Geogr. 2006;14(6):463–70. https://doi.org/10.1016/j.jtrangeo.2006.02.011.

30. Zelinková J. The anticipated impacts of road pricing on the quality of life in Prague City Centre. ToTS. 2011;4(1):11–8. https://doi.org/10.2478/V10158-011-0002-z.

31. Israel Schwarzlose AA, Mjelde JW, Dudensing RM, Jin Y, Cherrington LK, Chen J. Willingness to pay for public transportation options for improving the quality of life of the rural elderly. Transport Res Part A: Policy Pract. 2014;61:1–14. https://doi.org/10.1016/j.tra.2013.12.009.

32. Geurs KT, van Wee B. Accessibility evaluation of land-use and transport strategies: review and research directions. J Transport Geogr. 2004;12:127–40. https://doi.org/10.1016/j.jtrangeo.2003.10.005.

33. Ritsema van Eck J, Burghouwt G, Dijst M. Lifestyles, spatial configurations and quality of life in daily travel: an explorative simulation study. J Transport Geogr. 2005;13:123–34. https://doi.org/10.1016/j.jtrangeo.2004.04.013.

34. Lee RJ, Sener IN. Transportation planning and quality of life: where do they intersect? Transport Policy. 2016;48:146–55. https://doi.org/10.1016/j.tranpol.2016.03.004.

35. Nakamura K, Morita H, Vichiensan V, Togawa T, Hayashi Y. Comparative analysis of QOL in station areas between cities at different development stages, Bangkok and Nagoya. Transport Res Procedia. 2017;25:3188–202. https://doi.org/10.1016/j.trpro.2017.05.361.

36. Thompson I. The impact of transport connectivity on housing prices in London. A companion to transport, space and equity; 2019.

37. Haslauer E, Delmelle EC, Keul A, Blaschke T, Prinz T. Comparing subjective and objective quality of life criteria: a case study of green space and public transport in Vienna, Austria. Soc Indic Res. 2015;124(3):911–27. https://doi.org/10.1007/s11205-014-0810-8.
38. Douglas O, Russell P, Scott M. Positive perceptions of green and open space as predictors of neighbourhood quality of life: implications for urban planning across the city region. J Environ Plann Manage. 2019;62(4):626–46. https://doi.org/10.1080/09640568.2018.1439573.
39. Salvo G, Lashewicz BM, Doyle-Baker PK, McCormack GR. Neighbourhood built environment influences on physical activity among adults: a systematized review of qualitative evidence. Int J Environ Res Public Health. 2018;15(5):897. https://doi.org/10.3390/ijerph15050897.
40. Engel L, Chudyk AM, Ashe MC, McKay HA, Whitehurst DGT, Bryan S. Older adults' quality of life–exploring the role of the built environment and social cohesion in community-dwelling seniors on low income. Soc Sci Med. 2016;164:1–11.
41. Garin N, Olaya B, Miret M, Ayuso-Mateos JL, Power M, Bucciarelli P, Haro JM. Built environment and elderly population health: a comprehensive literature review. Clin Pract Epidemiol Mental Health. 2014;10:103.
42. Keul AG, Prinz T. The Salzburg quality of urban life study with GIS support. In: Marans RW, Stimson RJ, editors. Investigating quality of urban life: theory, methods, and empirical research. Dordrecht: Springer; 2011. p. 273–93.
43. Sallis JF, Saelens BE, Frank LD, Conway TL, Slymen DJ, Cain KL, Chapman JE, Kerr J. Neighborhood built environment and income: examining multiple health outcomes. Soc Sci Med. 2009;68(7):1285–93.
44. Sarmiento OL, Schmid TL, Parra DC, Díaz-del-Castillo A, Gómez LF, Pratt M, Jacoby E, Pinzón JD, Duperly J. Quality of life, physical activity, and built environment characteristics among Colombian adults. J Phys Activity Health. 2010;7(s2):S181–95.
45. Fernandes P, Vilaça M, Macedo E, Sampaio C, Bahmankhah B, Bandeira JM, Guarnaccia C, Rafael S, Fernandes AP, Relvas H, Borrego C, Coelho MC. Integrating road traffic externalities through a sustainability indicator. Sci Total Environ. 2019;691:483–98. https://doi.org/10.1016/j.scitotenv.2019.07.124.
46. Connelly LB, Supangan R. The economic costs of road traffic crashes: Australia, states and territories. Accident Anal Prevent. 2006;38(6):1087–93. https://doi.org/10.1016/j.aap.2006.04.015.
47. Kovačević J, Miškulin M, Ličanin MM, Barać J, Biuk D, Palenkić H, Matić S, Kristić M, Biuk E, Miškulin I. Quality of life in road traffic accident survivors. Slovenian J Public Health. 2020;59(4):202–10.
48. Rissanen R, Berg H-Y, Hasselberg M. Quality of life following road traffic injury: a systematic literature review. Accident Anal Prevent. 2017;108:308–20. https://doi.org/10.1016/j.aap.2017.09.013.
49. Shinar D. Traffic safety and human behavior. Emerald Group Publishing; 2017.
50. Putra K, Juwita ES. The effect of public transport services on quality of life in Medan city. In: AMER International Conference on Quality of Life, AicQoL. 2016 Medan Indonesia; 2016. p. 25–27.
51. Deegan M, Baker K. Quality of life survey results. Upside Allentown; 2017.
52. Pánek J, Pászto V, Šimáček P. Spatial and temporal comparison of safety perception in urban spaces. Case study of Olomouc, Opava and Jihlava. In: Proceedings of GIS Ostrava. Springer; 2017. p. 333–46
53. Orru K, Orru H, Maasikmets M, Hendrikson R, Ainsaar M. Well-being and environmental quality: does pollution affect life satisfaction? Qual Life Res. 2016;25(3):699–705.
54. Shepherd D, Dirks K, Welch D, McBride D, Landon J. The covariance between air pollution annoyance and noise annoyance, and its relationship with health-related quality of life. Int J Environ Res Public Health. 2016;13(8):792. https://doi.org/10.3390/ijerph13080792.
55. Sharmeen F, Arentze T, Timmermans H. An analysis of the dynamics of activity and travel needs in response to social network evolution and life-cycle events: a structural equation model. Transport Res Part A: Policy Pract. 2014;59:159–71. https://doi.org/10.1016/j.tra.2013.11.006.

56. Argan MT. Eskişehir, Turkey as a crossroads for leisure, travel and entertainment. Cities. 2016;56:74–84.
57. Liao XY, So S-I, Lam D. Residents' perceptions of the role of leisure satisfaction and quality of life in overall tourism development: case of a fast-growing tourism destination–Macao. Asia Pac J Tour Res. 2016;21(10):1100–13.
58. Uysal M, Sirgy MJ, Woo E, Kim HL. Quality of life (QOL) and well-being research in tourism. Tourism Manage. 2016;53:244–61.
59. Wang S. Leisure travel outcomes and life satisfaction: an integrative look. Annal Tour Res. 2017;63:169–82.
60. Yu GB, Sirgy MJ, Bosnjak M. The effects of holiday leisure travel on subjective well-being: the moderating role of experience sharing. J Travel Res. 2020;
61. Zhu M, Gao J, Zhang L, Jin S. Exploring tourists' stress and coping strategies in leisure travel. Tourism Manage. 2020;81:104167.
62. Chatterjee K, Chng S, Clark B, Davis A, Vos JD, Ettema D, Handy S, Martin A, Reardon L. Commuting and wellbeing: a critical overview of the literature with implications for policy and future research. Transport Rev. 2020;40(1):5–34. https://doi.org/10.1080/01441647.2019.1649317.
63. Gimenez-Nadal JI, Molina JA. Daily feelings of US workers and commuting time. J Transport Health. 2019;12:21–33. https://doi.org/10.1016/j.jth.2018.11.001.
64. Lorenz O. Does commuting matter to subjective well-being? J Transport Geogr. 2018;66:180–99. https://doi.org/10.1016/j.jtrangeo.2017.11.019.
65. Rüger H, Pfaff S, Weishaar H, Wiernik BM. Does perceived stress mediate the relationship between commuting and health-related quality of life? Transport Res Part F: Traffic Psychol Behav. 2017;50:100–8. https://doi.org/10.1016/j.trf.2017.07.005.
66. Bohte W, Maat K, van Wee B. Measuring attitudes in research on residential self-selection and travel behaviour: a review of theories and empirical research. Transport Rev. 2009;29:325–57. https://doi.org/10.1080/01441640902808441.
67. Choi J, Coughlin JF, D'Ambrosio L. Travel time and subjective wellbeing. Transport Res Rec. 2013;2357:100–8.
68. De Vos J, Schwanen T, Van Acker V, Witlox F. Travel and subjective wellbeing: a focus on findings, methods, and future research needs. Transport Rev. 2013;33:421–42. https://doi.org/10.1080/01441647.2013.815665.
69. Ettema D, Friman M, Gärling T, Olsson LE. Travel mode use, travel mode shift and subjective wellbeing: overview of theories, empirical findings and policy implications. In: Wang D, He S, editors. Mobility, sociability and wellbeing of urban living. Berlin: Springer; 2016.
70. Martin A, Goryakin Y, Suhrcke M. Does active commuting improve psychological wellbeing? Longitudinal evidence from eighteen waves of the British Household Panel Survey. Prevent Med. 2014;69:296–303.
71. St-Louis E, Manaugh K, Van Lierop D, El-Geneidy A. The happy commuter: a comparison of commuter satisfaction across modes. Transport Res Part F. 2014;26:160–70.
72. Hernández D. Public transport, well-being and inequality: coverage and affordability in the city of Montevideo; 2017.
73. Abou-Zeid M, Ben-Akiva M. Satisfaction and travel choices. In: Handbook of sustainable travel. Springer; 2014. p. 53–65
74. Ettema D, Gärling T, Olsson LE, Friman M. Out-of-home activities, daily travel, and subjective wellbeing. Transport Res Part A: Policy Pract. 2010;44:723–32.
75. Singleton PA. Validating the satisfaction with travel scale as a measure of hedonic subjective wellbeing for commuting in a U.S. city. Transport Res Part F: Traffic Psychol Behav. 2019a;60:399–414. https://doi.org/10.1016/j.trf.2018.10.029.
76. Ye R, Titheridge H. Satisfaction with the commute: the role of travel mode choice, built environment and attitudes. Transport Res Part D: Transport Environ. 2017;52:535–47.
77. Kanwal S, Rasheed MI, Pitafi AH, Pitafi A, Ren M. Road and transport infrastructure development and community support for tourism: the role of perceived benefits, and community satisfaction. Tourism Manage. 2020;77:104014. https://doi.org/10.1016/j.tourman.2019.104014.

78. Chidambaram B. Vehicle emission reduction - An experimental approach for analysing sustainable traffic strategies. Aachen: Shaker Verlag; 2015.
79. Downs A. Still stuck in traffic. Brookings Institution Press; 2004.
80. Churchill S, Smyth R. Transport poverty and subjective wellbeing. Transport Res Part A: Policy Pract. 2019;124:40–54. https://doi.org/10.1016/j.tra.2019.03.004.
81. Sirgy MJ, Gao T, Young RF. How does residents' satisfaction with community services influence quality of life (QOL) outcomes? Appl Res Qual Life. 2008;3:81. https://doi.org/10.1007/s11482-008-9048-4.
82. Zhang J, Timmermans H. Activity-travel behavior analysis for universal mobility design. Transportmetrica. 2012;8:149–56. https://doi.org/10.1080/18128602.2010.539412.
83. D'Orso G, Migliore M. A GIS-based method for evaluating the walkability of a pedestrian environment and prioritised investments. J Transport Geogr. 2020;82:102555. https://doi.org/10.1016/j.jtrangeo.2019.102555.
84. Banister D, Ban Bowling A. Quality of life for the elderly: the transport dimension. Transport Policy. 2004;11(2):105–15. https://doi.org/10.1016/S0967-070X(03)00052-0.
85. Mouratidis K. Built environment and social wellbeing: how does urban form affect social life and personal relationships? Cities. 2018;74:7–20. https://doi.org/10.1016/j.cities.2017.10.020.
86. Townley G, Brusilovskiy E, Snethen G, Salzer MS. Using geospatial research methods to examine resource accessibility and availability as it relates to community participation of individuals with serious mental illnesses. Am J Community Psychol. 2018;61:47–61. https://doi.org/10.1002/ajcp.12216.
87. Delbosc A, Currie G. Accessibility and exclusion related to well being. In: Friman M, Ettema D, Olsson LE, editors. Quality of life and daily travel. Cham: Springer International Publishing; 2018. p. 57–69.
88. Gomes E, Banos A, Abrantes P, Rocha J. Assessing the effect of spatial proximity on urban growth. Sustainability. 2018;10(5):1308.
89. Herbolsheimer F, Mahmood A, Ungar N, Michael YL, Oswald F, Chaudhury H. Perceptions of the neighborhood built environment for walking behavior in older adults living in close proximity. J Appl Gerontol; 2020.
90. Kasraian D, Maat K, van Wee B. The impact of urban proximity, transport accessibility and policy on urban growth: a longitudinal analysis over five decades. Environ Plann B: Urban Anal City Sci. 2019;46(6):1000–17.
91. Kramer A. The unaffordable city: housing and transit in North American cities. Cities. 2018;83:1–10.
92. Moran MR, Rodríguez DA, Cotinez-O'Ryan A, Miranda JJ. Park use, perceived park proximity, and neighborhood characteristics: evidence from 11 cities in Latin America. Cities. 2020;105:102817.
93. Levinson DM, Krizek KJ. Metropolitan transport and land use: planning for place and plexus. Routledge; 2018.
94. Dėdelė A, Miškinytė A, Andrušaitytė S, Nemaniūtė-Gužienė J. Dependence between travel distance, individual socioeconomic and health-related characteristics, and the choice of the travel mode: a cross-sectional study for Kaunas, Lithuania. J Transport Geogr. 2020;86 https://doi.org/10.1016/j.jtrangeo.2020.102762.
95. Cao X, Mokhtarian PL, Handy SL. Do changes in neighborhood characteristics lead to changes in travel behavior? A structural equations modeling approach. Transportation. 2007;34(5):535–56. https://doi.org/10.1007/s11116-007-9132-x.
96. World Health Organization (WHO). Global Status Report on Road Safety 2015. World Health Organization; 2015.
97. Hidalgo D, Huizenga C. Implementation of sustainable urban transport in Latin America. Res Transport Econ. 2013;40(1):66–77.
98. Chidambaram B, Janssen MA, Rommel J, Zikos D. Commuters' mode choice as a coordination problem: a framed field experiment on traffic policy in Hyderabad, India. Transport Res Part A: Policy Pract. 2014;65:9–22. https://doi.org/10.1016/j.tra.2014.03.014.

99. Benevenuto R, Caulfield B. Poverty and transport in the global south: an overview. Transport Policy. 2019;79:115–24.

100. Can A, Aumond P. Estimation of road traffic noise emissions: the influence of speed and acceleration. Transport Res Part D: Transport Environ. 2018;58:155–71. https://doi.org/10.1016/j.trd.2017.12.002.

101. da Schio N, Boussauw K, Sansen J. Accessibility versus air pollution: a geography of externalities in the Brussels agglomeration. Cities. 2019;84:178–89.

102. Gössling S, Choi AS. Transport transitions in Copenhagen: comparing the cost of cars and bicycles. Ecol Econ. 2015;113:106–13.

103. Grieco M. Poverty mapping and sustainable transport: a neglected dimension. Res Transport Econ. 2015;51:3–9.

104. Mackett RL, Thoreau R. Transport, social exclusion and health. J Transport Health. 2015;2(4):610–7.

105. Pereira RH, Schwanen T, Banister D. Distributive justice and equity in transportation. Transport Rev. 2017;37(2):170–91. https://doi.org/10.1080/01441647.2016.1257660.

106. Petrov AI. Road traffic accident rate as an indicator of the quality of life. Econ Soc Changes: Facts, Trends, Forecast. 2016;3:154–72.

107. Santos G, Behrendt H, Maconi L, Shirvani T, Teytelboym A. Part I: Externalities and economic policies in road transport. Res Transport Econ. 2010;28(1):2–45. https://doi.org/10.1016/j.retrec.2009.11.002.

108. Sovacool BK, Kester J, Noel L, de Rubens GZ. Energy injustice and Nordic electric mobility: inequality, elitism, and externalities in the electrification of vehicle-to-grid (V2G) transport. Ecol Econ. 2019;157:205–17.

109. Jackson SE, Firth JA, Firth J, Veronese N, Gorely T, Grabovac I, Yang L, Smith L. Social isolation and physical activity mediate associations between free bus travel and wellbeing among older adults in England. J Transport Health. 2019;13:274–84. https://doi.org/10.1016/j.jth.2019.03.006.

110. Jacob L, Tully MA, Barnett Y, Lopez-Sanchez GF, Butler L, Schuch F, López-Bueno R, McDermott D, Firth J, Grabovac I, Yakkundi A, Armstrong N, Young T, Smith L. The relationship between physical activity and mental health in a sample of the UK public: a cross-sectional study during the implementation of COVID-19 social distancing measures. Mental Health Phys Activity. 2020;19:100345. https://doi.org/10.1016/j.mhpa.2020.100345.

111. Knell G, Durand CP, Shuval K, Kohl I, Salvo D, Sener IN, Gabriel KP. Transit use and physical activity: findings from the Houston travel-related activity in neighborhoods (TRAIN) study. Prevent Med Rep. 2018;9:55–61. https://doi.org/10.1016/j.pmedr.2017.12.012.

112. Rantanen T, Ayräväinen I, Eronen J, Lyyra T, Törmäkangas T, Vaarama M, Rantakokko M. The effect of an outdoor activities' intervention delivered by older volunteers on the quality of life of older people with severe mobility limitations: a randomized controlled trial. Aging Clin Exp Res. 2014;27(2):161–9.

113. Brömmelhaus A, Feldhaus M, Schlegel M. Family, work, and spatial mobility: the influence of commuting on the subjective well-being of couples. Appl Res Qual Life. 2020;15(3):865–91. https://doi.org/10.1007/s11482-019-9710-z.

114. Chen HS. Travel well, road warriors: assessing business travelers' stressors. Tour Manage Perspect. 2017;22:1–6. https://doi.org/10.1016/j.tmp.2016.12.005.

115. Gustafson P. Work-related travel, gender and family obligations. Work, Employ Soc. 2006;20(3):513–30.

116. Gustafson P. Business travel from the traveller's perspective: stress, stimulation and normalization. null. 2014;9(1):63–83. https://doi.org/10.1080/17450101.2013.784539.

117. Jensen MT. Exploring business travel with work–family conflict and the emotional exhaustion component of burnout as outcome variables: the job demands–resources perspective. Eur J Work Org Psychol. 2014;23(4):497–510. https://doi.org/10.1080/1359432X.2013.787183.

118. Legrain A, Eluru N, El-Geneidy AM. Am stressed, must travel: the relationship between mode choice and commuting stress. Transport Res Part F: Traffic Psychol Behav. 2015;34:141–51. https://doi.org/10.1016/j.trf.2015.08.001.

119. Rüger H, Viry G. Work-related travel over the life course and its link to fertility: a comparison between four European countries. Eur Sociol Rev. 2017;33(5):645–60. https://doi.org/10.1093/esr/jcx064.

120. Chidambaram B, Scheiner J. Understanding relative commuting within dual-earner couples in Germany. Transp Res Part A Policy Pract. 2020;134:113–29. https://doi.org/10.1016/j.tra.2020.02.006.

121. Dunatchik A, Speight S. Re-examining how partner co-presence and multitasking affect parents' enjoyment of childcare and housework. Soc Sc. 2020;7:268–90. https://doi.org/10.15195/v7.a11.

122. Ho C, Mulley C. Tour-based mode choice of joint household travel patterns on weekend and weekday. Transportation. 2013;40(4):789–811. https://doi.org/10.1007/s11116-013-9479-0.

123. Oakil ATM, Nijland L, Dijst M. Rush hour commuting in the Netherlands: gender-specific household activities and personal attitudes towards responsibility sharing. Travel Behav Soc. 2016;4:79–87. https://doi.org/10.1016/j.tbs.2015.10.003.

124. Vovsha P, Petersen E. Explicit modeling of joint travel by household members: statistical evidence and applied approach. Transport Res Record. 1831;1:1–10. https://doi.org/10.3141/1831-01.

125. Russell M, Price R, Signal L, Stanley J, Gerring Z, Cumming J. What do passengers do during travel time? JPT. 2011;14:7.

126. Varghese V, Jana A. Impact of ICT on multitasking during travel and the value of travel time savings: empirical evidences from Mumbai, India. Travel Behav Soc. 2018;12:11–22. https://doi.org/10.1016/j.tbs.2018.03.003.

127. Kim MJ, Hall CM. A hedonic motivation model in virtual reality tourism: comparing visitors and non-visitors. Int J Inf Manage. 2019;46:236–49.

128. Sweet M, Kanaroglou P. Gender differences: the role of travel and time use in subjective wellbeing. Transport Res Part F: Traffic Psychol Behav. 2016;40:23–34. https://doi.org/10.1016/j.trf.2016.03.006.

129. Ettema D, Gärling T, Eriksson L, Friman M, Olsson LE, Fujii S. Satisfaction with travel and subjective wellbeing: development and test of a measurement tool. Transport Res Part F Traffic Psychol Behav. 2011;14(3):167–75.

130. Nordbakke S, Schwanen T. Transport, unmet activity needs and wellbeing in later life: exploring the links. Transportation. 2015;42(6):1129–51. https://doi.org/10.1007/s11116-014-9558-x.

131. Reardon L, Abdallah S. Well-being and transport: taking stock and looking forward. Transport Rev. 2013;33(6):634–57. https://doi.org/10.1080/01441647.2013.837117.

132. Singleton PA, Clifton KJ. Towards measures of affective and eudaimonic subjective wellbeing in the travel domain. Transportation. 2019; https://doi.org/10.1007/s11116-019-10055-1.

133. Singleton PA. Walking (and cycling) to wellbeing: modal and other determinants of subjective wellbeing during the commute. Travel Behav Soc. 2019b;16:249–61. https://doi.org/10.1016/j.tbs.2018.02.005.

134. Woodcock J, Tainio M, Cheshire J, O'Brien O, Goodman A. Health effects of the London bicycle sharing system: health impact modelling study. Br Med J. 2014;13:348–425. https://doi.org/10.1136/bmj.g425.

135. Vaitsis P, Basbas S, Nikiforiadis A. How Eudaimonic aspect of subjective wellbeing affect transport mode choice? The case of Thessaloniki, Greece. Soc Sci. 2019;8(1):1–19. https://doi.org/10.3390/socsci8010009.

136. Mann E, Abraham C. The role of affect in UK commuters' travel mode choices: an interpretative phenomenological analysis. Br J Psychol. 2006;97(2):155–76.

137. Bergstad CJ, Gamble A, Gärling T, Hagman O, Polk M, Ettema D, Friman M, Olsson LE. Subjective wellbeing related to satisfaction with daily travel. Transportation. 2011;38:1–15.

138. Croce AI, Musolino G, Rindone C, Vitetta A. Transport system models and big data: Zoning and graph building with traditional surveys, FCD and GIS. ISPRS Int J Geo-Inf. 2019;8:187. https://doi.org/10.3390/ijgi8040187.

139. Martens K, Bastiaanssen J, Lucas K. Measuring transport equity: key components, framings and metrics. In: Measuring transport equity. Elsevier; 2019. p. 13–36. https://doi.org/10.1016/B978-0-12-814818-1.00002-0

140. Aziz HMA, Nagle NN, Morton AM, Hilliard MR, White DA, Stewart RN. Exploring the impact of walk–bike infrastructure, safety perception, and built-environment on active transportation mode choice: a random parameter model using New York City commuter data. Transportation. 2018;45:1207–29. https://doi.org/10.1007/s11116-017-9760-8.

141. Conway MW, Salon D, King DA. Trends in taxi use and the advent of ridehailing, 1995–2017: evidence from the US National Household Travel Survey. Urban Sci. 2018;2:79. https://doi.org/10.3390/urbansci2030079.

142. Alsger A, Assemi B, Mesbah M, Ferreira L. Validating and improving public transport origin–destination estimation algorithm using smart card fare data. Transport Res Part C: Emerg Technol. 2016;68:490–506. https://doi.org/10.1016/j.trc.2016.05.004.

143. Poonawala H, Kolar V, Blandin S, Wynter L, Sahu S. Singapore in motion: insights on public transport service level through Farecard and mobile data analytics. In: Proceedings of the 22nd ACM SIGKDD international conference on knowledge discovery and data mining, KDD '16. Association for Computing Machinery. New York; 2016. p. 589–98.

144. Zhang Y. Future wireless networks and information systems. Springer Science & Business Media; 2012.

145. Panagopoulos T, Tampakis S, Karanikola P, Karipidou-Kanari A, Kantartzis A. The usage and perception of pedestrian and cycling streets on residents' wellbeing in Kalamaria, Greece. Land. 2018;7(3) https://doi.org/10.3390/land7030100.

146. Mondschein A, Taylor BD. Is traffic congestion overrated? Examining the highly variable effects of congestion on travel and accessibility. J Transport Geogr. 2017;64:65–76. https://doi.org/10.1016/j.jtrangeo.2017.08.007.

147. Frey HC, Unal A, Rouphail NM, Colyar JD. On-road measurement of vehicle tailpipe emissions using a portable instrument. J Air Waste Manage Assoc. 2003;53(8):992–1002. https://doi.org/10.1080/10473289.2003.10466245.

148. Smit R, Kingston P. Measuring on-road vehicle emissions with multiple instruments including remote sensing. Atmosphere. 2019;10(9):1–17. https://doi.org/10.3390/atmos10090516.

149. Chatzidiakou L, Krause A, Han Y, Chen W, Yan L, Popoola OAM, Jones RL. Using low-cost sensor technologies and advanced computational methods to improve dose estimations in health panel studies: results of the AIRLESS project. J Exposure Sci Environ Epidemiol. 2020; https://doi.org/10.1038/s41370-020-0259-6.

150. Litman T. Evaluating transportation equity: guidance for incorporating distributional impacts in transportation planning. Vic Transport Policy Instit Vic Br. 2005;8(2):50–65.

151. Offer S, Schneider B. Revisiting the gender gap in time-use patterns: multitasking and wellbeing among mothers and fathers in dual-earner families. Am Sociol Rev. 2011;76:809–33. https://doi.org/10.1177/0003122411425170.

152. Foley L, Dumuid D, Atkin AJ, Olds T, Ogilvie D. Patterns of health behavior associated with active travel: a compositional data analysis. Int J Behav Nutr Phys Act. 2018;15:26. https://doi.org/10.1186/s12966-018-0662-8.

153. Susilo YO, Liotopoulos FK. Measuring door-to-door journey travel satisfaction with a mobile phone app. In: Friman M, Ettema D, Olsson LE, editors. Quality of life and daily travel, applying quality of life research. Cham: Springer International Publishing; 2018. p. 119–38.

154. Kwon J, Lee H. Why travel prolongs happiness: longitudinal analysis using a latent growth model. Tour Manage. 2020;76:103944. https://doi.org/10.1016/j.tourman.2019.06.019.

155. Alshehri A, O'Keefe R. Analyzing social media to assess user satisfaction with transport for London's Oyster. Int J Hum–Comput Interact. 2019;35:1378–87. https://doi.org/10.1080/10447318.2018.1526442.

156. Friman M, Fellesson M. Service supply and customer satisfaction in public transportation: the quality paradox. J Public Transport. 2009;12 https://doi.org/10.5038/2375-0901.12.4.4.

157. Cordera R, Nogués S, González-González E, dell'Olio L. Intra-urban spatial disparities in user satisfaction with public transport services. Sustainability. 2019;11:5829. https://doi.org/10.3390/su11205829.
158. Banyard V, Hamby S, Grych J. Health effects of adverse childhood events: identifying promising protective factors at the intersection of mental and physical wellbeing. Child Abuse Neglect. 2017;65:88–98. https://doi.org/10.1016/j.chiabu.2017.01.011.
159. Giebel C, McIntyre JC, Alfirevic A, Corcoran R, Daras K, Downing J, Gabbay M, Pirmohamed M, Popay J, Wheeler P. The longitudinal NIHR ARC North West Coast Household Health Survey: exploring health inequalities in disadvantaged communities. BMC Public Health. 2020;20:1–11.
160. Bishwajit G, O'Leary DP, Ghosh S, Yaya S, Shangfeng T, Feng Z. Physical inactivity and self-reported depression among middle-and older-aged population in South Asia: World health survey. BMC Geriatrics. 2017;17:100.
161. Stubbs B, Koyanagi A, Hallgren M, Firth J, Richards J, Schuch F, Rosenbaum S, Mugisha J, Veronese N, Lahti J. Physical activity and anxiety: a perspective from the World Health Survey. J Affect Disorders. 2017;208:545–52.
162. Diener E, Chan MY. Happy people live longer: subjective wellbeing contributes to health and longevity. Appl Psychol. 2011;3:1–43. https://doi.org/10.1111/j.1758-0854.2010.01045.x.
163. Camporeale R, Caggiani L, Ottomanelli M. Modeling horizontal and vertical equity in the public transport design problem: a case study. Transport Res Part A: Policy Pract. 2019;125:184–206. https://doi.org/10.1016/j.tra.2018.04.006.
164. Di Ciommo F, Shiftan Y. Transport equity analysis. Taylor & Francis; 2017.
165. Litman T. Evaluating transportation equity: Guidance for incorporating distributional impacts in transportation planning. Victoria: Victoria Transport Policy Institute; 2007.
166. Ricciardi AM, Xia JC, Currie G. Exploring public transport equity between separate disadvantaged cohorts: a case study in Perth, Australia. J Transport Geogr. 2015;43:111–22.
167. Klier T, Linn J. Using taxes to reduce carbon dioxide emissions rates of new passenger vehicles: evidence from France, Germany, and Sweden. Am Econ J: Econ Policy. 2015;7(1):212–42. https://doi.org/10.1257/pol.20120256.
168. Borndörfer R, Grötschel M, Klostermeier F, Kittner C. Telebus Berlin: vehicle scheduling. In: Computer-aided transit scheduling: proceedings, Cambridge, MA; 2012. vol. 471, p. 391.
169. Lowe C, Stanley J, Stanley J. Transport industry adapting to change: an Australian case study. Res Transport Econ. 2020;83:100940.
170. Wac K. Quality of life technologies. In: Gellman M, editor. Encyclopedia of behavioral medicine. New York: Springer; 2020. https://doi.org/10.1007/978-1-4614-6439-6_102013-1.
171. Estrada-Galiñanes V, Wac K. Collecting, exploring and sharing personal data: why, how and where. Data Sci. 2019;3:79–106. https://doi.org/10.3233/ds-190025.
172. Wac K. From quantified self to quality of life. In: Digital health. Cham: Springer; 2018. p. 83–108. https://doi.org/10.1007/978-3-319-61446-5_7.
173. Sun W, Liu J, Zhang H. When smart wearables meet intelligent vehicles: challenges and future directions. IEEE Wireless Commun. 2017;24(3):58–65. https://doi.org/10.1109/MWC.2017.1600423.

Part VI
Conclusive Remarks

Chapter 25
The Future of Quantifying Behaviors, Health, and Quality of Life

Katarzyna Wac

Daily behaviors influence an individual's health and overall quality of life (QoL). Specific patterns of behavior such as smoking, having a poor diet, being physically inactive, or consuming alcohol influence the long-term development of chronic diseases such as type 2 diabetes, chronic obstructive pulmonary disease, or cardiovascular disease [1]. In the US alone, chronic diseases are becoming increasingly common, and about half of all deaths can be attributed to preventable behaviors and exposures [2, 3]. Figure 25.1 represents the proportional contribution of various factors to causes of death, including genetic, behavioral, and systemic aspects of one's life [2, 3].

The influence of daily behavior on the development of chronic diseases and, ultimately, mortality is a stark example of how behaviors influence QoL in the long

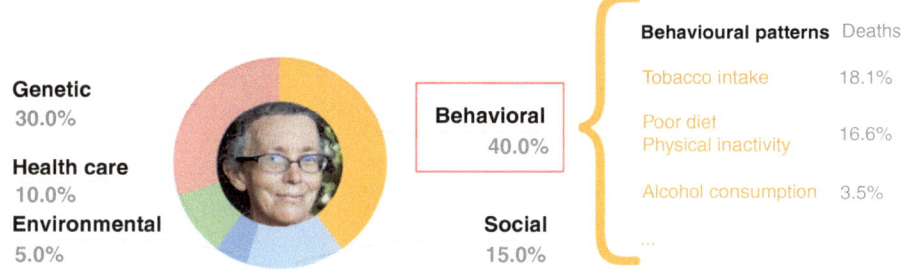

Fig. 25.1 Proportional contribution of various factors to mortality

K. Wac (✉)
Geneva School of Economics and Management, Center for Informatics, Quality of Life Technologies Lab, University of Geneva, Geneva, Switzerland

Quality of Life Technologies Lab, Department of Computer Science, University of Copenhagen, Copenhagen, Denmark
e-mail: katarzyna.wac@unige.ch

K. Wac, S. Wulfovich (eds.), *Quantifying Quality of Life*, Health Informatics,
https://doi.org/10.1007/978-3-030-94212-0_25

term. It is therefore important for health practitioners and researchers, as well individuals themselves to accurately to accurately quantify individuals' behaviors, and psychological states to improve their awareness of these factors, encourage necessary habit changes, and provide more individualized treatment approaches. Currently, individuals' behaviors and psychological states are mainly quantified via methods such as the use of self-reported measures, many of which consist of validated scales referred to as patient-reported outcomes (PROs) [4]. Among other behaviors, PROs may include self-reported measures of physical activity (e.g., IPAQ [5]), sleep (e.g., PSQI [6]), or nutritional habits (e.g., Mediterranean diet score [7]).

To characterize the current landscape of tools and methods used to quantify QoL, the chapters of this book discuss various PROs that have been used to assess behaviors relating to all 24 aspects of an individual's QoL as defined by the World Health Organization [8]. In particular, each chapter identifies methods and instruments for assessing a particular variable of QoL; some of the methods and instruments are considered gold standards in the field of QoL assessment. A summary of the existing self-reported QoL measures that are most commonly used in adult populations can also be found on the website of the QoL lab, pros.qol.unige.ch, which will be regularly updated in the future.

As is discussed extensively throughout the book, these self-reported PROs have multiple limitations, including the infrequency with which they are collected, the subjective nature of responses, their reliance on memory recall, and their susceptibility to the influence of social norms. Additionally, self-reported PROs are typically performed outside of the context in which the assessed behavior occurs. Therefore, the use of personalized and miniaturized technological innovations, including smartphones [9], mobile applications [10], and wearables [11, 12], is proposed in this book as an alternative means of quantifying QoL that can enable the continuous and more accurate assessment of daily life behaviors that contribute to or result from an individual's QoL.

The advantages of using these technologies include the fact that they allow more frequent measurements than PROs, provide longitudinal data, are objective, sensory-based, and non-judgmental, and permit a context-rich approach to the assessment of individual states and behaviors [13]. The behavioral and health outcomes captured by these personal technologies, the US Food and Drug Administration (FDA) calls "digital health technology tools (DHTTs) [14, 15], are referred to as technology-reported outcomes (TechROs) [4]. According to regulatory bodies such as the FDA, TechRO data is an example of real-world data (RWD). The RWD includes electronic health records, information from insurance claims and billing, registries, and data collected from medical devices used outside of clinical settings. The analysis of RWD produces a broad range of objective evidence that can be used for research and regulatory purposes, including information about individuals' behavior and the potential advantages or risks of specific behavioral or pharmacological interventions [15, 16]. The emergence of standardized definitions of TechROs and the growing use of TechROs as a data source are important to note in the context of the present book. The work of the QoL lab defined the larger field of "quality of life technologies" (QoLTs) to which TechROs belong. These technologies encompass a broad range of tools that can be leveraged to assess, maintain, improve, prevent decline in, or compensate for one's life quality, which can be applied at the individual, interpersonal, community, group, or population level [17].

Expanding on the research on QoLTs, the contributors to this book discuss and evaluate various TechROs that have been used to assess behaviors relating to different aspects of an individual's QoL, as well as technologies that could be applied to the collection of TechRO data in the future. Moreover, each chapter elaborates on the role of the emerging quantified self (QS) movement in the context of the QoL variable it focuses on. The QS movement[1] involves a group of highly motivated individuals who leverage personalized technologies to assess and intervene in their behaviors with the ultimate goal of realize the QS ideal of "knowing thyself." Engaging in the QS practices presented in this book may not be possible for the average individual today. However, the book's discussions highlight practices that may be commonplace in the future, as they demonstrate the feasibility of various technology-enabled approaches to behavior self-management in particular areas of one's daily life.

Before TechROs can be leveraged in clinical practice, many challenges need to be overcome. These include ensuring the accurate collection of technology-based data [18], interpreting the data in its appropriate context, and resolving potential ethical dilemmas in the ways data is collected and used [19, 20]. Ultimately, the data collected through use of a given technology must accurately represent its defined variable, and the individual must accept the use of these technologies in their daily life for their relevant purposes [21].

In spite of these, however, the book overall presents a positive outlook concerning the state of the art and current developments in the use of TechROs and the future potential for leveraging TechROs to supplement PROs in the assessment of individuals' behaviors and health and life quality outcomes. In one such scenario, individuals may be assessed using TechROs throughout their lifetime, with the results of assessments being leveraged to improve their health and levels of care, including self-care. Figure 25.2 presents a vision of the future quantification of individuals' behavioral data. Drawing upon comparisons to the genome (a term that denotes the entirety of an organism's genetic material) and its common application in current genetics testing, the collection of the behavioral data is called one's "behaviome." Ultimately, the mapping of this behaviome through behavioral assessments could improve diagnoses, treatments, and practices for the prevention of diseases and, if applied regularly over an individual's lifetime, improve their overall health outcomes.

The technologies used to collect TechRO data can empower individuals to become "co-producers" of and experts in their own health and QoL, both in the short and the long term [22]. Medicine may thereby be transformed into a more personalized, predictive, participatory, and preventative system. Meanwhile, quantifying one's QoL might become a new norm for individuals and populations at large, enabling the transition from healthcare to self-care that the world needs.

---

[1] <Footnote ID="Fn1"><Para ID="Par8">QuantifiedSelf, <ExternalRef><RefSource>quantifieds elf.com</RefSource><RefTarget TargetType="URL" Address="http://quantifiedself.com"/></ ExternalRef>, visited May 23, 2021.</Para></Footnote>

Fig. 25.2 Quality of life technologies: from genome to behaviome

References

1. Bauer UE, Briss PA, Goodman RA, Bowman BA. Prevention of chronic disease in the 21st century: elimination of the leading preventable causes of premature death and disability in the USA. Lancet. 2014;384(9937):45–52. https://doi.org/10.1016/S0140-6736(14)60648-6.
2. Naghavi M, Abajobir AA, Abbafati C, et al. Global, regional, and national age-sex specific mortality for 264 causes of death, 1980–2016: a systematic analysis for the global burden of disease study 2016. Lancet. 2017;390(10100):1151–210. https://doi.org/10.1016/S0140-6736(17)32152-9.
3. Mokdad AH, Marks JS, Stroup DF, Gerberding JL. Actual causes of death in the United States, 2000. JAMA. 2004;291(10):1238–45. https://doi.org/10.1001/jama.291.10.1238.
4. Mayo NE, Figueiredo S, Ahmed S, Bartlett SJ. Montreal accord on patient-reported outcomes (PROs) use series – paper 2: terminology proposed to measure what matters in health. J Clin Epidemiol. 2017;89:119–24. https://doi.org/10.1016/j.jclinepi.2017.04.013.
5. Hagströmer M, Oja P, Sjöström M. The international physical activity questionnaire (IPAQ): a study of concurrent and construct validity. Public Health Nutr. 2006;9(6):755–62. https://doi.org/10.1079/phn2005898.
6. Buysse DJ, Reynolds CF, Monk TH, Berman SR, Kupfer DJ. The Pittsburgh sleep quality index: a new instrument for psychiatric practice and research. Psychiatry Res. 1989;28(2):193–213. https://doi.org/10.1016/0165-1781(89)90047-4.
7. Yang J, Farioli A, Korre M, Kales SN. Modified Mediterranean diet score and cardiovascular risk in a north American working population. PLoS One. 2014;9(2):e87539. https://doi.org/10.1371/journal.pone.0087539.
8. Skevington SM, Lotfy M, O'Connell KA, WHOQOL Group. The World Health Organization's WHOQOL-BREF quality of life assessment: psychometric properties and results of the international field trial. A report from the WHOQOL group. Qual Life Res. 2004;13(2):299–310. https://doi.org/10.1023/B:QURE.0000018486.91360.00.
9. Dey AK, Wac K, Ferreira D, Tassini K, Hong J-H, Ramos J. Getting closer: an empirical investigation of the proximity of user to their smart phones. In: Proceedings of the 13th international

conference on ubiquitous computing - UbiComp '11. ACM Press;2011:163. https://doi.org/10.1145/2030112.2030135

10. Wac K. Smartphone as a personal, pervasive health informatics services platform: literature review. Yearb Med Inform. 2012;21(01):83–93. https://doi.org/10.1055/s-0038-1639436.

11. Wac K. From quantified self to quality of life. Digital Health. 2018:83–108. https://doi.org/10.1007/978-3-319-61446-5_7.

12. Boillat T, Rivas H, Wac K. "Healthcare on a Wrist": increasing compliance through checklists on wearables in obesity (self-)management programs. In: Rivas H, Wac K, editors. Digital health. Health informatics; 2018. p. 65–81. https://doi.org/10.1007/978-3-319-61446-5_6.

13. Berrocal A, Manea V, De MA, Wac K. mQoL lab: step-by-step creation of a flexible platform to conduct studies using interactive, mobile, wearable and ubiquitous devices. Procedia Comput Sci. 2020;175:221–9. https://doi.org/10.1016/j.procs.2020.07.033.

14. Taylor KI, Staunton H, Lipsmeier F, Nobbs D, Lindemann M. Outcome measures based on digital health technology sensor data: data- and patient-centric approaches. npj Digit Med. 2020;3(1):97. https://doi.org/10.1038/s41746-020-0305-8.

15. Food and Drug Administration. Patient-focused drug development: methods to identify what is important to patients guidance for industry. US Dep Heal Hum Serv: Published online; 2019.

16. Sherman RE, Anderson SA, Dal Pan GJ, et al. Real-world evidence — what is it and what can it tell us? N Engl J Med. 2016;375(23):2293–7. https://doi.org/10.1056/NEJMsb1609216.

17. Wac K. Quality of life technologies. In: Encyclopedia of behavioral medicine. New York: Springer; 2020. p. 1–2. https://doi.org/10.1007/978-1-4614-6439-6_102013-1.

18. van Berkel N, Goncalves J, Wac K, Hosio S, Cox AL. Human accuracy in mobile data collection. Int J Hum Comput Stud. 2020;137:102396. https://doi.org/10.1016/j.ijhcs.2020.102396.

19. Manea V, Schnoor Hansen M, Elbeyi SE, Wac K. Towards personalizing participation in health studies. In: Proceedings of the 4th international workshop on multimedia for personal health & health care – HealthMedia '19. ACM Press; 2019. p. 32–39. https://doi.org/10.1145/3347444.3356241

20. Estrada-Galiñanes V, Wac K. Collecting, exploring and sharing personal data: why, how and where. Hoehndorf R, ed. Data Sci. 2020;3(2):79–106. https://doi.org/10.3233/DS-190025.

21. Wulfovich S, Fiordelli M, Rivas H, Concepcion W, Wac K. "I must try harder": Design implications for mobile apps and wearables contributing to self-efficacy of patients with chronic conditions. Front Psychol. 2019;2388. https://doi.org/10.3389/fpsyg.2019.02388.

22. Gauthier T, Wac K. A foresight analysis of pervasive healthcare technologies. J Futur Stud. Published online. 2015. https://arodes.hes-so.ch/record/1529/files/Gauthier_2015_foresight_analysis.pdf.

Correction to: Quantifying Quality of Life

Katarzyna Wac and Sharon Wulfovich

Correction to:
Quantifying Quality of Life, © The Author(s) 2022
K. Wac, S. Wulfovich (eds.), Health Informatics,
https://doi.org/10.1007/978-3-030-94212-0

The chapter 12 was inadvertently published with incorrect citation to figure 12.5 in page 306. This has now been corrected and correct citation has been placed in page 302. The figure 12.5 has been moved from page 306 to page 302.

The chapter 18 was inadvertently published with incorrect family name. This has now been updated to "S. Mercuri Chapuis".

The updated version of these chapters can be found at:
https://doi.org/10.1007/978-3-030-94212-0_12
https://doi.org/10.1007/978-3-030-94212-0_18

© The Author(s) 2022   C1
K. Wac, S. Wulfovich (eds.), *Quantifying Quality of Life*, Health Informatics,
https://doi.org/10.1007/978-3-030-94212-0_26

Index